High-Resolution X-Ray Scattering

Springer
New York
Berlin
Heidelberg
Hong Kong
London
Milan
Paris
Tokyo

Advanced Texts in Physics

This program of advanced texts covers a broad spectrum of topics that are of current and emerging interest in physics. Each book provides a comprehensive and yet accessible introduction to a field at the forefront of modern research. As such, these texts are intended for senior undergraduate and graduate students at the M.S. and Ph.D. levels; however, research scientists seeking an introduction to particular areas of physics will also benefit from the titles in this collection.

Ullrich Pietsch
Václav Holý
Tilo Baumbach

High-Resolution X-Ray Scattering

From Thin Films to Lateral Nanostructures

Second Edition

With 240 Figures

Springer

Ullrich Pietsch
Institute of Physics
University of Potsdam
Potsdam 14469
Germany

Václav Holý
Department of Solid State Physics
Masaryk University
Brno 61137
Czech Republic

Tilo Baumbach
Institut fuer Synchrotronstrahlung
Forschungszentrum Karlsruhe in der
Helmholtz-Gemeinschaft
76021 Karlsruhe
Germany

Cover illustration: Reciprocal space mapping in vicinity of the 224 reflection of multi-layer structure consisting in five SiGe layers of different composition grown on Si substrate (adapted from Figure 9.13, page 192).

Library of Congress Cataloging-in-Publication Data
Pietsch, Ullrich, 1952–
High-resolution X-ray scattering: from thin films to lateral nanostructures / Ullrich Pietsch, Václav Holý, Tilo Baumbach.
p. cm.
Includes bibliographical references and index.
ISBN 0-387-40092-3 (acid-free paper)
1. Thin films Optical properties. 2. X-rays—Scattering. 3. Xprays Diffraction.
4. Nanostructure materials. I. Holy, Václav, 1953– II. Baumbach, Tilo, 1961– III. Title.
QC176.84.O7P54 2004
530.4′175—dc22 2003070360

ISBN 0-387-40092-3 Printed on acid-free paper.

Printed in the United States of America. (EB)

9 8 7 6 5 4 3 2 1 SPIN 10928905

www.springer-ny.com

Springer-Verlag is a part of *Springer Science+Business Media*

springeronline.com

This book is dedicated to Norio Kato (1923-2002), one of the most prominent scientists in the field of x-ray dynamical theory

Preface

During the last 20 years interest in high-resolution x-ray diffractometry and reflectivity has grown as a result of the development of the semiconductor industry and the increasing interest in material research of thin layers of magnetic, organic, and other materials. For example, optoelectronics requires a subsequent epitaxy of thin layers of different semiconductor materials. Here, the individual layer thicknesses are scaled down to a few atomic layers in order to exploit quantum effects. For reasons of electronic and optical confinement, these thin layers are embedded within much thicker cladding layers or stacks of multilayers of slightly different chemical composition. It is evident that the interface quality of those quantum wells is quite important for the function of devices.

Thin metallic layers often show magnetic properties which do not appear for thick layers or in bulk material. The investigation of the mutual interaction of magnetic and non-magnetic layers leads to the discovery of colossal magnetoresistance, for example. This property is strongly related to the thickness and interface roughness of covered layers.

The properties of supramolecular structures made from organic thin films can differ entirely from those of individual layers. Supramolecular structures can be composed by amphiphilic layers, for example, which have the capability for lateral self-organization. Particular head groups are attached to these molecules to give them proper functionality. These layers are separated by layers of another material either carrying a second functionality or simply for spatial separation of the first kind of layers. The different sublayers can exhibit different degrees of perfection and crystallinity.

Thin layers and multilayers can be grown using molecular-beam-epitaxy (MBE) or metal-organic-vapor-deposition epitaxy (MOVPE), for example. Layer-by-layer deposition of particular semiconductor material allows for the composition of tailored stacks of sublayers with monolayer and submonolayer accuracy. The functionality of the respective devices requires perfect lattice matching between the sublayers and smooth interfaces. Under conditions of crystal growth, several real-structure effects appear that may reduce the efficiency of the electronic or optoelectronic device. For example, coherent hetero-epitaxy of highly mismatched material combinations is possible only up to a critical layer thickness only. When the thickness increases, misfit dislocations are created at the interfaces in order to reduce the strain energy of the system. Another process of relaxation is the growth of strain-reduced islands, which disturbs the smoothness of interfaces. In order to prevent this, slightly misoriented substrates are used, and the epitaxy provides terraces, i.e., locally smooth but macroscopically rough interfaces. Additionally, the statistical character of the growing process gives rise to various local fluctuations of the layer thickness and to waviness of the interfaces on a mesoscopic and nanoscopic scale.

Nowadays the Stranski-Krastanov growth process is used especially to create quantum dots. Here, the growth switches from layer-by-layer into an island-like growth mode. The dots can show quantum size effects which can be used for exciting a particular emission wavelength or reducing the laser threshold of an optoelectronic device. Unfortunately, these dots can hardly be arranged in an ordered array and are still not monodisperse in size, which requires the use of statistical methods of data evaluation. Ordered arrays of lateral nanostructures can be obtained by selective etching of semiconductor quantum-well structures or metallic multilayers, respectively. Because the distance of the defined surface dots or stripes is of the order of several 10–100 nm, this technological process defines a one- or two-dimensional mesoscopic lattice which is best accessed by the x-ray scattering process.

Metallic layers are often prepared using sputtering techniques. In contrast to MBE, this process runs far from thermodynamic equilibrium and the deposited films sometimes become amorphous or polycrystalline. On the other hand, the deposition rate is higher and thus the statistical character of the atomic deposition is more pronounced than in the MBE process. The structural parameters controlling the application of the respective multilayer films for x-ray optics, for example, are the waviness and roughness of the interfaces and the thickness of the individual sublayers. Depending on the growing mode, the initial long-range waviness and the interface roughness of the substrate may be either replicated layer-by-layer or smoothed out during the deposition process. The lateral and vertical correlation of these parameters is worthy of detailed investigation because the parameters provide insights into the growing process itself.

Organic layers were deposited by rather crude methods, such as spin-coating or Langmuir-Blodgett transfer from a liquid surface. The structure of these layers depends on whether self-organization takes place or not, so they can be deposited layer-by-layer often accompanied by reducing layer perfection. The in-plane structure of lipid layers, for example, behaves like a two-dimensional powder. During heating liquid crystals undergo several structural phase-transitions, changing the lateral and vertical ordering of molecules. Lipid layers can be combined with polymers or with layers containing inorganic species to supramolecular structures. Characterization of the three-dimensional structure is important for understanding the functionality of these supramolecular structures, which may differ from the individual functions of the constituents. Considering the different size and ordering of molecules in lateral and vertical directions, different methods of structural investigation are required, i.e., x-ray reflectivity in the vertical, but x-ray diffraction in the lateral, direction.

Individual layers and interfaces within a layering heterostructure or a multilayer are identified rather indirectly. Due to the Fourier character of x-ray scattering, the particular layer or interface is measured as a scattering peak appearing at a certain scattering angle with definite intensity and definite peak shape. The peak shape, in particular, depends on the correlation length of structural attributes, i.e. the distance and number of structural features along the interface contributing to the scattering peak with equal phase. As in conventional optics the angular position and the peak width is inversely proportional to the number and distance of these scattering objects. This reciprocity is the reason for describing x-ray scattering in *reciprocal space* instead of direct space. However, in the best case, the correlation length of structural features equals that of the whole crystal. The corresponding peak shape can be entirely described by the dynamical theory of x-ray diffraction [270]. Any reduction of the correlation length, whether it is caused by the finite size of the crystal or by structural defects within a large crystal, reduces the correlation length and subsequently increases the peak width. These cases can be described in terms of the kinematic scattering theory as in Warren [386], for example. Although the basic theories are well known [175, 213, 270, 396], their application on thin layer and multilayer analysis requires a number of modifications and extensions of the theory, which demands a separate treatment.

For many problems x-ray scattering methods are complementary to scanning-probe techniques. Whereas scanning techniques, ergo atomic-force microscopy, provide a direct picture of a particular surface area, x-ray scattering gives a measure of the reciprocal space of the sample, which represents structural information averaged over a large sample volume. In addition, x-ray techniques are non-destructive and give access to internal interfaces, and they can be fast when using highly intense x-ray sources. Because the x-ray refraction index for matter is smaller than unity, x-ray methods can be depth selective if the beam strikes the sample surface under a shallow angle of incidence. Thus very thin layers and surface elevations of a few nanometers can be measured.

At present, x-ray techniques are routine methods in scientific material science laboratories and one can find more and more applications to routine probes of industrial processes. The wide use of x-ray diffraction techniques is based on new developments in x-ray scattering equipment. Nowadays x-ray tubes can provide an incident intensity which is comparable to that of a bending magnet at a synchrotron facility. Modern x-ray diffractometers and reflectometers are equipped with optimized optical elements which give the user flexibility in the choice of experimental conditions. Therefore these instruments can be used for high-resolution experiments as well as for the measurement of imperfect materials.

This book is a second edition of our previous monograph, *High-Resolution X-Ray Scattering from Thin Films and Multilayers* [167], which appeared in 1999. Before this, there was no particular monograph in the literature dealing with the problem of x-ray diffractometry and reflectometry from thin films. At nearly the same time other books appeared describing one [8, 120, 376] or both topics [61]. Despite this competition our book experienced good acceptance by readers and was sold out after two years. This might have been caused by the general architecture of the book, which can be helpful for newcomers as well as experienced researchers.

Our book displays a synergetic structure of four main parts - *Experimental Realization, Basic Principles, Solution of Experimental Problems*, and *X-Ray Scattering by Laterally Structured Semiconductor Nanostructures* - presented in mutual connection to one another. Nevertheless, each part is organized in such a way that it can be understood separately to a large extent. The first part will introduce the reader to the general set-up of an x-ray reflection and x-ray diffraction experiment and in the function and arrangement of optical elements to achieve the experimental resolution *necessary* to solve a particular problem.

The second part describes the underlying theory in an extensive way; it describes all the formulas necessary to interpret an x-ray reflection or diffraction experiment and considers the kinematic and dynamical theories of x-ray diffraction which are used to simulate the scattering curves presented in the experimental chapters.

Finally, the third and fourth parts, which are the longest ones, present a lot of experimental problems, which have been solved by the authors and in other laboratories in the past. There are problems of the thickness determination of layers with thicknesses of less than 5 nm, the thickness determination of multilayer structures, the measurement of misfits and interface strains, the determination of the interface roughness, and correlation properties of the interface roughness and of crystal defects. As in the previous edition we focus our interest on actual problems of x-ray diffraction analysis and the investigation of lateral nanostructures, such as surface gratings and quantum dots. The examples presented were taken from scientific projects dealing with either semiconductors, metals, or organic materials. Thus we believe that the experience transmitted by us is of a general nature and not restricted to sophisticated problems. The third and fourth parts are self-explanatory but refer to expressions of general theory given in Part II.

All the chapters in the first edition were critically reviewed. Obvious misprints and errors were corrected and ambiguous phrases were reformulated. In the first part we considered new commercial instruments that appeared on the market during the last years, providing an increasing x-ray flux for home laboratory apparatus. Also the new challenge in the production of synchrotron radiation, the free-electron laser, is addressed. The theoretic part is completed by the formalism of *Distorted-Wave Born Approximation.* This became necessary for the description of statistical properties of surfaces and interfaces and, in particular, of semiconductor quantum dot structures.

The examples of the third part also were revised. Many figures were replaced by actual examples taken in our laboratories and from laboratories of the Ferdinand-Braun Institut (Berlin); the universities of Würzburg, Linz, and Magdeburg, and the European Synchrotron Radiation Facility in Grenoble (ESRF), the Hamburger Synchrotron Strahlungslabor (HASYLAB), and the company Bruker AXS. The chapter about gratings and dots was rewritten and much extended. It is now divided into three chapters (12 - 14) because of the many new results received by the authors over the last years: Chapter 12, dealing with periodic surface nanostructures, is completed by examples of organic and metallic systems. Chapter 13 considers the problem of strain analysis in artificially structurized lateral nanostructures. The strain and correlation properties of quantum dot structures are described in chapter 14.

Finally, the authors thank several people who have helped us during the preparation of the manuscript. First of all, there are all our co-workers and Ph.D. students who have worked with us during the last years in our laboratories. In additon, besides many others, we thank Bernd Husemann for technical assistance while preparing the LATEX-style manuscript and Daniel Lübbert and Birgit-Marina Pietsch for support with the graphics design. Finally we thank the Bruker AXS GmbH, Karlsruhe for providing schemes of new technical solutions.

Potsdam, Germany — *Ullrich Pietsch*
Brno, Czech Republic — *Vaclav Holý*
Karlsruhe, Germany — *Tilo Baumbach*
February 2004

Contents

Part I Experimental Realization

Part II Basic Principles

Part I

Experimental Realization

The first part of this book acquaints the reader with the x-ray equipment necessary to analyze thin films and surface nanostructures. After we describe the particular components of the equipment, such as conventional and synchrotron x-ray sources and different kinds of monochromators, analyzers, detectors (Chap. 1), we combine them to build diffractometers and reflectometers with certain experimental properties exploiting certain scattering geometries (Chap. 2). X-ray reflectometers are equipped to measure the scattering intensity at very small angles measured with respect to the sample surface. Here one has to distinguish between apparatus for angle-resolved measurements with monochromatic radiation and those for energy-dispersive recording using a white incident beam. Both methods provide different advantages, either for multilayer characterization on the absolute scale or for fast measurements on relative scale. A diffractometer can be designed for relatively fast measurements of samples with small layer thickness, i.e., broad scattering peaks, and with large lattice mismatch and for high-resolution, i.e., for multilayers with small lattice mismatch and for inspecting lateral nanostructures. Finally the grazing-incidence diffraction method combines in-plane diffraction with out-of-plane reflectivity. This method has the advantage of probing the change of the structure parameters from the near surface region down to the substrate. The last chapter of this part deals with the problem of coherence. Here one has to distinguish the coherence of the source and its modification by the various optical elements used for the experiment. This coherence is necessary in order to probe the correlation properties of the sample under investigation. Each experiment requires a certain degree of resolution in angular and reciprocal space and its variation during the experiment (Chap. 3).

1 Elements for Designing an X-Ray Diffraction Experiment

The aim of this chapter is to discuss the basic elements of the apparatus necessary for performing x-ray diffraction and x-ray reflectometry experiments. These are the sources providing the x-rays, the optical elements to monochromate and collimate the incident beam for an optimum interaction with the sample and the detection units, recording the diffracted/reflected radiation as a function of a spatial angle or of energy. The particular design will be different, of course, depending on whether we are performing the experiment in a home laboratory or at a beamline of a storage ring facility. Although the arrangement of elements may differ, the functions of the elements will be the same.

Here the basic properties will be discussed on a qualitative level only. A detailed explanation in terms of kinematic or dynamical theory will be given in Part II of this book. However, a knowledge of the basic elements should be helpful for the readers designing their own experiments.

1.1 X-Ray Sources

Sealed x-ray tubes are the most widely used x-ray sources. The radiation is created by an electron beam accelerated within a vacuum tube toward a metal target. Two types of radiation are excited (Fig. 1.1) due to the interaction of the accelerated electrons with the electrons of the target. The inelastic interaction with the outer electron cloud of an atom gives rise to a nearly continuous spectrum, the *bremsstrahlung*, with an onset wavelength $\lambda_{\min} = 123.9/\mathrm{V}$, where V is the acceleration voltage of the x-ray generator, measured in keV; $\lambda_{\min}$ is measured in nanometers.

The second type of radiation is the *characteristic radiation.* It is created whenever the energy of incident electrons is large enough to excite a K electron of the target material. The hole created will be filled by an electron of neighboring shells, creating a photon of an energy which corresponds to the energy difference between the energy states concerned. The most intense line is the $K\alpha$ doublet; it corresponds to the energy difference between the $2p_{x,y} - 1s$ and the $2p_z - 1s$ states, respectively. Because the energy levels of $2p_x$ and $2p_y$ equal but differ from that of the $2p_z$ one the intensity ratio between the $K\alpha_1$ and $K\alpha_2$ line is $2:1$. Seldom does one make use of the $K\beta$ line, which mainly is due to the $3p-1s$ electron transition and less intense compared to the $K\alpha$ line. An angular dispersive experiment mostly makes use of one of the $K\alpha$ lines, because it can exceed the intensity of *bremsstrahlung* by about 4 orders of magnitude. The maximum input elec-

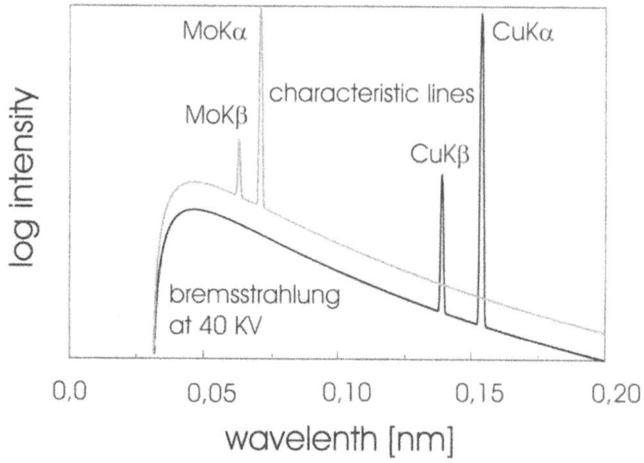

Fig. 1.1. Schematic emission spectrum of a sealed x-ray tube with a copper and molybdenum *anticathode*, respectively, and an acceleration voltage of $V = 40kV$. The radiation consists of the continuous *bremsstrahlung* radiation and the characteristic $K\alpha$ doublet (not resolved here) at $\lambda_1 = 0.154056$ nm and $\lambda_2 = 0.154439$ nm, and $\lambda_1 = 0.07093$ nm and $\lambda_2 = 0.071359$ nm, for copper and molybdenum, respectively. Additionally, there is the $K\beta$ line, which is composed from several sublines with the center wavelength at $\lambda_\beta = 0.1396$ nm and 0.06323 nm, respectively.

trical power of common sealed tubes is limited to about 2–3 kW, depending of the focus size and the effectivity of the cooling system. Less than 0.3 % of this power is transformed into x-ray radiation, leaving the tube through four beryllium windows under an exit angle of 84° with respect to the tube axis. Using an emission area at the anticathode of $1 \times 10\,\mathrm{mm}^2$, one can exploit the point focus of $1 \times 1\,\mathrm{mm}^2$ or the long focus with $1 \times 10\,\mathrm{mm}^2$. Fine-focus tubes deliver a point focus of $0.5 \times 1\,\mathrm{mm}^2$ and a long focus of $0.05 \times 10\,\mathrm{mm}^2$, respectively. When performing an angle dispersive experiment under air conditions a typical generator voltage of 40 to 50 kV and the use of copper $K\alpha$ radiation is recommended, because it provides maximum intensity at the sample site, compared to other target materials. This is a compromise between loss of intensity by air scattering and absorption, on one hand, and the λ^2 or λ^3 dependence of sample scattering power, on the other hand. For white-beam experiments a molybdenum or tungsten target is preferred, when using a similar generator voltage as mentioned above, because of its higher *bremsstrahlung* radiation output.

The usable intensity is increased by a factor of 10, using rotation-anode generators. The heat load at the target is reduced by frequently changing the focal point of the electron beam on the target; the target material is coated on the surface of a cylinder which rotates with high angular velocity around its main axis. Rotational-anode generators typically consume about 18–25 kW of electrical power. Because of the high-speed rotation, this type of equipment requires more frequent service than sealed tube systems do.

The intensity of emitted radiation can also be increased by focusing the electron beam before hitting the anticathode. This can be achieved by electron optics installed within the electron beam path. The advantage of this optics is the reduction of the focus down to several micrometers. In addition the power of consumption is reduced to a few 100 W. The drawback of this source lies in a complicated vacuum system and a stabilized electronics for feeding the electron optics [43].

Recently a new type of microfocus tube came on the market operating at a power of less than 100 W but providing similar intensity on the sample site compared with rotational anodes without using any electron optics. This is possible by focusing the emitted x-rays within the tube by using small ellipsoidal mirrors or polycapillary optics [136, 365].

Storage ring facilities provide the most intensive x–rays, i.e., the synchrotron radiation. Extensive reviews have been published on the properties and application of synchrotron radiation in several monographs [8, 76, 198, 209]. Here we will concentrate on selected topics which are absolutely necessary for synchrotron radiation users.

High-energy electrons or positrons are moving within a storage ring and emit radiation whenever their path is curved. Second-generation storage rings produce radiation by the use of bending magnets. Here the curvature is induced by high magnetic fields, providing a radiation fan with a small opening angle within the ring plane. Within the ring plane the radiation is highly collimated and entirely linearly polarized, where the $\boldsymbol{E}$-vector is in the ring plane. Above and below the ring plane the radiation is elliptically polarized.

The emission spectrum of synchrotron radiation is well defined [323]. Its calculation is based on a knowledge of the bending radius R and the energy E_{el} of accelerated electrons or positrons. The spectrum is characterized by the critical energy E_c of the storage ring, given by

$$E_c = \frac{3hc}{4\pi R}\left(\frac{E_{\mathrm{el}}}{m_0 c^2}\right)^3 , \tag{1.1}$$

where h is the Planck constant, c the velocity of light, and m_0c^2 the rest mass energy of the electrons or positrons. In the ring plane the emitted radiation distribution, i.e., the photon rate per energy interval $\delta E/E$, is calculated by [390]

$$\frac{\mathrm{d}N/\mathrm{d}t}{\delta E/E} = \frac{e^2 c N_{\mathrm{el}}}{6\pi\epsilon_0 E_c R^2}\left(\frac{E_{\mathrm{el}}}{m_0 c^2}\right)^4 S(\xi) \tag{1.2}$$

with the spectral function $S \leq 1$,

$$S(\xi) = \frac{9\sqrt{3}}{8\pi}\xi \int_{E/E_c}^{\infty} K_{5/3}(\xi)\mathrm{d}\xi,$$

where $\xi = E/E_c$ and $K_{5/3}(\xi)$ is a modified Bessel function. N_{el} is the number of stored electron or positrons. As seen in Fig. 1.2, the emission spectrum of bending magnets and wigglers increases continuously up to the critical energy E_c but decreases very fast beyond. To compare different radiation sources the quality of the emitted radiation is characterized by its brilliance:

$$\text{brilliance} = \frac{\mathrm{d}N/\mathrm{d}t}{\text{mrad}^2 \times \text{mm}^2 \times 0.1\,\%\ \text{energy interval}}\,. \tag{1.3}$$

The brilliance describes the number of photons, N, emitted in one second from a source area of 1 mm^2, into a radiation cone defined by an spatial opening angle of 1 mrad^2 and normalized on a spectral bandwidth of 0.1% [233]. Another general quality parameter for characterizing the radiation source is its brightness, defined by

$$\text{brightness} = \frac{\mathrm{d}N/\mathrm{d}t}{\text{mrad}^2 \times \text{mA}^2 \times 0.1\,\%\ \text{energy interval}}\,. \tag{1.4}$$

This contains the beam current, measured in mA, as normalization parameter instead of the source area.

In Fig. 1.2 we compare the brilliance of several sources including x-ray tubes and various insertion devices at the example of the *Hamburger Synchrotronstrahlungslabor* (HASYLAB at DESY). In contrast to the line spectrum provided by an x–ray tube the radiation spectrum of a bending magnet is smooth over a large range of energy. The wavelength can be chosen freely, depending on the experimental problem in question. Due to the much higher intensity and better collimation of the probing x-ray beam a diffraction experiment can be performed much faster and with higher angle resolution at a bending magnet beamline than in a home laboratory.

In third-generation synchrotron sources so-called insertion devices are installed to improve the source properties. The electrons/positrons pass through a periodic magnetic field in order to increase radiation power in a particular direction (see top part of Fig. 1.3). Depending on the number of antiparallel arranged magnets and the spacing between them, the individual emission acts are in phase (undulator) or without a definite phase relation (wiggler). As shown in Fig. 1.2, the gain of intensity between bending magnets and insertion devices is about six orders of magnitude larger. This intensity is so high that the beam–sample interaction cannot be neglected, and even organic materials can be destroyed.

Besides the high intensity, one of the most promising properties of synchrotron radiation is its coherence. Similar to lasers in the visible spectral range, synchrotron radiation can be used to detect tiny phase differences between partial waves scattered at the same time from different spatial positions

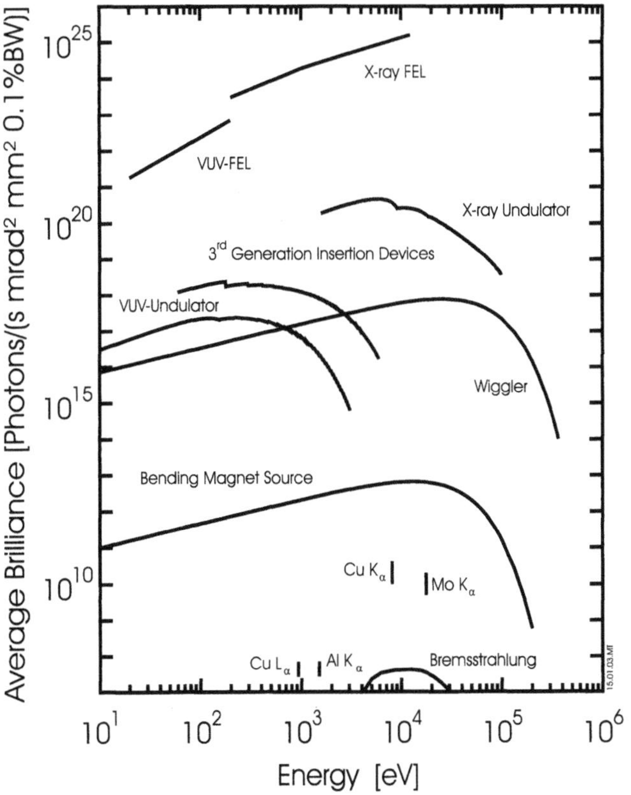

Fig. 1.2. The average brilliance of various radiation sources. Comparing the characteristic radiation from sealed tubes with the radiation of bending magnets gives a gain of 2–3 orders of magnitude. Wigglers and undulators increase the brilliance by a further 5–6 orders of magnitude. A similar hump is expected after bringing free-electron laser facilities in service. Meanwhile the TESLA-TEST-Facility (VUV-FEL) is in operation at the *Deutsches Electron-Synchrotron (DESY)* lasing at $\lambda \leq 85\,\mathrm{nm}$. The TTF-FEL is the prototype of an extended version (x-ray FEL) which will operate in the hard x-ray range [370].

of the sample (speckles) or at different times from the same sample position (time-correlation) in the nanometer range.

Radiation at storage rings exhibits a particular time structure. This is caused by the fact that charge carriers are stored in bunches. The bunch length is measured in terms of the time necessary to pass the focal point of the experiment; it is on the order of about 100 ps. The time interval between two bunches corresponds to several ns. This time structure makes it possible to run time-resolved experiments.

A third characteristic time constant of a storage ring is the lifetime of the stored charge carriers. It is limited by collisions of electrons or positrons with residual gas atoms within the ring and at the ring walls or by the interaction among the charge carriers themselves. Thus the number of electrons or positrons, i.e., the beam current stored, decreases approximately exponentially with time, continuously reducing the radiation power available. The lifetime depends primarily on the quality of the ultra-high-vacuum system of the storage ring. The lifetime varies between about 10 hours at HASYLAB and 48 hours at the ESRF, as typical examples. At a certain lower limit of ring current, the radiation process is stopped and new charge carriers are injected for a new synchrotron radiation run.

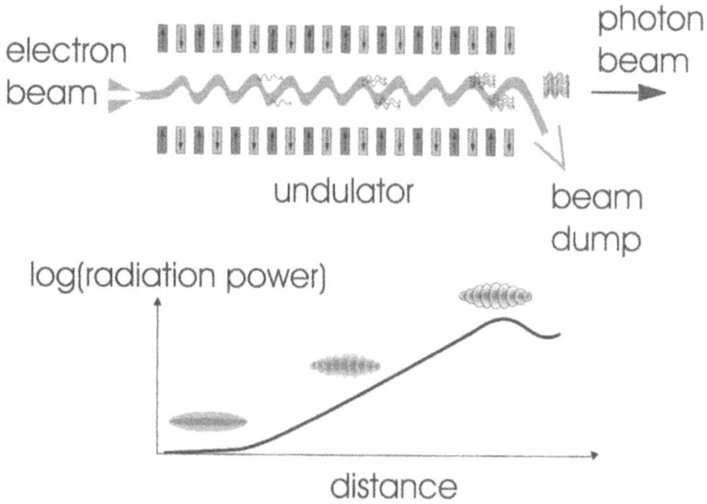

Fig. 1.3. Scheme of the SASE principle. A highly accelerated electron pulse becomes bunched while traveling throughout the high-field undulator. The self-organized density wave of electrons equals the wavelength of the emitted radiation. The SASE-principle enables an FEL to lase.

Fourth-generation synchrotron sources will become available for users in the near future. They are based on the principle of a free-electron laser (FEL). Here an electron bunch is accelerated by a linear accelerator (LINAC). During the passage of these relativistic electrons throughout a very long undulator the electron bunch undergoes self-organization to a traveling density wave with a wavelength equal to that of the emitted radiation. Due to this bunching any spontaneous photon emission becomes self-amplified (SASE principle) which accelerates the emitted density wave in proportion to the number square of bunching electrons. The general scheme of an SASE-FEL is shown in Fig. 1.3. LINAC-driven SASE-FELs are expected to provide extremely coherent x-ray pulses of less than 100 fs with a peak brilliance beyond

10^{33}photons/s/mm^2/mrad^2/0.1%BW. Fig. 1.2 shows the average brilliance of the FELs. The gain in intensity will be much larger than the gain which happens between undulators and sealed tubes. These new kinds of sources are under development worldwide. Meanwhile the minimum wavelength emitted by an FEL is below 75 nm [19]. A project for installation of an X-FEL with emission wavelengths in the range of 0.1 nm is planned at Deutsches Elektron-Synchrotron DESY in Hamburg [370] and at SLAC in Stanford [334]. Compared to third-generation sources free-electron lasers will initiate a completely new type of experiment exploiting fs-pulses and extremely high coherence. Whether these sources can be used for diffraction experiments in a manner described in this book is not clear yet. In this sense the denotation of LINAC-driven FELs as fourth-generation synchrotrons might be misleading.

Generally, there is not continuous access to synchrotron radiation sources. Several synchrotron radiation laboratories allocate the access via a proposal system. Thus the time for synchrotron radiation experiments seldom exceeds more than 1–2 weeks per year and project, which restricts the usage of synchrotron radiation to a few selected experiments. Even in the future, the routine characterizations of samples will be realized mainly at experimenters home laboratories.

1.2 Optical Elements

To perform a high-resolution experiment one needs a well-collimated and monochromatized incident beam of sufficient intensity that strikes the sample. After interaction with the sample, that part of the incident beam needs to be recorded by a detector that scatters into a definite direction in space. Often one is interested in evaluating the coherent part of the scattering event only. This can be realized by optical elements installed between the beam path, i.e., between source and sample and/or between sample and detector. A simple way to design the probing beam is to use various slit systems. They consist of either a circular aperture or two blade pairs arranged perpendicularly to each other, both symmetrically versatile with respect to the center of the beam. Both restrict the incident/reflected beam to a proper area suitable to probe the sample under investigation (see Fig. 1.4). In addition, the beam divergence becomes reduced (as long as the aperture diameter or the blade distances are on the order of the coherence length of radiation – see later). Usually the slit width is by many order of magnitude larger than the wavelength so that the Fraunhofer diffraction from the slits is negligible. To avoid fluorescence radiation in the energy range of about 10 keV, the slits have to be made from tungsten or platinum.

Additional efforts for designing the scattering experiment involve the implementation of more sophisticated optical elements. Here, in general, one has to distinguish between optical elements improving the beam quality and those increasing the intensity at the sample site. The elements of the first

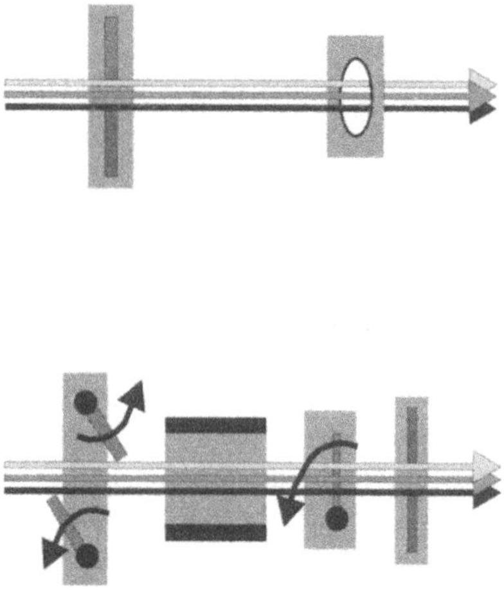

Fig. 1.4. Optical element consisting of a combination of a slit and a circular aperture (top); or a combination of two separated slits, a Soller slit (a set of parallel metal sheets), and flexible absorbers (bottom) to limit the illuminated sample area.

group are referred to as monochromators and analyzers, respectively, since they reduce the energy or angular range accepted by the sample or the detector. An optical element used for parallelizing the beam is referred to as the collimator. The second group of optical elements are parabolically or elliptically bent mirrors, mainly bent in one or two directions with different bending radii. These mirrors collect the photons oncoming from the incident beam and deflect them toward the sample, at the cost of an increased angular divergence. Similar effects are achieved by beam compressors constructed by a non parallely arranged double crystal arrangement (see later).

In order to understand the function of optical elements we have to anticipate a few formulas from the theoretical chapters presented in Part II of this book. Tailoring of the beam can be performed by Bragg diffraction at a single crystal. Depending on its lattice spacing and crystal perfection, the angular width and the energy band pass, respectively, of the Bragg-diffracted beam becomes reduced after passing the optical element. From the dynamical theory of x-ray diffraction (see Chap. 6) its half-width $\Delta\eta_h$ in angular space varies as

$$\Delta\eta_{\boldsymbol{h}} = 2C|\chi_{\boldsymbol{h}}|\frac{1}{\sqrt{|b|}\sin(2\Theta_B)} . \tag{1.5}$$

C is the linear polarization factor defined in Sect. 5.1, Θ_B is the Bragg angle introduced in Chap. 4, and $\chi_{\boldsymbol{h}}$ is the $\boldsymbol{h}$-th Fourier coefficient of the crystal polarizability $\chi(\boldsymbol{r})$. This coefficient depends on the atomic scattering factors $f_s(\boldsymbol{h})$ of the atoms in positions $\boldsymbol{r}_s$ in the crystal unit cell (see Sect. 5.2):

$$\chi_{\boldsymbol{h}} = -\frac{\lambda^2 r_{\mathrm{el}}}{\pi V_{\mathrm{el}}} \sum_s f_s(\boldsymbol{h}) \mathrm{e}^{-\mathrm{i}\boldsymbol{h}.\boldsymbol{r}_s}, \tag{1.6}$$

where r_{el} is the classical electron radius defined in Sect. (5.1) and V_{el} is the volume of the unit cell. $f_s(\boldsymbol{h})$ varies as a function of the length of the scattering vector $\boldsymbol{h}$. It equals the ordinary number of the element for $\boldsymbol{h} = 0$ but decreases rapidly with increasing $\boldsymbol{h}$. The asymmetry factor b is defined by

$$b = -\frac{\sin(\Theta_B + \phi)}{\sin(\Theta_B - \phi)}, \tag{1.7}$$

in which ϕ means the angle between the diffracting lattice plane and the sample surface (asymmetry angle – see Sect. 4.2 for more details).

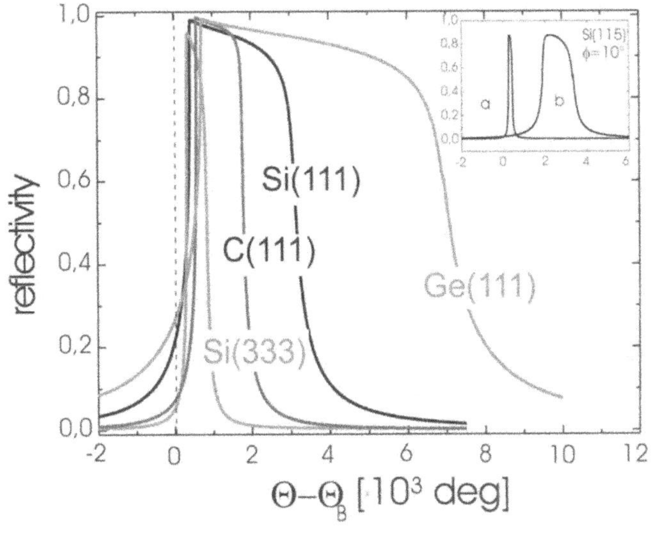

Fig. 1.5. Normalized dynamical diffraction curves of the silicon (111) and (333) Bragg diffractions from [111] oriented crystal (symmetric diffraction geometry). The $2\theta = 27.6°$ corresponds to a wavelength of $\lambda_{111} = 0.150\,\mathrm{nm}$ and $\lambda_{333} = 0.05\,\mathrm{nm}$, respectively. As seen, both curves coincide at the low-angle side. The intrinsic half-width is reduced additionally if the diffracting lattice planes do not coincide with the surface (asymmetric diffraction geometry). Here the silicon (115) net plane is artificially inclined by 10° with respect to the crystal surface. The different curves shown in the inset correspond to the situation of grazing exit (a) and grazing incidence (b). For comparison the (111) Bragg diffraction curves of diamond and germanium are shown additionally, appearing at the same Bragg angle for $\lambda = 0.102\,\mathrm{nm}$ and $0.156\,\mathrm{nm}$, respectively. The calculations have been performed for a p-polarized incident beam.

From arguments given in Sect. 6.4 and as shown in Fig. 1.5 the angular width $\Delta\eta$ of the dynamical diffraction curve of the silicon (111) Bragg diffraction is several times greater than its third harmonic, i.e., the (333) diffracting at $\lambda/3$. This is mainly caused by the λ^2 dependence of χ_h and is due to the smaller f_s in (1.6). This behavior has been experimentally verified by Bonse et al. [58]. $\Delta\eta$ can be reduced further when the diffracting lattice plane does not coincide with the crystal surface. In that case the diffraction geometry becomes asymmetric with respect to the sample surface. This is shown in the inset of Fig. 1.5 for the example of the asymmetrically cut silicon (115) lattice plane. Compared with the symmetrical (333) diffraction, the diffracted beam becomes narrower if the beam leaves the surface at a grazing angle $(\Theta_B + \phi)$, but it becomes wider in the opposite case. Additionally, the angle position of the peak maximum differs.

The different angular acceptance of a particular Bragg diffraction can be used to select a certain energy band pass from a continuous spectrum. At fixed Θ_B the band pass can be reduced using high-indexed net planes (as 333 in Fig.1.5 instead of 111). A further reduction of the band pass is achieved by increasing the asymmetry of the diffraction. From a derivative of the Bragg law and substituting the energy for the wavelength one obtains the relation $\Delta E/E = \Delta\eta/\Theta_B$. Using the examples presented in Fig. 1.5, we find the selected energy band pass to be 20×10^{-5}, 3.5×10^{-5}, and 1.5×10^{-5} for Si (111) and (333) in symmetric and (115) in grazing exit geometry, respectively.

Another way to narrow the angular width in angular space or to reduce the energy band pass is to diminish the electron density of monochromator material. The lower the electron density, the smaller χ_h is, and subsequently the curve width, but at the cost of the diffraction power. Integrated over the entire diffraction curve, the diffracted power decreases in the same direction. As shown in Fig. 1.5, the angular width of the diamond (111) diffraction is narrower than the silicon (111) calculated at an equal Bragg angle. Thus C(111) accepts a wavelength range which is approximately half of that accepted by Si(111). On the other hand, the respective diffraction curve of germanium (111) is twice as large as Si(111). Thus germanium yields a higher flux at the sample site compared to silicon. Compared with silicon, diamond shows better heat transport properties and a smaller thermal expansion coefficient. This makes the material preferable for synchrotron radiation use [227]. Therefore, diamond crystals become more and more popular as monochromators and are implemented at several beamlines at the ESRF [140]. However, on account of the crystal perfection and the low material price, silicon is still the most widely used material for monochromators.

The combination of two or more crystals into one unique optical element results in a double-crystal or multiple-crystal arrangement. The properties of a double-crystal arrangement, i.e., the properties of a particular combination of a monochromator crystal and the crystalline sample, were already discussed by von Laue [213]. Supposing a parallel setting of the sample and

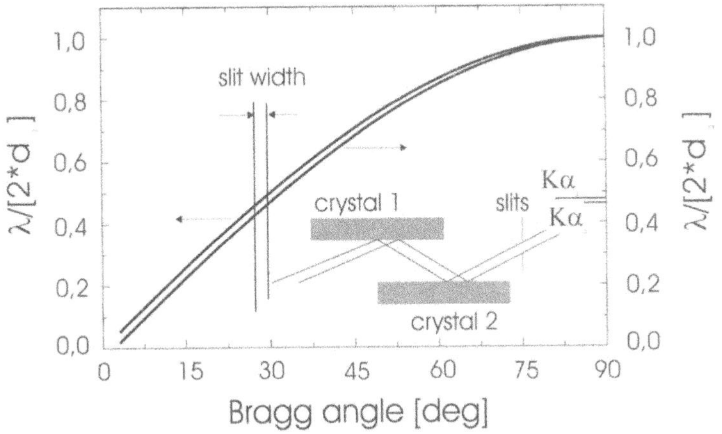

Fig. 1.6. DuMond diagram and set-up of a non-dispersive double-crystal arrangement.

monochromator, the so called (+−) setting [105] is non-dispersive if both crystals are equal and similar lattice planes are used for diffraction. That means that the second crystal accepts the same range of wavelengths accepted by the the first crystal supposing the Bragg angles of both crystals, $\Theta_{B,1}$ and $\Theta_{B,2}$, are equal. If $\mathcal{R}_j(\eta_{i,j})$ is the diffraction power of the crystal j at the incident angle $\alpha_{i,j}$ and β is the rotation angle of the second crystal with respect to the first one, the intensity measured after the second crystal is

$$I(\beta) = \int \mathrm{d}\eta_{i,1} \mathcal{R}_1(\eta_{i,1}) \otimes \mathcal{R}_2(\eta_{i,2}), \tag{1.8}$$

where

$$\eta_{i,2} = \pm\left(-\frac{\eta_{i,1}}{b_1} + \beta\right) \tag{1.9}$$

is the angular deviation of the incidence angle at the second crystal. b_1 is the asymmetry factor of the first crystal; the sign $\pm$ refers to the non-dispersive and dispersive arrangements, respectively. Eq. (1.8) is valid for a single polarization. For a non-polarized x-ray source the intensity results from a incoherent superposition of S- and P-components. Similar formulas can be derived for a multiple setup of crystals.

For a particular energy the convolution product in Eq. (1.8) gives a reflection with a curve width of about $\sqrt{2}\Delta\eta$ as defined in Eq. (1.6) since the Bragg angles of both crystals are equal. This curve is much narrower than the divergence of an incident beam leaving a sealed tube and can be used as a monochromator and collimator. This can be visualized by a DuMond diagram

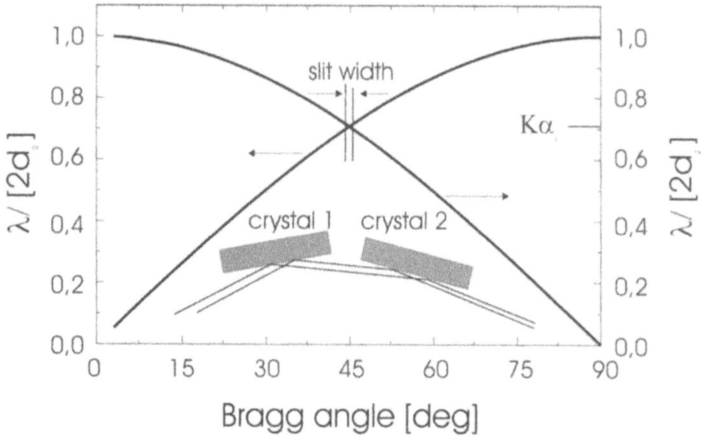

Fig. 1.7. DuMond diagram and experimental set-up for a dispersive double-crystal arrangement.

(Fig. 1.6). It shows the functional dependence of the Bragg law for both crystals. The co-ordinate axes for both crystals are arranged parallelly, but the horizontal (angular) axes of both are displaced by the angle β against each other. For crystals with identical Bragg angles $\Theta_{B1,2}$ and a divergent incident beam, both curves are parallel; i.e., all energies diffracted by the first crystal are diffracted by the second crystal as well. After diffraction, the intensities corresponding to the $K\alpha_1$ and $K\alpha_2$ lines of an x-ray tube still are strong and appear simultaneously. The other energies are not Bragg-diffracted toward the selected direction, and subsequently they are extremely reduced in intensity. If $\Theta_{B1} \neq \Theta_{B2}$, the first and second crystal diffract at a slightly different wavelength, each of which gives rise to a *dispersion enlargement* of the convoluted curve. This enlargement can be approximated by

$$\Delta\eta(\Delta\lambda) = \frac{\Delta\lambda}{\lambda}(\tan\Theta_{B1} - \tan\Theta_{B2}), \quad (1.10)$$

where $\frac{\Delta\lambda}{\lambda}$ is the energy band pass approaching the experiment. The energy band pass can be further reduced by narrow slits. In particular the selection of the $K\alpha_1$ line requires that a slit be placed between sample and monochromator, as shown in the inset of Fig. 1.6. If the monochromator material differs from that of the sample, the arrangement is not perfectly non-dispersive. This slightly increases the acceptable band pass but at the expense of the achievable angular resolution.

A strictly dispersive arrangement is realized by an antiparallel setting of monochromator and crystal; the (++) arrangement is shown in Fig. 1.7 [105]. The DuMond graphs of both crystal reflections are aligned in opposite directions. Both reflection curves overlap at one particular angle only. Since this overlap on the energy axes is much smaller than the $K\alpha_1$–$K\alpha_2$ separation, the optical element accepts one of both lines only. This arrangement has been developed first for spectroscopic applications and is implemented now in four-bounced monochromators (see later). Manufactured as a monolithic optical element, the (+−) and (++) set–up can be arranged in front of or after the sample to realize high-resolution conditions (see Chap. 3). Nowadays the non-dispersive arrangement is realized by channel-cut monochromators [145]. Here the *channel* is grooved into a perfect silicon block aligned parallel to a definite lattice plane, as (110), for example. Depending on the angle of incidence with respect to this plane, the incident beam is diffracted two or more times inside before leaving the groove. The beam now available has the property of an improved peak-to-background ratio since the final rocking curve is the product of these two or four intrinsic Bragg diffractions. Supposing a peak reflectivity of 90% and a reflectivity in the tails of 0.01 %, the final reflectivity amounts to 60% at the peak but 10^{-8} at the tails after four reflections. Note, however, a channel-cut crystal is not able to separate the $K\alpha_1$ from the $K\alpha_2$ line or to suppress higher harmonics . This requires a slit system again or a combination of dispersive and non-dispersive elements. Several solutions are discussed by Hashizume et al. [150]. In particular they proposed a triple-diffraction monolithic silicon monochromator which delivers a strictly monochromatic beam for a single definite wavelength. Other monolithic arrangements have been published by Kohra [200].

The four-reflection monochromator consists of two channel-cut crystals in (++) setting [27]. It combines the advantage of the low tail intensity of a channel-cut (+−) setting with the highly reduced energy band pass of the (++) arrangement. This is the most effective way to separate the $K\alpha_1$ line from the emission spectrum of an x-ray tube. Additionally, it maintains the incident beam direction. Figure 1.8 shows such a device for implementation to a high-resolution diffractometer. The (022) and (044) settings are pre-adjusted and can be used alternatively.

However, each monochromatization and collimation reduces the total radiation power at the sample position. Thus extreme conditions are not necessary and are not desirable, in general. As a *golden rule*, the divergence of the incident beam should not be smaller than the width of the intrinsic diffraction curve of the sample under investigation. Otherwise the convolution product between both the monochromator curve and the sample is too small to achieve sufficient flux; the counting time per angular step needs to be increased in order to guarantee good counting statistics. Under laboratory conditions, extreme collimation as delivered by a four-bounce monochromator is necessary only for the investigation of very narrow rocking curves, i.e., highly perfect materials such as crystalline silicon or GaAlAs/GaAs heterostructures. Another reason for using high resolution is the *reciprocal space mapping*, where the intensity distribution is recorded in the vicinity of a reciprocal lattice point (see Chap. 8).

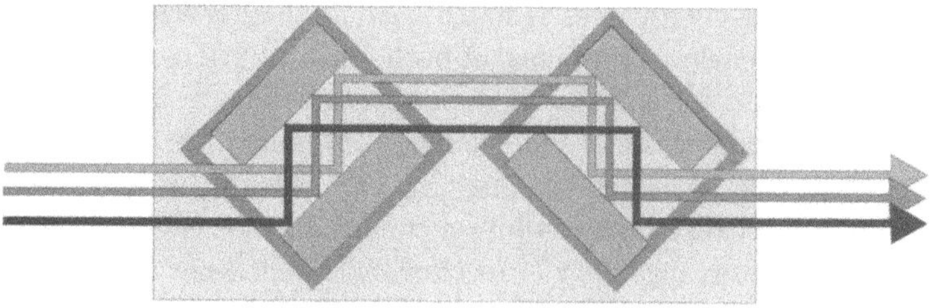

Fig. 1.8. Four-bounce monochromator made from two counter– rotating monolithic germanium crystals, cut nearly parallel to the (022) crystal plane. The (044) reflection can be attuned after counter–rotation of both crystals to the respective Bragg angles.

For measuring crystal powders or materials with a mosaic spread, a monochromator with even large mosaicity (pyrolithic graphite, for example) is recommended in order to minimize the counting time.

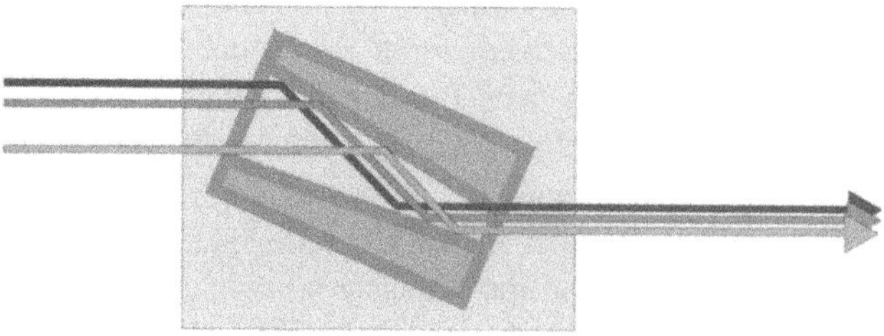

Fig. 1.9. Beam compressor made by a conic channel-cut crystal. It collects the incoming intensity and produces an increased power density at the sample position at the cost of an increased beam divergence. The use of a beam compressor is recommended when the divergence of the incident beam is nearly completely accepted by the optical element.

An alternative way to increase the photon flux at the sample site is to use a beam compressor (see Fig. 1.9). A beam compressor is a channel-cut crystal with non-parallel surfaces narrowing the channel width along the beam path. The diffracting lattice planes of both crystals and their miscuts with respect to the inner surfaces are chosen in such a way that the incoming photons being collected at the entrance face become concentrated toward the exit face by multiple inner reflection. The price of an increased power density is a slightly increased beam divergence, compared with the angular acceptance of the compressor. The compressor is useless if the incoming beam divergence is much larger than the compressor acceptance.

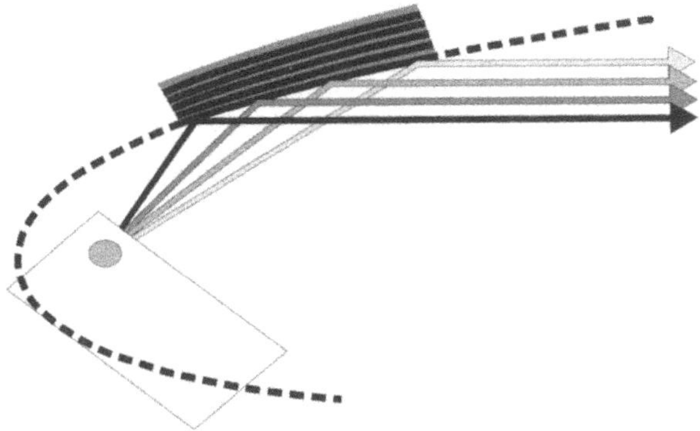

Fig. 1.10. Schematic setup of a parabolically bent multilayer mirror with laterally varying lattice spacing. It collects the divergent radiation emitted by the sealed tube and produces a nearly parallel intense beam.

A home laboratory experiment always suffers from the inefficient exploitation of the radiation produced by the x-ray tube. Because the radiation is emitted into the whole upper half-space of the tube, the photon flux toward a particular direction is relatively low. This general problem cannot be overcome using flat monochromators or collimators but it can partially solved using a parabolically bent multilayer mirror [322] with additionally varying lattice spacing across the beam pass (Fig. 1.10). This mirror accepts a divergent incident beam but reflects a rather parallel beam. The rest divergence amounts to about 80 arcsec which is larger than the divergence of a germanium monochromator (12 arcsec). However, a parabolically bent mirror (*Göbel mirror*) increases the intensity at the sample site by a factor of about ten compared to a beam which is conventionally defined by a collimation line of two slits. The *Göbel mirror* attached to a sealed tube yields an intensity at the sample site that is comparable with that of a slit arrangement and a rotation-anode tube [355].

Synchrotron radiation facilities supply a nearly parallel and highly intense beam of radiation. Here the problems of designing an x-ray diffraction or reflectivity experiment are rarely related to achieve a gain of intensity. The main problems are related to a proper choice of the energy band pass and of the angular resolution.

An intensity problem may arise if the beam divergence is much narrower than the intrinsic reflectivity curve of the sample. As already mentioned, this may happen when measuring samples with large mosaicity spread. In this case the convolution of monochromator and sample (see Eq. (1.8)) provides a flat curve of low intensity, because the incident beam always probes a very narrow part of the sample curve. Sometimes the intensity measured at a synchrotron facility can be lower than that recorded from the same sample using home equipment and a rather divergent incident beam.

For the efficient measurement of imperfect samples, the angular divergence of the incident beam has to be increased. This can be achieved by using bent mirrors like those used for spectroscopy applications. Unfortunately, these mirrors are bent into two directions and the bending radius is often too large for our application. A better method is the use of high-reflectivity multilayer mirrors [321]. They are composed of thin layers of at least two materials providing high electron density contrast; often they are made by double layers from two metals with a large difference in the ordinary number, such as C/W or Si/Ta, for example. The vertical lattice spacing is chosen in such a way that the first-order Bragg peak reflects a particular energy band of interest which can be exploited for the experiment. As shown

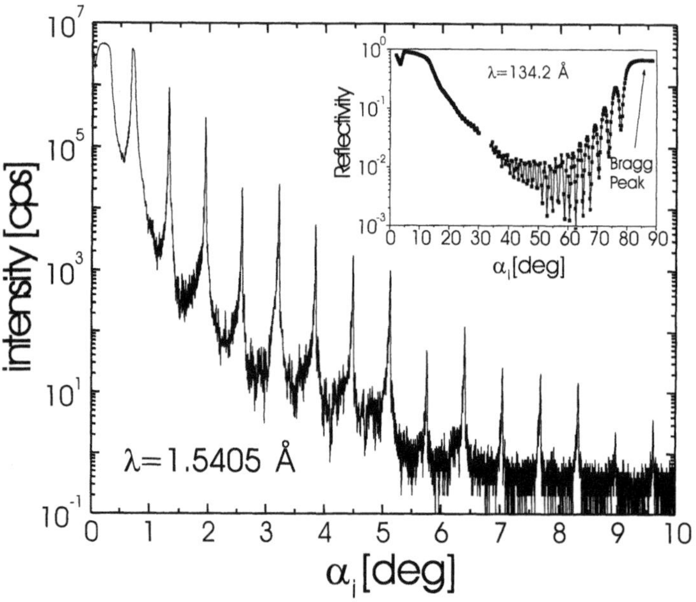

Fig. 1.11. X-ray reflectivity of a 50-period Mo/Si multilayer made for application as a mirror for soft x-rays probed with CuK_α radiation. The inset shows the same sample measured at an energy of 92.4 eV. Here the 1st- order Bragg peaks appears under normal incidence. The band pass achieved amounts to about 4% [377].

in Sec. 5.4, of the reflection maximum of a multilayer depends on the number of covered layers. For instance, about 100 double layers of 5 nm each give a beam divergence of about 1 mrad or a band pass on the order of 10^{-2}. This possibility is often applied at various synchrotron radiation laboratories. Figure 1.11 shows an example of an Mo/Si multilayer grown on polished quartz substrate. It is particularly designed for application in EUV photithography, i.e., at photon energies below 92.4 eV. In this case a single element of the 50-period multilayer consists of four sublayers: 1.6 nm Mo + 0.8 nm Mo_xSi_{1-x} + 3.7 nm Si + 0.8 nm Mo_xSi_{1-x}. The perfection achieved is probed by an x-ray specular reflectivity experiment. Sharp Bragg peaks appear up to an incident angle of about 10°. At the 1st-order Bragg peak the reflectivity is 83% and the band pass amounts to 4%. The reflectivity for an energy of 92.4 eV is shown in the inset of Fig. 1.11. Here the 1st-order Bragg peak appears under normal incidence with an reflectivity of about 70%.

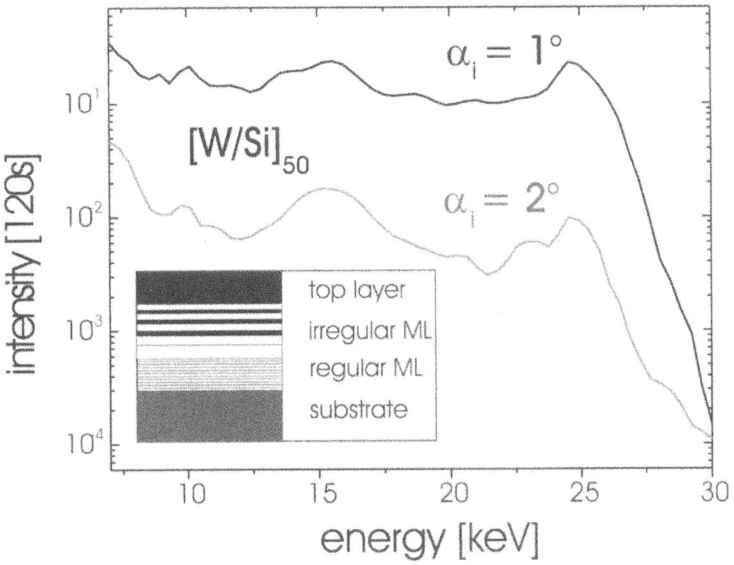

Fig. 1.12. Reflection spectrum of a super-mirror made by 50periods of an Si/W multilayer (ML). A regular bi–layer spacing is followed by an irregular spacing in order to achieve high reflectivity within the angular range between the angle of total external reflection and the first-order Bragg peak of the regular multilayer.

A novel design of those multilayers is a *super-mirror*. Here, the lattice spacing of the a bi-layer varies with the number of covered layers in such a way that the angular range between the first-order Bragg peak and the edge of total external reflection reflects with approximately equal intensity. Consequently, this mirror accepts a wide range of energies but rejects all energies above a critical one. This optical element can be used for energy-dispersive experiments. Figure 1.12 shows a reflectivity curve of a super-mirror, designed for application at BESSY II [112].

A crystal monochromator, as shown in Fig. 1.5, always accepts higher harmonics of a particularly selected Bragg reflection. Due to their spectral characteristics the higher harmonics show very similar peak intensities compared with the main reflection and may excite higher harmonics of Bragg reflection of the sample as well. In order to measure a *clean* signal of a single Bragg reflection, one has to to suppress these higher harmonics from the incident beam.

This can be done by inserting a plane mirror in front of the monochromator. If it is illuminated under a fixed grazing angle α_i, those energies E become totally reflected that are smaller than the critical angle of total external reflection, which depends on the mirror material (Fig. 1.13). The high order harmonics fall out due to the E^{-4} dependence of the Fresnel reflectivity. The energy band pass varies with α_i^{-1}. This energy selection is most effective using gold-coated mirrors because of its high electron density and chemical stability. A true but rather large band pass is obtained if the high-energy cut-off is combined with an absorption unit suppressing a particular low-energy part of the incoming white spectrum. Due to the E^{-3} dependence of photo absorption [8], attenuation of low energies is very effective, whereas high energies penetrates a thin metallic foil nearly unaffected. For energies less than about 5 keV, air is already a very effective absorber.

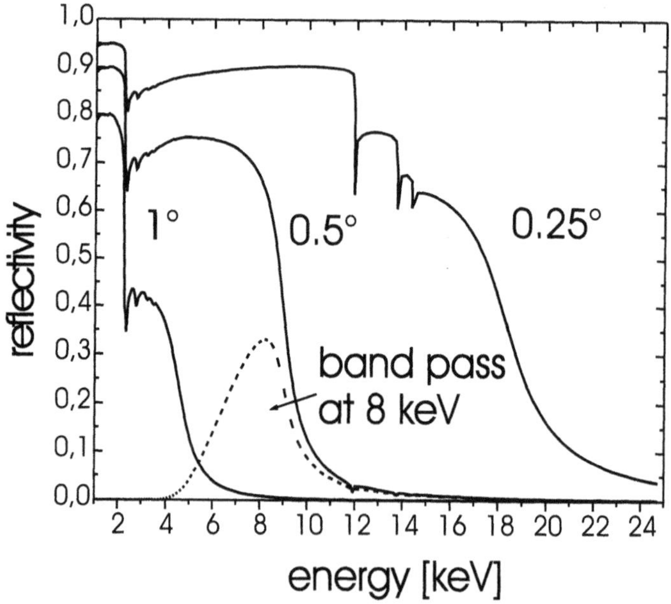

Fig. 1.13. Energy characteristics of a plane gold mirror for different angles of incidence α_i. An energy band pass of several percent is achieved by insertion of a thin aluminum foil within the beam path.

Another way to attenuate higher harmonics is to detune one crystal of a double crystal monochromator by a small amount. In most cases the double crystal arrangement is realized by two separate plane monochromators. Then, the first crystal is cooled by flowing water or liquid nitrogen in order to dissipate the heat load caused by the highly intense white synchrotron radiation. The absorbed heat alters the lattice parameter of the first crystal and the second monochromator has to be re-adjusted to maintain maximum reflectivity of the whole optical element. The chance for detuning is based on the different angular widths of the Bragg diffraction curves as shown in Fig. 1.5. Whereas the low-angle sides of the higher harmonics coincide with that of the main diffraction, the angular position of the high-angle sides differ substantially. Using an Si(111) monochromator, for example, the third harmonics can be suppressed if the second crystal of the monochromator is detuned by about $\Theta - \Theta_B) \approx 0.003°$, which still fulfills the (111) diffraction but not the (333) one.

1.3 Detectors

Detectors have to register the diffracted or reflected intensity at a particular position in space and time. It may be realized by moving a point detector through the diffraction curve or using a one-dimensional or two-dimensional detection unit at a fixed angular position. Whereas the area of the detector window and the precision of the detector movement determines the resolution in the first case, the spatial resolution of a single recording unit is essential in the second case. Sometimes the detector also has to select a particular energy. The required energy resolution depends on whether one performs an angular- or energy–dispersive experiment. However, the analysis of thin films and multilayers requires the registration of scattering intensity over several orders of magnitude. This necessitates high dynamics of the detection process.

Commercially available x-ray diffractomers or reflectometers are often equipped with proportional or scintillation counters [260]. A proportional counter consists of a metal cylinder with a central wire anode capped by a beryllium or mylar side window. The tube is typically filled with xenon gas along with a small amount of CO_2 or CH_4 for discharging the charge cloud created by a single photon. An incoming x-ray photon ionizes a number of noble gas molecules proportional to its photon energy. The ionization energy is on the order of 30 eV. Thus an 8 keV photon produces about 240 electron-ion pairs. Under an applied high voltage of 1.4–1.6 keV, both electrons and ions ionize a lot of additional gas molecules. Although the multiplication (avalanche) factor can reach up to several thousand, the number of electron-ion pairs is strictly proportional to the photon energy. The positive ions need a longer time to approach the cathode. Meanwhile it is enough time to form an electron pulse close to the anode. This pulse is differentiated from noise using a pulse high analyzer and stored as a single count using a scaling circuit.

After this, a certain time (dead time τ) is necessary to quench the foregoing charge carrier cloud in the vicinity to the central wire. This limits the maximum number of photons which can be detected in a given time. If the photon flux is higher not all the photons are registered. The detected number of photons N approximately follows the relation [199]

$$N = \frac{N_0}{1 + N_0 \tau} . \tag{1.11}$$

Here N_0 is the true number of photons entering the detector per second. The number of registered photons depends on the dead time τ, which is on the order of several microseconds. If $1/\tau = N_0$, then 50% of the incoming photons are detected.

Proportional counters are used for photon fluxes up to several 10^5 *counts per second* (cps). The low energy resolution of about 20% makes it possible to attenuate higher harmonics but it is insufficient to suppress the Kβ line.

In a scintillation counter the incoming photons strike a fluorescent crystal, which emits photons of visible light. For hard x-rays NaI crystals doped with 1 % of Tl are effective scintillators. A single 8-keV photon produces about 250 photons. About 50% of them leave the scintillation crystal toward the photo-multiplier tube to form an electron pulse close to the anode, registered as a single count by the pulse-hight analyzer. As shown in Fig. 1.14, one selected photon of visible light excites one or more photo electrons at the first dynode which become multiplied by high-voltage acceleration toward the next dynode. Using a cascade of dynodes, a multiplication factor of about 10^6 is typically achieved. Scintillation counters behave linearly within 1% up to a counting rate of about 10^5 cps. They cannot be used for energy selection, because their energy resolution is not better than 40%.

So-called *plastic counters* are scintillator counters where the fluorescence crystal is replaced by an organic material, like polyvinyltoluene [52]. Compared with conventional scintillation detectors both the rise time and decay time are on the order of nanoseconds, which gives an up to two orders of magnitude higher count rate.

For high photon fluxes, low-efficiency counters are recommended [372]. Ion chambers are constructed similarly to proportional counters. For an operation voltage of 200–300 V the avalanche process becomes negligible and the detection efficiency depends primarily on the number of absorbed photons and the photoionization yield of the respective counting gas used. In synchrotron radiation laboratories ionization chambers are often used for monitoring the primary incident photon flux. For that purpose the x-ray beam passes through the ion chamber and creates a certain number of electron-ion pairs within the gas volume. Although the total number of absorbed photons is low, the induced discharge current registered at the outer circuit is strictly proportional to the incoming photon flux.

Experimental arrangements exploiting a white primary x-ray beam require an energy-dispersive detection system, which can be realized by a solid-

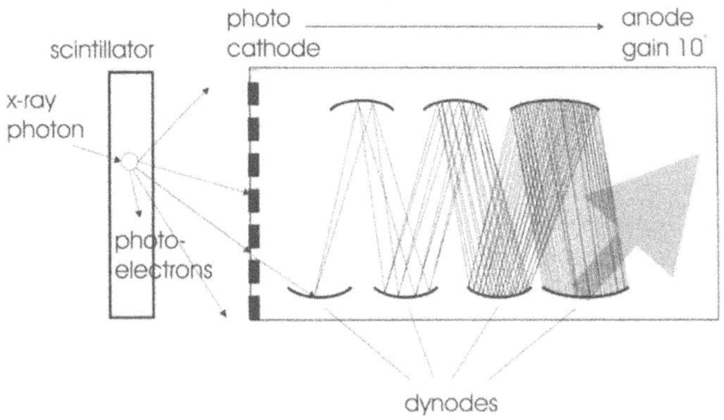

Fig. 1.14. Functional scheme of a scintillation detector. X-ray photons create photons in the visible spectral range within the scintillator, which themselves excite photo electrons at the first dynode which become multiplied by acceleration within a high electric field between several dynodes (photo multiplier).

state detector [70]. There are pure germanium and lithium drifted silicon diodes which reach a sufficient efficiency for detecting x-ray photons. Germanium crystals are more efficient than Si:Li but they show additional *escape* peaks within the spectrum (see below). The principle of detection is simple: each x-ray photon creates a definite number of electron-hole pairs within the intrinsic (i) region of a semiconductor p-i-n-diode. Taking into account the energy necessary for creating a single electron-hole pair (3.3 eV for silicon), the number of created charge-carriers pairs is proportional to the energy of the incoming photon. The resulting electrical pulse is analyzed by use of a pulse height analyzer and stored into a certain channel of a multi-channel analyzer.

These detectors operate under liquid nitrogen cooling to prevent thermal creation of charge carriers and the drift of Li^+ ions across the silicon lattice. Furthermore, the preamplifier operates at liquid nitrogen temperatures as well. Fortunately the purity and perfection of crystals now available makes it unnecessary to keep the detectors under liquid nitrogen cooling when they are not in use.

In a hard x-ray regime the achievable energy resolution is smaller than 200 eV, that is, less than 2.5% at 5.6 keV. The relative resolution $\frac{\Delta E}{E} = -\frac{\Delta\Theta_B}{\Theta_B}$ of solid state detectors is up to two orders of magnitude worse than the corresponding resolution evaluated from the intrinsic angular width of a perfect single-crystal rocking curve. However, the resolution of a solid-state detector is sufficient for thin film analysis, especially for measurements on the relative scale.

Solid-state detectors become especially advantageous for non-perfect or artificially layered samples and in the case of powder diffraction. In multilayer systems, for example, one can detect the entire scattering curve of the sample in a few seconds which is most attractive in the study of time-dependent processes (see Chap. 8).

An additional advantage is the simple separation of higher harmonics and the better peak-to-background ratio compared with scintillation counters. The disadvantage of an energy-dispersive system is the relatively small flux which can be detected simultaneously. The counting loss becomes important as it is already between 10,000 cps and 15,000 cps for Si:Li's and close to 20,000 cps for Ge-detectors.

A second disadvantage is the appearance of additional fluorescence lines at the spectrum emitted from elements of the sample and equipment and the so-called *escape peaks.* Because there is a definite probability that an incoming photon excites a K electron of a germanium or silicon atom one finds additional peaks in the spectrum displaced from highly intensive peaks by 9 keV or 1.8 keV, respectively, to lower energies. Their identification becomes crucial especially in strongly structurized spectra.

A third problem arises due to the probability that the system may create two photons of the same energy entering the detector at the same time. It produces a number of electron-hole pairs which is equivalent to that number of photons created by a single photon of twice energy. It produces *pile-up* peaks at twofold energy. The pile-up probability is on the order of about 0.1 % for the twofold and 10^{-4} to 10^{-5} for the threefold energy with respect to the incident one. Figure 1.15 demonstrates this effect using the example of a CuK-fluorescence line measured with a Si:Li detector at the energy–dispersive beamline at BESSY II. Here, several orders of pile-up and escape peaks are visible on the log-scale of intensity. The pile-up effect is essential whenever the integrated intensity is much larger than the inverse of the dead time of the detector system. Using a Si:Li detector, the effect is negligible only if the integrated intensity is less then 10^4 cps.

Another energy-dispersive system is based on a silicon p-i-n diode operating under Peltier cooling [9]. Since it is small, the system is flexible in its application. The energy resolution is slightly lower than that of a Si:Li but both systems are equal in the maximum detectable count rate.

Recently, an improved p-i-n diode came on market, also equipped with Peltier cooling. An much faster pre-amplifier enables the detection of a photon rate above 10^5 cps with low counting loss [289]. Again the energy resolution is same as for a Si:Li.

There are many technical solutions of position-sensitive detectors (PSD) [14]. One example is a linear PSD consisting of a metal or quartz fiber coated with carbon [59, 62]. This low- or high-resistance wire, respectively, is centered within a straight or bent metal tube, which may be aligned perpendicular or parallel to the x-ray scattering plane. The tube is continuously flushed by a noble gas (argon 95%) with the addition of quenching-gas molecules (methane 5%). In the case of the PSD sold by Comp. BRAUN (see Fig. 1.16),

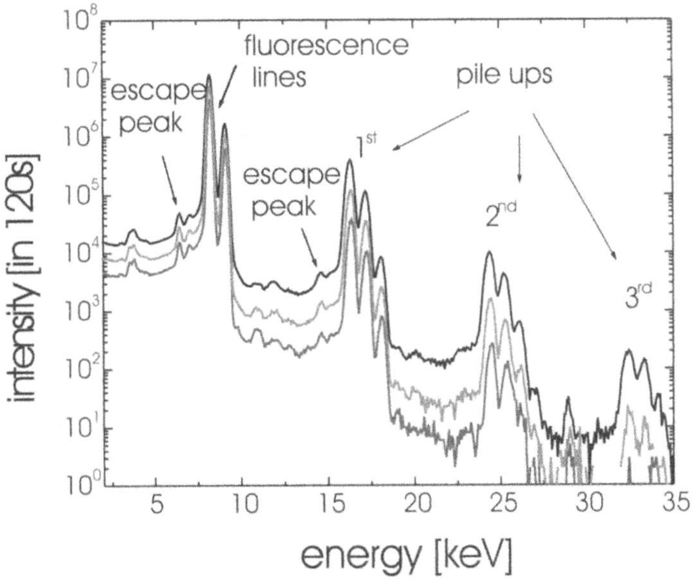

Fig. 1.15. Pile-up effect of an intense copper fluorescence line measured by an Si:Li detector at three different incident intensities. The escape peaks are shown at the low-energy side of each main peak. Note the log-scale of intensity.

the incoming photons ionize the gas molecules, which themselves are amplified toward the cathode wire at a certain position Z. The created charge pulse is registered at counters 1 and 2 installed at opposite ends of the wire and connected via separate circuits. The running time of the charge pulse between Z and a counter is proportional to the respective spacing. Consequently counters 1 and 2 will register a different charge Q_1 and Q_2 which give a measure for the position Z.

$$Z = L\frac{Q_1}{Q_1 + Q_2}. \tag{1.12}$$

This detection mode corresponds to the situation of so-called high-resistance wires. Using low-resistance wires, the spatial position of the event along the wire axis is measured by the time difference between electron and hole pulses, respectively, detected by identical amplifiers at both ends of the wire [59]. The spatial resolution depends on the gas pressure, voltage, and gas composition. In practice the achievable spatial resolution is about 40 μm, which allows for an angular resolution of about 0.005° if the detector is placed about 45 cm from the sample.

Besides the detectors running with continuous gas stream, there are other systems which store the gas for a definite time (one week) [391]. A disadvan-

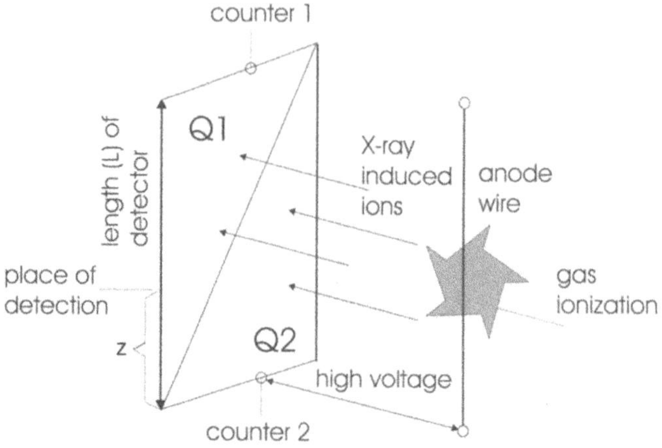

Fig. 1.16. Schematic set-up of a linear position-sensitive detector. An x-ray photon creates ions near the anode wire which are accelerated toward the two half-cathodes. The impact position Z on the wire is determined by evaluating the charges Q_1 and Q_2 at two counters 1 and 2. (with permission of MBRAUN GmbH).

tage for all these systems is the relatively low integral photon flux that can be recorded simultaneously. About 10^5 cps can be detected with a counting loss smaller than 10%. The maximum counting rate is limited to about 5×10^5 cps or smaller. Otherwise the coating of the wire becomes destroyed.

Area detectors are important for simultaneous registration of reciprocal-space maps. Besides very expensive systems of two-dimensional grids of wires based on principles as explained above [118], there are image plates and CCD systems in operation. Image plates are sheets coated with europium-activated barium halides (BaFX: Eu^{2+}, X=Cl or Br) [335]. They are exposed like photographic films. A single photon excites an electron into a metastable F-center. The spatial resolution of this process is on the order of 200×200 μm^2. Macroscopically, it creates a latent picture of the scattering event. After illumination the image plate has to be read out immediately because the life-time of metastable levels is limited to several minutes. This is performed by a laser system scanning the image plate line by line; it takes about 2–3 minutes. Under the influence of green laser light, the stored electron becomes free and relaxes into the ground state accompanied by the emission of visible (blue light) photons, which are recorded as a function of spatial position on the image plate. Under these conditions the intensity of the blue light is a function of the initially absorbed x-ray photons. The scanned two-dimensional intensity distribution is stored and it can be used later in further data processing. The scanning read-out equipment is the most expensive part of the system; the price for a single image plate is very cheap (US \$10).

Two other modern area detector systems are photodiode arrays or charge-coupled devices (CCDs). In general, the created charge carriers are stored in the depletion region of a p-i-n diode. A CCD system [170] consists of an ar-

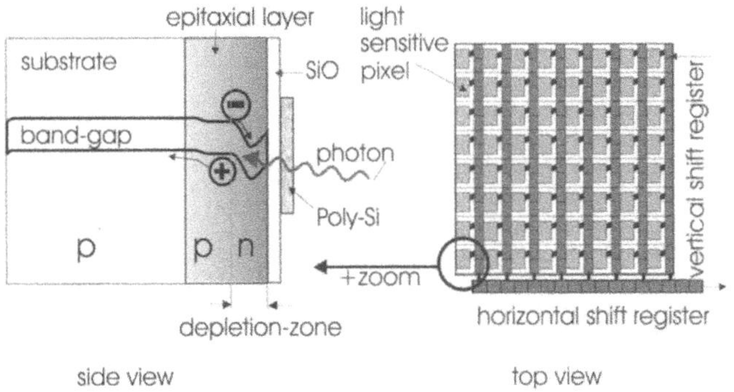

Fig. 1.17. Scheme of a CCD. The x-ray photons create a charge at the depletion zone of a single capacity. The read-out of information is performed like a scoop chain via vertical shift registers downward and finally along the horizontal shift register. The left figure shows a side view of a single capacity; the right figure shows a top view of the capacitor array connected to the shift registers.

ray of light sensitive pixels made by MOS capacitors. Incoming x-ray photons create electron-hole pairs within the depletion zone which are stored at the position of the pn-junction (see left part of Fig. 1.17). The read-out of the stored charges takes place like a scoop chain via horizontally and vertically arranged shift registers (see right part of Fig. 1.17). After one interval of illumination (typically 20 ms) all created charges were transferred via a transfer channel into the light-insensitive vertical shift register. During the next illumination time each of these charges becomes shifted by one step downward. At the same time all charges of lowermost registers, the horizontal shift register, are read out and can be stored electronically. The next line of charges is read out in the next illumination time, etc.

The dynamics of a CCD suffers from the remaining dark current of the individual MOS capacitors. This dark noise can be reduced efficiently by Peltier cooling. At a transfer frequency of about 1 MHz, the read-out noise corresponds to a charge of 10 electron charges where one pixel is able to store 10^5 electrons. The pixel size amounts to about 20×20 m. For x-ray photons the probability of absorption is lower than for visible light photons. Therefore the storage process is more efficient if the x-ray photons are transformed into visible ones by means of a scintillator crystal attached in front of the CCD. The efficiency is further increased by amplifying the visible light output, improving the signal-to-noise ratio.

2 Diffractometers and Reflectometers

In the previous chapter we described the basic elements of x-ray equipment, namely, x-ray sources, optical elements to define the beam pass and a proper band pass of energy, and various recording units to detect the x-rays. In this chapter we deal with the experimental arrangement as a whole. There are general aspects to consider in setting-up an x-ray experiment: The sample is illuminated by an incident beam striking the sample surface under a definite angle of incidence α_i. The incident beam may be characterized by its incident divergence $\Delta\alpha_i$ and its energy spread $\Delta\lambda$. After interaction of the beam with the sample, the scattering intensity escaping from the surface at a take-off angle α_f has to be recorded. Owing to the finite size of the detector window, all the photons leaving the sample surface in a take-off angle range $(\alpha_f - \Delta\alpha_f/2, \alpha_f + \Delta\alpha_f/2)$ are registered by the detector simultaneously. The plane containing the source and the surface normal of the sample is called the *scattering plane*. Since the scattering happens within the scattering plane, the scattering geometry is called *coplanar*. In the so-called *not coplanar* or *off-plane* scattering geometry (Sect. 4.3), the directions of the incident and the scattered beams are also characterized by the azimuthal angles θ_i and θ_f and by the widths of the corresponding angular intervals $\Delta\theta_i$ and $\Delta\theta_f$.

Different problems of measurement require slightly different diffractometer arrangements. X-ray reflectometry uses scans at very small α_i. If one is interested in thin-film analysis, high-resolution in angular space is not necessary. As shown later, a β-filtering of the white spectrum of an x-ray tube gives sufficient energy resolution so long as the layer thickness to be analyzed does not approach the micron range and so long as one is not interested in recording reciprocal–space maps.

The situation is different for wide-angle diffractometry. Here high resolution in energy and angular space is necessary for analyzing the lattice misfit between epilayer and substrate even for nearly lattice-matched heterostructures. High resolution also guarantees the low noise also necessary for analyzing very thin films. High resolution is obtained by using optical elements. As shown in Section 1.2 these elements reduce $\Delta\alpha_i$ and/or $\Delta\alpha_f$ at the cost of decreasing the integrated scattering power. This has to be compensated for by insertion of photon collectors as beam compressors or Göbel mirrors or otherwise by increasing the recording time per angular step. Often one has to compromise between the degree of resolution and the time necessary for measurement. In this chapter we will show that high resolution is not required in general. Samples with mosaic spread, heterostructures with a large lattice mismatch or epilayers with a thickness of a few 100 nm can be investigated under slightly relaxed conditions of resolution without loss of information.

2.1 X-Ray Reflectometers

X-ray specular reflectometry is used to measure the thickness of individual thin layers, the vertical spacing of a multilayer stacking, the surface and interface roughnesses, and the average density of a layered system. According to the α_i^{-4} law of Fresnel reflectivity (Sects. 5.2 and 6.4), the intensity leaving a smooth surface decreases very rapidly as the angle of incidence increases. To record the reflected intensity over more than 6 orders of magnitude one needs a highly intensive incident beam and/or a detector with low noise.

The apparative requirements should be demonstrated by the following examples. In case of layered samples the layer thickness is determined from the angular distance between the thickness oscillations (Kiessig maxima, see Sect. 8.1). The required angular resolution depends on the total thickness of film. Film thicknesses of about 50 nm provide an angular distance between Kiessig maxima of about 0.1°. Such a film can be investigated using the Cu$K\alpha$-doublet of a sealed tube collimated by two spatially separated located slits before the sample and and an additional slit placed before the detector.

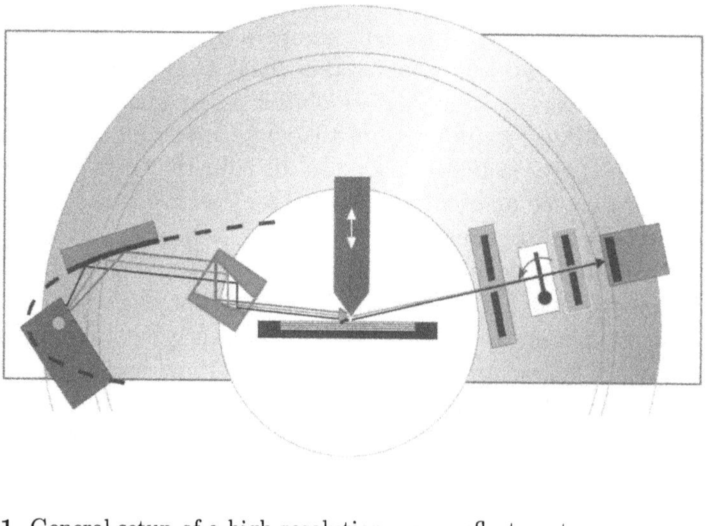

Fig. 2.1. General setup of a high-resolution x-ray reflectometer.

The precise determination of the critical angle of total external reflection, α_c, which is necessary for electron density analysis, demands a much better angular resolution. Often it is sufficient to match α_c approximately with the angular position of half-intensity compared with the primary beam (Sect. 8.1). The average density is obtained with an accuracy of 5% if α_c is measured to an accuracy of 2.3 %. It requires precise angle adjustment on the order of 0.001°.

It follows from these previous estimates that a reflectivity experiment needs high flux at the sample site, moderate angular resolution in most cases, but accurate angle adjustment between sample and detector circle. At home-laboratories an angle-dispersive reflectometer (Fig. 2.1) should consist of the source attached to a photon-collecting system as Göbel-mirror (completed by beam compressor), horizontal and vertical slits to define the beam size, and a detector with a large dynamical range. The slits can be inserted either before or after the sample. The angular resolution of the experiment is additionally controlled by a knife-edge attached close to the center of goniometer. Both the sample circle, denoted by ω, and the detector circle, 2θ, move with an accuracy of $\Delta\omega = \Delta 2\theta \leq 0.001°$.

The specular reflectivity is recorded by running an $\omega/2\Theta$ scan, where ω corresponds to the true angle of incidence α_i and 2Θ is the angular position of the detector measured with respect to the incident beam direction. Both α_i and the exit angle $\alpha_f = 2\Theta - \alpha_i$ are moving simultaneously by the same amount. With so-called θ–θ reflectometers, where the sample stays fixed, α_i and α_f vary directly instead of $\omega = \alpha_i$ and $2\Theta = \alpha_i + \alpha_f$ which is the case with at equipment, where the x-ray tube is fixed. The properties of the $\omega/2\Theta$ scan and others with different angular ratios will be described in Sect. 3.2 in detail.

The function of the knife-edge, shown in Fig. 2.1, can be explained as follows. For geometrical and intensity reasons, both $\Delta\alpha_i$ and $\Delta\alpha_f$ cannot be reduced too much. Because the incident beam divergence and the detector acceptance are large, the irradiated sample area must be reduced to achieve sufficient angular resolution. This is achieved by sinking the knife-edge very close to the axis of goniometer rotation, which additionally helps to bring the sample surface to the rotation center (see below). Under these conditions, only those parts of the beam reach the detector, which are escaping the sample straight beneath the knife-edge. The extreme limitation of the scattering area reduces the detectable intensity by several orders of magnitude. This disadvantage can be compensated for by insertion of a photon collector in front of the sample (see Fig 1.9).

An alternative setup is shown in Fig. 2.2. The first Göbel-mirror is sufficient to suppress the $K\beta$ line. High resolution still can be achieved by replacing the beam compressor in Fig. 2.1 by a Bartels monochromator (Sect. 1.2 and Fig. 1.8). Relaxed resolution but higher flux is achieved by removing the beam compressor but setting a second Göbel mirror in front of the

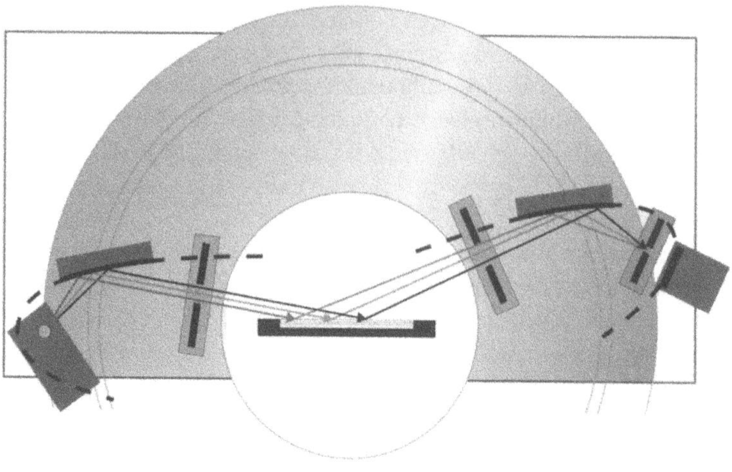

Fig. 2.2. Scheme of a powder diffractometer applied for x-ray reflectometry with relaxed angular resolution.

detector. The gain of detected intensity is due to the fact that the second Göbel mirror increases the beam cross section. Specularly and non-specularly scattering photons are deflected toward the center of a point detector. This set-up corresponds to that of a powder diffractometer [167].

At synchrotron facilities the incident beam is parallel enough. The reflectometer has to be equipped vertically in order to make use of the extreme collimation of the incident beam in this direction (see Sect 1.1). For angular dispersive experiments one has to insert an optical element for monochromazing the incident x-ray spectrum. This can be done either before or after the sample.

A completely different arrangement is necessary to perform an energy-dispersive experiment (Fig. 2.3) [57]. Here the white beam strikes the sample surface under a fixed $\omega = \alpha_i$ and the reflected beam is recorded by an energy–dispersive detector at a fixed angle $2\Theta = \alpha_i + \alpha_f$. In this case the only requirement is an approximately parallel incident beam provided at a synchrotron facility without optical elements except slits. In a home laboratory a sufficiently parallel beam is prepared by passing the beam through two slits fixed at both ends of an about 1-meter-long tube, which should be evacuated to reduce air scattering and absorption by air. Only low air absorbance supplies the whole white spectrum of an x-ray tube [238]. The resolution of an energy-dispersive scattering experiment is determined by the energy resolution of the detector. It amounts to ($\Delta E/E \leq 2.5\,\%$), which is sufficient for thin film analysis. The advantage of this set-up is the possibility to record time-dependent processes. The scattering spectrum always is available and can be controlled directly by the user. The appearance of intense lines in the incident spectrum of a sealed tube, the energy dependence of sample absorbance, and the detector response (see Sect. 1.3) make it difficult to interpret the spectrum quantitatively. For quantitative analysis one has to remove the Kα and Kβ lines from the *bremsstrahlung* spectrum as well as several fluorescence lines excited in the equipment by the incident photons. Therefore the energy-dispersive set-up is recommended for measurements on a relative scale, than for absolute measurements. However, using laboratory sources and choosing $\alpha_i \cong 1°$, about 5 minutes sufficed to collect a spectrum of an organic multilayer sample with satisfying counting statistics [229].

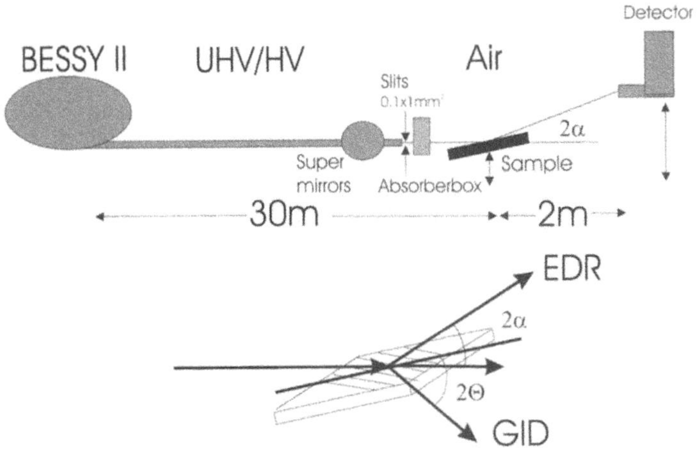

Fig. 2.3. Setup of the energy-dispersive reflectometer installed at the EDR beamline at BESSY II. It is equipped to measure simultaneously the reflectivity and grazing-incidence diffraction of a sample.

At the EDR beamline of BESSY II the same sample was measured in about 10 seconds with much better statistics [57]. Here energy-dispersive reflectometry is very efficient for rapid sample analysis and even for routine measurements.

Sample alignment is a crucial problem for accurate reflectivity measurements. In particular, the main error of density determination via measurement of the critical angle of total external reflection is the inaccurate sample alignment. Also a true specular scan can be recorded only when 2Θ is exactly twice ω. The procedure of alignment is same for an angle- and energy-dispersive set-up and will be explained in the following.

In order to measure the angles correctly, the rotational axis of the sample circle (ω-circle) has to be aligned exactly with the sample surface (Fig. 2.4). Additionally, one has to make sure that the position of the rotation axes of both circles coincides with the center of the incident beam; that means that the sample has to shadow half of the incident beam. With commercial

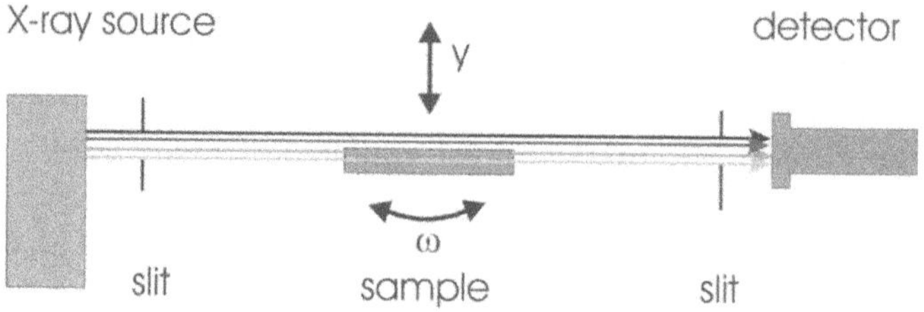

Fig. 2.4. Procedure to align the sample surface normal with the rotation axis of the reflectometer. This procedure is general for all types of reflectometers and diffractometers.

diffractometers it is a guarantee that the rotation axes of sample and detector coincide. However, the procedure to adjust the sample surface consists of an iterative movement and rocking of the sample across the incident beam (ω-scan). Both scans have to repeat until the peak intensity of the ω–scan is half of the intensity of the incident beam, measured without sample. In this case the ω-axis lies exactly in the sample surface and this surface is parallel to the incident-beam direction.

After this adjustment the angular position of the sample, however, may not coincide with the zero mark of the ω-circle. This might be caused by surface damage or by a miscut of the surface with respect to the bottom plane of the sample. Additional tests are necessary to redefine the ω-scale. To do this with sufficient accuracy one has to choose an incidence angle in the range $0 < \alpha_i < \alpha_c$ and find the angular position of specularly reflected beam at an angle 2θ. If 2Θ does not coincide with $2\alpha_i$, the zero point of the ω-circle needs to be re-scaled by $(2\Theta/2 - \alpha_i)$. Repetition of the procedure at various values of α_i improves the precision of sample alignment.

The reflectivity experiments should be optimized in such a way that the specular reflectivity can be measured over a wide range of α_i This is necessary to show typical features as Bragg peaks and Kiessig fringes characterizing the sample. Due to the α^{-4} dependence of Fresnel reflectivity the intensity drops over 6 to 8 orders of magnitude. Sometimes it is helpful to record parts of a single reflectivity curve at different conditions of angular resolution and counting time. So the angular range close to α_c can be detected with the highest resolution available, but the wide angle range should be recorded using a relaxed resolution. Under home laboratory conditions one needs to change the angular resolution, i.e., to increase or decrease $\Delta\alpha_i$ and $\Delta\alpha_f$. When synchrotron radiation is used, separate detection of the low-angle and wide-angle range is necessary due to the limitation of detector sensitivity.

2.2 High-Resolution Diffractometer

In the semiconductor industry, in particular, the necessity to analyze epitaxially grown highly perfect multilayer materials encouraged the development of new types of diffractometers. They are well adapted to measure layer thickness, lattice mismatches and lattice strains of heterostructures. The investigation of quantum well structures, i.e., thin layers with thicknesses of less than 10 nm embedded within much thicker cladding layers, requires measurement of rocking curves over a wide angular range left and right with respect to the Bragg peak of the substrate. The method of reciprocal-space mapping makes it necessary to have good resolution in two directions of reciprocal space. All this requires a highly intense but parallel incident beam and a low background associated with a good as possible angular resolution. These needs should be satisfied by a high-resolution set-up at a synchrotron beamline. Unfortunately, due to the limited access, such experimental stations cannot be used for routine characterization. Thus, most of measurements have to be performed at home laboratories.

Modern high-resolution diffractometers are equipped with a four-bounce monochromator (Fig. 2.5). The intensity of the x-ray tube is increased by attaching a Göbel mirror. Maximum resolulution, that is necessary for reciprocal-space mapping, is achieved by use of a channel-cut analyzer before the detector. As for reflectometers, the diffractometer makes motor-controlled angular steps $\Delta\omega = \Delta 2\Theta \leq 0.001°$ on both sample and detector circle. The equipment is suited to record reciprocal-space maps in the vicinity of a particular Bragg peak of the sample in several hours.

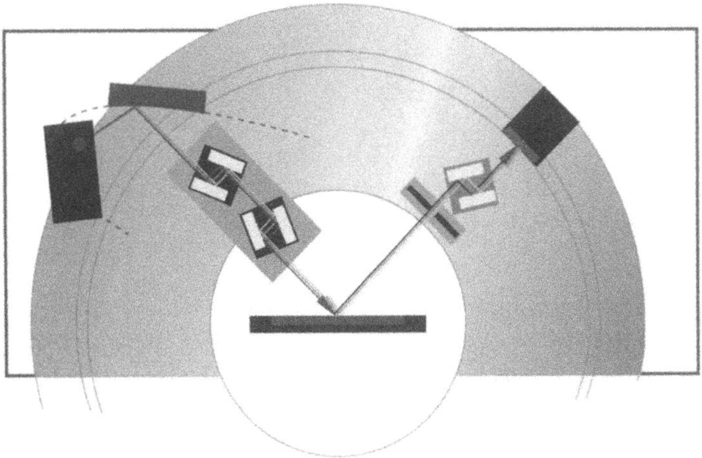

Fig. 2.5. High-resolution diffractometer.

The equipment shown in Fig. 2.5 should be simplified if the sample is not highly perfect. In this case the Bartels monochromator can be replaced by a single- or double-crystal arrangement. The analyzer crystal may be replaced by a slit system for simple rocking curve analysis. The achievable angular and energy resolution can be estimated using the DuMond diagrams shown in Sect. 1.2. Using (+−) set-up, the *dispersion enlargement* of each peak of the rocking curves follows from Eq. (1.10) identifying $\eta_{1,2}$ by the angular deviation from the diffraction maximum of the monochromator η_M and of sample η_S, respectively. The broadening of the diffraction curve can reach several hundred seconds of arc if the monochromator/analyzer and sample Bragg angles differ. In a strict sense, high resolution is achieved only if monochromator, sample, and analyzer are made from same material and scatter at exactly the same Bragg angles. The use of a four-bounce monochromator overcomes this problem. As demonstrated in Sect. 1.2, it combines the +− and ++ set-up and allows one to measure always the intrinsic rocking curve of sample. Because the apparative broadening is very small, the the exploitable intensity is small compared with the usage of a double-crystal arrangement.

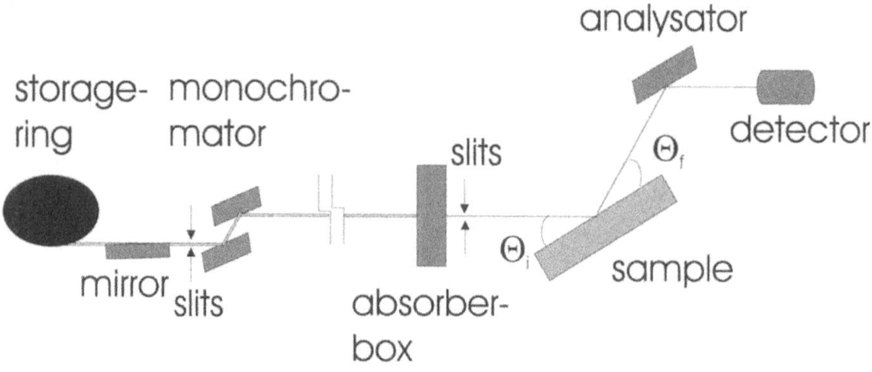

Fig. 2.6. General set-up of a high-resolution diffractometer equipped at a beamline of a storage ring facility.

All parts of the equipment have to be mechanically stable on a time scale of several hours or days so that reciprocal-space maps can be recorded under constant conditions (see Sect. 3.2). Reciprocal-space maps are obtained running ω-scans for different 2Θ or running $(\omega + \omega_0) : 2\Theta$ scans for different off-set angles ω_0 and plotting the recorded intensities in a 2D frame.

To run a particular scan across the reciprocal space, the sample and the detector circles have to move in an arbitrary ratio which differs from 1:2. These scans must be supported by the diffractometer software. In fact, computer software is an essential component of modern diffractometers. Fast access to a code of rocking curve simulation enables interpretation of the experimental curves straight after the measurement and interactive action. However, the user has to make sure beforehand that the basic assumptions of the simulation software agrees with those of the actual experiment.

Finally we propose an optimum arrangement for high-resolution diffraction which can be installed at a beamline at a synchrotron radiation facility. A schematic set-up is shown in fig. 2.6. Because the demand for intensity is not as high (about 8 orders of magnititude are sufficient) the high photon flux delivered by an undulator can be used to design a setup with very good angular resolution. A heavyweight goniometer can help to guarantee reproducible steps of less than 0.0005°. The small divergence of the synchrotron radiation in vertical direction should not be enlarged by bent mirrors. Plane monochromator crystals performed by a double-crystal arrangement supply an energy band pass with a resolution on the order of 10^{-4}. A collimation line defined by two pairs of slits, a guiding slit pair straight after the monochromator and a defining slit pair close before the sample, reduce the divergence of the incident beam. Supposing a source height of 100 μm and a length of the beamline of 20 m, the slit height of 1 mm provides a divergence of less than one second of arc, which is much smaller than the intrinsic half-width of the silicon monochromator crystal (see Fig. 1.5). Despite these constrains the flux at the sample site will exceed 10^{12} photons/sec if we choose a typical undulator beamline of ESRF. Finally a plane monochromator should be attached close before the point detector to guarantee high-resolution of the exit angle α_f, as well.

2.3 Limits of the Use of Powder Diffractometers

Several problems of x-ray characterization can be solved by using low-resolution diffractometers, i.e., a powder diffractometer. If the layered material is damaged, the Bragg peaks become broad and low in intensity. Following Eq. (1.8), the divergence of the incident beam should not be much smaller than the angular width of the diffraction curve of the sample under investigation. In the case of poly-crystalline material or material with a mosaic spread the intrinsic curve width may increase to several minutes of arc. Here focusing beam arrangements are preferred. A bent crystal-monochromator or analyzer in front of the sample or the detector, respectively, provides increased intensity and a relaxed angular resolution. A disadvantage of this arrangement is the simultaneous appearance of the Kα doublet in the rocking curve, which has to be taken into account for the simulation of the diffraction curves.

Figure 2.7 shows the (422) diffraction curves of a partially relaxed 40 period $In_{0.2}Ga_{0.8}As/GaAs$ multilayer recorded, first, with a graphite-analyzer-equipped powder diffractometer, similar to that shown in Fig. 2.2 and, second, under high-resolution conditions (see Fig. 2.5) using a four-bounce monochromator without Göbel mirror. The time taken to record the high-resolution

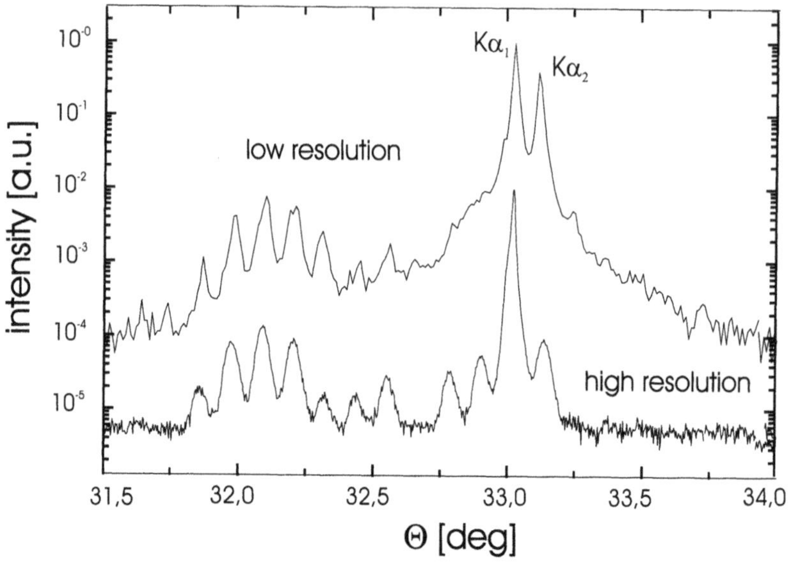

Fig. 2.7. Rocking curve of an $In_{0.2}Ga_{0.8}As/GaAs$ multilayer recorded with a powder diffractometer, similar to that shown in Fig. 2.2, and with high-resolution conditions, using a four-bounce monochromator.

curve was about five times longer than that for the powder diffractometer. The main difference between both curves is the appearance of the $K\alpha_2$ line at the powder curve. Except in the region close to the substrate peak, the superlattice peaks are clearly visible. Here, they are smeared out because of the lower angular resolution compared with the high-resolution curve. Because the measured angular width of superlattice peaks is determined by the mosaicity, the structure parameters of the multilayer sample can be estimated on the basis of this powder curve [290].

2.4 Grazing-Incidence Diffraction

A schematic illustration of the set-up of a grazing-incidence diffraction (GID) experiment is shown in Fig. 2.8 [89, 281]. Here one has to distinguish between the *plane of incidence* containing α_i and α_f and the surface normal and the *scattering plane* lying approximately perpendicular to the plane of incidence and containing the angles θ_i and θ_f. The latter ones are measured with respect to the diffracting lattice plane. The method unifies in-plane Bragg diffraction and out-of-plane reflectivity combined with the feasibility for depth resolution.

The experimental set-up is the following: a monochromatic and parallel x-ray beam as defined in Fig. 2.6 strikes the sample surface at an angle α_i close

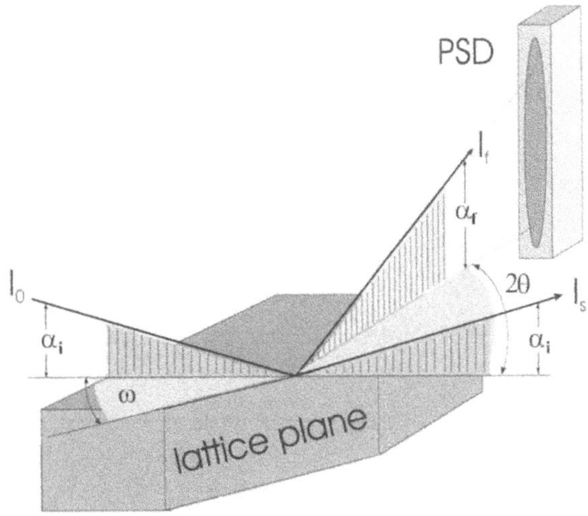

Fig. 2.8. Schematic set-up of a grazing-incidence diffraction experiment.

to the critical angle α_c. The sample is rotated around the surface normally until a particular lattice plane lying perpendicular to the surface fulfills the Bragg condition under an in-plane Bragg angle $2\Theta_{B||} = \theta_i + \theta_f$ (see Sect. 3.3).

Owing to refraction of the incoming beam at the air–sample interface the penetration depth of the probing x-ray can be controlled by choosing α_i to be smaller or larger than α_c (Fig. 2.9). In the first case the incoming beam becomes evanescent and propagates parallel to and close below the surface. The minimum penetration depth is on the order of 4–10 nm, depending on the density of material. On increasing α_i, the penetration depth within the sample increases up to about 400–600 nm.

The GID geometry requires collimation in both directions, perpendicular and parallel to the plane of incidence. In general, a set-up like that shown in Fig. 2.6 can be used but with additional efforts to gain the in-plane resolution. In fact, one needs a second monochromator installed perpendicular to the first one; this reduces the photon flux at the sample site again. For practical reasons the divergence with respect to α_i should be one order of magnitude smaller than that with respect to $2\Theta_{B||}$. On the other hand, a moderate divergence regarding $2\Theta_{B||}$ is necessary to excite the crystal truncation rod (CTR) (see Sect. 3.2). The intensity distribution along the crystal truncation rod can be recorded simultaneously as a function of α_f using a position-sensitive detector (Sect. 1.3). The measurement of reciprocal-space maps requires the insertion of an analyzer crystal. Because the reflectivity of the truncation rod is on the order of 10^{-8}, the intensity of a sealed tube is not enough to record in–plane scattering curves. The intensity of a rotational-

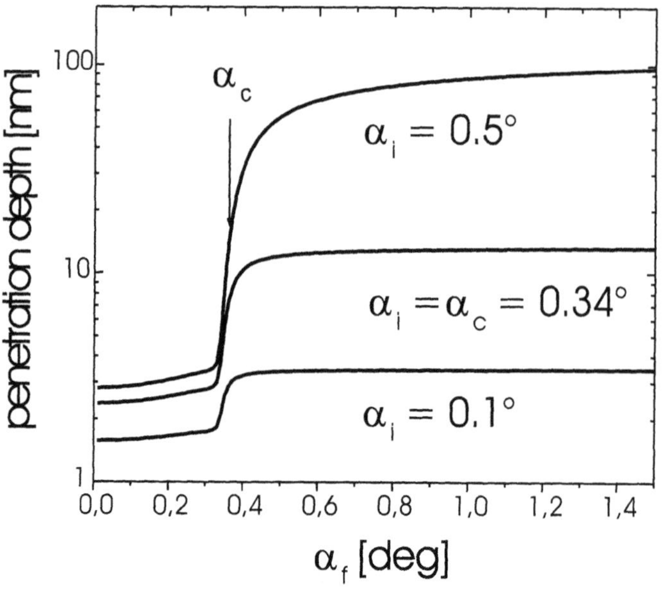

Fig. 2.9. The effective penetration depth below a GaAs surface for a GID experiment, calculated for different incidence angles α_i and exit angle α_f.

anode is sufficient for the detection of in-plane rocking curves integrated over the whole α_f range, i.e., with a wide open detector window. The measurement of α_f-resolved curves with good angular resolution is only possible with use of synchrotron radiation. α_f integrated measurement can also be performed using the energy-dispersive arrangement shown in Fig. 2.3.

3 Scans and Resolution in Angular and Reciprocal Space

In this and in the next chapters we show that the scattered intensity can be expressed as a function of the *scattering vector*

$$\boldsymbol{Q} = \boldsymbol{K} - \boldsymbol{K}_i, \tag{3.1}$$

if the the direction of the perfectly monochromatic incident plane wave is described by its wave vector $\boldsymbol{K}_i$ and that of scattered wave by $\boldsymbol{K}$. Therefore, the angular distribution of the scattered intensity, representing the properties of the sample, can be described by its distribution in reciprocal space (*reciprocal-space map*). Strictly speaking, this expression is applicable only if the wave vectors $\boldsymbol{K}$ and $\boldsymbol{K}_i$ and the surface normal vector $\boldsymbol{n}$ lie in the same plane (*scattering plane*). This scattering geometry is called *coplanar.* As we show later, in the non-coplanar scattering geometry (in particular, in the grazing-incidence diffraction – GID, see Sect. 4.3) the scattered intensity depends on both vectors $\boldsymbol{K}_i$ and $\boldsymbol{K}$ independently.

A non-zero divergence of the primary beam, its non-zero spectral width, as well as a non-zero angular width of the acceptance of the detection system broadens all features of the reciprocal-space within the detection plane. Then, the measured intensity corresponds to an average of a definite area in reciprocal space. This particular area is called *resolution area* or *resolution element.*

A quantitative description of an x-ray scattering experiment requires knowledge of this area and the interference with the apparatus function caused by the optical elements used in the experiment. The experimental resolution may be considered on two levels:

First, a simple estimate can be given by calculating the *coherence lengths* of radiation and their projections onto the sample surface. Then, this coherent radiation interacts with the sample. The degree of sample perfection may be characterized by a *correlation length.* The result of these interactions defines the correlation length of the scattering process. It is measured from the half-width of the rocking curve. Using this quantity, one can make an estimate as to whether the coherence of the incoming radiation or the sample imperfection determines the resolution of the experiment. An extreme situation arises either if a perfect crystal is measured with a powder diffractometer or if a powdered sample is investigated with a high-resolution diffractometer. In the first case the measured rocking curve is not smaller than the angular divergence of the incident beam; in the second case it reflects the mosaic spread of the sample.

The coherence of third-generation synchrotron radiation can be measured directly by recording the Fraunhofer diffraction figure after passage of the beam through a pinhole. This general proof of coherence can be obtained with monochromatic or white radiation.

The second view on experimental resolution is based on direct measurement or calculation of the resolution area in reciprocal space. The measurement can be performed via reciprocal-space mapping in the vicinity of a Bragg peak of a perfect crystal. This map reflects the actually achieved resolution of the experiment before insertion of the sample in the beam path. From calculations one can show that the resolution element changes during the experiment. Under the complex conditions of the GID-geometry, it is impossible to define the resolution area exactly. Here, the experimental resolution can discussed on an approximate level only. For more details about the relations between coherence and resolution we recommend the paper published by Sinha *et al.* [333].

3.1 Coherence of Radiation and Correlation of Sample Properties

As we know from the optics of visible light, an interference pattern appears under conditions of coherence. This requires not only the coherence of the light, which means the capability of different partial waves to interfere, but also the coherence of the scattering lattice; i.e., the definite periodicity of the scattering centers across the irradiated sample area. The coherence properties of an x-ray wave field are described by a *mutual coherence function* introduced in Chap. 4. The coherence of the lattice is often described by the *correlation length*, which is the average value of the lateral distance between two scattering centers where the scattered waves are in phase. In general, the experiment probes the smaller of the two types of coherence.

The coherence is restricted either by sample inhomogeneities or by the phase mismatches in the probing radiation. Any restriction of coherence is measured by a broadening of the scattering curve. In order to make sure that the broadening indicates a property of sample, it is necessary to estimate the coherence length of the incoming x-ray beam first.

To do this, one has to focus on the scattering plane. No optical element has to be placed between the source, sample, and detector, except for various slits. The direction perpendicular to the scattering plane is ignored. One has to distinguish between lateral and vertical coherence lengths, i.e., the coherence in time and space, respectively, and also their projections on the irradiated sample area (see Fig 3.1). The first coherence length, the temporal or longitudinal correlation length L_{p0}, depends on the spectral purity of the radiation, given by $\frac{\Delta\lambda}{\lambda}$:

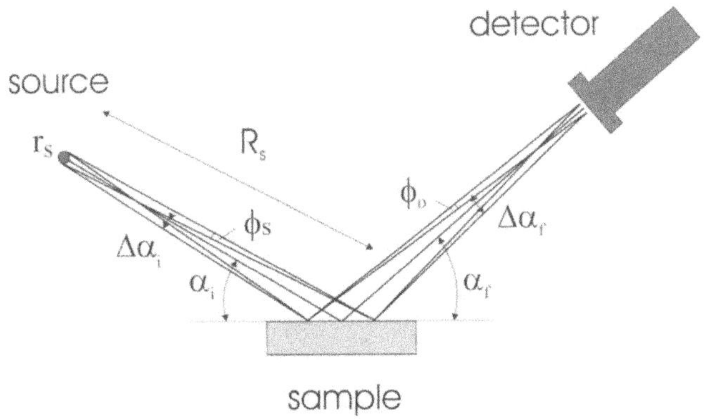

Fig. 3.1. Definition of parameters used to estimate the coherence length.

$$L_{p0} = \frac{\lambda}{2}\frac{\lambda}{\Delta\lambda} \ . \tag{3.2}$$

L_{p0} is the length of the wave train along the propagation direction having the spectral width $\Delta\lambda$. Two wave trains can interfere if the difference in their arrival times at the detector is smaller than $t = L_{p0}/c$ (*temporal coherence*), where c is the velocity of light.

The spatial or lateral coherence length L_{s0} is determined entirely by geometrical considerations:

$$L_{s0} = \lambda\frac{R}{2r_s} \approx \frac{\lambda}{2\phi_s} \approx \frac{\lambda}{2\Delta\alpha_i}, \tag{3.3}$$

in which R is the distance between the source and the sample, r_s is the source size, and ϕ_s is the angle under which any sample point *sees* the source. L_{s0} is the maximum spacing between two points of an extended source emitting photons which can interfere at any given sample point. This is equivalent to the assumption that a single source point emits coherent photons under the angle ϕ_s which can interfere at two different sample points separated by the length L_{s0}.

Assuming the incoming wave field is a superposition of plane waves, ϕ_s is the true divergence of the incoming beam. Every optical element that changes the divergence of the wave changes the coherence width as well. The plane wave concept is equivalent to the assumption of a homogeneous wave defined in 4.1. There we show that the coherence function of such a wave is equal to the Fourier transformation of the intensity distribution $J(\boldsymbol{K})$ in reciprocal space, given in Eq. (4.9).

For the scattering process, the projections of the two coherence lengths onto the sample surface are important. They are given by

$$L_p = \frac{L_{p0}}{\cos\alpha_i} \tag{3.4}$$

$$L_s = \frac{L_{s0}}{\sin\alpha_i}. \tag{3.5}$$

It is straightforward to see that L_s becomes very large under grazing incidence even when a sealed x-ray tube is used. Using the line focus ($0.1 \times 10\,\text{mm}^2$) of a copper-tube and a sample-to-source distance $R = 30\,\text{cm}$, one obtains $L_{s0} \approx 250\,\text{nm}$ and $L_{p0} = 30\,\text{nm}$, taking into account the difference between the two lines of the Kα doublet. For grazing angles $\alpha_i \approx 1°$, L_s exceeds $15\,\mu\text{m}$. When synchrotron radiation is used, R becomes very long (30–40 m) and r_s very narrow (40–100 μm). Under grazing-incidence conditions L_s can exceed 1 mm.

Similar relations exist with respect to the detector arrangement. The coherence length L_D evaluated for the detector depends on the detector slit width r_D and the sample and detector distance R_D, i.e., the detector acceptance angle ϕ_D or the beam divergence $\Delta\alpha_f$. Taking into account both the incident-beam and the exit-beam parts, the resulting coherence length is

$$\begin{aligned} L_{\text{total}} = {} & \frac{\lambda}{2}\left(\frac{1}{\Delta\alpha_i \sin\alpha_i} + \frac{1}{\Delta\alpha_f \sin\alpha_f}\right) + \\ & + \frac{\lambda^2}{2\Delta\lambda}\left(\frac{1}{\cos\alpha_i} + \frac{1}{\cos\alpha_f}\right). \end{aligned} \tag{3.6}$$

If the geometric condition $L_D \leq L_s$ is satisfied, the detector acceptance determines the effective coherence length .

The insertion of an optical element into the beam path changes both the spatial and angular distribution of radiation approaching the sample. This modifies the coherence properties of the beam because the respective optical element reduces the spectral width or the angular divergence of the incident and reflected beam. In general the smallest value of each quantity has to be used in Eqs. (3.2) and (3.3) [354].

Both lengths become very large when using four-bounce monochromators or the highly parallel synchrotron radiation. Here L_p and L_s approach several microns. Under these conditions the correlation properties of the sample determine the scattering process. On the other hand, for bent-crystal monochromators the coherence length is smaller than 100 nm, because the bending of the monochromator acts like an divergent optical element, diminishing the coherence length at the sample position.

The coherence length of a beam can be measured directly by recording the diffraction figure excited by a pinhole inserted into the beam path. As in optics of visible light, one expects a concentric ring pattern if the coherence length is larger than the pinhole diameter. Such interference figures of x-rays were published first by Abernathy et al. received at a beamline of a third-generation synchrotron sources [1] and using monochromatic radiation. One experimental arrangement to measure the coherence length and subsequentely the *effective source size* of the experiment is shown in Fig. 3.2. The diffraction figure is recorded at a distance z apart from the entrance slit system built by two slits separated by the spacing D in Fig. 3.2. Meanwhile a similar pattern has been recorded during a white-beam experiment. As

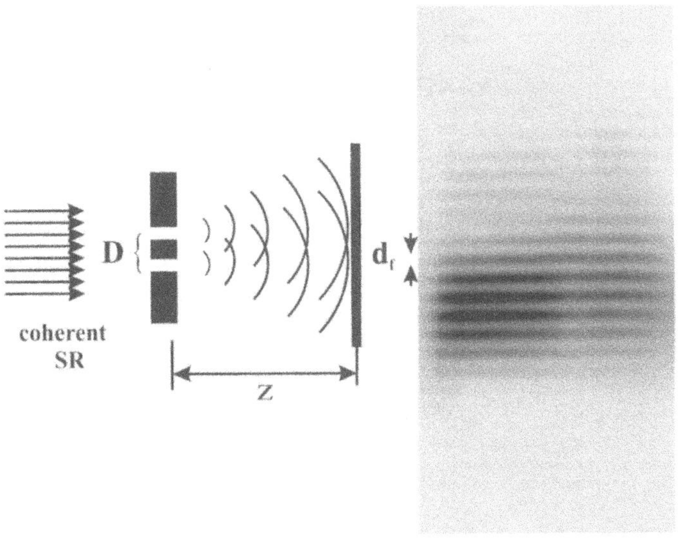

Fig. 3.2. Fraunhofer diffraction figure measured from a double-slit experiment using a monochromatic beam. The coherent beam passes two slits separated by the distance D. At a distance z apart one observes a set of diffraction maxima with spacing d_f. The experiment was performed at an undulator beamline of the ESRF providing a source size of 160 μm [215].

can be seen in Fig. 3.3, the Fraunhofer oscillations are well resolved within an energy range $5 \leq E \leq 15$ keV using an energy dispersive detector, although the temporal coherence is only $L_p \leq 50$ nm. To record the pattern one has to scan the detector vertically across the diffracted beam taking a full energy spectrum at each detector position z. Because the angular separation between the diffraction maxima is on the order of minutes of arc, the Fraunhofer diffraction patterns have to be recorded at large sample-detector distance.

3.2 Scans Across the Reciprocal Space

X-ray scattering experiments can be simply understood using the model of the reciprocal space. The reciprocal space is an alternative representation of the crystal system [196]. Its symmetry properties are the same as those of the direct lattice. For the description of x-ray scattering one has to consider a second co-ordinate system, the laboratory system, which describes the directions of the incident and exit beam with respect to the sample. In contrast to the reciprocal lattice of the crystal, which can be oblique, one chooses a laboratory system with Cartesian coordinates Q_x, Q_y, and Q_z centered at the origin of the reciprocal lattice of the crystal. The vectors of the incident

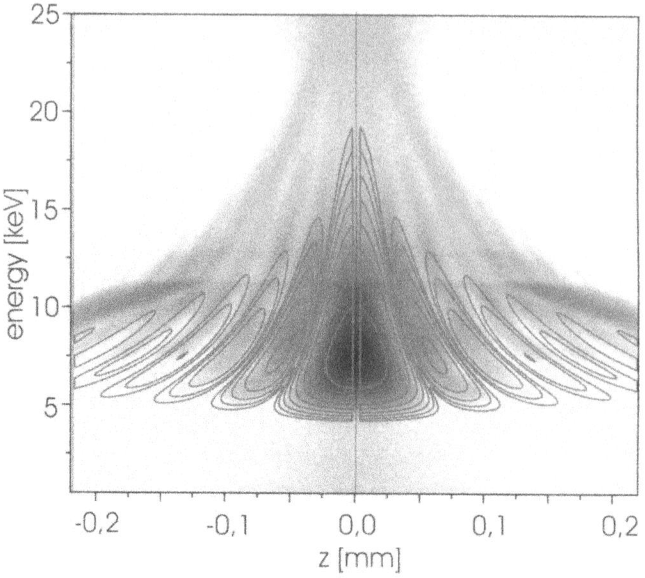

Fig. 3.3. Fraunhofer diffraction pattern recorded at the EDR beamline at BESSY II. Fraunhofer oscillations appears simulatenous for energies between 5–15 keV. Solid lines show a fit using Airy functions [258].

and diffracted beams $\boldsymbol{K}_i$ and $\boldsymbol{K}$ are vectors of the laboratory system. The difference corresponds with the scattering vector $\boldsymbol{Q}$. The length of each vector is $K = \frac{2\pi}{\lambda} \approx 0.5E$ [keV, nm^{-1}], where λ is the wavelength in 1/nm and E the energy of radiation in [keV].

For coplanar scattering geometries $\boldsymbol{K}_i$ and $\boldsymbol{K}$ and the normal of the sample surface lie in a common scattering plane. For a proper description of scattering, one aligns the Q_z-axis parallel to the normal of the sample surface in outward direction and Q_x is parallel to the in-plane component of $\boldsymbol{K}_i$. Then the Q_x, Q_z-plane describes the scattering plane. For these conditions the coordinates of the scattering vector $\boldsymbol{Q}$ in Eq. (3.1) can be described by (see also Sec. 4.3):

$$Q_z = K(\sin\alpha_i + \sin\alpha_f) \ , \tag{3.7}$$

$$Q_x = K(\cos\alpha_f - \cos\alpha_i) \ . \tag{3.8}$$

These formulas connect the coordinates Q_x, Q_z of the reciprocal space with the angular coordinates α_i, α_f in direct space measured with respect to the sample surface/lattice plane.

On the goniometer, the rotation of the sample circle is denoted by ω, and that of the detector by 2Θ. These angles are related to α_i and α_f by

$$\omega = \alpha_i \quad \text{and} \quad 2\Theta = \alpha_i + \alpha_f \ . \tag{3.9}$$

We will show later that in high-resolution geometry the resolution in angular space can be good enough to inspect experimentally different directions in reciprocal space. As Fig. 3.4 shows, one can distinguish four particular scans which are often performed on a goniometer:

1. The radial- or offset scan is directed parallel or inclined to Q_z. It requires a goniometer movement in a ratio $\Delta\omega/\Delta 2\Theta = 1/2$. Generally ω differs from $2\Theta/2$.
2. One particular case of a radial-scan is the specular , or Q_z, scan, which is directed parallel to the surface normal (positive Q_z direction). It is performed by using $\Delta\omega/\Delta 2\Theta = 1/2$ with $\omega = 2\Theta/2$. With respect to the surface, this means $\alpha_i = \alpha_f$.
3. The angular or rocking or constant ω scan rotates the sample across a fixed 2Θ position. It runs approximately perpendicular to Q_z if α_i and α_f differ only slightly from each other. A strict Q_x scan needs a correction to keep the length of the scattering vector constant.
4. The α_f or detector or 2Θ scan changes the detector angle α_f for fixed angle of incidence α_i. It describes a circle within the Q_x, Q_z plane.
5. In special cases one has to consider the $\boldsymbol{Q}$ scan , which is directed along an arbitrary direction in reciprocal space with $Q_x \neq 0$.

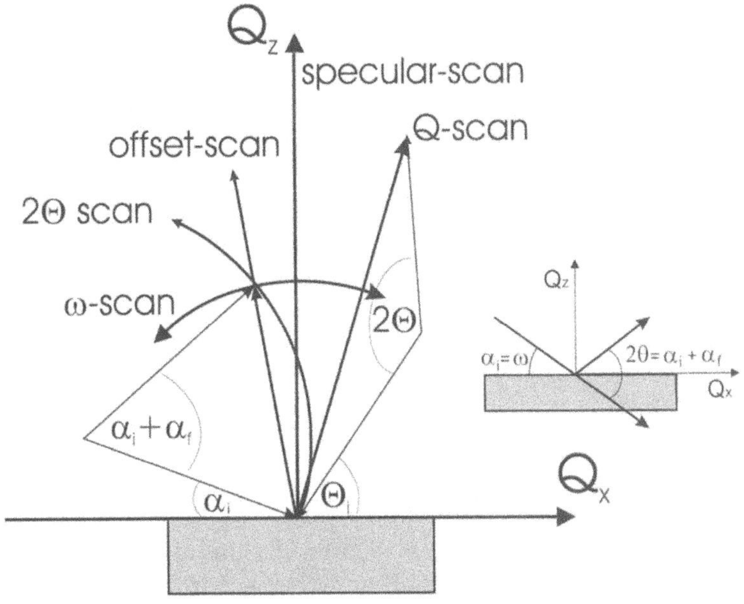

Fig. 3.4. Scans in reciprocal space using an angle-dispersive set-up.

For symmetrical Bragg diffractions or in the case of probing specular reflectivity, the radial scan is directed exactly parallel to Q_z, i.e., along the *crystal truncation rod* (see Sect. 4.3). This is not the case for asymmetric Bragg reflections (Q scan). Here α_i and α_f or ω and 2Θ must be varied in a ratio different from 1:1 or 1:2, respectively. The correct scan conditions are kept for

$$\frac{\Delta\alpha_f}{\Delta\alpha_i} = -\frac{1}{b} = \frac{\sin(\Theta_B - \phi)}{\sin(\Theta_B + \phi)}, \tag{3.10}$$

which is equivalent to

$$\frac{\Delta\omega}{\Delta(2\Theta)} = \frac{1}{1 - 1/b}, \tag{3.11}$$

where b is the asymmetry factor defined in Eq. (1.7) (see Sect. 4.3). The intensity distribution within the Q_x, Q_z plane of reciprocal space (*reciprocal-space map*) can be recorded by a set of offset scans at fixed 2Θ or other angular combinations.

On plotting such reciprocal-space maps alternatively as a function of the goniometer angles ω and 2Θ, the crystal truncation rod appears parallel to the 2Θ-axis, probing symmetrical reflections, but inclined with respect to 2Θ-axis recording asymmetric ones.

However, any scan in reciprocal space can only be performed within the angular interval $-2\Theta/2 < \omega < 2\Theta/2$, which means from $\alpha_i \geq 0$ up to $\alpha_f \geq 0$. Some areas in reciprocal space can never be inspected by scans running coplanar diffraction geometries (Fig. 3.4). This becomes important whenever α_i or α_f is very small, which means close to (000), or when probing strongly asymmetric Bragg reflections. These shadowed areas become accessible when using transmission geometry [283] or when using a non-coplanar scattering scheme, as shown in Fig. 3.9. The GID geometry is very important for recording small-angle x-ray diffuse scattering [300] because it has the advantage that the component Q_y perpendicular to the Q_x, Q_z plane is probed at fixed α_i and α_f.

For the energy-dispersive set-up, variation of the vector $\boldsymbol{Q}$ is achieved by changing the energy for fixed α_i. The coordinates Q_z and Q_x are described by

$$Q_z = \frac{2\pi E}{hc}(\sin\alpha_i + \sin\alpha_f) \approx \Theta E\,, \tag{3.12}$$

$$Q_x = \frac{2\pi E}{hc}(\cos\alpha_f - \cos\alpha_i) \approx (\alpha_i - \alpha_f)\Theta E\,, \tag{3.13}$$

where $2\Theta = \alpha_i + \alpha_f$ and $4\pi/hc \approx 1 \times 10^{-10}$ [1/m] if E is given in [keV]. The scan direction in reciprocal space is determined by the ratio of the fixed angles α_i and α_f. As shown in Fig. 3.5 one can perform two scans only:

1. The radial or offset scan is directed along a straight line non-parallel to Q_z. It is fixed by the difference $\alpha_i - \alpha_f \neq 0$. Here Q_x increases as a function of E.

2. The specular scan records the intensity distribution as a function of energy for $\alpha_i = \alpha_f$, i.e. for $Q_x = 0$.
3. The intensity distribution along Q_x can be reconstructed from a set of different offset scans, plotting the intensities at a particular E as a function of the offset angle $\Delta\alpha = \alpha_i - \alpha_f$.

All other scans being possible in the angle-dispersive scheme can be realized under energy-dispersive condition only by the simultaneous variation of angles and energy. The Q_x scan has some importance which has to be constructed from a set of several offset scans [268].

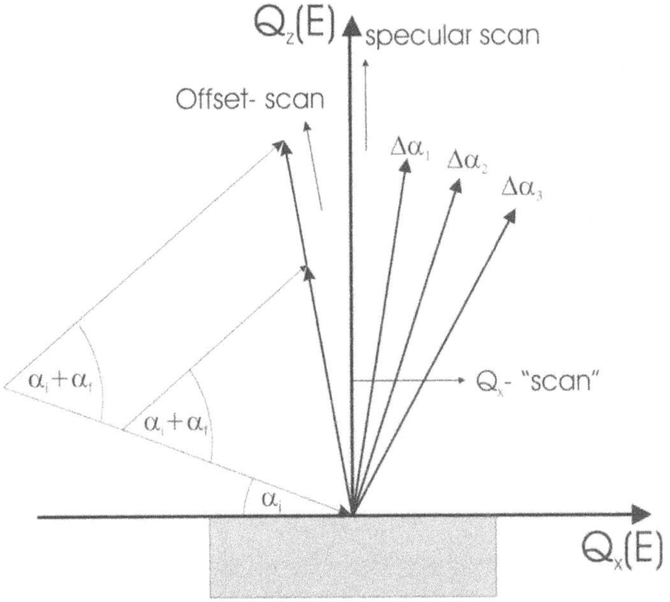

Fig. 3.5. Scans across the reciprocal space using an energy-dispersive set-up. The intensity distribution along Q_x can be reconstructed from a set of offset scans with different $\Delta\alpha$.

3.3 Resolution Elements

The possible angular resolution of the scattering experiment can be estimated by calculating the resolution area A_E. This measures the area *in reciprocal space* that is simultaneously illuminated by the incident beam and accepted by the detector under given geometrical and spectral conditions. At any given point of the rocking curve the recorded intensity is an average across that area.

A sketch of A_E and its variation in reciprocal space is shown in Fig. 3.6. In general it is an obliquely shaped face given by

$$A_E = \Delta Q_x \Delta Q_z \tag{3.14}$$

and it is inclined with respect to the Q_x- and Q_z-axes. The derivatives δQ_z and δQ_x of Q_z and Q_x, respectively, follow from Eq. (3.7) :

$$\delta Q_z = K(\delta\alpha_f \cos\alpha_f + \delta\alpha_i \cos\alpha_i) + \delta K(\sin\alpha_i + \sin\alpha_f) \tag{3.15}$$

$$\delta Q_x = -K(\delta\alpha_i \sin\alpha_i - \delta\alpha_f \sin\alpha_f) + \delta K(\cos\alpha_f - \cos\alpha_i)\ ;. \tag{3.16}$$

These derivatives contain the differentials of the angles and the **K** vector. If

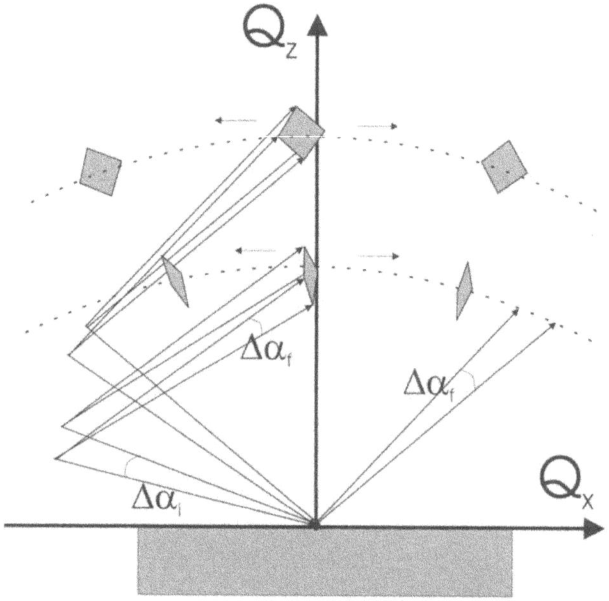

Fig. 3.6. The shape of the resolution area in reciprocal space and its variation with changing geometric conditions of the experiment.

one identifies $\delta\alpha_{i,f}$ with the angular divergence $\Delta\alpha_{i,f}$, then neglecting δK and assuming statistical independence of the parameters used, A_E can be calculated approximately by Gaussian superposition of δQ_z and δQ_x

$$\Delta Q_x = K\sqrt{\sin^2\alpha_i \Delta\alpha_i^2 + \sin^2\alpha_f \Delta\alpha_f^2} \tag{3.17}$$

$$\Delta Q_z = K\sqrt{\Delta\alpha_i^2 \cos^2\alpha_i + \Delta\alpha_f^2 \cos^2\alpha_f}\ . \tag{3.18}$$

Using the small-angle approximation ($\sin\alpha \approx \alpha$ and $\cos\alpha \approx 1$), ΔQ_z and ΔQ_x can be approximated by

$$\Delta Q_z \approx \sqrt{2}K\Delta\alpha_i \quad \text{and} \quad \Delta Q_x \approx \sqrt{2}K\alpha_i\Delta\alpha_i \tag{3.19}$$

for scans close to Q_z axis ($\alpha_i \approx \alpha_f$) [103]. In this case ΔQ_z is nearly independent of α_i , but ΔQ_x depends on, α_i. In the angular system of the goniometer $\Delta Q_x \approx \Delta\omega Q_z$ holds.

As shown in Fig. 3.6, the orientation of A_E on Q_x changes with ω. When running non-specular scans, the projection of A_E on Q_x is given approximately by

$$\Delta Q_x \approx \begin{cases} K\omega\Delta\alpha_i & \text{if} \quad \omega \approx 2\Theta \\ K2\Theta\Delta\alpha_f & \text{if} \quad \omega \ll 2\Theta \end{cases} . \tag{3.20}$$

That means that the resolution with respect to Q_x changes significantly when $\alpha_i \equiv \omega$ changes. For incident angles of several degrees (wide-angle measurements) the resolution area stays approximately unchanged in shape and in its orientation with respect to Q_x and Q_z. This is due to the fact that the angular differences remain small. A_E develops additional features when using

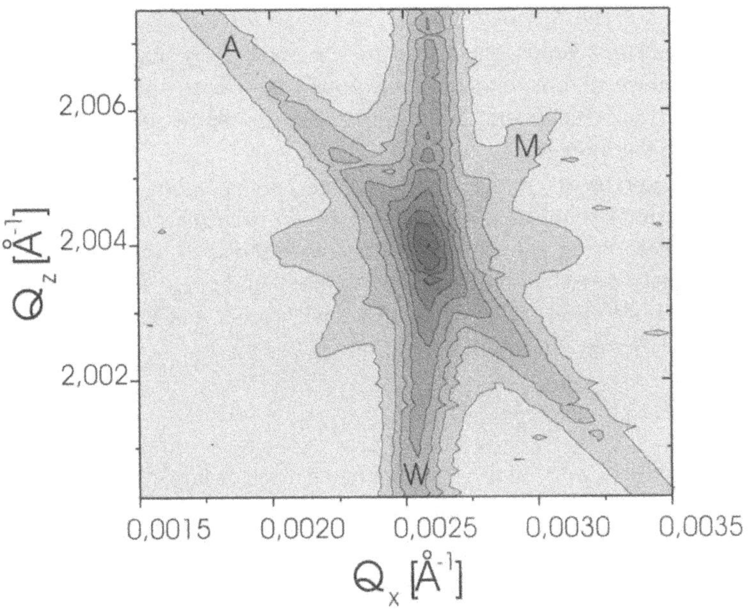

Fig. 3.7. The shape of the resolution element close to the Si (111) Bragg diffraction measured from FZ Silicon wafer with an x-ray tube but using a Bartels monochromator and a single crystal analyzer. A, W, and M denote the direction of the analyzer- , wavelength- and monochromator streak, respectively.

further optical elements as in a high-resolution arrangements (see Fig. 3.7).

Now both $\Delta\alpha_i$ and $\Delta\alpha_f$ depend on the angular acceptance of the monochromator and analyzer crystal. Assuming finite divergence of the incident beam but a very small acceptance of the detector, for example, the resolution area is a very small stripe (*analyzer streak*) inclined by an angle $2\Theta_{\mathrm{analyzer}}/2$ with respect to Q_z (see 3.7 and 3.15). In a similar way a finite acceptance of the detector and a parallel incident beam create the *monochromator streak*. It is inclined by $2\Theta_{\mathrm{mono}}/2$ with respect to Q_z but in the opposite direction compared with the analyzer streak. Often there is a third streak pointing to the origin of reciprocal space, the *wavelength streak*. This is caused by the finite energy resolution ΔK of the particular Bragg diffraction used.

The intensity distribution of the streaks becomes very narrow using multiple-reflection optical elements. Using the four-bounce monochromator, (see Sect. 2.2) the monochromator streaks becomes very small, as shown in the example of Fig. 3.7.

The resolution element of any experimental set-up can be measured by recording the reciprocal space map in the vicinity of a sharp Bragg reflection of a perfect silicon wafer. Figure 3.8 shows an experimentally determined resolution element measured under GID geometry at a beamline of a synchrotron facility. Besides the monochromator and analyzer streaks, the figure shows additional features which are caused by the improperly aligned focusing mirrors. To achieve high-resolution conditions one has to take care that these additional streaks disappear. This can be done by subsequent realignment of all beamline components.

The measurement of the intensity distribution along a truncation rod (Q_x = constant) requires good resolution of one of the angles α_i or α_f. If the diffuse scattering can be neglected, the intensity of the truncation rod can be measured without any analyzing element ($\Delta\alpha_f \to \infty$), i.e., using a double- crystal diffractometer with an open detector window. In this case the resolution in the Q_z direction is

$$\Delta Q_z = K\Delta\alpha_i \frac{\sin(\alpha_i + \alpha_f)}{\sin\alpha_f}, \tag{3.21}$$

but the angles α_f and α_i are dependent on each other in order to hold Q_x fixed. If an analyzer crystal with an angular resolution $\Delta\alpha_f$ is used, the Q_z-resolution yields

$$\Delta Q_z = K\frac{\sin 2\Theta}{\sin\alpha_f}\Delta \quad \text{where} \quad \Delta = \min(\Delta\alpha_i, \frac{\sin\alpha_f}{\sin\alpha_i}\Delta\alpha_f), \tag{3.22}$$

considering Eq. (3.15). This shows that the resolution can be improved only if $\Delta\alpha_f$ is smaller than $\Delta\alpha_i$.

High resolution is absolutely necessary for diffuse scattering experiments, i.e., if one wishes to measure the intensity far from the crystal truncation rod. In this case the resolution element in reciprocal space is entirely described by Eq. (3.17).

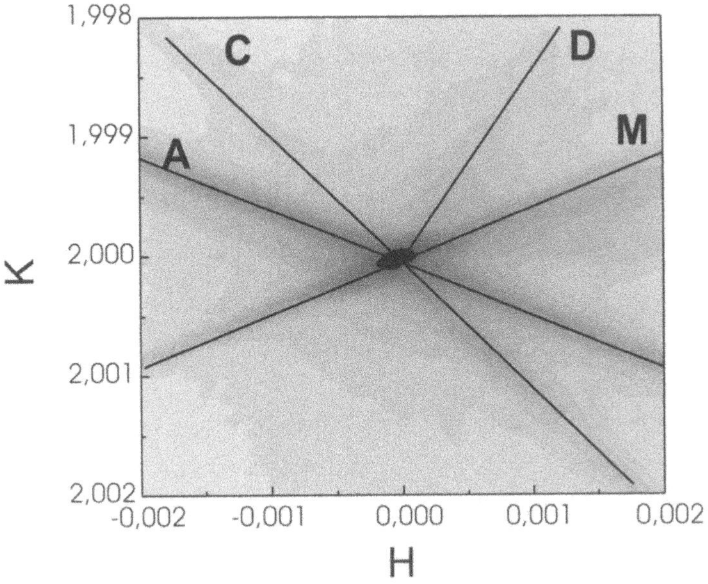

Fig. 3.8. Experimentally determined resolution element close to the Si(220) reflection measured in GID geometry at a beamline of a synchrotron facility. M and A denote the monochromator and analyzer streak, respectively. The streaks C and D originate from other beamline components of an improperly aligned experimental set-up. The axes are denoted in the laboratory system of the beamline.

The resolution area has to be rescaled when using an energy-dispersive experiment. Now the reciprocal space is not unique. Each energy defines another Ewald sphere in reciprocal space (see Fig. 3.5). The crystal truncation rod is examined for fixed angles but for different energies. Nevertheless, the incident beam is divergent and the detector has a finite angular acceptance. To determine A_E one has to consider the relative low-energy resolution of the solid-state detector given by ΔK (see Eq. (3.15)), resulting in [238]:

$$\Delta Q_z = K\sqrt{\Delta\alpha_f^2 + \Delta\alpha_i^2 + (\frac{\Delta K}{K})^2(\alpha_i + \alpha_f)^2} \tag{3.23}$$

$$\Delta Q_x = K\sqrt{\Delta\alpha_f^2\alpha_f^2 + \Delta\alpha_i^2\alpha_i^2 + (\frac{\Delta K}{2K})^2(\alpha_f^2 - \alpha_i^2)^2}\ . \tag{3.24}$$

Since $\Delta K/K = \Delta E/E \approx 2\,\%$ and α_i and α_f are small, the last term in the expression for ΔQ_x is very small and can be neglected. In contrast to the angular-dispersive set-up, ΔQ_x remains approximately independent, but ΔQ_z becomes angular dependent and changes with energy.

For grazing-incidence diffraction the definition of a resolution area seems to be meaningless because the beam becomes scattered out of the plane of

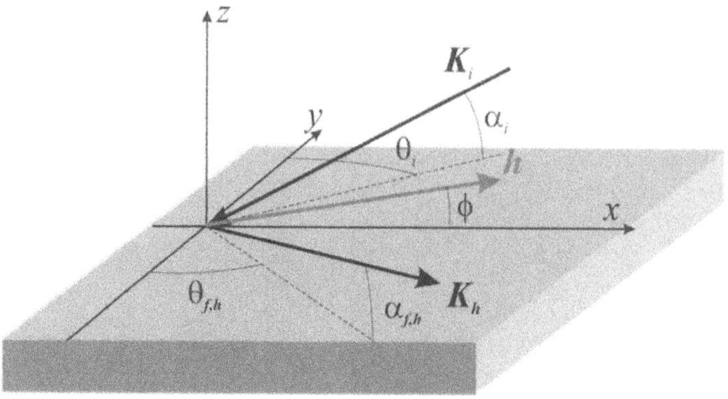

Fig. 3.9. Definition of angles for the GID set-up.

incidence. Additionally, Q_x and Q_y, as shown in Fig. 3.9, depend on the four angles $\alpha_i, \alpha_f, \theta_i$, and θ_f, simultaneously. For the description of a GID experiment one needs three Q-vectors given by

$$Q_x = K(\cos\alpha_f \sin\theta_f + \cos\alpha_i \sin\theta_i) \quad (3.25)$$
$$Q_y = -K(\cos\alpha_f \cos\theta_f - \cos\alpha_i \cos\theta_i) \quad (3.26)$$
$$Q_z = K(\sin\alpha_f + \sin\alpha_i). \quad (3.27)$$

If one of the four angles remains fixed, one is able to scan across the reciprocal space in three dimensions changing the other three angles. To make use of the possibility for depth resolution (see 2.4), α_i is often selected as the fixed angle.

Only when $\alpha_i = 0$ and $\alpha_f = 0$ will the Q_z component be zero and the *diffraction plane* the Q_x, Q_y-plane. Otherwise it lies obliquely in reciprocal space. The Q_x, Q_y plane contains the parallel component of the scattering vector, $Q_\| = \sqrt{Q_x^2 + Q_y^2}$.

The recording of truncation rods, i.e. the intensity distribution along Q_z at fixed $Q_\| = Q_x = 0$, requires the variation of all four angles mentioned above. The direction of the diffracted beam and, consequently, the angles α_f and θ_f, are entirely determined by the other both angles. The relations among these four angles are entirely described in Sect. 4.3 (Eq. (4.32)).

As seen from Eq. (3.25), the resolution element may be defined within the knowledge of the four-dimensional angular space. That means that the resolution element is not a unique function of $\boldsymbol{Q}$ but depends on the particular choice of the four angles $\alpha_{i,f}, \theta_{i,f}$. On the other hand, one can express the resolution element as a function of $\boldsymbol{Q}$ and one of the angles, α_i, for instance. This is meaningful if α_i is used to tune the penetration depth of the incoming beam (see Fig. 2.9). Using this approach the resolution area $A_{E,GID} = \Delta Q_\| \Delta Q_z$ can be determined from

$$\Delta Q_{\|} = K\sqrt{\sin^2 \Theta_{B\|}[\alpha_f^2(\Delta\alpha_f)^2 + \alpha_i^2(\Delta\alpha_i)^2] + \cos^2 \Theta_{B\|}[(\Delta\theta_i)^2 + (\Delta\theta_f)^2]} \tag{3.28}$$

$$\Delta Q_z = K\sqrt{(\Delta\alpha_f)^2 + (\Delta\alpha_i)^2}. \tag{3.29}$$

This projected *resolution area* is tilted with respect to Q_z. The tilt angle changes when α_f and α_i are changed [119]. For fixed α_i the tilt angle increases with increasing α_f (see Fig. 3.10). Therefore, the best resolution in reciprocal space is realized for small α_i and small α_f. As a consequence of this effect a α_f scan for fixed α_i is always a scan through the (Q_x, Q_z)-plane. Depending on the deviation from the exact in-plane Bragg condition $\Delta\Theta$, the scan cuts the truncation rod at different Q_z.

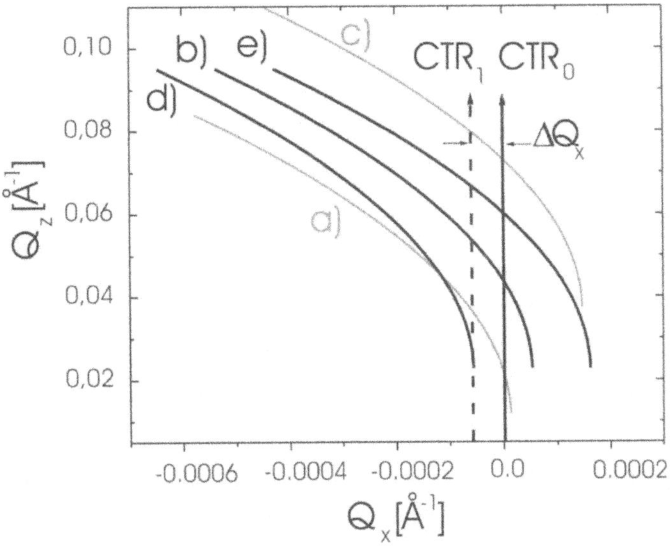

Fig. 3.10. Trajectories of different α_f scans across the reciprocal space calculated in the interval $0.01 \leq \alpha_f \leq 1°$ for different but fixed α_i. ($\Delta\Theta = 0$ but $\alpha_i = 0.15°$ (a), 0.30° (b), 0.50° (c), for $\alpha_i = 0.30°$ with $\Delta\Theta = +0.002°$ (d) and -0.002° (e)). Depending on $\Delta\Theta$ the scans cut the crystal truncation rod (CTR) at different Q_z. In case of a thin layer grown on substrate with slightly different in-plane lattice parameters the scan may cut both truncation rods.

Using the theory presented in Sect. 4.3, one can determine the angular resolution necessary to record a crystal truncation rod. For fixed α_i, the finite divergence of the other three angles creates an illumination of a part of the crystal truncation rod. Moreover, the illumination of the entire crystal truncation rod requires a non-vanishing in-plane divergence $\Delta\theta_{i,f}$. The smaller θ_i is, the shorter the illuminated part of the crystal truncation rod. Similar considerations hold for fixed α_f. Both can be summarized by the conditions presented in (4.33). These conditions are similar to the *master formulas* of GID (see Sect. 4.3) which applied for the evaluation of surface-scattering experiments [119, 281].

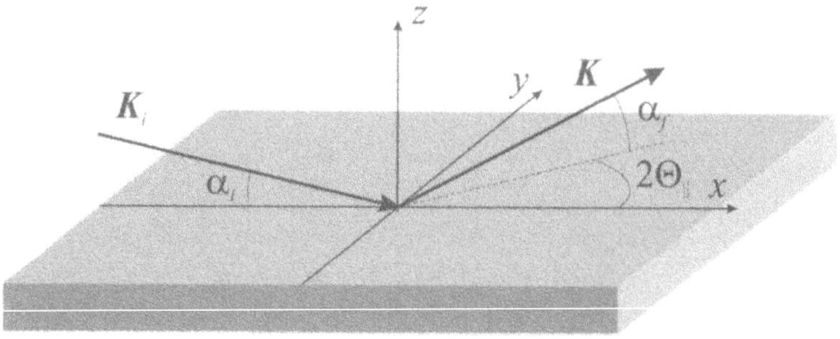

Fig. 3.11. Grazing-incidence small-angle scattering arrangement (GISAXS).

A special case of grazing-incidence geometry is the grazing-incidence small-angle x-ray scattering arrangement (GISAXS) shown schematically in Fig. 3.11. In this arrangement, the incidence and take-off angles $\alpha_{i,f}$ as well as the in-plane scattering angle $2\Theta_{\|}$ are small, so that the coordinates of the scattering vector $\boldsymbol{Q}$ are given by the following approximate expression:

$$\begin{aligned}\boldsymbol{Q} &= K(\cos\alpha_f\cos(2\Theta_{\|}) - \cos\alpha_i, \cos\alpha_f\sin(2\Theta_{\|}), \sin\alpha_f + \sin\alpha_i)\\ &\approx K((\alpha_i^2 - \alpha_f^2)/2 - 2\Theta_{\|}^2, 2\Theta_{\|}, \alpha_i + \alpha_f). \end{aligned} \tag{3.30}$$

Usually, the scattered intensity is measured as a function of Q_y, Q_z for a constant incidence angle α_i. Using the derivatives

$$\Delta Q_z = K\sqrt{\Delta\alpha_f^2 + \Delta\alpha_i^2 + (\frac{\Delta K}{K})^2(\alpha_i + \alpha_f)^2} \tag{3.31}$$

$$\Delta Q_y = K\sqrt{\Delta(2\Theta)^2 + (\frac{\Delta K}{K})^2(2\Theta)^2}. \tag{3.32}$$

and $\Delta K \approx 0$, the resolution element A_E is

$$A_E = K^2\Delta 2\Theta\Delta\alpha_f, \tag{3.33}$$

simply a function of the angular resolution in both directions.

Part II

Basic Principles

X-ray scattering methods discussed in this book measure reciprocal-space distribution of x-ray intensity scattered from a sample. In order to achieve a resolution in reciprocal space, the primary beam irradiating the sample has to be very well monochromatized and collimated, and the angular acceptance of the detector must be as narrow as possible. The resolution in reciprocal space is connected with the resolution in real space by an uncertainity principle analogous to the Heisenberg uncertainty principle, which is well known from quantum mechanics. Therefore, high-resolution x-ray scattering *in reciprocal space* implies a very bad resolution *in real space* and vice versa. High-resolution x-ray scattering methods are therefore non-local and the measured reciprocal-space distribution of the scattered intensity is collected from a relatively large volume of the scattering sample. On the other hand, several attempts have been published, where both reciprocal-space and real-space resolutions were achieved (see [104], among others); this compromise, however, inevitably deteriorates both resolutions.

High-resolution scattering methods are in principle non-local, but usually they are used in order to obtain information on local structure of a sample. For instance, x-ray reflectometry is a suitable tool for determining thicknesses of layers in a multilayer; diffuse x-ray scattering probes the local chemical composition and its inhomogeneities in a quantum dot. Owing to a large interaction volume, this data is averaged over a large amount of quantum dots or over a large volume of the multilayer so that its statistical relevance is much better than in the case of local methods (transmission electron microscopy, for instance). On the other hand, the analysis of the scattering data is much more complicated than in the case of the local methods, since we cannot "see" the quantum dot immediately in the intensity pattern. In most cases, there is no straightforward path from the experimental data to the local structure of the sample under investigation and the experimental data should be compared numerically with a suitable structure model. Generally, this comparison consists of the following steps:

1. Formulation of the structure model. This model should reflect all the preliminary knowledge of the sample, including the growth mechanism, influence of a post-growth treatment, etc. The formulation of a structure model is not the subject of this book and we present here only simple models often used in the analysis of the scattering data.
2. Numerical simulation of x-ray scattering based on the structure model. This step will be discussed in Part II of this book. Depending on the structure model and on the scattering method used, a suitable theoretical formulation of the scattering problem has to be used. In many experimentally important cases, the kinematical approximation of the scattering theory is fully sufficient. In this approach, described thoroughly in Chap. 5, we neglect multiple scattering of x-ray photons in the sample, and the resulting scattered wave can be expressed by means of a Fourier transformation of the electron density. The kinematical approximation can be

used even for relatively complicated structure models. This approach is applicable for a sample with relatively small scattering volume and/or for the waves far away from the diffraction maximum. The effects of multiple x-ray scattering in thin layers occur often in small-angle scattering geometry; in wide-angle scattering they are relatively rare. These effects have to be described using much more complicated dynamical scattering theory, briefly discussed in Chapter 6. Owing to its mathematical complexity, the application of this theory is limited only to simple structure models. Chapter 7 describes the semikinematical scattering theory. This approximative method is a good choice if the dynamical effects cannot be neglected and the structure model is not simple.

3. Comparison of the measured data with the theoretical simulations. This step can be performed only qualitatively with the naked eye or by a numerical fitting method. The description of several fitting methods lies outside the scope of this book.

In Parts III and IV we demonstrate this general approach by numerous experimental examples.

4 Basic Principles

In this part, we present a formalism for the description of the wavefields in vacuum. This formalism is independent of the scattering theory used.

The properties of the scattered waves derived in this part, follow only from the symmetry of the scattering objects and from the assumption of elastic scattering.

4.1 Description of the X-Ray Wavefield in Vacuum

For the *elastic* scattering of x-rays, the frequencies ω of the primary and scattered waves are equal; thus the wave vectors of these waves (in vacuum) have the same lengths,

$$|\boldsymbol{K}_i| = |\boldsymbol{K}| = \frac{\omega}{c} = \frac{2\pi}{\lambda} \equiv K. \tag{4.1}$$

We assume here that *the primary wave* irradiating the sample is plane and monochromatic (*plane wave approach*). The influence of its divergence and non-monochromaticity is included in the resolution function described in Chapter 3.3. The plane primary wave is described by

$$\boldsymbol{E}_i(\boldsymbol{r}, t) = \boldsymbol{E}_i \mathrm{e}^{-\mathrm{i}(\omega t - \boldsymbol{K}_i . \boldsymbol{r})}. \tag{4.2}$$

In strict quantum mechanical notation, formula (4.2) is the space representation $\langle \boldsymbol{r} | \boldsymbol{K}_i \rangle$ of the eigenstate $|\boldsymbol{K}_i\rangle$ of the momentum operator. Since all waves considered here have the same frequency, their time dependencies are given by the same factor $\exp(-\mathrm{i}\omega t)$; we omit this factor in the following.

The scattered wave is described as a coherent superposition of plane wave components with equal frequencies, i.e., with wave vectors of constant lengths but various directions,

$$\boldsymbol{E}(\boldsymbol{r}) = \int \mathrm{d}\Omega \boldsymbol{E}(\Omega) \mathrm{e}^{\mathrm{i}\boldsymbol{K}.\boldsymbol{r}}, \tag{4.3}$$

where $\mathrm{d}\Omega \equiv \mathrm{d}\phi \mathrm{d}\theta \sin\theta$ and the spherical coordinate angles θ, ϕ determine the direction of $\boldsymbol{K}$. We rewrite the last expression into a more convenient form using the integration over the components of $\boldsymbol{K}$. In the following, we use the coordinate system with x and y-axes in the sample surface and the z-axis parallel to the *outward* surface normal. Then $\mathrm{d}\Omega = \mathrm{d}K_x \mathrm{d}K_y/(KK_z) \equiv$

$d^2\boldsymbol{K}_{\|}/(KK_z)$, where $K_z = \sqrt{K^2 - K_x^2 - K_y^2} > 0$, i.e., the scattered wave propagates in the half-space above the sample surface. Then from (4.3) one obtains

$$\boldsymbol{E}(\boldsymbol{r}) = \int \frac{d^2\boldsymbol{K}_{\|}}{KK_z} \boldsymbol{E}(\boldsymbol{K}) e^{i\boldsymbol{K}.\boldsymbol{r}}. \tag{4.4}$$

In an actual structure, the atomic positions are not exactly defined, they can be described by random functions. For instance, the morphology of a sample with a rough surface is described by random function $z(\boldsymbol{r}_{\|}) \equiv z(x, y)$ of the local surface. The waves scattered by such a sample are random as well. We will not deal with coherent scattering methods, where the microstructure of these random waves are investigated (see, for instance, [285]). Instead, we study the properties of the random waves *averaged* over a statistical ensemble of all microscopic configurations of the sample that are not resolvable macroscopically. In the example with the rough surface, the scattered wave is averaged over all the shapes of the surface morphology that correspond to the same macroscopic quantities such as root mean square roughness, roughness correlation length, etc. If the sample size is much larger than a characteristic size of the sample disorder (the correlation length of the surface roughness, in the example), the sample contains all possible microscopic configurations and the ensemble average can be replaced by the average over the sample volume or sample surface.

The properties of the ensemble-averaged scattered waves can be studied using the *mutual coherence function* (MCF) of the the scattered wave defined by [60]

$$\Gamma(\boldsymbol{r}, \boldsymbol{r}') = \langle E(\boldsymbol{r}) E^*(\boldsymbol{r}') \rangle . \tag{4.5}$$

This definition considers a scalar amplitude E only; thus we do not investigate here the polarization properties of the waves scattered by random samples. In contrast to classical optics, where $\langle\ \rangle$ usually denotes the averaging over the statistical ensemble of the wavefield, here the same symbol is used for the averaging over the statistical ensemble of the sample microstructures.

From the MCF, the intensity of the wavefield in point $\boldsymbol{r}$ is

$$I(\boldsymbol{r}) = \Gamma(\boldsymbol{r}, \boldsymbol{r}). \tag{4.6}$$

Using the Fourier transformation of the scattered wave in Eq. (4.4) we obtain the MCF of this wave in the form

$$\Gamma(\boldsymbol{r}, \boldsymbol{r}') = \frac{1}{K^2} \int \frac{d^2\boldsymbol{K}_{\|}}{K_z} \int \frac{d^2\boldsymbol{K}'_{\|}}{K'_z} \Gamma(\boldsymbol{K}, \boldsymbol{K}') e^{i(\boldsymbol{K}.\boldsymbol{r} - \boldsymbol{K}'.\boldsymbol{r}')}, \tag{4.7}$$

where

$$\Gamma(\boldsymbol{K}, \boldsymbol{K}') \equiv \langle E(\boldsymbol{K}) E^*(\boldsymbol{K}') \rangle \tag{4.8}$$

is the Fourier transformation of the MCF.

If the MCF $\Gamma(\boldsymbol{r}, \boldsymbol{r}')$ depends only on the *relative* position $\boldsymbol{r}-\boldsymbol{r}'$ of the two points, the wavefield is called *homogeneous*. The intensity of the homogeneous wavefield is independent of the position $\boldsymbol{r}$: $I = \Gamma(\boldsymbol{r}, \boldsymbol{r}) \equiv \Gamma(0)$. The Fourier transformation of the MCF of a homogeneous wavefield is

$$\Gamma(\boldsymbol{K}, \boldsymbol{K}') = \delta^{(2)}(\boldsymbol{K}_{\parallel} - \boldsymbol{K}'_{\parallel})J(\boldsymbol{K}), \tag{4.9}$$

where $\delta^{(2)}$ is the two-dimensional Dirac delta-function and

$$J(\boldsymbol{K}) = \frac{K^2K_z^2}{4\pi^2}\int \mathrm{d}^2(\boldsymbol{r}_{\parallel} - \boldsymbol{r}'_{\parallel})\Gamma(\boldsymbol{r} - \boldsymbol{r}')\mathrm{e}^{-\mathrm{i}\boldsymbol{K}.(\boldsymbol{r}-\boldsymbol{r}')} \tag{4.10}$$

is the intensity of the plane component of the homogeneous wavefield having the wave vector $\boldsymbol{K}$ (i.e., the intensity distribution in reciprocal space). In the last expression the value of the integral over $\mathrm{d}^2\boldsymbol{r}_{\parallel} \equiv \mathrm{d}x\mathrm{d}y$ does not depend on the vertical coordinate z. For a homogeneous wavefield, the function $J(\boldsymbol{K})$ will be used for its description, instead of its MCF.

4.2 General Description of the Scattering Process

Any scattering process considered in this book will be described by the scalar wave equation (or by a system of such equations)

$$(\Delta + K^2)E(\boldsymbol{r}) = \hat{\mathbf{V}}(\boldsymbol{r})E(\boldsymbol{r}), \tag{4.11}$$

where $\hat{\mathbf{V}}(\boldsymbol{r})$ is the operator of the *scattering potential.* If we restrict ourselves to a non-magnetic material (the sample magnetic permeability equals the vacuum permeability: $\mu = \mu_0$), from the Maxwell equations the following form of the scattering potential can be derived [18, 60, 270],

$$\hat{\mathbf{V}}(\boldsymbol{r}) = \mathrm{graddiv} - K^2\chi(\boldsymbol{r}), \tag{4.12}$$

where $\chi(\boldsymbol{r}) = \varepsilon_{\mathrm{rel}}(\boldsymbol{r}) - 1$ is the dielectric susceptibility (polarizability) of the material, $\varepsilon_{\mathrm{rel}}$ is the relative permittivity, and the first term graddiv in the formula expresses the non-transversality of the E-wave in the material.

The differential wave equation (4.11) can be rewritten in the integral form

$$E(\boldsymbol{r}) = E_i(\boldsymbol{r}) + \int \mathrm{d}^3\boldsymbol{r}' G_0(\boldsymbol{r} - \boldsymbol{r}')\hat{\mathbf{V}}(\boldsymbol{r}')E(\boldsymbol{r}'), \tag{4.13}$$

where E_i (the incident wave) is the solution of the vacuum wave equation

$$(\Delta + K^2)E_i(\boldsymbol{r}) = 0 \tag{4.14}$$

and $G_0(\boldsymbol{r} - \boldsymbol{r}')$ is the *Green function* of a free particle, i.e., the solution of the wave equation with a delta-like right-hand side:

$$(\Delta + K^2)G_0(\boldsymbol{r} - \boldsymbol{r}') = \delta^{(3)}(\boldsymbol{r} - \boldsymbol{r}').$$

Solving this equation we obtain the Green function of a free particle in two equivalent forms [94, 288]:

$$G_0(\boldsymbol{r}-\boldsymbol{r}') = -\frac{1}{4\pi}\frac{\mathrm{e}^{\mathrm{i}K|\boldsymbol{r}-\boldsymbol{r}'|}}{|\boldsymbol{r}-\boldsymbol{r}'|} = -\frac{\mathrm{i}}{8\pi^2}\int \frac{\mathrm{d}^2\boldsymbol{K}_\parallel}{K_z}\mathrm{e}^{\mathrm{i}\boldsymbol{K}.(\boldsymbol{r}-\boldsymbol{r}')}. \tag{4.15}$$

In the second expression (the Weyl plane-wave representation), the Green function is constructed as a superposition of plane waves with the same frequency.

In the integral form of the wave equation (4.13), the second term on the right-hand side is the scattered wave. The solution of this wave equation can be written in the symbolic form (see any textbook on quantum mechanics, e.g., [92])

$$E(\boldsymbol{r}) = E_i(\boldsymbol{r}) + \int \mathrm{d}^3\boldsymbol{r}' G_0(\boldsymbol{r}-\boldsymbol{r}')\hat{\mathbf{T}}(\boldsymbol{r}')E_i(\boldsymbol{r}'), \tag{4.16}$$

where

$$\hat{\mathbf{T}} = \hat{\mathbf{V}} + \hat{\mathbf{V}}\hat{\mathbf{G}}_0\hat{\mathbf{V}} + \hat{\mathbf{V}}\hat{\mathbf{G}}_0\hat{\mathbf{V}}\hat{\mathbf{G}}_0\hat{\mathbf{V}} + \cdots \tag{4.17}$$

is the symbolic expression of the *scattering operator* $\hat{\mathbf{T}}(\boldsymbol{r})$. In the *kinematic* approximation $\hat{\mathbf{T}} \approx \hat{\mathbf{V}}$, i.e., we neglect multiple scattering processes and Eq.(4.17) gives an explicit expression for the scattered wavefield. The validity of the kinematic approximation has to be tested for a particular sample and experimental conditions used.

Using the plane-wave representation of the Green function, the scattered wave can be expressed as

$$\begin{aligned} E_s(\boldsymbol{r}) &= -\frac{\mathrm{i}}{8\pi^2}E_i \int \frac{\mathrm{d}^2\boldsymbol{K}_\parallel}{K_z}\mathrm{e}^{\mathrm{i}\boldsymbol{K}.\boldsymbol{r}} \int \mathrm{d}^3\boldsymbol{r}'\,\hat{\mathbf{T}}(\boldsymbol{r}')\mathrm{e}^{-\mathrm{i}(\boldsymbol{K}-\boldsymbol{K}_i).\boldsymbol{r}'} \\ &\equiv -\frac{\mathrm{i}}{8\pi^2}E_i \int \frac{\mathrm{d}^2\boldsymbol{K}_\parallel}{K_z}\mathrm{e}^{\mathrm{i}\boldsymbol{K}.\boldsymbol{r}}\langle \boldsymbol{K}|\hat{\mathbf{T}}|\boldsymbol{K}_i\rangle, \end{aligned} \tag{4.18}$$

where we have assumed the primary wave according to Eq. (4.2) and we have expressed the integral over $\boldsymbol{r}'$ in the Dirac notation used in quantum mechanics.

In most textbooks on x-ray scattering, the *Fraunhofer approximation* is assumed. In this approach, the region where $\hat{\mathbf{T}}(\boldsymbol{r})$ differs substantially from zero (i.e., the sample volume) is much smaller than the sample-detector distance. Then the sample acts as a point scatterer and the scattered wave has a spherical wavefront. The wave vector of the scattered wavefield is then fully determined by the relative position of the sample and detector. Using this approach, the expression for the Green function of a free particle can be simplified using the approximation [60],

$$K|\boldsymbol{r}-\boldsymbol{r}'| \approx Kr - \boldsymbol{K}_s.\boldsymbol{r}'$$

where $\boldsymbol{K}_s = K\boldsymbol{r}/r$ is the wave vector pointing from the (point-like) sample to the detector and having the length K. The scattered wave is then

$$E_s(\boldsymbol{r}) \approx -\frac{1}{4\pi}\frac{\mathrm{e}^{\mathrm{i}Kr}}{r}\int \mathrm{d}^3\boldsymbol{r}'\hat{\mathbf{T}}(\boldsymbol{r}')\mathrm{e}^{-\mathrm{i}(\boldsymbol{K}_s-\boldsymbol{K}_i).\boldsymbol{r}'} \equiv$$

$$\equiv -\frac{1}{4\pi}\frac{e^{iKr}}{r}\langle \boldsymbol{K}_s|\hat{\mathrm{T}}|\boldsymbol{K}_i\rangle, \tag{4.19}$$

assuming the incident wave in the form (4.2). Comparing with Eq. (4.18) we find that the scattered wave in the Fraunhofer approximation is proportional to the Fourier transformation of the scattering operator $\hat{\mathrm{T}}$, the argument of which is the scattering vector (wave vector transfer),

$$\boldsymbol{Q} = \boldsymbol{K}_s - \boldsymbol{K}_i, \tag{4.20}$$

where the direction of the wave vector $\boldsymbol{K}_s$ of the scattered radiation is *fully determined* by the detector position. In the general formula (4.18), the scattered wave is a superposition of the Fourier transformations of $\hat{\mathrm{T}}$ with *various* wave vectors $\boldsymbol{K}$ with constant lengths K and different directions.

The scattering process is usually described by the *differential cross section* $d\sigma$ determining the flux of the scattered photons into an elementary solid angle $d\Omega$ pointing towards the detector, and normalized to the unit flux *density* of the incident photons. From Eq. (4.19) the differential cross section follows in the form [92]

$$d\sigma = \frac{1}{16\pi^2}\left|\langle \boldsymbol{K}_s|\hat{\mathrm{T}}|\boldsymbol{K}_i\rangle\right|^2 d\Omega. \tag{4.21}$$

For laterally large samples, like perfect single crystals, thin layers, and multilayers, the Fraunhofer approximation is usually not valid. From classical optics (see [60], for instance) it follows that the Fraunhofer approximation is applicable if the scattering sample is much smaller than the circular first Fresnel zone having the radius $\sqrt{\lambda r}$, where r is the distance sample-detector and λ is the x-ray wavelength. In most of the laboratory experimental arrangements, the radius of the first Fresnel zone is a few μm, i.e., much smaller than the sample size. Another approach must be used in this case. In contrast to the Fraunhofer approximation, let us assume that the sample is laterally *infinite* and the scattered wavefield is homogeneous. Then, the intensity of a plane component of this wavefield given by Eq. (4.10) is

$$J(\boldsymbol{K}) = \frac{K^2 I_i}{16\pi^2 A}\left|\langle \boldsymbol{K}|\hat{\mathrm{T}}|\boldsymbol{K}_i\rangle\right|^2, \tag{4.22}$$

where $I_i = |E_i|^2$ is the intensity of the incident wave and A is the area of the irradiated sample surface. Using the approximation of the homogeneous wavefield, the concept of the differential cross section can be preserved, and we *define* the differential cross section in the form

$$d\sigma = \frac{A}{K^2 I_i}J(\boldsymbol{K})d\Omega = \frac{1}{16\pi^2}\left|\langle \boldsymbol{K}|\hat{\mathrm{T}}|\boldsymbol{K}_i\rangle\right|^2 d\Omega. \tag{4.23}$$

In this approach, however, the direction of the wave vector $\boldsymbol{K}$ is arbitrary and, in contrast to the Fraunhofer approximation, it is *not* determined by the position of the detector. Here, in the approximation of the homogeneous wavefield, the sample acts as a point-like scatterer *in reciprocal space* and the differential cross section determines the flux of the scattered photons into an elementary solid angle $d\Omega$, whose direction is determined by the arbitrary wave vector $\boldsymbol{K}$.

Using the assumption of the wave homogeneity, the intensity of the scattered wave does not depend on the position $\boldsymbol{r}$ of the observer. It can be derived using Eqs. (4.7), (4.9) and (4.22),

$$I = \frac{I_i}{16\pi^2 A} \int \frac{d^2 \boldsymbol{K}_\parallel}{K_z^2} \left| \langle \boldsymbol{K} | \hat{\mathrm{T}} | \boldsymbol{K}_i \rangle \right|^2 , \tag{4.24}$$

as a superposition of the contributions of all plane components constituting the homogeneous wave.

4.3 Direction of Scattered Waves

The direction of the waves scattered from a sample can be derived using the condition $|\boldsymbol{K}| = |\boldsymbol{K}_i|$ and considering the symmetry of the sample structure. Let us assume in this chapter that the sample is irradiated by an ideally plane wave, the sample surface is ideally flat, and the sample structure has a translational symmetry. For an amorphous sample, a displacement of the sample by any vector $\boldsymbol{R}$ parallel to the sample surface is a symmetry operation, for perfect crystalline samples the translational symmetry operations are the translations by the lattice vectors $\boldsymbol{R}_{\boldsymbol{n}} = n_1 \boldsymbol{a}_1 + n_2 \boldsymbol{a}_2$, where $\boldsymbol{a}_{1,2}$ are the basis vectors of the crystal lattice parallel to the sample surface and $n_{1,2}$ are integers. Then, the scattering potential $\hat{\mathrm{V}}(\boldsymbol{r})$ occurring in the wave equation (4.11) exhibits the same translational symmetry and the solution of the wave equation can be expressed using the Bloch theorem, which is well known from textbooks (see [195], for instance). For crystalline samples the Bloch theorem predicts the solution of the wave equation outside the sample in the form

$$E(\boldsymbol{r}) = \sum_{\boldsymbol{g}_\parallel} E_{\boldsymbol{g}} e^{i \boldsymbol{K}_{\boldsymbol{g}} \cdot \boldsymbol{r}} , \tag{4.25}$$

where $\boldsymbol{g}$ is the vector of the lattice reciprocal to the crystal lattice of the sample and $\parallel$ denotes a component parallel to the sample surface. Independent of the nature of the scattering sample, the solution of the wave equation outside the sample obeys the vacuum dispersion condition (4.1); therefore the wave vectors $\boldsymbol{K}_{\boldsymbol{g}}$ of the plane components of the vacuum wave are

$$\boldsymbol{K}_{\boldsymbol{g}} = \left(\boldsymbol{K}_{i\parallel} + \boldsymbol{g}_\parallel , \sqrt{K^2 - |\boldsymbol{K}_{i\parallel} + \boldsymbol{g}_\parallel|^2} \right) . \tag{4.26}$$

The in-plane components of these wave vectors (parallel to the sample surface) follow from the *lateral diffraction condition*

$$\boldsymbol{K}_{\boldsymbol{g}\parallel} = \boldsymbol{K}_{i\parallel} + \boldsymbol{g}_\parallel ; \tag{4.27}$$

therefore, the lateral component $\boldsymbol{Q}_{\parallel}$ of the scattering vector equals a lateral component of any vector $\boldsymbol{g}$ of the reciprocal lattice. Their vertical components can be obtained from the elasticity condition (4.1). The meaning of Eq. (4.25) is obvious; if a plane monochromatic wave irradiates an ideal crystalline sample, the scattered wave consists of a series of plane waves, the in-plane components of their wave vectors differ by $\boldsymbol{g}_{\parallel}$, and the lengths of these wave vectors correspond to the constant wave frequency.

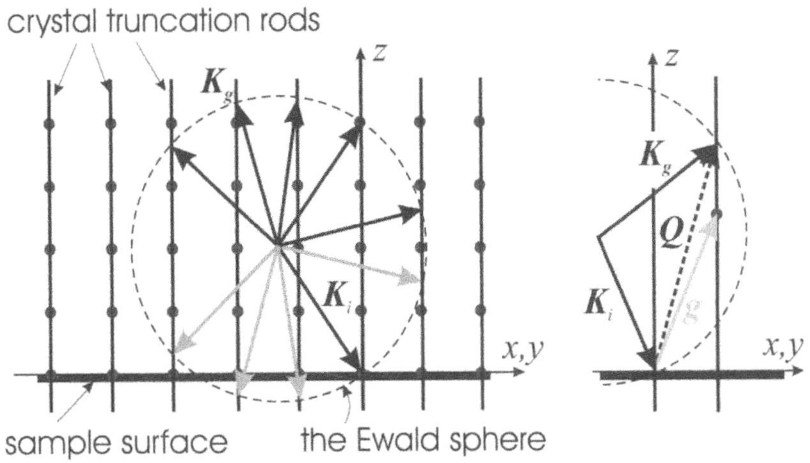

Fig. 4.1. The Ewald construction. The wave vector of the primary wave $\boldsymbol{K}_i$ points to the origin of reciprocal space, the wave vectors of the diffracted waves $\boldsymbol{K}_{\boldsymbol{g}}$ are given by Eq. (4.27). The wave vectors drawn in gray represent the diffracted waves penetrating into the sample; these waves are not measurable in the reflection geometry. In the right panel, the detail of the Ewald construction is drawn, for a given reciprocal lattice vector $\boldsymbol{g}$.

The lateral diffraction condition and the energy conservation in the scattering process are expressed graphically by means of the *Ewald construction* shown in Fig. 4.1. If the starting points of all the wave vectors $\boldsymbol{K}_i$ and $\boldsymbol{K}$ coincide, the end-points must lie at a spherical surface with radius $K = 2\pi/\lambda$ (the Ewald sphere). Since the difference of the in-plane components of $\boldsymbol{K}_i$ and $\boldsymbol{K}$ is an in-plane component of a reciprocal lattice vector, the end-points of the wave vectors $\boldsymbol{K}_{\boldsymbol{g}}$ lie in the cross sections of the Ewald sphere with the straight lines perpendicular to the surface and going through the reciprocal lattice points (the *crystal truncation rods* – CTR). For amorphous samples, only one truncation rod exists, intersecting the origin of reciprocal space, so that the scattered wavefield consists of a single plane component with the wave vector

$$\boldsymbol{K}_0 \equiv \boldsymbol{K}_R = (\boldsymbol{K}_{i\parallel}, \sqrt{K^2 - K_i^2}) \equiv (\boldsymbol{K}_{i\parallel}, -K_{iz}). \tag{4.28}$$

This is the specularly reflected wave. The vertical components of the primary and reflected waves have opposite signs $K_{Rz} = -K_{iz} > 0$.

The lateral diffraction conditions (4.27) and (4.28) for a crystalline and an amorphous sample do not depend on the scattering mechanism; so that they are valid both in the kinematical and dynamical approaches. Of course, various scattering theories yield different intensity distributions along a given truncation rod.

Let us deal now with the particular plane component of the scattered wavefield corresponding to the reciprocal lattice vector $\boldsymbol{g} = \boldsymbol{h}$. We call this component the *diffracted wave* with the diffraction vector $\boldsymbol{h}$. The wave vector $\boldsymbol{K}_{\boldsymbol{h}}$ of this diffracted wave obeys the lateral diffraction condition (4.27) in the form $\boldsymbol{K}_{\boldsymbol{h}\|} = \boldsymbol{K}_{i\|} + \boldsymbol{h}_{\|}$ and the condition of the energy conservation $|\boldsymbol{K}_{\boldsymbol{h}}| = K$. From these two conditions it is possible to determine the direction of $\boldsymbol{K}_{\boldsymbol{h}}$ if $\boldsymbol{K}_i$ is known. We distinguish two different geometries (see also Sec. 3.2):

1. The coplanar geometry – the common plane of the vectors $\boldsymbol{K}_i$ and $\boldsymbol{K}_{\boldsymbol{h}}$ (the *scattering plane*) is perpendicular to the sample surface, and therefore all the vectors $\boldsymbol{K}_i$, $\boldsymbol{K}_{\boldsymbol{h}}$, $\boldsymbol{h}$ and $\boldsymbol{n}$ (the unit vector of the surface normal) lie in the same plane. The angle ϕ between the diffraction vector $\boldsymbol{h}$ and $\boldsymbol{n}$ is the *asymmetry angle* ($\phi > 0$ for $\alpha_i < \alpha_{\boldsymbol{h}}$), see Fig. 4.2. Since the vectors $\boldsymbol{K}_{i,\boldsymbol{h}}$ are each determined by only one angular variable, the scattered intensity $J(\boldsymbol{Q})$ is determined by two angular variables, or equivalently by two components of the scattering vector $\boldsymbol{Q}$ in the scattering plane.
2. The non-coplanar geometry – the vectors $\boldsymbol{K}_i$, $\boldsymbol{K}_{\boldsymbol{h}}$, and $\boldsymbol{n}$ do not lie in the same plane. In this case, the scattered intensity J depends on four angular coordinates determining the directions of the vectors $\boldsymbol{K}_i$ and $\boldsymbol{K}_{\boldsymbol{h}}$. Therefore, in the non-coplanar case the scattered intensity can be represented as a function of the scattering vector $\boldsymbol{Q}$ (three coordinates) and the angle of incidence α_i of the primary wave.

The direction of the scattered wave follows from the Ewald construction shown in Fig. 4.1. Let us deal with this direction in a more detail.

In the *coplanar case* the lateral diffraction condition has the form

$$\cos\alpha_{\boldsymbol{h}} = \cos\alpha_i - 2\sin\Theta_B\cos\phi \tag{4.29}$$

(see Fig. 4.2). In the reflection geometry (the Bragg case) the diffracted wave is emitted into the upper half-space and $\alpha_{\boldsymbol{h}} > 0$. In the transmission geometry (the Laue case), the diffracted wave penetrates into the sample and $\alpha_{\boldsymbol{h}} < 0$. The angle between $\boldsymbol{K}_i$ and $\boldsymbol{K}_{\boldsymbol{h}}$ is the *scattering angle* $2\Theta = \alpha_i + \alpha_{\boldsymbol{h}}$, Θ_B is the *Bragg angle* defined as

$$\sin\Theta_B = \frac{|\boldsymbol{h}|}{2K}.$$

As we show later, kinematical diffraction exhibits a maximum of the diffracted intensity, if the scattering angle equals $2\Theta_B$ (the Bragg condition). Let us

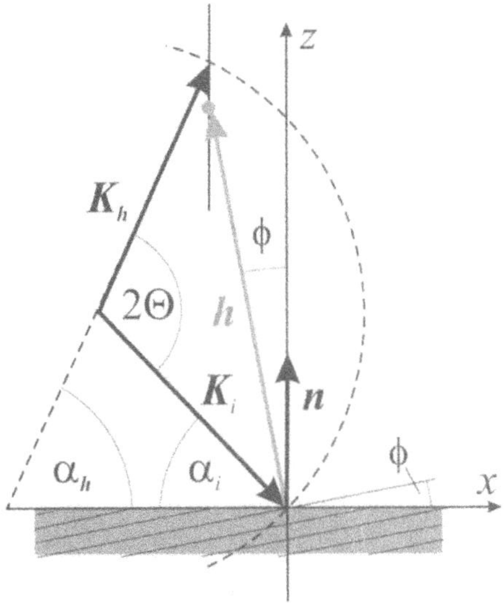

Fig. 4.2. The sketch of the coplanar scattering geometry.

assume that the incidence angle α_i slightly differs from its value α_{iB} in the kinematical Bragg maximum, i.e., $\alpha_i = \alpha_{iB} + \eta_i$, $|\eta_i| \ll \alpha_{iB}$ and similarly, $\alpha_{\boldsymbol{h}} = \alpha_{\boldsymbol{h}B} + \eta_{\boldsymbol{h}}$, $|\eta_{\boldsymbol{h}}| \ll \alpha_{\boldsymbol{h}B}$. Then, differentiating Eq. (4.29), we obtain

$$\frac{\eta_{\boldsymbol{h}}}{\eta_i} = -\frac{\gamma_0}{\gamma_{\boldsymbol{h}}}, \tag{4.30}$$

where γ_0 and $\gamma_{\boldsymbol{h}}$ denote the direction cosines of the primary and diffracted waves in the Bragg maximum with respect to the *internal* surface normal:

$$\gamma_0 = \sin\alpha_{iB} = \sin(\Theta_B - \phi), \ \gamma_{\boldsymbol{h}} = -\sin\alpha_{\boldsymbol{h}B} = -\sin(\Theta_B + \phi).$$

The ratio of the direction cosines is also called the diffraction asymmetry factor (or b-factor): $b = \gamma_{\boldsymbol{h}}/\gamma_0$.

From Eq. (4.30) it follows that any x-ray beam can be collimated by a diffraction from a perfect crystal. Let us assume that the crystal is irradiated by a divergent wave with the divergence angle $\Delta\alpha_i$. Then the divergence of the diffracted beam, $\Delta\alpha_{\boldsymbol{h}}$, is

$$\Delta\alpha_{\boldsymbol{h}} = \frac{1}{|b|}\Delta\alpha_i.$$

In the Bragg case with $\alpha_{iB} < \alpha_{\boldsymbol{h}B}$ (low-incidence asymmetry, Fig. 4.3) the divergence of the beam is *reduced* due to the diffraction, i.e., $\Delta\alpha_{\boldsymbol{h}B} < \Delta\alpha_i$, if $\alpha_{iB} > \alpha_{\boldsymbol{h}B}$ (high-incidence asymmetry, Fig. 4.3), $\Delta\alpha_{\boldsymbol{h}B} > \Delta\alpha_i$ holds and the divergence of the beam *increases* by the diffraction. If we express the asymmetry factor by the ratio $A_{\boldsymbol{h}}/A_i$ of the cross sections of the diffracted and primary beams, we find

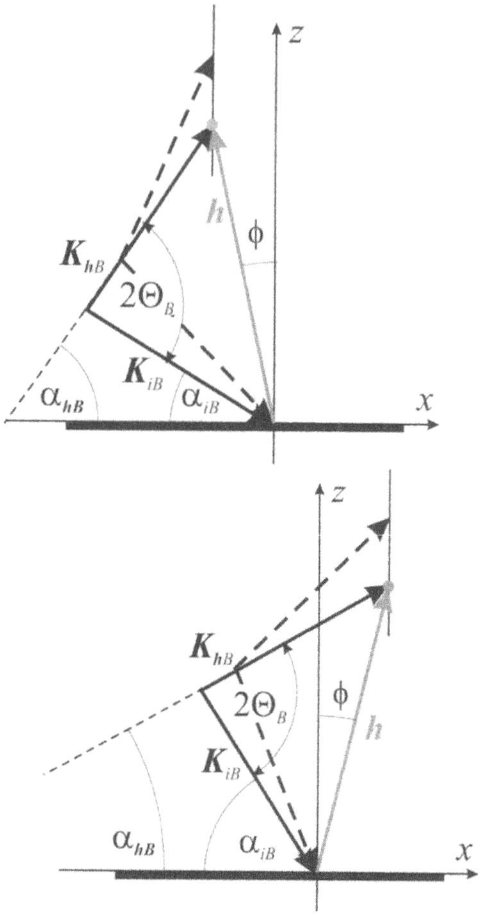

Fig. 4.3. The low-incidence asymmetry (left) and high-incidence asymmetry (right) in the reflection case. The full arrows denote the wave vectors $\boldsymbol{K}_{iB}$ and $\boldsymbol{K}_{\boldsymbol{h}B}$ of the primary and diffracted waves, respectively, in the Bragg position, the dashed arrows represent these vectors for a general direction of the primary wave.

$$A_i \Delta\alpha_i = A_{\boldsymbol{h}} \Delta\alpha_{\boldsymbol{h}}; \tag{4.31}$$

therefore the decrease of the divergence of the beam is accompanied with an increase of the beam cross section and vice versa. In the *symmetric* Bragg-case diffraction (and in x-ray reflection) $\boldsymbol{h}_{\|} = 0$, $\alpha_{\boldsymbol{h}B} = \alpha_{iB}$ and $\eta_{\boldsymbol{h}} = \eta_i$ hold so that the divergence and the cross section of the beams are not changed.

In the transmission geometry, the situation is different. In the symmetrial Laue case, where $\boldsymbol{h}$ is parallel to the sample surface, $\alpha_{\boldsymbol{h}B} = \alpha_{iB} + \pi$ (see Fig. 4.4), $b = 1$, and $\eta_{\boldsymbol{h}} = -\eta_i$. In this arrangement, the scattering angle 2Θ equals $2\Theta_B$ for *any small* deviation η_i from the kinematical Bragg position.

In the *non-coplanar geometry* we restrict ourselves to the most typical case, where the scattering plane defined by the vectors $\boldsymbol{K}_i$ and $\boldsymbol{K}_{\boldsymbol{h}}$ is nearly parallel to the sample surface (*grazing-incidence arrangement* – *GID*, see Fig. 3.9 in Sec. 3.3). From the lateral diffraction condition (4.27), we obtain the following equations

$$\begin{aligned} \cos\alpha_{\boldsymbol{h}} \sin\theta_{\boldsymbol{h}} + \cos\alpha_i \sin\theta_i &= 2\sin\Theta_B \cos\phi \equiv 2\sin\Theta_{B\parallel}, \\ \cos\alpha_{\boldsymbol{h}} \cos\theta_{\boldsymbol{h}} &= \cos\alpha_i \cos\theta_i, \end{aligned} \tag{4.32}$$

where $\theta_{i,\boldsymbol{h}}$ are the azimuthal angles of $\boldsymbol{K}_{i,\boldsymbol{h}}$ and ϕ is the asymmetry angle of the grazing-incidence geometry defined in Fig. 3.9. Therefore, in the incidence geometry the direction of the primary wave is determined by two angles α_i and θ_i; the two angles $\alpha_{\boldsymbol{h}}$ and $\theta_{\boldsymbol{h}}$ of the scattered beam are determined by two equations (4.32). Thus a plane wave is produced by a grazing-incidence diffraction of a plane monochromatic primary wave.

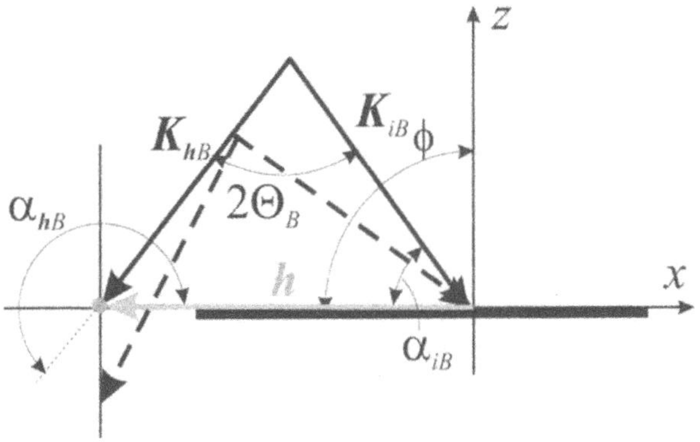

Fig. 4.4. Sketch of the symmetric transmission (Laue) geometry ($b = 1$). Like the previous picture, the full arrows denote the wave vectors $\boldsymbol{K}_{iB}$ and $\boldsymbol{K}_{\boldsymbol{h}B}$ of the primary and diffracted waves, respectively, in the Bragg position, the dashed arrows represent these vectors for a general direction of the primary wave. In the symmetrical Laue case, the scattering angle 2Θ remains unchanged for small deviations from the kinematical Bragg position.

As in the coplanar case, we assume that the direction of the primary wave differs only slightly from the kinematical Bragg position. We denote the small deviations

$$\delta\theta_s = \theta_s - \theta_{sB}, \;\; s = i, \boldsymbol{h},$$

and we assume that the incidence and exit angles are very small, i.e., $\alpha_{i,\boldsymbol{h}} \ll 1$. Then, the angular parameters of the diffracted wave are

$$\begin{aligned} \alpha_{\boldsymbol{h}}^2 &= \alpha_i^2 \cos(2\Theta_{B\parallel}) + 2\delta\theta_i \sin(2\Theta_{B\parallel}) \\ \delta\theta_{\boldsymbol{h}} &= \tfrac{1}{2}\alpha_i^2 \sin(2\Theta_{B\parallel}) - \delta\theta_i \cos(2\Theta_{B\parallel}). \end{aligned} \tag{4.33}$$

These formulas have an interesting consequence. Let us irradiate the sample by a monochromatic wave with a small non-zero horizontal divergence $\Delta(\delta\theta_i)$ and no vertical divergence $\Delta\alpha_i = 0$. Differentiating the first equation in (4.33), we obtain

$$\Delta\alpha_{\boldsymbol{h}} = \frac{\Delta(\delta\theta_i)}{\alpha_{\boldsymbol{h}}} \sin(2\Theta_{B\|}). \tag{4.34}$$

Since, in the grazing-incidence geometry, the exit angle $\alpha_{\boldsymbol{h}}$ is rather small, a small horizontal divergence $\Delta(\delta\theta_i)$ of the incident wave is transformed by a *large* vertical divergence $\Delta\alpha_{\boldsymbol{h}}$ of the diffracted wave.

5 Kinematical Theory

The essence of the kinematical x-ray scattering theory lies in the assumption that an x-ray photon, after being scattered by an electron, cannot be scattered by another electron again. Thus, only one scattering act can take place on a single ray. Intuitively, this assumption is more likely to be fulfilled for thinner layers. As we demonstrate in this and the following chapters, the validity of the kinematical approximation depends not only on the layer thickness, but also on other parameters (e.g., strength of the scattering, crystalline perfection, geometrical arrangement). Therefore, it may happen that the scattering process in a relatively thick layer can be described kinematically, while in other situations, the dynamical theory is necessary even for extremely thin layers. Kinematical diffraction theory has been described in many textbooks and monographs, for instance [20, 77, 142, 386].

5.1 Scattering From a Perfect Layer

We start from the scalar wave equation in the integral form defined in the previous chapter in Eq. (4.16). We have shown in the previous chapter that, in the kinematical approximation, the scattering operator $\hat{\mathbf{T}}$ is replaced by the scattering potential $\hat{\mathbf{V}}$, Eq. (4.16) represents an explicit expression for the scattered wave, and the scattering potential has the form (4.12). The actual expression for the polarizability depends on the polarization process assumed. For instance, if classical scattering from electrons is considered,

$$\chi(\boldsymbol{r}) = -\frac{\lambda^2}{\pi} r_{\mathrm{el}} C \varrho(\boldsymbol{r}), \tag{5.1}$$

where $r_{\mathrm{el}} = e^2/(4\pi\epsilon_0 m_0 c^2) \approx 2.82 \times 10^{-15}$ m is the classical electron radius, $\varrho(\mathrm{r})$ is the electron density, and $C = \sin\xi$ is the linear polarization factor. Here ξ denotes the angle between the polarization vector of the primary radiation and the position vector $\boldsymbol{r}$ of the observation point.

There are two limiting cases of the polarization of the primary wave with respect to the scattering plane. In the S-polarization, the polarization vectors of the primary and scattered waves are perpendicular to the scattering plane (see Fig. 5.1) and $C = 1$. In the P-polarization, both the polarization vectors lie in the scattering plane and $C = \cos(2\Theta)$, where 2Θ is the scattering angle defined in the previous chapter. Strictly speaking, the angle ξ and the linear polarization factor C are not constant in the scattering sample; however, this dependence will be neglected and we will consider C as a constant characterizing the scattering geometry.

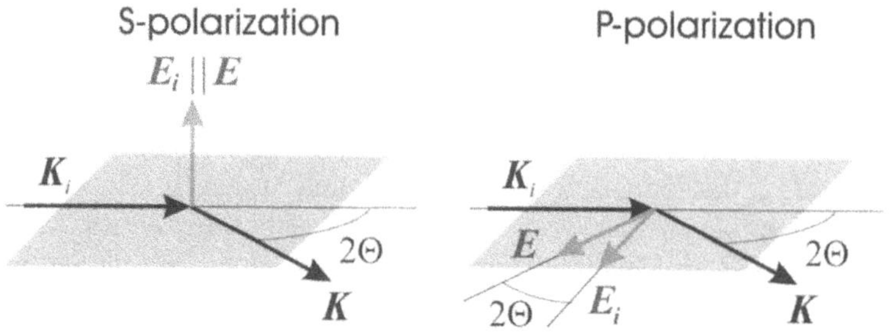

Fig. 5.1. S- and P-polarizations.

Within the kinematic approximation, the amplitude of the scattered wave is

$$E_s(\boldsymbol{r}) = \int \mathrm{d}^3\boldsymbol{r}' G_0(\boldsymbol{r} - \boldsymbol{r}')\hat{\mathbf{V}}(\boldsymbol{r}')E_i(\boldsymbol{r}'). \tag{5.2}$$

In the following we omit the subscript s in the scattered wave. The incident wave $E_i(\boldsymbol{r}')$ propagates in vacuum; therefore $\mathrm{graddiv}\boldsymbol{E}_i(\boldsymbol{r}') = 0$. Using the expressions (4.12, 5.1) for the scattering potential and the plane wave expression (4.2) for the primary wave, for the scattered wave we obtain the formula

$$E(\boldsymbol{r}) = 4\pi r_{\mathrm{el}} C E_i \int \mathrm{d}^3\boldsymbol{r}' G_0(\boldsymbol{r} - \boldsymbol{r}')\varrho(\boldsymbol{r}')\mathrm{e}^{\mathrm{i}\boldsymbol{K}_i.\boldsymbol{r}'}. \tag{5.3}$$

Since we assume that the sample has a perfect single crystalline structure, the electron density $\varrho(\boldsymbol{r})$ can be written as a superposition of *identical* unit cells

$$\varrho(\boldsymbol{r}) = \sum_{\boldsymbol{R} \in \Omega_{\mathrm{cryst}}} \varrho_{\mathrm{cell}}(\boldsymbol{r} - \boldsymbol{R}), \tag{5.4}$$

where $\boldsymbol{R}$ are the position vectors of the unit cells and Ω_{cryst} represents the crystal volume. We introduce the shape function of the crystal $\Omega_{\mathrm{cryst}}(\boldsymbol{r})$ that equals unity in the crystal and zero outside it. Using the well-known properties of periodic functions, the electron density of the crystal can be expressed as

$$\varrho(\boldsymbol{r}) = \frac{1}{8\pi^3 V_{\mathrm{cell}}} \sum_{\boldsymbol{g}} \int \mathrm{d}^3\boldsymbol{Q} F_{\mathrm{cell}}(\boldsymbol{Q}) \Omega_{\mathrm{cryst}}^{\mathrm{FT}}(\boldsymbol{Q} - \boldsymbol{g})\mathrm{e}^{\mathrm{i}\boldsymbol{Q}.\boldsymbol{r}}, \tag{5.5}$$

where V_{cell} is the volume of the crystal elementary unit cell, the superscript ^{FT} means the Fourier transformation, and

$$F_{\text{cell}}(\boldsymbol{Q}) = \int \mathrm{d}^3\boldsymbol{r}\,\varrho_{\text{cell}}(\boldsymbol{r})\mathrm{e}^{-\mathrm{i}\boldsymbol{Q}.\boldsymbol{r}} \tag{5.6}$$

is the Fourier transformation of the contribution of a single elementary unit cell to the electron density; this Fourier transformation is called *structure factor of the unit cell.* The sum in Eq. (5.5) runs over all vectors $\boldsymbol{g}$ of the reciprocal lattice.

Using the expression (4.15) for the plane-wave representation of the Green function and assuming that the crystal has the form of a laterally infinite planparallel slab with the thickness T and the upper (irradiated) surface at $z = 0$ we obtain from Eqs. (5.3,5.5) the final formula for the amplitude of the scattered wave

$$E(\boldsymbol{r}) = -2\pi\mathrm{i}\frac{r_{\text{el}}}{V_{\text{cell}}}CE_i\sum_{\boldsymbol{g}}\frac{\mathrm{e}^{\mathrm{i}\boldsymbol{K}_{\boldsymbol{g}}.\boldsymbol{r}}}{K_{\boldsymbol{g}z}}F_{\text{cell}}(\boldsymbol{K}_{\boldsymbol{g}} - \boldsymbol{K}_i)G_{\text{cryst}}(K_{\boldsymbol{g}z} - \tilde{K}_{\boldsymbol{g}z}). \tag{5.7}$$

We have used Eq. (4.26) for the components of the vector $\boldsymbol{K}_{\boldsymbol{g}}$ of the plane component of the scattered wave obeying the lateral diffraction condition (4.27) and

$$\tilde{\boldsymbol{K}}_{\boldsymbol{g}} \equiv (\boldsymbol{K}_{\boldsymbol{g}\|}, \tilde{K}_{\boldsymbol{g}z}) = \boldsymbol{K}_i + \boldsymbol{g}. \tag{5.8}$$

In a general case, this vector does *not* correspond to any wavefield, since its length differs from K (see also Fig. 5.2). The ending point of $\tilde{\boldsymbol{K}}_{\boldsymbol{g}}$ always coincides with the reciprocal lattice point $\boldsymbol{g}$; therefore, the difference between $\boldsymbol{K}_{\boldsymbol{g}}$ and $\tilde{\boldsymbol{K}}_{\boldsymbol{g}}$ represents the deviation from the the reciprocal lattice point for a given vector $\boldsymbol{K}_i$.

We define $\Omega_{\text{cryst}}(\boldsymbol{r})$ the shape function of a crystal (unity in the crystal volume and zero outside it). For a planparallel crystal, its Fourier transformation $\Omega^{\text{FT}}_{\text{cryst}}(\boldsymbol{Q} - \boldsymbol{g})$

$$\Omega^{\text{FT}}_{\text{cryst}}(\boldsymbol{Q} - \boldsymbol{g}) = 4\pi^2\delta^{(2)}(\boldsymbol{Q}_\| - \boldsymbol{g}_\|)G_{\text{cryst}}(Q_z - g_z)$$

contains the one-dimensional geometrical factor of the crystal

$$G_{\text{cryst}}(q_z) = \int_{-T}^{0}\mathrm{d}z\mathrm{e}^{-\mathrm{i}q_z z} = \frac{\mathrm{i}}{q_z}\left(1 - \mathrm{e}^{-\mathrm{i}q_z T}\right) \equiv$$

$$\mathrm{e}^{-\mathrm{i}q_z T/2}T\mathrm{sinc}\left(\frac{q_z T}{2}\right), \tag{5.9}$$

where $\mathrm{sinc}(x) = \sin(x)/x$ and $q_z = Q_z - g_z$. The main maximum of G_{cryst} lies at $q_z = 0$. $\boldsymbol{q} = \boldsymbol{Q} - \boldsymbol{g}$ is the *reduced scattering vector.*

The structure factor F_{cell} of the unit cell can be calculated approximately, replacing the actual electron density of the atoms in a unit cell $\varrho_{\text{cell}}(\boldsymbol{r})$ by a sum of the contributions of *isolated* atoms, neglecting the influence of chemical bonds on the total electron density $\varrho_{\text{cell}}(\boldsymbol{r})$. Then, the structure factor equals a sum of the scattering factors of the individual atoms of the cell; the atomic scattering factor $f_s(\boldsymbol{Q})$ is the Fourier transformation of the electron density of an individual atom s, and its values for various $Q \equiv |\boldsymbol{Q}|$ can be found in the literature [254]. The tabulated $f_s(Q)$ have to be corrected by the contribution of the inelastic scattering (anomalous dispersion), which depends on the wavelength used in the experiment. The anomalous dispersion can be large for the wavelength close to the fundamental absorption edge of the constituting elements.

Using these $f_s(\boldsymbol{Q})$, the structure factor of the unit cell is given by the well-known formula

$$F_{\text{cell}}(\boldsymbol{Q}) = \sum_s f_s(\boldsymbol{Q}) \mathrm{e}^{-\mathrm{i}\boldsymbol{Q}.\boldsymbol{r}_s}, \tag{5.10}$$

where the sum runs over the atoms belonging to the same elementary unit cell.

Considering Eq. (5.10), the meaning of Eq. (5.7) is obvious. The scattered wave is a coherent superposition of plane waves with wave vectors $\boldsymbol{K}_g$; the amplitude of a plane component is proportional to the structure factor $F_{\text{cell}}(\boldsymbol{K}_g - \boldsymbol{K}_i)$ and to the geometrical factor $G_{\text{cryst}}(K_{gz} - \tilde{K}_{gz})$. The structure factor changes slowly with the scattering vector $\boldsymbol{K}_g - \boldsymbol{K}_i$, while the geometrical factor changes rapidly with the difference $K_{gz} - \tilde{K}_{gz}$. If this difference is zero, the corresponding plane component has a maximum amplitude (see Fig. 5.2).

Close to the maximum, the wave vector $\boldsymbol{K}_g$ can be replaced by $\tilde{\boldsymbol{K}}_g$ in the argument of the structure factor. Then, the structure factor is $F_{\text{cell}}(\boldsymbol{g})$ and its value can be calculated easily if the positions of the atoms in the elementary unit cell are known. For a diamond lattice for instance, whose unit cell consists of 8 identical atoms with the scattering factors $f(\boldsymbol{g})$ (in the approximation of isolated atoms), the structure factor is zero if the coordinates $g_{1,2,3}$ are of nonequal parity. If all these coordinates are even and their sum is divisible by 4, $F_{\text{cell}}(\boldsymbol{g}) = 8f(\boldsymbol{g})$ holds; if they are odd $F_{\text{cell}}(\boldsymbol{g}) = 4(1 \pm \mathrm{i})f(\boldsymbol{g})$.

From Eq. (5.7) it follows that the intensity of the scattered wave is a coherent sum of several contributions corresponding to the diffracted waves with different diffraction vectors $\boldsymbol{g}$. These waves obey the lateral diffraction condition (4.27) and their wave vectors $\boldsymbol{K}_g$ follow also from the Ewald construction shown in Fig. 4.1. If the reciprocal lattice vectors $\boldsymbol{g}$ entering the sum in Eq. (5.7) are not coplanar, the different plane components of $E(\boldsymbol{r})$ have different linear polarization factors C. We do not consider these differences; they can be included easily in the expression describing the wavefield.

From Eq. (5.7), the MCF of the scattered wave can be obtained using the definition formula (4.5). The resulting expression will contain a double sum $\sum_{\boldsymbol{g}} \sum_{\boldsymbol{g}'}$; therefore, the scattered wave is not homogeneous and its intensity depends on the position due to the oscillatory term $\exp[\mathrm{i}(\boldsymbol{K}_g - \boldsymbol{K}_{g'}).\boldsymbol{r}]$. The period of these oscillations is comparable to the parameters of the crystal lattice. These very rapid oscillations can be detected only by means of special

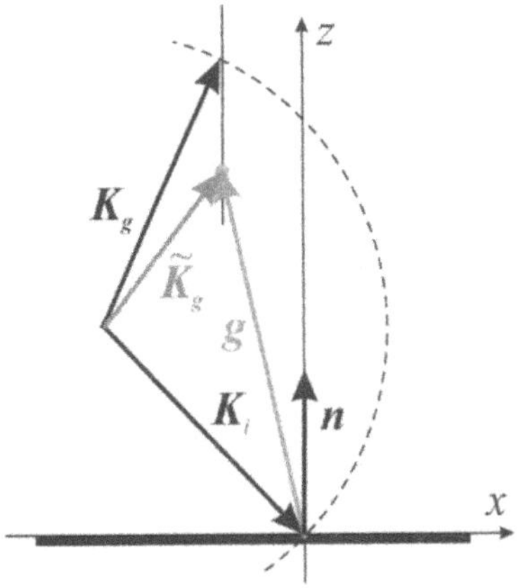

Fig. 5.2. The definition of the wave vectors $\boldsymbol{K_g}$ and $\tilde{\boldsymbol{K}}_{\boldsymbol{g}}$.

methods (for instance, using the x-ray standing-wave method – XRSW, see [383]), any usual detector averages these oscillations. The mutual coherence function of this averaged wave can be obtained if we neglect the terms $\boldsymbol{g}_{\parallel} \neq \boldsymbol{g'}_{\parallel}$. Thus, the double sum $\sum_{\boldsymbol{g}} \sum_{\boldsymbol{g'}}$ is replaced by the sum $\sum_{\boldsymbol{g}_{\parallel}} \sum_{g_z} \sum_{g'_z}$. The resulting wavefield is a coherent superposition of the waves with different g_z, but an *incoherent* superposition of the waves with different $\boldsymbol{g}_{\parallel}$. Considering Fig. 4.1, it means that the intensity of the scattered radiation is an incoherent superposition of the the contributions of individual crystal truncation rods (CTR), where each CTR is a coherent superposition of the contributions of individual reciprocal lattice points lying on this rod. This wavefield is homogeneous and we can easily express its intensity I and the reciprocal space distribution of intensity $J(\boldsymbol{Q})$

$$I = I_i \frac{4(\pi r_{\mathrm{el}} C)^2}{V_{\mathrm{cell}}^2} \sum_{\boldsymbol{g}_{\parallel}} \left| \frac{F_{\mathrm{cell}}(\boldsymbol{g}_{\parallel}, K_{\boldsymbol{g}z} - K_{iz})}{K_{\boldsymbol{g}z}} \sum_{g_z} G_{\mathrm{cryst}}(K_{\boldsymbol{g}z} - \tilde{K}_{\boldsymbol{g}z}) \right|^2 \quad (5.11)$$

and

$$J(\boldsymbol{Q}) = I_i K^2 \frac{4(\pi r_{\mathrm{el}} C)^2}{V_{\mathrm{cell}}^2} \sum_{\boldsymbol{g}_{\parallel}} \delta^{(2)}(\boldsymbol{Q}_{\parallel} - \boldsymbol{g}_{\parallel}) \times$$

$$\times \left| F_{\mathrm{cell}}(\boldsymbol{Q}) \sum_{g_z} G_{\mathrm{cryst}}(Q_z - g_z) \right|^2 . \quad (5.12)$$

In Eq. (5.11), the vectors $\boldsymbol{K_g}$ and $\tilde{\boldsymbol{K}}_{\boldsymbol{g}}$ are determined by the wave vector of the incident wave $\boldsymbol{K}_i$ by the formulas (4.26) and (5.8); the scattering vector $\boldsymbol{Q} = \boldsymbol{K} - \boldsymbol{K}_i$ in Eq. (5.12) is arbitrary and it denotes a position in reciprocal space.

The distribution of the scattered intensity in reciprocal space consists of a periodic sequence of infinitely narrow maxima (crystal truncation rods) given by the delta-peaks $\delta^{(2)}(\boldsymbol{Q}_{\|} - \boldsymbol{g}_{\|})$. The position of these rods has been derived from the translational symmetry of the sample in the previous chapter (see Fig. 4.1). In a measured reciprocal-space map, the lateral width of the crystal truncation rods is determined only by the resolution function of the experimental arrangement, i.e., by the coherence of the primary wave and angular resolution of the detector. The intensity distribution along the rods is given by a coherent superposition of the contributions of different reciprocal lattice points lying on a given rod. Each contribution is a product of the actual value of the structure factor $F_{\mathrm{cell}}(\boldsymbol{Q})$ and the geometrical factor $G_{\mathrm{cryst}}(Q_z - g_z)$. Since the crystal is usually much thicker than one elementary unit cell, the geometrical factor $G_{\mathrm{cryst}}(Q_z - g_z)$ changes much more rapidly than the structure factor $F_{\mathrm{cell}}(\boldsymbol{g}_{\|}, Q_z)$. Therefore, in the region of the main maximum of G_{cryst}, the actual structure factor can be replaced by its value $F_{\mathrm{cell}}(\boldsymbol{g})$ in the reciprocal lattice point. This simplification is used in the two-beam approximation discussed in the next section.

The position of the intensity maxima on the CTR is determined by the maximum of the geometrical factor G_{cryst}. The intensity maxima therefore coincide with the reciprocal lattice points and they are given by the vertical diffraction condition

$$Q_z = g_z \text{ or equivalently } K_{\boldsymbol{g}z} = \tilde{K}_{\boldsymbol{g}z}. \tag{5.13}$$

In contrast to the lateral diffraction condition (4.27), whose validity is not restricted to the kinematical diffraction, the vertical diffraction condition is valid *only* within the kinematic approximation. Dynamical effects caused by multiple scattering processes modify this condition; x-ray refraction, for instance, shifts the intensity maximum to larger Q_z; this will be shown in Chapter 7. Combining the lateral and vertical diffraction conditions we obtain the (full) diffraction condition (the Bragg condition)

$$\boldsymbol{Q} = \boldsymbol{g}, \tag{5.14}$$

the validity of which is also restricted to the kinematic approximation only.

The width of the intensity maximum on a CTR is given by the width of the geometrical factor G_{cryst}; and it equals approximately $\delta Q_z \approx 2\pi/T$. Thus, the width of the diffraction maxima do not depend on the diffraction vector $\boldsymbol{g}$. The geometrical factor exhibits also subsidiary maxima (see Figs. 5.3 and 5.4). Their period is $2\pi/T$ again (*thickness fringes*) and it does not depend on $\boldsymbol{g}$.

The general formula (5.3) can also be used for the calculation of the wave scattered from a homogeneous amorphous sample, where the electron density $\varrho(\boldsymbol{r})$ is constant. Repeating the procedure leading to Eq. (5.7), we obtain

$$E(\boldsymbol{r}) = -2\pi \mathrm{i} r_{\mathrm{el}} C E_i \varrho \frac{\mathrm{e}^{\mathrm{i}\boldsymbol{K}_R \cdot \boldsymbol{r}}}{K_{Rz}} G_{\mathrm{cryst}}(K_{Rz} - K_{iz}), \tag{5.15}$$

where

$$\boldsymbol{K}_R = (\boldsymbol{K}_{i\|}, \sqrt{K^2 - |\boldsymbol{K}_{i\|}|^2})$$

is the wave vector of the *specularly reflected* wave. If we express the (constant) electron density by the polarizability using Eq. (5.1), the intensity of the reflected wave is

$$I_R = I_i |r_{\text{kin}}|^2, \tag{5.16}$$

where r_{kin} is the kinematical approximation of the Fresnel reflectivity coefficient

$$r_{\text{kin}} = -\frac{K^2 \chi}{4K_{Rz}^2}. \tag{5.17}$$

The exact (dynamical expression) for the Fresnel reflectivity coefficient will be presented in chapter 6.

5.2 Two-Beam Approximation

As we have shown in the previous section, the wavefield scattered from a perfect crystal is a coherent superposition of the contributions of various reciprocal lattice points lying on the same CTR and an incoherent superpostion different CTR's. In the following we take into account only one CTR and one reciprocal lattice point on it in addition to the origin (0,0,0). This is so-called *two beam approximation.*

For a given position $\boldsymbol{Q}$ in reciprocal space, the contribution of various reciprocal lattice points to the scattered intensity depends on the vertical distance $q_z = Q_z - g_z$ between the point $\boldsymbol{Q}$ in reciprocal space and a reciprocal lattice point $\boldsymbol{g}$. If $\boldsymbol{Q}$ lies close to a reciprocal lattice point $\boldsymbol{g} = \boldsymbol{h}$, the contributions of the other points are neglected. Moreover, in this close vicinity the actual value of the structure factor $F_{\text{cell}}(\boldsymbol{Q})$ is replaced by its value $F_{\text{cryst}}(\boldsymbol{h})$ in the chosen reciprocal lattice point. The resulting wavefield is homogeneous and its intensity and reciprocal space distribution follow from Eqs. (5.11) and (5.12)

$$I_{\boldsymbol{h}} = I_i \frac{4(\pi r_{\text{el}} C)^2}{V_{\text{cell}}^2} \left| \frac{F_{\text{cell}}(\boldsymbol{h})}{K_{\boldsymbol{h}z}} G_{\text{cryst}}(K_{\boldsymbol{h}z} - \tilde{K}_{\boldsymbol{h}z}) \right|^2 \tag{5.18}$$

and

$$J(\boldsymbol{Q}) = I_i K^2 \frac{4(\pi r_{\text{el}} C)^2}{V_{\text{cell}}^2} \delta^{(2)}(\boldsymbol{Q}_\| - \boldsymbol{h}_\|) \left| F_{\text{cell}}(\boldsymbol{h}) G_{\text{cryst}}(Q_z - h_z) \right|^2. \tag{5.19}$$

The structure factor $F_{\text{cell}}(\boldsymbol{h})$ can be expressed using the Fourier coefficients of the crystal polarizability $\chi(\boldsymbol{r})$. According to Eq. (5.1) the polarizability is proportional to the electron density and for an *infinite* crystal it can be written in a form of the Fourier series (see also Eq. (1.7))

$$\chi(\boldsymbol{r}) = \sum_{\boldsymbol{g}} \chi_{\boldsymbol{g}} \mathrm{e}^{\mathrm{i}\boldsymbol{g}.\boldsymbol{r}}. \tag{5.20}$$

Its coefficients are proportional to the Fourier coefficients of the electron density $\chi_{\boldsymbol{g}} = -r_{\mathrm{el}} C \varrho_{\boldsymbol{g}} \lambda^2/\pi$, and they are given by the integral

$$\varrho_{\boldsymbol{g}} = \frac{1}{V_{\mathrm{cell}}} \int_{V_{\mathrm{cell}}} \mathrm{d}^3\boldsymbol{r}\, \varrho(\boldsymbol{r}) \mathrm{e}^{-\mathrm{i}\boldsymbol{g}.\boldsymbol{r}} \approx \frac{1}{V_{\mathrm{cell}}} \int \mathrm{d}^3\boldsymbol{r}\, \varrho_{\mathrm{cell}}(\boldsymbol{r}) \mathrm{e}^{-\mathrm{i}\boldsymbol{g}.\boldsymbol{r}} = \frac{F_{\mathrm{cell}}(\boldsymbol{g})}{V_{\mathrm{cell}}}.$$

In the last expression, we have approximated the integral of the total electron density over the unit cell by the integral of the contribution ϱ_{cell} of a single unit cell over the *infinite* range of $\boldsymbol{r}$. Then, the $\boldsymbol{g}$-th Fourier coefficient $\chi_{\boldsymbol{g}}$ of the crystal polarizability can be expressed using the atomic scattering factors as follows (see also Eq. (5.10):

$$\chi_{\boldsymbol{g}} = -\frac{r_{\mathrm{el}} C \lambda^2}{\pi V_{\mathrm{cell}}} \sum_s f_s(\boldsymbol{g}) \mathrm{e}^{-\mathrm{i}\boldsymbol{g}.\boldsymbol{r}_s}. \tag{5.21}$$

Using this approximation, the two-beam expressions for the intensity and the reciprocal space distribution are

$$I_{\boldsymbol{h}} = I_i \frac{K^4}{4K_{\boldsymbol{h}z}^2} \left| \chi_{\boldsymbol{h}} G_{\mathrm{cryst}}(K_{\boldsymbol{h}z} - \tilde{K}_{\boldsymbol{h}z}) \right|^2 \tag{5.22}$$

and

$$J_{\boldsymbol{h}}(\boldsymbol{Q}) = I_i \frac{K^6}{4} \delta^{(2)}(\boldsymbol{Q}_{\|} - \boldsymbol{g}) \left| \chi_{\boldsymbol{h}} G_{\mathrm{cryst}}(Q_z - h_z) \right|^2. \tag{5.23}$$

The two-beam approximation is used in most parts of this book. This approximation allows us to express in a very simple way the reciprocal distribution of the scattered intensity $J(\boldsymbol{Q})$. Using Eqs. (5.7), (5.21) and (4.22) and assuming a laterally homogeneous sample (not necessarily in the form of a homogeneous planparallel slab), we obtain the following simple formula:

$$J(\boldsymbol{Q}) = \frac{I_i}{A} \frac{K^6}{16\pi^2} \left\langle \left| S_{\boldsymbol{h}}(\boldsymbol{Q} - \boldsymbol{h}) \right|^2 \right\rangle, \tag{5.24}$$

where

$$S_{\boldsymbol{h}}(\boldsymbol{Q}) = \chi_{\boldsymbol{h}} \Omega_{\mathrm{cryst}}^{\mathrm{FT}}(\boldsymbol{Q}) \tag{5.25}$$

is the *structure amplitude (structure factor)* of diffraction $\boldsymbol{h}$ and A is the irradiated sample surface. In many cases, the shape of the diffracting crystal expressed by the shape function $\Omega_{\mathrm{cryst}}(\boldsymbol{r})$ is included in the space dependence of the polarizability coefficient $\chi_{\boldsymbol{h}}(\boldsymbol{r})$. Then, the structure amplitude is expressed simply by the Fourier transformation of this coefficient:

$$S_{\boldsymbol{h}}(\boldsymbol{Q}) = \int \mathrm{d}^3\boldsymbol{r}\, \chi_{\boldsymbol{h}}(\boldsymbol{r}) \mathrm{e}^{-\mathrm{i}\boldsymbol{Q}.\boldsymbol{r}}. \tag{5.26}$$

This formalism, however can be used *only* in the two-beam approximation, where the true structure factor of the unit cell $F_{\mathrm{cell}}(\boldsymbol{Q})$ is replaced by its value $F_{\mathrm{cell}}(\boldsymbol{h})$ in the reciprocal lattice point $\boldsymbol{h}$.

The validity of the two beam approximation is illustrated in the following example. In Fig. 5.3 we compare the intensities diffracted from a planparallel GaAs single crystal with (001) surface calculated kinematically using the many-beam formula (5.11) and its two-beam approximation (5.22). The intensities are plotted as functions of Q_z; in the calculation we have assumed $\boldsymbol{g}_{\|} = 0$ (symmetric diffraction $00l$), so that Q_z is connected with the incidence angle of the primary beam by the formula $Q_z = 2K \sin \alpha_i$. The crystal thickness was $T = 10a_{\mathrm{GaAs}} = 56.53$ Å, wavelength $\lambda = 1.54$ Å. The fig-

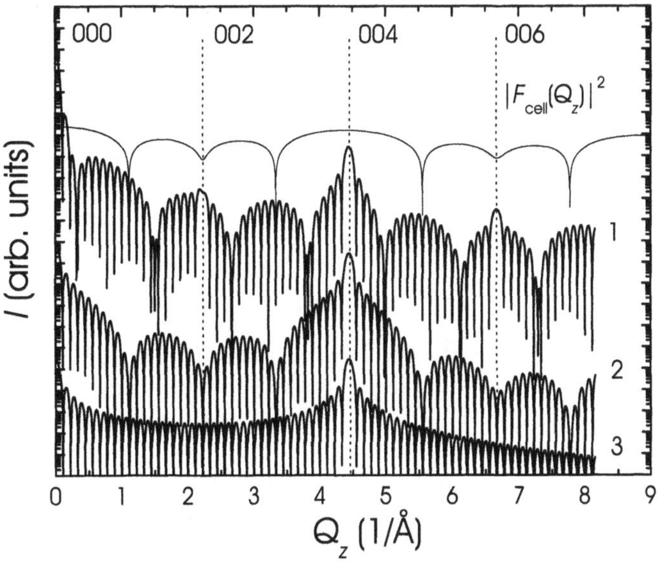

Fig. 5.3. The diffraction curves of a thin GaAs crystal. The curve (1) has been calculated using the many-beam formula (5.11); for the calculation (2) we have used the same formula but we took only one plane component $\boldsymbol{g} = (004)$ into account, i.e., we used a two-beam approximation with an *exact* value of $F_{\mathrm{cell}}(\boldsymbol{Q})$. Curve (3) follows from the two-beam expression (5.22) for $\boldsymbol{h} = (004)$. In addition the square of the structure factor F_{cell} is also plotted.

ure shows the importance of the many-beam approach, the approximations yielding the curves (2) and (3) cannot be used in the regions outside the diffraction maximum 004. Using the many-beam approach, we obtain the intensity distribution along the whole CTR including all the diffraction maxima 000, 002, 004, and 006. In the diffraction maxima 002 and 006 the structure factor has local minima; these maxima are weaker than the strong maximum 004. In the reciprocal lattice points 001, 003, and 005 the structure factor is zero (forbidden diffractions). Since the crystal is very thin, the kinematical approximation is valid, except for very small Q_z, where the kinematical intensity exceeds I_i. In this range, where the total external reflection of x-rays becomes important (see Chap. 6) the dynamical diffraction theory must be used.

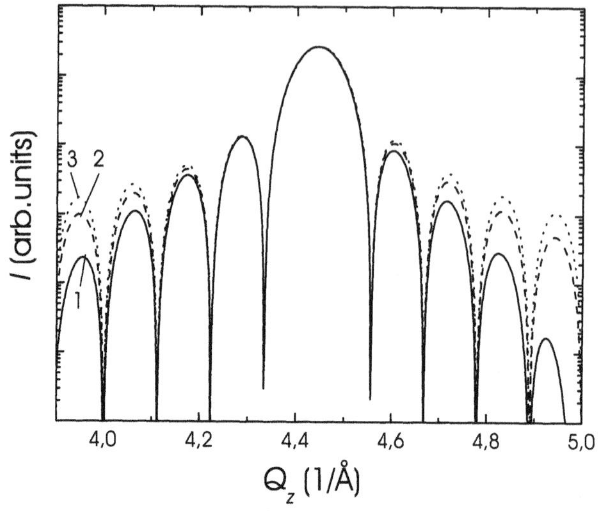

Fig. 5.4. The vicinity of the 004 diffraction maximum in Fig. 5.3. The meaning of the symbols 1, 2, and 3 is the same as in the previous figure.

In Fig. 5.4 we have enlarged the Q_z range in the vicinity of the 004 diffraction maximum, the differences between the full many-beam kinematical approach and the approximations presently used are clearly visible. For thicker crystals, the diffraction maxima are narrower and the intensities outside the maxima are lower than for thinner crystals. However, the many-beam approach is still important for the calculation of the intensity on a CTR far away from a maximum.

From Eq. (5.22) it follows that the intensity in the diffraction maximum is

$$I_{\boldsymbol{h}\,\max} = I_i \frac{K^4 T^2}{4K_{\boldsymbol{h}z}^2} \left|\chi_{\boldsymbol{h}}\right|^2 ,$$

i.e., it is proportional to the square of the crystal thickness T. Since the width of the diffraction maximum is inversely proportional to T, the *integrated intensity* $\mathcal{I}_{\boldsymbol{h}} = \int \mathrm{d}\alpha_i I_{\boldsymbol{h}}$ is proportional to T. These thickness dependencies of the maximum intensity and integrated intensity are characteristic for the kinematical approximation, where absorption and multiple scattering are neglected. Since the flux in the scattered beam must be smaller than (or equal to) the primary flux, the validity of the kinematical approximation is limited to very thin crystals. A more exact estimation of the range of validity of the kinematical theory will be presented in Chapter 7.

5.3 Kinematical Scattering From Deformed Crystals

The previous mathematical formalism can easily be modified for the kinematical description of x-ray scattering from a deformed crystal. In the following we assume that the deformation of the crystal *does not affect* the atomic scattering factors $f_s(\boldsymbol{Q})$ of individual atoms. Then, the structure factor of a deformed crystal unit cell lying in point $\boldsymbol{R}_n$ can be written as

$$F_{n,\text{cell}}(\boldsymbol{Q}) = \mathrm{e}^{-\mathrm{i}\boldsymbol{Q}.\boldsymbol{u}_n} \sum_s f_s(\boldsymbol{Q}) \mathrm{e}^{-\mathrm{i}\boldsymbol{Q}.(\boldsymbol{r}_s+\boldsymbol{u}_{ns})}, \tag{5.27}$$

where $\boldsymbol{u}_n$ is the displacement of the n-th unit cell *as a whole* and $\boldsymbol{u}_{ns}$ is the displacement of the s-th atom in the n-th unit cell with respect to a cell edge. In this so called *Takagi approximation* [357], we assume that the n-th unit cell is not modified due to the crystal deformation and it is only shifted by $\boldsymbol{u}_n$ as a rigid body. In this case, the structure factor differs from the structure factor of a non-deformed cell only by a phase factor:

$$F_{n,\text{cell}}(\boldsymbol{Q}) = \mathrm{e}^{-\mathrm{i}\boldsymbol{Q}.\boldsymbol{u}_n} F_{\text{cell}}(\boldsymbol{Q}). \tag{5.28}$$

This approximation is valid, if the components of the strain tensor are much smaller than unity and if no atoms are removed or added into the unit cell. In this section, we will use this approximation.

For a generally deformed crystal, the scattered wave is inhomogeneous even if we average the oscillatory terms in Eq.(5.7). The amplitude of the scattered wave can be obtained from Eqs. (5.3) and (5.28),

$$E(\boldsymbol{r}) = -\mathrm{i}E_i \frac{r_{\text{el}}C}{2\pi V_{\text{cell}}} \int \frac{\mathrm{d}^2\boldsymbol{K}_{\|}}{K_z} \mathrm{e}^{\mathrm{i}\boldsymbol{K}.\boldsymbol{r}} F_{\text{cell}}(\boldsymbol{Q}) \times$$

$$\times \sum_{\boldsymbol{g}} \int_{V_{\text{cryst}}} \mathrm{d}^3\boldsymbol{r}' \mathrm{e}^{-\mathrm{i}(\boldsymbol{Q}-\boldsymbol{g}).\boldsymbol{r}'} \mathrm{e}^{-\mathrm{i}\boldsymbol{Q}.\boldsymbol{u}(\boldsymbol{r}')}, \ \boldsymbol{Q} = \boldsymbol{K} - \boldsymbol{K}_i; \tag{5.29}$$

the integral on the right-hand side is calculated over the crystal volume V_{cryst}. Equation (5.29) enables us to calculate the scattered wavefield if the displacement field $\boldsymbol{u}(\boldsymbol{r})$ in the crystal is known.

The simplest deformation of a lattice is a homogeneous deformation, having constant components of the strain tensor ϵ_{jk}. Then, the displacement field is a linear function of the position

$$u_j(\boldsymbol{r}) = \epsilon_{jk} x_k. \tag{5.30}$$

Putting this expression in Eq. (5.29) we find that the mutual coherence function of the scattered wavefield is given by the same formula as for the non-deformed crystal (Eq. (5.7)); the wave vector $\boldsymbol{K}_g$ is replaced by the vector

$$\boldsymbol{K}_{\boldsymbol{g}}^{\text{def}} = \left(\boldsymbol{K}_{i\|} + \boldsymbol{g}_{\|}^{\text{def}}, \sqrt{K^2 - |\boldsymbol{K}_{i\|} + \boldsymbol{g}_{\|}^{\text{def}}|^2}\right), \tag{5.31}$$

where

$$g_j^{\text{def}} = (\delta_{jk} + \epsilon_{jk})^{-1} g_k \approx g_j - \epsilon_{jk} g_k. \tag{5.32}$$

Repeating the procedure from the previous section leading to the intensity and reciprocal space distribution of the homogeneous wavefield (Eqs. (5.11) and (5.12)) we find that the maxima of the scattered intensity in reciprocal space are shifted due to the homogeneous crystal deformation by $-\epsilon_{jk} g_k$; this shift is a linear function of the diffraction vector. This will be investigated in more detail in Chapter 9. Within the Takagi approximation and assuming a homogeneous deformation, the shape of the diffraction maxima remains unchanged.

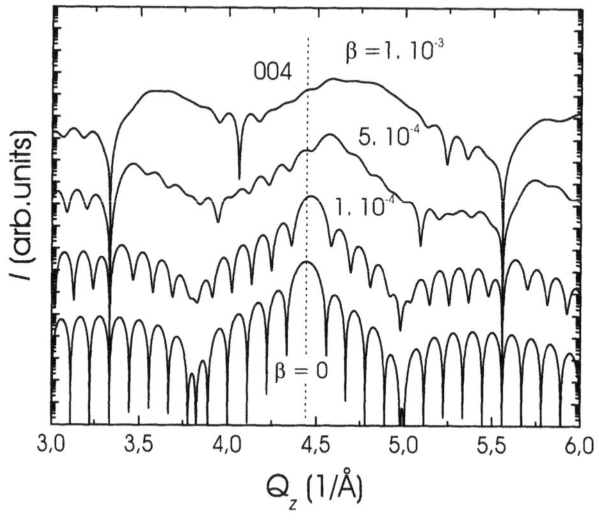

Fig. 5.5. The diffraction curves of a thin GaAs crystal with quadratic deformation along the z-axis, $u_z = \beta z^2$. The values of β are given in 1/Å. The many-beam kinematical approach is used.

Let us deal now with a special case of an inhomogeneous deformation field; namely we assume that the displacement is a nonlinear function of the coordinate z perpendicular to the surface. In this case, we obtain an expression for the mutual coherence function analogous to Eq. (5.7); the geometrical factor G_{cryst} of the crystal is replaced by

$$G_{\text{cryst}}^{\text{def}}(Q_z - g_z) = \int_{-T}^{0} \mathrm{d}z \mathrm{e}^{-\mathrm{i}(Q_z - g_z)z} \mathrm{e}^{-\mathrm{i}\boldsymbol{Q}.\boldsymbol{u}(z)}. \tag{5.33}$$

As in the non-deformed case, we can neglect the rapidly oscillating terms in the expression for the mutual coherence function. The resulting wavefield is homogeneous again, and its intensity and reciprocal space distribution are given by Eqs. (5.11) and (5.12), where the geometrical factor is replaced by the expression in Eq. (5.33). In Fig. 5.5 we show examples of the diffraction curves changing due to this deformation. As with Figs. 5.3 and 5.4, we assume a GaAs crystal with the surface (001)and thickness of $10a_{\mathrm{GaAs}}$; the x-ray wavelength is 1.54 Å. The displacement $\boldsymbol{u}(z)$ is parallel to the z-axis and it is given by a quadratic function $\boldsymbol{u}(z) = \beta z^2$. In the figure, we present the curves calculated for various values of β using the many-beam formula (5.11) with the geometrical factor (5.33). It can be seen that even a very small deformation ($\beta = 10^{-4}$ Å^{-1}) affects not only the position but also the shape of the diffraction maximum. For larger β, the diffraction maximum becomes broader and asymmetric and the thickness oscillations disappear.

5.4 Kinematical Scattering From Multilayers

In this section, we analyze a special type of a deformed crystal, namely, an epitaxial layered system, consisting of a sequence of N layers with different chemical compositions and different thicknesses; the multilayer stack is sketched in Fig. 5.6. We assume that the structure has no defects such as dislocations at the interfaces; therefore, the crystal lattices of individual layers have the *same* lateral lattice constants, i.e., the multilayer structure is pseudomorph. For the sake of simplicity, we restrict ourselves to the layers with cubic lattices, the interfaces in the multilayer are parallel to (001). Then, elementary unit cells in the layers are tetragonally distorted, having the same lattice constants a_x and a_y and different a_z. The detailed description of the deformation status of epitaxially grown layers will be given in Sect. 9.1. In an experimentally relevant case, the multilayer structure is deposited onto a substrate, much thicker than the multilayer stack. Then, the lateral lattice constants of the layers equal the lateral lattice constant of the substrate; the vertical lattice constants can be derived knowing the elasticity constants of the layers (see Chap. 9.1). However, in this chapter we do not include the substrate in the calculations since x-ray diffraction in the substrate is dynamic and it would require us to apply the dynamical diffraction theory. The kinematical scattering formalism can be applied also for any crystal symmetry and any orientation of the interfaces; in this general case, however, the calculation of the displacement field in the layers is more complicated.

A pseudomorph multilayer can be treated as a single crystal with a special kind of deformation that also affects the values of the structure factors; therefore, the Takagi approximation is not valid. In a pseudomorph structure, the structure factor F_{cell} and the displacement field $\boldsymbol{u}(\boldsymbol{r})$ depend only on the coordinate z perpendicular to the surface. If we neglect the oscillatory terms in the many-beam expression (5.7) we obtain for the intensity scattered from a multilayer the expression analogous to Eq. (5.11):

$$I = I_i \frac{4(\pi r_{\mathrm{el}} C)^2}{V_{\mathrm{cell}}^2} \sum_{\boldsymbol{g}_\parallel} \left| \frac{1}{K_{\boldsymbol{g}z}} \sum_{g_z} \sum_{n=1}^{N} F_{\mathrm{cell}}^{(n)}(\boldsymbol{g}_\parallel, K_{\boldsymbol{g}z} - K_{iz}) \times \right.$$

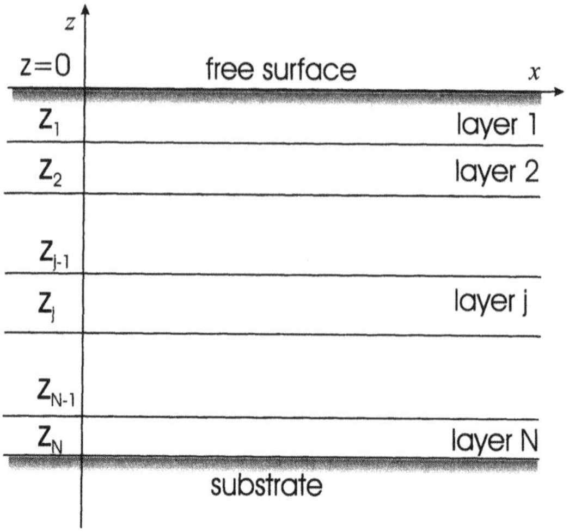

Fig. 5.6. The sketch of a multilayer structure.

$$\times \, G^{(n)}_{\text{layer}}(K_{gz} - \tilde{K}_{gz})\Big|^2 , \tag{5.34}$$

where the wave vectors $\boldsymbol{K_g}$ and $\tilde{\boldsymbol{K}}_{\boldsymbol{g}}$ are given by Eqs. (4.26) and (5.8); i.e., they are not affected by the displacement field. $F^{(n)}_{\text{cell}}(\boldsymbol{Q})$ is the structure factor of the unit cell in the n-th layer, and

$$G^{(n)}_{\text{layer}}(Q_z) = \int_{z_{n+1}}^{z_n} \mathrm{d}z e^{-\mathrm{i}Q_z(z+u(z))} \tag{5.35}$$

is the geometrical factor of the layer n lying between the interfaces $n+1$ and n. The deformation field $u(z)$ is defined with respect to *any* reference lattice; this lattice defines also the reciprocal lattice vectors $\boldsymbol{g}$. The lateral lattice parameter of the reference lattice must equal the common lateral lattice parameter of the layers. Usually, two equivalent choices are used for the reference lattice, namely, the lattice of the substrate or an averaged lattice of the multilayer.

An important case of a multilayer structure is a periodic multilayer with a pseudomorph structure. This multilayer consist of M identical bilayers; each bilayer is composed of a layer A with thickness $n_A a_A$ and the structure factor $F^{(A)}_{\text{cell}}(\boldsymbol{Q})$ and the layer B with thickness $n_B a_B$ and structure factor $F^{(B)}_{\text{cell}}(\boldsymbol{Q})$. $a_{A,B}$ are the (tetragonally distorted) vertical lattice parameters of the layers A and B (see Fig. 5.7). The period of the multilayer is $D = n_A a_A + n_B a_B$; the whole multilayer thickness is $T = MD$. In this case, we choose the averaged lattice of the multilayer as a reference lattice. The lateral lattice constant of the reference lattice equals that of the substrate and its value is not important for the intensity calculation. Its vertical lattice constant is

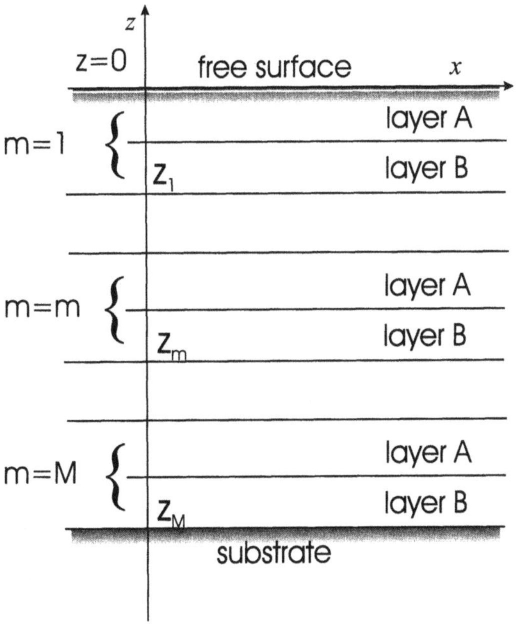

Fig. 5.7. The sketch of a periodic multilayer.

$$\langle a \rangle = \frac{n_A a_A + n_B a_B}{n_A + n_B}. \tag{5.36}$$

Using this definition, the displacement $u(z)$ is a periodic function of z with the period D. We denote $\delta_{A,B} = (a_{A,B} - \langle a \rangle)/\langle a \rangle$ the vertical mismatch of the layers A and B with respect to the reference lattice. Then, the displacement in the B-layers is $u_B(z) = \delta_B \Delta z$ and in the A-layers $u_A(z) = \delta_A \Delta z + (\delta_B - \delta_A) n_B \langle a \rangle$, where $\Delta z = z - z_m$ and z_m is the z-coordinate of the bottom of the m-th multilayer period.

From Eq. (5.34) we obtain the intensity scattered from a periodic multilayer

$$I = I_i \frac{4(\pi r_{\mathrm{el}} C)^2}{V_{\mathrm{cell}}^2} \sum_{\boldsymbol{g}_\parallel} \left| \frac{1}{K_{\boldsymbol{g}z}} \sum_{g_z} F_{\mathrm{period}}(\boldsymbol{g}_\parallel, K_{\boldsymbol{g}z} - K_{iz}) \times \right.$$

$$\left. \times G_{\mathrm{multilayer}}(K_{\boldsymbol{g}z} - \tilde{K}_{\boldsymbol{g}z}) \right|^2, \tag{5.37}$$

where

$$F_{\mathrm{period}}(\boldsymbol{Q}) = F_{\mathrm{cell}}^{(B)}(\boldsymbol{Q}) \frac{\mathrm{e}^{-\mathrm{i}q_B T_B} - 1}{-\mathrm{i}q_B} + F_{\mathrm{cell}}^{(A)}(\boldsymbol{Q}) \mathrm{e}^{-\mathrm{i}q_B T_B} \frac{\mathrm{e}^{-\mathrm{i}q_A T_A} - 1}{-\mathrm{i}q_A} \tag{5.38}$$

is the structure factor of a multilayer period. Here we have denoted $q_{A,B} = Q_z - g_z + Q_z \delta_{A,B}$ and $T_{A,B} = n_{A,B} \langle a \rangle$ are the thicknesses of the layers in the non-deformed (reference) lattice. In contrast to F_{cell}, the structure factor of the period depends also on g_z, which occurs in the factors $q_{A,B}$.

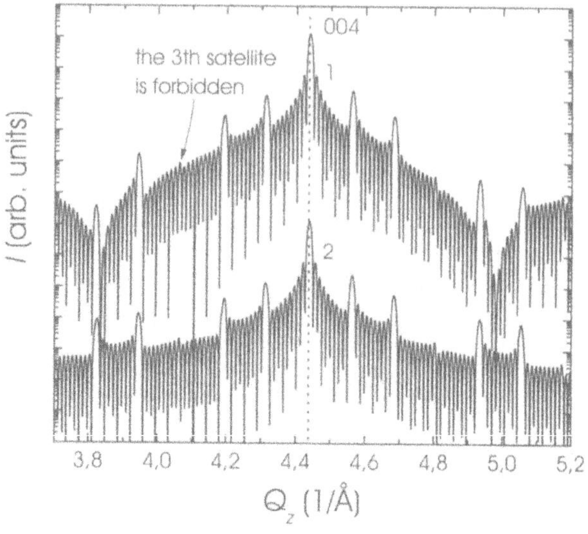

Fig. 5.8. The diffraction curves of a GaAs/AlAs multilayer; see the text for the multilayer structure. The curve (1) was calculated kinematically using the exact many-beam formula (5.37) and the two-beam approach (2).

The geometrical factor of the periodic multilayer is

$$G_{\mathrm{multilayer}}(q) = \mathrm{e}^{\mathrm{i}qD}\frac{\mathrm{e}^{\mathrm{i}qT}-1}{\mathrm{e}^{\mathrm{i}qD}-1}, \quad q = K_{\boldsymbol{g}z} - \tilde{K}_{\boldsymbol{g}z}. \tag{5.39}$$

If the multilayer consists of several bilayers, the geometrical factor changes more rapidly with Q_z than the structure factor of a single bilayer. Therefore, the intensity maxima on a CTR correspond to local maxima of the geometrical factor. Several types of the maxima of $G_{\mathrm{multilayer}}$ exist. The main maxima occur in the points $K_{\boldsymbol{g}z} = \tilde{K}_{\boldsymbol{g}z}$, i.e., in the lattice points of the lattice reciprocal to the reference lattice. These are the usual *diffraction maxima* analogous to the diffraction from a single layer, whose structure corresponds to the reference lattice. The maxima of the second type occur in the points, where $\exp(\mathrm{i}qD) = 1$, thus for $Q_z = g_z + 2p\pi/D$, where p is an integer. These maxima are called *satellites* or *superlattice peaks*; their distance is inversely proportional to the multilayer period. Between two superlattice satellites, $M-1$ zero points of the intensity exist, where $\exp(\mathrm{i}qT) = 1$; therefore between two satellites $M-2$ intensity maxima of the third type occur. These maxima are called *thickness fringes* and the distance between the adjacent zero-points on the intensity curve is $2\pi/T$.

The intensity maxima are modulated by the structure factor F_{period} that changes much slowly than the geometrical factor. If we neglect the influence of the deformation on the structure factor, i.e., if we replace $q_{A,B}$ by $q_z = Q_z - g_z$, the expression for the structure factor becomes simpler. If, in addition we put $q_z = 2p\pi/D$, i.e., we calculate the structure factor for the p-th satellite, the structure factor is

$$F_{\text{period}} = \frac{\mathrm{i}D}{2\pi p}(F_{\text{cell}}^{(B)} - F_{\text{cell}}^{(A)})\left(\mathrm{e}^{-2\pi\mathrm{i}pT_B/D} - 1\right). \tag{5.40}$$

The factor of the p-th satellite is zero, if $\exp(-2\pi\mathrm{i}pT_B/D) = 1$, i.e., if $n_B/(n_A + n_B) = m/p$, where m is an integer. For instance, if $n_B = 3n_A$, every 4th satellite disappears. However, this simple rule is valid only if we neglect the strain in the structure factor.

In Fig. 5.8 we have plotted the diffraction curves of a periodic multilayer consisting of 10 periods; each period is composed of a GaAs layer three lattice parameters thick and a AlAs layer six lattice parameters thick. The wavelength was 1.54 Å and for the calculation we have used both the many-beam approach (Eq. (5.37)) and the two-beam approximation. From the simplified expression (5.40) for the structure factor it follows that every third satellite should disappear. In fact, if we calculate the diffraction curve exactly, we can observe an indication of the third satellites. Between two satellites, 8 maxima of the third type can be resolved. From the figure the importance of the many-beam approach is obvious. The two-beam approximation provides incorrect intensities of higher-order superlattice satellites.

5.5 Kinematical Scattering From Randomly Deformed Crystals

Most of the samples investigated by x-ray scattering methods are random; i.e., their actual microstructure is not known. The scattering operator $\hat{\mathbf{T}}(\boldsymbol{r})$ of a random sample is a random functional of the coordinates. The wave scattered from such a sample is given by an integral of the scattering operator over the sample volume. If the sample volume is large enough so that it contains all possible configurations of the sample microstructure, the mutual coherence function of the scattered radiation can be *averaged* over the statistical ensemble of all microstructure configurations. This averaged wave is measured by conventional x-ray methods; in recent years, *coherent* x-ray scattering has been developed, where the exact (non-averaged) scattered wave is measured [285]. These methods will not be discussed here.

In this section we present the formulas for the mutual coherence function, scattered intensity, and its reciprocal space distribution for a wave scattered from a random sample averaged over the statistical ensemble. In order to simplify the formulas, we restrict ourselves to the two-beam case only, and we will use the polarizability coefficient χ_h instead of an exact value of the structure factor $F_{\text{cell}}(\boldsymbol{Q})$ (see the discussion preceding Eqs. (5.20) and (5.23)). In a deformed lattice, the polarizability coefficients are functions of the coordinates and the polarizability can be expressed as a *distorted* Fourier series

$$\chi^{\text{def}}(\boldsymbol{r}) = \sum_{\boldsymbol{g}} [\chi_{\boldsymbol{g}} + \delta\chi_{\boldsymbol{g}}(\boldsymbol{r})] \mathrm{e}^{\mathrm{i}\boldsymbol{g}.(\boldsymbol{r}-\boldsymbol{u}(\boldsymbol{r}))} \equiv \sum_{\boldsymbol{g}} \chi_{\boldsymbol{g}}^{\text{def}}(\boldsymbol{r}) \mathrm{e}^{\mathrm{i}\boldsymbol{g}.(\boldsymbol{r}-\boldsymbol{u}(\boldsymbol{r}))}. \tag{5.41}$$

The factor $\delta\chi_{\boldsymbol{g}}(\boldsymbol{r})$ includes both the relative displacements of the atoms within an elementary unit cell, and a possible change in the chemical composition of the cell by removing or adding the atoms. In the Takagi approximation, $\delta\chi_{\boldsymbol{g}}(\boldsymbol{r}) = 0$ and the polarizability of a deformed crystal in point $\boldsymbol{r}$ equals the polarizability of a non-deformed crystal in point $\boldsymbol{r} - \boldsymbol{u}(\boldsymbol{r})$, we do not use this approximation in this section. We divide the random displacement into two parts $\boldsymbol{u}(\boldsymbol{r}) = \langle \boldsymbol{u}(\boldsymbol{r}) \rangle + \delta\boldsymbol{u}(\boldsymbol{r})$; the first part is the averaged displacement and the second is the random deviation.

In order to simplify the following formulas, we will consider only *statistically homogeneous* samples, where all the macroscopic quantities describing the random crystal deformation, such as defect densities, correlation lengths etc. are constant in the entire crystal volume. In a statistically homogeneous crystal, the averaged displacement is a linear function of position, i.e., the average strain in the crystal is constant, i.e., $\langle u_j(\boldsymbol{r}) \rangle = \langle \epsilon_{jk} \rangle x_k$.

Using Eq. (5.28) for the amplitude of the wave scattered from a deformed crystal and the two-beam approximation, we obtain the mutual coherence function of the scattered wave in the form

$$\Gamma(\boldsymbol{r}, \boldsymbol{r}') = I_i \frac{K^4}{64\pi^4} \int \frac{\mathrm{d}^2 \boldsymbol{K}_{\|}}{K_z} \int \frac{\mathrm{d}^2 \boldsymbol{K}'_{\|}}{K'_z} \mathrm{e}^{\mathrm{i}(\boldsymbol{K}.\boldsymbol{r} - \boldsymbol{K}'.\boldsymbol{r}')} \times$$
$$\times \int \mathrm{d}^3 \boldsymbol{r}'' \int \mathrm{d}^3 \boldsymbol{r}''' \mathrm{e}^{-\mathrm{i}(\boldsymbol{K} - \tilde{\boldsymbol{K}}_{\boldsymbol{h}}^{\text{def}}).\boldsymbol{r}''} \mathrm{e}^{\mathrm{i}(\boldsymbol{K}' - \tilde{\boldsymbol{K}}_{\boldsymbol{h}}^{\text{def}}).\boldsymbol{r}'''} C(\boldsymbol{r}'' - \boldsymbol{r}'''). \tag{5.42}$$

Here we have denoted

$$\tilde{\boldsymbol{K}}_{\boldsymbol{h}}^{\text{def}} = \boldsymbol{K}_i + \boldsymbol{h}^{\text{def}}, \ h_j^{\text{def}} = h_j - \langle \epsilon_{jk} \rangle h_k;$$

$\boldsymbol{h}^{\text{def}}$ is the vector of the reciprocal lattice constructed to the *averaged* crystal lattice. Due to the averaged deformation, the reciprocal lattice points are shifted by $-\langle \epsilon_{jk} \rangle h_k$. Then, we have defined the *correlation function of the crystal deformation*

$$C(\boldsymbol{r}'', \boldsymbol{r}''') \equiv C(\boldsymbol{r}'' - \boldsymbol{r}''') =$$
$$= \left\langle \chi_{\boldsymbol{h}}^{\text{def}}(\boldsymbol{r}'') (\chi_{\boldsymbol{h}}^{\text{def}}(\boldsymbol{r}'''))^* \mathrm{e}^{-\mathrm{i}\boldsymbol{h}.(\delta\boldsymbol{u}(\boldsymbol{r}'') - \delta\boldsymbol{u}(\boldsymbol{r}'''))} \right\rangle. \tag{5.43}$$

In a statistically homogeneous crystal, this correlation function depends only on the difference $\boldsymbol{r}'' - \boldsymbol{r}'''$.

From the practical point of view it is useful to divide the correlation function C into two parts, the coherent and incoherent, as

$$C(\boldsymbol{r}''-\boldsymbol{r}''')=\left\langle\chi_{\boldsymbol{h}}^{\mathrm{def}}(\boldsymbol{r}'')\mathrm{e}^{-\mathrm{i}\boldsymbol{h}.\delta\boldsymbol{u}(\boldsymbol{r}'')}\right\rangle\left\langle\chi_{\boldsymbol{h}}^{\mathrm{def}}(\boldsymbol{r}''')\mathrm{e}^{-\mathrm{i}\boldsymbol{h}.\delta\boldsymbol{u}(\boldsymbol{r}''')}\right\rangle^{*}+$$

$$+M(\boldsymbol{r}''-\boldsymbol{r}'''). \tag{5.44}$$

In a statistically homogeneous crystal, the coherent part is constant and the second part is a function of $\boldsymbol{r}''-\boldsymbol{r}'''$. The function M used in the incoherent part is defined as

$$M(\boldsymbol{r}''-\boldsymbol{r}''')=\mathrm{Cov}\left(\chi_{\boldsymbol{h}}^{\mathrm{def}}(\boldsymbol{r}'')\mathrm{e}^{-\mathrm{i}\boldsymbol{h}.\delta\boldsymbol{u}(\boldsymbol{r}'')},\chi_{\boldsymbol{h}}^{\mathrm{def}}(\boldsymbol{r}''')\mathrm{e}^{-\mathrm{i}\boldsymbol{h}.\delta\boldsymbol{u}(\boldsymbol{r}''')}\right), \tag{5.45}$$

where $\mathrm{Cov}(a,b)$ denotes the covariance of two random quantities $\mathrm{Cov}(a,b)=\langle ab^*\rangle-\langle a\rangle\langle b\rangle^*$.

It can be easily shown that the scattered wave is homogeneous; i.e., the mutual coherence function is a function of $\boldsymbol{r}-\boldsymbol{r}'$. As we have shown in the previous chapter, this field can be described by the reciprocal space distribution $J(\boldsymbol{Q})$ of the scattered wave and by the (constant) field intensity I. These quantities can be also divided into the coherent and incoherent parts:

$$J_{\mathrm{coh}}(\boldsymbol{Q})=I_i\frac{K^6}{4}\left|\left\langle\chi_{\boldsymbol{h}}^{\mathrm{def}}\mathrm{e}^{-\mathrm{i}\boldsymbol{h}.\delta\boldsymbol{u}}\right\rangle\right|^2\delta^{(2)}(\boldsymbol{Q}_{\parallel}-\boldsymbol{h}_{\parallel}^{\mathrm{def}})\times$$

$$\times\left|G_{\mathrm{cryst}}(Q_z-h_z^{\mathrm{def}})\right|^2, \tag{5.46}$$

$$I_{\mathrm{coh}}=I_i\frac{K^4}{4(K_{\boldsymbol{h}z}^{\mathrm{def}})^2}\left|\left\langle\chi_{\boldsymbol{h}}^{\mathrm{def}}\mathrm{e}^{-\mathrm{i}\boldsymbol{h}.\delta\boldsymbol{u}}\right\rangle\right|^2\left|G_{\mathrm{cryst}}(K_{\boldsymbol{h}z}^{\mathrm{def}}-\tilde{K}_{\boldsymbol{h}z}^{\mathrm{def}})\right|^2, \tag{5.47}$$

where $\boldsymbol{K}_{\boldsymbol{h}}^{\mathrm{def}}$ is defined in Eq. (5.31). The coherent part of the scattered radiation is described by the formulas analogous to those of a perfect sample (Eqs. (5.22) and (5.23)), where the diffraction vector $\boldsymbol{h}$ has been replaced by that of an averaged lattice $\boldsymbol{h}^{\mathrm{def}}$ and instead of the polarizability coefficient $\chi_{\boldsymbol{h}}$ we used the expression

$$\left\langle\chi_{\boldsymbol{h}}^{\mathrm{def}}\mathrm{e}^{-\mathrm{i}\boldsymbol{h}.\delta\boldsymbol{u}}\right\rangle. \tag{5.48}$$

Within the Takagi approximation, this expression is a product of $\chi_{\boldsymbol{h}}$ with the factor $\mathcal{D}=\langle\exp(-\mathrm{i}\boldsymbol{h}.\delta\boldsymbol{u})\rangle$. This factor, called *static Debye-Waller factor*, diminishes the diffracted intensity. Therefore, the coherent diffraction curve of a statistically homogeneous sample, where the Takagi approximation is valid, has the same shape as that of a perfect sample, it is only shifted by $\Delta Q_j=-\langle\epsilon_{jk}\rangle h_k$ and diminished by $\mathcal{D}^2$. Like a perfect crystal, the reciprocal space distribution of the coherent part of the radiation consists of a sequence of truncation rods; therefore, if the sample is irradiated by a plane wave, the coherent part of the scattered beam is plane, too.

The reciprocal space distribution and the intensity of the incoherent part of the scattered wavefield are

$$J_{\mathrm{incoh}}(\boldsymbol{Q})=I_i\frac{K^6}{16\pi^2}\int\mathrm{d}^2(\boldsymbol{r}_{\parallel}-\boldsymbol{r'}_{\parallel})\int_{-T}^{0}\mathrm{d}z\times$$

$$\times \int_{-T}^{0} \mathrm{d}z' \mathrm{e}^{-\mathrm{i}(\boldsymbol{Q}-\boldsymbol{h}^{\mathrm{def}}).(\boldsymbol{r}-\boldsymbol{r}')} M(\boldsymbol{r}-\boldsymbol{r}') \tag{5.49}$$

and

$$I_{\mathrm{incoh}} = I_i \frac{K^4}{16\pi^2} \int \frac{\mathrm{d}^2 \boldsymbol{K}_{\parallel}}{K_z^2} \int \mathrm{d}^2(\boldsymbol{r}_{\parallel} - \boldsymbol{r}'_{\parallel}) \int_{-T}^{0} \mathrm{d}z \times$$

$$\times \int_{-T}^{0} \mathrm{d}z' \mathrm{e}^{-\mathrm{i}(\boldsymbol{K}-\tilde{\boldsymbol{K}}_{\boldsymbol{h}}^{\mathrm{def}}).(\boldsymbol{r}-\boldsymbol{r}')} M(\boldsymbol{r}-\boldsymbol{r}'). \tag{5.50}$$

The reciprocal space distribution of the incoherent part is not concentrated to distinct narrow crystal truncation rods; therefore if a randomly deformed sample is irradiated by a plane wave, the incoherent part of the scattered wave is *not* plane, this wave is usually called *diffusely scattered wave.* This wave depends on the covariance M of the deformation, thus a sample would not produce any incoherent wave if its deformation in two points were not correlated.

Usually, the characteristic size of the defects; i.e., the size of the region, where the covariance M differs substantially from zero, is much smaller than the sample thickness T. In this case, the formulas (5.49) and (5.50) can be simplified:

$$J_{\mathrm{incoh}}(\boldsymbol{Q}) \approx I_i \frac{K^6 T}{16\pi^2} M^{\mathrm{FT}}(\boldsymbol{Q} - \boldsymbol{h}^{\mathrm{def}}) \tag{5.51}$$

and

$$I_{\mathrm{incoh}} \approx I_i \frac{K^4 T}{16\pi^2} \int \frac{\mathrm{d}^2 \boldsymbol{K}_{\parallel}}{K_z^2} M^{\mathrm{FT}}(\boldsymbol{K} - \tilde{\boldsymbol{K}}_{\boldsymbol{h}}^{\mathrm{def}}), \tag{5.52}$$

where ^{FT} denotes the Fourier transformation.

The physical meaning of these formulae can be explained by Fig. 5.9. The wave vector of the coherent part of the scattered wave, $\boldsymbol{K}_{\boldsymbol{h}}^{\mathrm{def}}$, is determined by the cross section point of the Ewald sphere (its position depends on the direction of the wave vector $\boldsymbol{K}_i$ of the incident beam) with the CTR crossing the reciprocal lattice point of the averaged lattice. The Fourier transformation M^{FT} of the covariance of the deformation field is non-zero in the neighborhood of this reciprocal lattice point. The ending points of possible wave vectors $\boldsymbol{K}$ of the incoherent component of the scattered wave are in the cross section of the Ewald sphere with the "cloud" of M^{FT}. The intensity of the incoherent part is a superposition of the intensities of all these plane components, this superposition is calculated by an integral over the Ewald sphere, this is the integral $\int \mathrm{d}^2\boldsymbol{K}_{\parallel}/K_z^2$ in Eq. (5.52). The reciprocal space distribution of the incoherent wave is simply proportional to $M^{\mathrm{FT}}(\boldsymbol{q})$, where $\boldsymbol{q} = \boldsymbol{Q} - \boldsymbol{h}^{\mathrm{def}}$ is the position vector in reciprocal space with respect to the reciprocal lattice point of the averaged lattice (reduced scattering vector). Therefore the inverse Fourier transformation of a measured reciprocal space map $J_{\mathrm{incoh}}(\boldsymbol{Q})$ yields directly the covariance of the deformation field

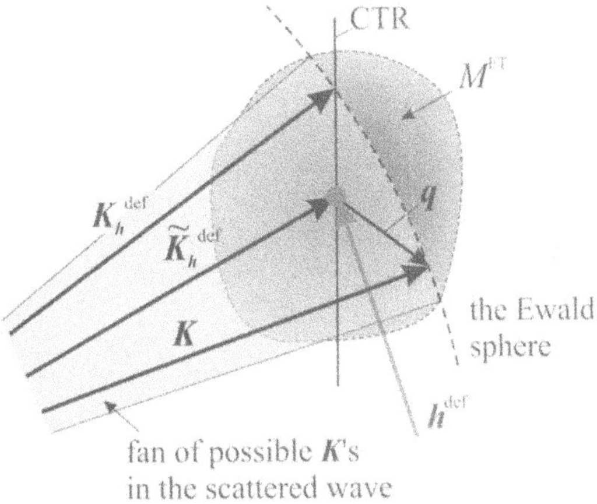

Fig. 5.9. The wave vectors of the coherent and incoherent parts of the scattered wave.

$M(\boldsymbol{r} - \boldsymbol{r}')$ without any a priori assumption. However, this analysis requires the experimental data to be measured in a rather large area in reciprocal space in order to achieve the necessary resolution in real space.

6 Dynamical Theory

So far, we have described the scattering of x-rays from thin layers using the kinematical approximation. We have neglected multiple scattering; i.e., we have neglected the influence of the scattering process on the transmitted beam.

The kinematical approximation is valid if the scattered wave represents a small disturbance of the transmitted wave. In Sect. 5.2 we found that the intensity of the wave diffracted kinematically from a perfect crystalline layer is in diffraction maximum proportional to the square of the layer thickness T. Therefore, we can expect that for thicker layers the diffracted intensity might be large and the kinematical approximation would not be valid.

The scattered amplitude also increases with decreasing incidence angle α_i; and therefore, for small α_i, where x-ray reflection from the interfaces is important, a dynamical description of the scattering process would be more appropriate. In this chapter, we formulate basic equations of x-ray scattering, taking the multiple scattering processes into account (*dynamical x-ray scattering*). In the *one-beam case* (x-ray reflection), we derive the dynamical reflectivity of a crystal with parallel surfaces and of a multilayer, and we compare them with the corresponding kinematical formulae. In the *two-beam case* (dynamical x-ray diffraction), we obtain expressions for the intensities diffracted dynamically from a perfect crystal. Further, we also deal with a many-beam case of diffraction from thin layers and we show its experimental relevance especially for x-ray diffraction from multilayers.

Dynamical diffraction theory is described thoroughly elsewhere in monographs [18, 20, 270]; here we present only the basic ideas relevant to the dynamical calculation of the scattering from crystals with parallel surfaces. We derive general expressions that can be used both in the one-beam and in many-beam cases.

6.1 The Wave Equation for a Periodic Medium

In our description of the dynamical theory we start from the wave equation used already in Section 4.2 in Eq. (4.11) containing the scattering potential for a classical scattering from electrons (4.12). For a non-magnetic material, the crystal polarizability $\chi(\boldsymbol{r}) = \varepsilon_{\rm rel}(\boldsymbol{r}) - 1$ is proportional to the electron density $\varrho(\boldsymbol{r})$ (see Eq. (5.1)).

In Section 4.2 we have rewritten the wave equation (4.11) into an integral form (4.13) and (4.16) that can be solved by successive iterations. The first iteration of its solution (the first Born approximation) describes the kinematical scattering, this was the subject of Chapter 5. The obvious improvement of the kinematical approach could be the second or higher iterations. However, it can be shown easily that the series of successive iterations converges very slowly, and moreover, the expressions for higher iterations of the wavefield are extremely complicated. Therefore, in order to include multiple scattering in our calculations, another method must be used.

For the solution of the wave equation we use the fact that, for a perfect crystal, the electron density and the crystal polarizability are periodic functions of the position. Strictly speaking, this is valid only for an infinite crystal; in any finite crystal, the polarizability is not periodic. We ignore this objection for the moment and solve the wave equation for an infinite ideal crystal. The finiteness of the crystal will be expressed by boundary conditions at the crystal surface that connect the wavefield outside the crystal with the wavefields in the *infinite* crystal. In the kinematical approximation, the finiteness of the crystal has been taken into account correctly. As we show later, the dynamical theory is valid for larger crystals, where we can assume that the equations for wavefield in the crystal are not affected by the crystal size. A detailed analysis of this problem can be found in the literature [371].

Since the polarizability of an infinite crystal and its scattering potential are periodic, the corresponding solution of the wave equation (4.11) has the form of the Bloch wave [195]

$$\boldsymbol{E}(\boldsymbol{r}) = \sum_{\boldsymbol{g}} \boldsymbol{E}_{\boldsymbol{g}} \mathrm{e}^{\mathrm{i}\boldsymbol{k}_{\boldsymbol{g}}.\boldsymbol{r}}, \tag{6.1}$$

where $\boldsymbol{k}_{\boldsymbol{g}} = \boldsymbol{k}_0 + \boldsymbol{g}$ and $\boldsymbol{k}_0$ is a wave vector from the first Brillouin zone that is unknown for the time. The wavefield in a crystal is also represented by a superposition of plane components with amplitudes $\boldsymbol{E}_{\boldsymbol{g}}$ and wave vectors $\boldsymbol{k}_{\boldsymbol{g}}$. Putting this expression into (4.11) and expressing the polarizability by the Fourier series (5.20), we obtain an infinite system of *algebraic* equations instead of one differential equation:

$$K^2 \boldsymbol{E}_{\boldsymbol{g}} - k_{\boldsymbol{g}} \times (\boldsymbol{E}_{\boldsymbol{g}} \times \boldsymbol{k}_{\boldsymbol{g}}) = -K^2 \sum_{\boldsymbol{p}} \chi_{\boldsymbol{p}} \boldsymbol{E}_{\boldsymbol{g}-\boldsymbol{p}}. \tag{6.2}$$

This system of equations can be solved, if we restrict the number of equations, i.e., the number of the reciprocal lattice vectors $\boldsymbol{g}$ in the series (6.1). If we assume that this series contains only one term, only one plane waves propagates in the crystal (*one-beam approximation*). This approximation is valid if the crystal does not diffract and this plane wave is the transmitted wave.

If the series (6.1) contains two non-zero terms, two plane waves propagate in the crystal (*two-beam approximation*); this approximation is valid close to the diffraction position of the crystal, these waves are the transmitted and the diffracted waves. In thin layers, a *many-beam* case can also be important. In this case, the wavefield in the crystal consists of one transmitted wave and two or more diffracted waves.

The amplitude equations (6.2) can be found in the literature also in a different form using the electric displacement vector $\boldsymbol{D}(\boldsymbol{r}) = \varepsilon_0(1+\chi(\boldsymbol{r}))\boldsymbol{E}(\boldsymbol{r})$ instead of the electric intensity $\boldsymbol{E}(\boldsymbol{r})$ [270] (ε_0 is the vacuum permittivity):

$$(K^2 - k_g^2)\boldsymbol{D}_g = \boldsymbol{k}_g \times (\boldsymbol{k}_g \times \sum_{\boldsymbol{p}} \chi_{\boldsymbol{p}} \boldsymbol{D}_{g-\boldsymbol{p}}). \tag{6.3}$$

The advantage of this approach is that the $\boldsymbol{D}$-waves in the crystal are always transversal, since div$\boldsymbol{D} = 0$ in any medium. However, deriving Eq. (6.3), we have used the approximation $\chi(\boldsymbol{r})/(1+\chi(\boldsymbol{r})) \approx \chi(\boldsymbol{r})$, which may not be correct for soft x-rays. In this book we use only the $\boldsymbol{E}$-description.

Equation (6.2) represents a system of linear homogeneous algebraic equations for the unknown amplitudes $\boldsymbol{E}_g$. A non-zero solution of the system exists, if the determinant of the system matrix is zero (*dispersion condition*). From this condition, possible vectors $\boldsymbol{k}_0$ follow; the ending points of these vectors can be expressed in reciprocal space by a *dispersion surface*. Generally, in an n-beam case, the dispersion surface is of order $2n$. The particular forms of the dispersion surface will be discussed later.

6.2 Boundary Conditions

In this section we formulate the boundary conditions on a plane interface lying in the xy-plane. The z-axis is parallel to the outward surface normal. Our considerations will include not only the free crystal surface but also an internal boundary in a layered sample. From classical electrodynamics it follows that the in-plane components $\boldsymbol{E}_\parallel$ and $\boldsymbol{H}_\parallel$ as well as the vertical components $\boldsymbol{D}_z$ and $\boldsymbol{B}_z$ are continuous; the fulfillment of the first two conditions implies the validity of the other two.

The wavefields in a laterally large perfect sample with parallel surfaces can be expressed as a superposition of plane waves. The amplitudes of the electric and magnetic components of those plane waves are connected by the Maxwell equation

$$\boldsymbol{H} = \frac{c\varepsilon_0}{K}\boldsymbol{k} \times \boldsymbol{E}; \tag{6.4}$$

and therefore, if the electric component of the wave is plane, the magnetic component is plane, too.

In Chapter 5 we introduced two polarization states of the scattered radiation. In the S-polarization, the polarization vectors $\boldsymbol{E}_i$ and $\boldsymbol{E}$ of the primary and scattered waves are perpendicular to the plane of the wave vectors $\boldsymbol{K}_i$ and $\boldsymbol{K}$ (the scattering plane). Since these waves are vacuum waves, they are purely transversal and in the S-polarization the vectors $\boldsymbol{E}_i$ and $\boldsymbol{E}$ are parallel (see Fig. 6.1). In the P-polarization, the vectors $\boldsymbol{E}_i$ and $\boldsymbol{E}$ lie in the scattering plane, and the angle between them is the scattering angle 2Θ. Within the dynamical description, the situation is more complicated, since the $\boldsymbol{E}$-vectors in the material are not transversal.

Let us deal in detail with two limiting geometries introduced already in Sect. 4.3, namely, the coplanar geometry, where the wave vectors of all waves lie in the same scattering plane perpendicular to the surface, and the grazing-incidence geometry, where the scattering plane is nearly parallel to the surface. We introduce the unit vectors $\boldsymbol{t}$, $\boldsymbol{n}$, and $\boldsymbol{s}$, as shown in Fig. 6.1. For the S-polarization in the coplanar case and the P-polarization in the

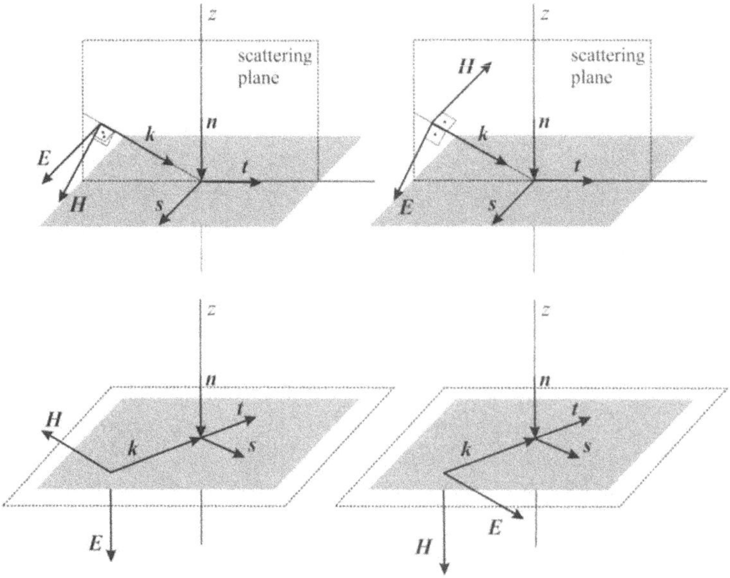

Fig. 6.1. Geometry of the wavefield irradiating a plane interface: S-polarization, coplanar case (*upper left*); P-polarization, coplanar case (*upper right*); S-polarization, grazing incidence case (*bottom left*), and P-polarization, grazing incidence case (*bottom right*).

grazing-incidence case, the in-plane components of the $\boldsymbol{E}$ and $\boldsymbol{H}$ waves are

$$\boldsymbol{E}_{\parallel} = E\boldsymbol{s}, \ \boldsymbol{H}_{\parallel} = \frac{k_z}{K}\varepsilon_0 cE\boldsymbol{t}. \tag{6.5}$$

In the opposite cases (P-polarization in the coplanar and S-polarization in the grazing-incidence cases), the in-plane components are

$$\boldsymbol{E}_{\|} \approx \frac{k_z}{K} E\boldsymbol{t}, \; \boldsymbol{H}_{\|} \approx -\varepsilon_0 c E \boldsymbol{s}. \tag{6.6}$$

The formulae (6.5) and (6.6) are exact only for the S-polarization in the coplanar case; in the other three cases, they are only approximate. The smaller the value of χ and the smaller the angle of the wave vector plane with the surface in the grazing-incidence case, the better the approximation is.

The boundary conditions at the interface S can be written in the following general form:

$$\begin{aligned} \sum_n E_n^{\mathrm{a}} \boldsymbol{s}_n \mathrm{e}^{\mathrm{i}\boldsymbol{k}_n^{\mathrm{a}} \cdot \boldsymbol{r}}\Big|_{\boldsymbol{r}\in S} &= \sum_n E_n^{\mathrm{b}} \boldsymbol{s}_n \mathrm{e}^{\mathrm{i}\boldsymbol{k}_n^{\mathrm{b}} \cdot \boldsymbol{r}}\Big|_{\boldsymbol{r}\in S}, \\ \sum_n E_n^{\mathrm{a}} k_{zn}^{\mathrm{a}} \boldsymbol{t}_n \mathrm{e}^{\mathrm{i}\boldsymbol{k}_n^{\mathrm{a}} \cdot \boldsymbol{r}}\Big|_{\boldsymbol{r}\in S} &= \sum_n E_n^{\mathrm{b}} k_{zn}^{\mathrm{b}} \boldsymbol{t}_n \mathrm{e}^{\mathrm{i}\boldsymbol{k}_n^{\mathrm{b}} \cdot \boldsymbol{r}}\Big|_{\boldsymbol{r}\in S}, \end{aligned} \tag{6.7}$$

where the sums are calculated over all plane components irradiating the interface, and $\boldsymbol{E}_n^{\mathrm{a,b}}$ and $\boldsymbol{k}_n^{\mathrm{a,b}}$ are the $\boldsymbol{E}$ vectors and the wave vectors, respectively, of the nth wave above (a) and below (b) the interface.

The boundary conditions (6.7) have the same form for both of the polarization states and in both the coplanar and the grazing-incidence geometries. Close to the diffraction maximum and for a small asymmetry angle ϕ (so-called *conventional diffraction*), the z components k_{zn} are nearly the same and the vectors $\boldsymbol{t}_n$ coincide. Thus, the second equation in (6.7) follows from the first one and the boundary conditions have the simplified approximate form

$$\sum_n E_n^{\mathrm{a}} \mathrm{e}^{\mathrm{i}\boldsymbol{k}_n^{\mathrm{a}} \cdot \boldsymbol{r}}\Bigg|_{\boldsymbol{r}\in S} = \sum_n E_n^{\mathrm{b}} \mathrm{e}^{\mathrm{i}\boldsymbol{k}_n^{\mathrm{b}} \cdot \boldsymbol{r}}\Bigg|_{\boldsymbol{r}\in S}. \tag{6.8}$$

Thus, within this approximation, both the in-plane and the normal components of $\boldsymbol{E}$ are continuous across the interface.

The boundary conditions (6.7) or (6.8) must be valid at every point $\boldsymbol{r} = (x, y, z = \text{const})$ in the interface. This can only be fulfilled if the in-plane components $\boldsymbol{k}_{n\|}^{\mathrm{a,b}}$ of the wave vectors on both sides of the interface are the same. The same condition must be fulfilled at the free sample surface $z = 0$; the wave vectors of the internal wavefields must have the same in-plane components as the primary and scattered wavefields. This condition and the requirement that the end-point of the wave vector must lie on the dispersion surface determine the wave vectors of the waves in the crystal.

If we put the start points of all the wave vectors in the crystal at the same point, their end-points must lie at the intersections of the dispersion surface with the common interface normal (the tie-points). This construction is equivalent to the Ewald construction introduced in Sect. 4.3; in the kinematical theory the actual dispersion surface has been replaced by the Ewald spherical surface.

6.3 X-Ray Reflection

As the simplest case, we assume that the wavefield in the crystal is a single plane wave. Then, the sum in (6.1) contains only one term,

$$\boldsymbol{E}_0 \mathrm{e}^{\mathrm{i}\boldsymbol{k}_0.\boldsymbol{r}},$$

and the system (6.2) of the amplitude equations has only one equation,

$$(k^2 - k_0^2)\boldsymbol{E}_0 = 0. \tag{6.9}$$

The condition for the existence of a non-trivial solution is the dispersion equation

$$k_0 = k \equiv nK = K\sqrt{1+\chi_0} \approx K(1+\chi_0/2). \tag{6.10}$$

Therefore, the end-points of all admissible wave vectors in the crystal lie on a spherical dispersion surface (the *Ewald sphere*) of radius nK, where $n = 1 + \chi_0/2$ is the refractive index. For *any* material, $\mathrm{Re}(\chi_0) < 0$ holds in the x-ray region, and $n < 1$. Therefore, a critical angle of incidence α_c exists, so that for angles $\alpha_i < \alpha_c$ total *external* reflection takes place. Since $|\chi_0| \ll 1$, the critical angle can be expressed as

$$\alpha_c \approx \sqrt{-\mathrm{Re}(\chi_0)} = \lambda\sqrt{\frac{r_{el}}{\pi}\langle\varrho(\boldsymbol{r})\rangle}; \tag{6.11}$$

thus the critical angle depends on the *mean* electron density, and it does not depend on the arrangement of the electrons in the unit cell and on the crystal structure. For usual materials and the wavelengths of few Å, the critical angle is smaller than one degree. The values of refraction indexes and critical angles of various materials can be found in many databases; see [254], for instance.

Since the dispersion surface is spherical, there are two tie-points in general of the dispersion surface with the interface normal. The dispersion surface and the Ewald construction for the wave vectors are shown in Fig. 6.2. In the sample, two plane waves (transmitted and reflected) propagate. The amplitudes and the wave vectors of these waves are $\boldsymbol{E}_R$, $\boldsymbol{k}_R$ (reflected) and $\boldsymbol{E}_T$, $\boldsymbol{k}_T$ (transmitted).

Figure 6.2 shows two equivalent forms of the Ewald construction. The left panel shows the form usual in the kinematical theory: the wave vectors of the primary and transmitted beams are drawn with the same end-points, and the starting points for the beams outside and inside the crystal coincide. In dynamical scattering theory, another, equivalent, way of drawing the Ewald construction is adopted (Fig. 6.2, the right drawing). In this form of the construction, we draw the wave vectors of the primary, reflected and transmitted beams with the same end-points. We will show later that in the two-beam case, if the diffracted and reflected-diffracted beams in both materials are plotted with the same end-points, these end-points coincide with the reciprocal-lattice point H.

The boundary conditions at the interface $z = 0$ follow from (6.7):

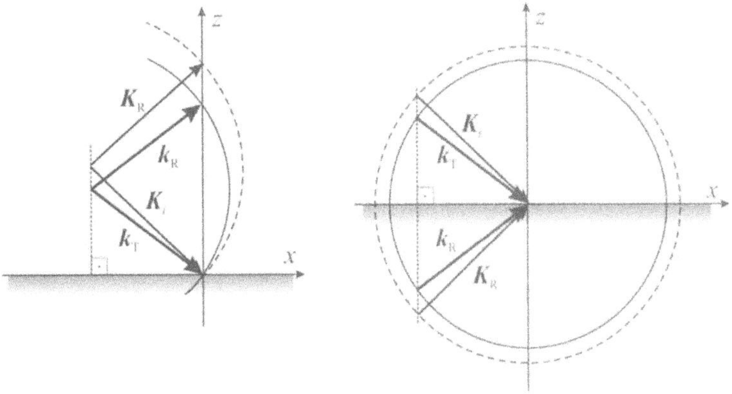

Fig. 6.2. The Ewald construction for the refracted wave, one-beam case. In the left panel, the Ewald construction has been drawn in the same way as in the kinematical theory; the equivalent form of the Ewald construction used in the dynamical theory is shown in the right. Full and dashed arcs denote the Ewald spheres in the material (with radius $k = nK$) and in vacuum (radius K), respectively.

$$E_{\mathrm{T}}^{\mathrm{a}} + E_{\mathrm{R}}^{\mathrm{a}} = E_{\mathrm{T}}^{\mathrm{b}} + E_{\mathrm{R}}^{\mathrm{b}},\; k_{\mathrm{T}z}^{\mathrm{a}}(E_{\mathrm{T}}^{\mathrm{a}} - E_{\mathrm{R}}^{\mathrm{a}}) = k_{\mathrm{T}z}^{\mathrm{b}}(E_{\mathrm{T}}^{\mathrm{b}} - E_{\mathrm{R}}^{\mathrm{b}}), \tag{6.12}$$

where the superscripts "a" and "b" denote the quantities above and below the interface, respectively. These conditions, as well as the dispersion equation (6.9), do not depend on the polarization, and therefore both the amplitudes and the wave vectors of the waves propagating in the crystal are independent of the polarization.

If the interface is the free surface of a semi-infinite sample irradiated by a plane primary wave $\boldsymbol{E}_i \exp(\mathrm{i}\boldsymbol{K}_i.\boldsymbol{r})$, the boundary conditions are

$$E_i + E_{\mathrm{R}} = E_{\mathrm{T}},\; K_{iz}(E_i - E_{\mathrm{R}}) = k_{\mathrm{T}z} E_{\mathrm{T}}, \tag{6.13}$$

since there is no interface lying below the free surface that could emit a reflected wave into the sample. From (6.13), the well-known Fresnel reflectivity and transmittivity coefficients of a crystal surface can easily be deduced (see e.g., [60, 259])

$$r = \frac{E_{\mathrm{R}}}{E_i} = \frac{K_{iz} - k_{0z}}{K_{iz} + k_{0z}},\; t = \frac{E_{\mathrm{T}}}{E_i} = \frac{2K_{iz}}{K_{iz} + k_{0z}}. \tag{6.14}$$

Since the boundary conditions derived above (6.7) are exact only for the S-polarization, these Fresnel coefficients for the P-polarization are slightly different. However, for the angles α_i of few degrees the difference is negligibly small. For the x-ray wavelengths of few Å's, the reflectivity of any sample for larger angles of incidence is very small and the Fresnel coefficients for the S-polarization can be used also for P-polarized primary beam. However, x-ray reflection for longer wavelengths is measurable also for larger α_i's. Then, the exact Fresnel coefficients of both polarizations must be used.

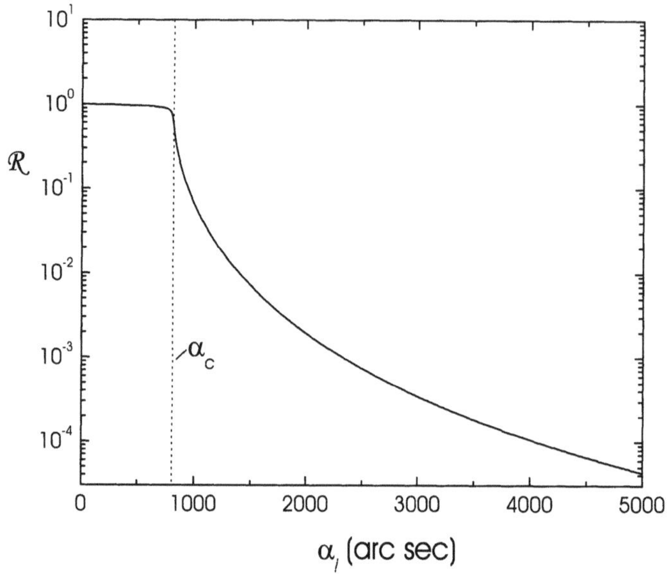

Fig. 6.3. Reflectivity curve of a semiinfinite Si sample with flat surface.

In Fig. 6.3 we show the reflectivity $\mathcal{R}_R = |r|^2$ of a Si surface calculated for the wavelength $\lambda = 1.54$ Å. Below the critical angle α_c, the reflectivity exhibits a plateau corresponding to the total external reflection. Due to the absorption (i.e., due to the imaginary part of χ_0), the reflectivity in this region is smaller than unity. Above the critical angle, the reflectivity decreases as Q_z^{-4}, i.e., as α_i^{-4}. X-ray reflection is often used for the determination of the thickness and density of thin layers; this is described in detail in Chapter 8.

6.4 Two-Beam Diffraction

In this case we take two terms in the series in (6.1) into account:

$$\boldsymbol{E}(\boldsymbol{r}) = \boldsymbol{E}_0 \mathrm{e}^{\mathrm{i}\boldsymbol{k}_0 . \boldsymbol{r}} + \boldsymbol{E}_{\boldsymbol{h}} \mathrm{e}^{\mathrm{i}\boldsymbol{k}_{\boldsymbol{h}} . \boldsymbol{r}}, \ \boldsymbol{k}_{\boldsymbol{h}} = \boldsymbol{k}_0 + \boldsymbol{h},$$

where $\boldsymbol{h}$ is the diffraction vector. For the S-polarization, from the amplitude equations (6.2), we derive the amplitude equations

$$\begin{aligned} \left(k_0^2 - k^2\right) E_0 &= K^2 \chi_{-\boldsymbol{h}} E_{\boldsymbol{h}} \\ \left(k_{\boldsymbol{h}}^2 - k^2\right) E_{\boldsymbol{h}} &= K^2 \chi_0 E_{\boldsymbol{h}}, \end{aligned} \tag{6.15}$$

where $k = nK$ is the radius of the one-beam dispersion surface.

In the P-polarization, the formulas are complicated by the fact that the $\boldsymbol{E}$-waves in the material are not transversal. If we define the coordinate system

$x_\alpha x_\beta$ in the scattering plane according to Fig. 6.4, the amplitude equations are

$$\begin{aligned}
(k_0^2 - k^2)E_{0\alpha} - k_{0\alpha}(k_{0\alpha}E_{0\alpha} + k_{0\beta}E_{0\beta}) &= K^2\chi_{-\boldsymbol{h}}E_{\boldsymbol{h}\alpha} \\
(k_0^2 - k^2)E_{0\beta} - k_{0\beta}(k_{0\alpha}E_{0\alpha} + k_{0\beta}E_{0\beta}) &= K^2\chi_{-\boldsymbol{h}}E_{\boldsymbol{h}\beta} \\
(k_{\boldsymbol{h}}^2 - k^2)E_{\boldsymbol{h}\alpha} - k_{\boldsymbol{h}\alpha}(k_{\boldsymbol{h}\alpha}E_{\boldsymbol{h}\alpha} + k_{\boldsymbol{h}\beta}E_{\boldsymbol{h}\beta}) &= K^2\chi_{\boldsymbol{h}}E_{0\alpha} \\
(k_{\boldsymbol{h}}^2 - k^2)E_{\boldsymbol{h}\beta} - k_{\boldsymbol{h}\beta}(k_{\boldsymbol{h}\alpha}E_{\boldsymbol{h}\alpha} + k_{\boldsymbol{h}\beta}E_{\boldsymbol{h}\beta}) &= K^2\chi_{\boldsymbol{h}}E_{0\beta}
\end{aligned} \tag{6.16}$$

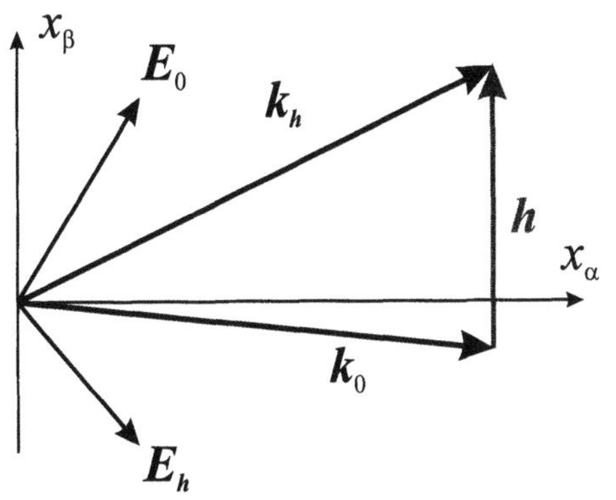

Fig. 6.4. The coordinate system used in the P-polarization, the two-beam case.

The condition for the existence of a non-zero solution yields the dispersion equation

$$\left(k_0^2 - k^2\right)\left(k_{\boldsymbol{h}}^2 - k^2\right) = K^4\chi_{\boldsymbol{h}}\chi_{-\boldsymbol{h}}\left(1 - P\frac{k_{0\alpha}^2 h^2}{k^4 - K^4\chi_{\boldsymbol{h}}\chi_{-\boldsymbol{h}}}\right), \tag{6.17}$$

where $P = 0$ for the S-polarization and $P = 1$ for the P-polarization. The dispersion equation determines the shape of the dispersion surface. In Fig. 6.5 we have plotted the intersection of the dispersion surface with the scattering plane for the S-polarization in the coplanar two-beam case. Far from the Laue points La, the dispersion surface can be approximated by spheres with radius $k = nK$ centered at O and H. Therefore, far from La, the two-beam case turns into the one-beam case, where only one strong wave $\boldsymbol{E}_0$ or $\boldsymbol{E}_{\boldsymbol{h}}$ can propagate only.

In the vicinity of La, both $\boldsymbol{E}_0$ and $\boldsymbol{E}_{\boldsymbol{h}}$ are strong. In Fig. 6.5 the full lines denote the dispersion surface, whereas the dashed lines represent the intersection of the *vacuum* dispersion surfaces with the scattering plane. These surfaces are spheres with radius K and centers O and H again.

The dispersion equation (6.17) contains the polarizability coefficients $\chi_{\boldsymbol{h}}$, $\chi_{-\boldsymbol{h}}$, corresponding to both $\boldsymbol{h}$ and $-\boldsymbol{h}$. Thus, the dynamical description accounts for the diffraction processes with diffraction vectors $\boldsymbol{h}$ and $-\boldsymbol{h}$. The

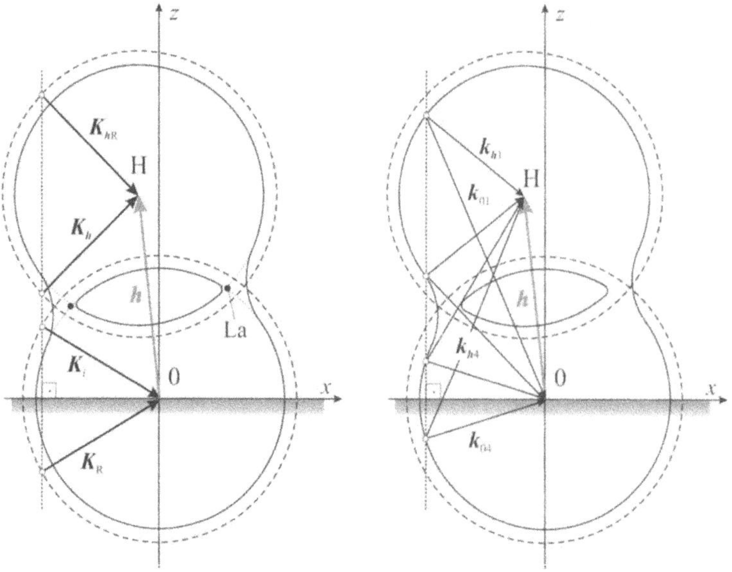

Fig. 6.5. Dispersion surface of the coplanar two-beam case: *left* the wave vectors of the vacuum waves, *right* the wave vectors of the waves in the crystal. In the right panel, the tie-points are denoted by small empty circles. La denotes the Laue point.

dispersion surface is a surface of the fourth order, and therefore four tie-points of the dispersion surface exist with the interface normal. The shape of the dispersion surface depends on the polarization; for a given polarization, eight plane waves (four waves $\boldsymbol{E}_0$ and four waves $\boldsymbol{E}_{\boldsymbol{h}}$) propagate through the crystal.

The shape of the dispersion surface for the P-polarization differs significantly from that for S-polarization only in the close vicinity of the Laue points. In this region, the dispersion condition (6.17) can be simplified by writing

$$1 - p\frac{k_{0\alpha}^2 h^2}{k^4 - K^4 \chi_{\boldsymbol{h}} \chi_{-\boldsymbol{h}}} \approx C^2, \tag{6.18}$$

where C is the linear polarization factor defined in Sect. 5.1.

Knowing the position of the tie-points on the dispersion surface, we can calculate the wave vectors $\boldsymbol{k}_{0\boldsymbol{h}n}$, $n = 1, \ldots, 4$, of the plane waves. These wave vectors are depicted in the right-hand part of Fig. 6.5. Putting the lengths of the vectors $\boldsymbol{k}_{0\boldsymbol{h}n}$, $n = 1, \ldots, 4$, into (6.15) or (6.16), we can calculate the amplitudes of these waves. However, because of the condition (6.17), the amplitude equations are linearly dependent, and in the S-polarization we can only obtain the ratios

$$c_n = \frac{E_{\boldsymbol{h}n}}{E_{0n}}. \tag{6.19}$$

In the P-polarization, from the system of equations (6.16) we can calculate the ratios

$$p_n = \frac{E_{0\alpha n}}{E_{0\beta n}}, \; c_n = \frac{E_{\boldsymbol{h}\alpha n}}{E_{0\beta n}}, \; d_n = \frac{E_{\boldsymbol{h}\beta n}}{E_{0\beta n}}. \tag{6.20}$$

In the two-beam case, the application of the continuity condition of the in-plane components of the wave vectors is a little more complicated than in the one-beam case. Since $\boldsymbol{k_h} = \boldsymbol{k}_0 + \boldsymbol{h}$ and, in general, $\boldsymbol{h}_{\|} \neq 0$, the in-plane components of $\boldsymbol{k}_0$ and $\boldsymbol{k_h}$ *cannot* be the same. Therefore, the boundary conditions (6.7) can be fulfilled only if the amplitudes of the $\boldsymbol{E}_0$ and $\boldsymbol{E_h}$ waves below and above the interface equal separately. Therefore, the boundary conditions in the two-beam case are (S-polarization)

$$\begin{aligned} &\textstyle\sum_{n=1,4} E_{0n} = \text{const}, \qquad \textstyle\sum_{n=1,4} c_n E_{0n} = \text{const}, \\ &\textstyle\sum_{n=1,4} k_{0nz} E_{0n} = \text{const}, \; \textstyle\sum_{n=1,4} c_n k_{\boldsymbol{h}zn} E_{0n} = \text{const}. \end{aligned} \tag{6.21}$$

and

$$\begin{aligned} &\textstyle\sum_{n=1,4} \frac{1}{K}(k_{0\alpha n} - k_{0\beta n} p_n) E_{0\beta n} = \text{const}, \\ &\textstyle\sum_{n=1,4} \frac{1}{K}(k_{0\alpha n} d_n - (k_{0\beta n} + h) c_n) E_{0\beta n} = \text{const}, \\ &\textstyle\sum_{n=1,4} \frac{K}{h}(-h_z p_n - h_x) E_{0\beta n} = \text{const}, \\ &\textstyle\sum_{n=1,4} \frac{K}{h}(-h_z c_n - h_x d_n) E_{0\beta n} = \text{const} \end{aligned} \tag{6.22}$$

for the P polarization. These forms of the boundary conditions are valid both in the coplanar and in the grazing incidence geometry. The theory for the coplanar arrangement has been published in [4, 44, 69, 146, 147, 148, 194], the explicit formulas for the amplitudes and the wave vectors in the GID geometry can be found in [3, 5, 6].

The wave vectors of the vacuum wave can also be constructed using the Ewald construction, as shown in Fig. 6.5 (the left panel). Let us construct these vectors so that the end-points of the wave vectors of the primary and reflected beams coincide with origin O and those of the diffracted and diffracted-reflected beams with the reciprocal-lattice point H. Then, the starting points of these vectors lie on the same surface normal. The common surface normal intersects the pair of spherical vacuum dispersion surfaces in four tie-points, corresponding to the primary beam $\boldsymbol{K}_i$, the reflected beam $\boldsymbol{K}_{\mathrm{R}}$, the diffracted beam $\boldsymbol{K_h}$ and the diffracted-reflected beam $\boldsymbol{K}_{\boldsymbol{h}\mathrm{R}}$.

Numerical calculation shows that the intensity of the reflected and/or diffracted-reflected beam is not negligible only if the incidence angle α_i and/or the exit angle $\alpha_{\boldsymbol{h}}$ is very small. In the coplanar case, this can be realized using strongly asymmetrical diffraction; then, usually, only one of these waves is strong. In the grazing-incidence geometry, both waves can be strong, since the input and exit angles are small simultaneously.

From the Ewald construction it follows that

$$\boldsymbol{K}_{\mathrm{R}\|} = \boldsymbol{K}_{0\|}, \ \boldsymbol{K}_{\boldsymbol{h}\|} = \boldsymbol{K}_{\boldsymbol{h}\mathrm{R}\|} = \boldsymbol{K}_{0\|} + \boldsymbol{h}_{\|}. \tag{6.23}$$

Thus, as in the kinematical case, the reflected intensity is concentrated along the crystal truncation rod through the origin O and the diffracted intensity in a crystal truncation rod through H. The lateral diffraction condition (6.23) is equivalent to (4.27), derived in Sect. 4.3 using only basic symmetry considerations.

If the sample is semi-infinite, only the Bragg reflection geometry can be realized. Then, not all of the four tie-points are physically relevant. Namely, two tie-points represent forbidden waves with $\mathrm{Im}(k_z) > 0$, i.e., with negative absorption (for $z \to -\infty$). Therefore, for a given polarization, only two $\boldsymbol{E}_0$ waves and two $\boldsymbol{E}_{\boldsymbol{h}}$ waves propagate in a semi-infinite sample.

Let us assume that the permitted waves are labelled by $n = 1, 2$. The boundary conditions at the free surface are (S-polarization)

$$\begin{aligned} E_i + E_{\mathrm{R}} &= E_{01} + E_{02}, \qquad K_{0z}(E_i - E_{\mathrm{R}}) = k_{0z1}E_{01} + k_{0z2}E_{02} \\ E_{\boldsymbol{h}} &= c_1 E_{01} + c_2 E_{02}, \qquad K_{\boldsymbol{h}z}E_{\boldsymbol{h}} = k_{\boldsymbol{h}z1}c_1E_{01} + k_{\boldsymbol{h}z2}c_2E_{02}, \end{aligned} \tag{6.24}$$

and

$$\begin{aligned} E_i + E_{\mathrm{R}} &= \textstyle\sum_{n=1,2}(k_{0\alpha n} - k_{0\beta n}p_n)E_{0\beta n}, \\ K_{0z}(E_i - E_{\mathrm{R}}) &= \textstyle\sum_{n=1,2} \frac{K}{h}(-h_z p_n - h_x)E_{0\beta n}, \\ E_{\boldsymbol{h}} &= \textstyle\sum_{n=1,2} \frac{1}{K}(k_{0\alpha n}d_n - (k_{0\beta n} + h)c_n)E_{0\beta n}, \\ K_{\boldsymbol{h}z}E_{\boldsymbol{h}} &= \textstyle\sum_{n=1,2} \frac{K}{h}(-h_z c_n - h_x d_n)E_{0\beta n} \end{aligned} \tag{6.25}$$

for P polarization. A sample irradiated by a plane wave E_i emits the diffracted wave $E_{\boldsymbol{h}}$ and the reflected wave E_{R}; the diffracted-reflected wave cannot be emitted into the upper half-space in the Bragg geometry.

The amplitude equations and the boundary conditions can be substantially simplified in the *conventional diffraction* case, where the tie-points lie close to the Laue point La [270]. In this case, the asymptotic Ewald spheres with radius nK and centers at the reciprocal-lattice points O and H can be replaced by tangential planes (Fig. 6.6). Then, $k_0 \approx k_{\boldsymbol{h}} \approx k$; and the dispersion equation (6.17) can be simplified to

$$(k_0 - k)(k_{\boldsymbol{h}} - k) = \frac{1}{4}K^2C^2\chi_{\boldsymbol{h}}\chi_{-\boldsymbol{h}}. \tag{6.26}$$

This equation is of order 2, and therefore two tie-points and, consequently, two waves $\boldsymbol{E}_0$ and two waves $\boldsymbol{E}_{\boldsymbol{h}}$ exist in the diffracting crystal for each polarization state. The corresponding amplitude equations are

$$\begin{aligned} (k_0 - k)E_0 &= \tfrac{1}{2}KC\chi_{-\boldsymbol{h}}E_{\boldsymbol{h}}, \\ (k_{\boldsymbol{h}} - k)E_{\boldsymbol{h}} &= \tfrac{1}{2}KC\chi_{\boldsymbol{h}}E_0. \end{aligned} \tag{6.27}$$

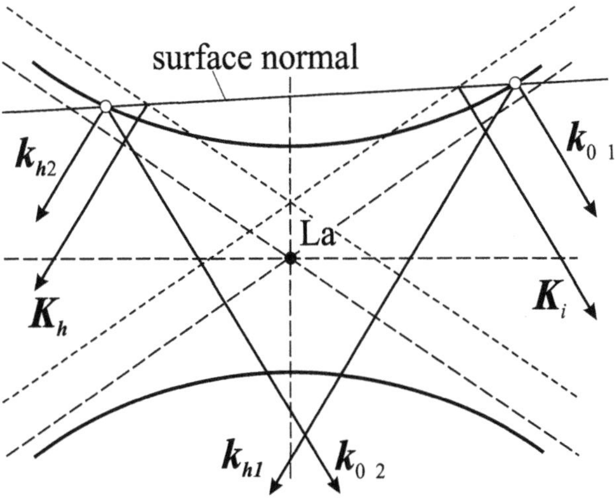

Fig. 6.6. Two-beam dispersion surface close to the Laue point; only two tie-points and two pairs of the internal waves $\boldsymbol{k}_{01,2}$ and $\boldsymbol{k}_{\boldsymbol{h}1,2}$ are considered.

These equations are valid both for S- and P-polarizations; the formulae for c_n and $\boldsymbol{k}_{0,\boldsymbol{h}n}$ for the conventional diffraction are much simpler than in the general case. Analogously, in the conventional diffraction case, the common normal intersects the vacuum Ewald sphere in only two tie-points (see Fig. 6.6), which correspond to the primary wave $\boldsymbol{K}_i$ and the diffracted wave $\boldsymbol{K}_{\boldsymbol{h}}$. In the Laue transmission case for a crystal with parallel surfaces, the wave vector of the vacuum transmitted wave is identical to $\boldsymbol{K}_i$. No reflected or diffracted-reflected waves are present in conventional diffraction. The conventional diffraction formalism is valid only for small angular deviations from the diffraction maximum, where the curvature of the asymptotic Ewald spheres can be neglected, and for small asymmetries, where the angles of incidence and exit $\alpha_{i,f}$ are sufficiently large.

The boundary conditions are very simple in the case of conventional diffraction by a semi-infinite sample. In this case, only one of the two tie-points is physically relevant (say, the first one) and we obtain

$$E_i = E_{01}, E_{\boldsymbol{h}} = c_1 E_{01}. \tag{6.28}$$

The intensity of the diffracted wave is $I_{\boldsymbol{h}} = I_i |c_1|^2$. A detector in a usual experimental arrangement does not measure the intensity of the diffracted wave, but its flux $\Psi_{\boldsymbol{h}} = I_{\boldsymbol{h}} A_{\boldsymbol{h}}$, where $A_{\boldsymbol{h}}$ is the area of the cross section of the diffracted beam. Therefore, we define the quantity *diffractivity* as a ratio of the fluxes of the diffracted and primary beams

$$\mathcal{R}_{\boldsymbol{h}} = \frac{\Psi_{\boldsymbol{h}}}{\Psi_i} = \frac{|\gamma_{\boldsymbol{h}}|}{\gamma_0} |c_1|^2, \tag{6.29}$$

where $\gamma_{0,h}$ are the direction cosines of the primary and diffracted waves, defined in Sect. 4.3. Within the conventional approach, the diffracting crystal emits only one wave (the diffracted wave); the reflected wave is completely neglected.

Several attempts have been published in the literature introducing corrections to the conventional dispersion formula (6.26) in order to extend its validity range also for larger deviations from the diffraction maximum [121, 398]. These corrections are always valid only in a certain angular range, but the corrected dispersion equation has order 2, so that its numerical solution is very easy.

As a simple example, we show the diffraction curves calculated for a semi-infinite Si crystal, diffraction 004, the S-polarization, in the Bragg reflection case with various asymmetry angles ϕ, and for a crystal with finite thickness (Fig. 6.7). Since the asymmetry angles used here are rather small, these curves have been calculated using the conventional diffraction formulas (6.27) and (6.28). The diffraction curve exhibits a distinct plateau; its asymmetry is due

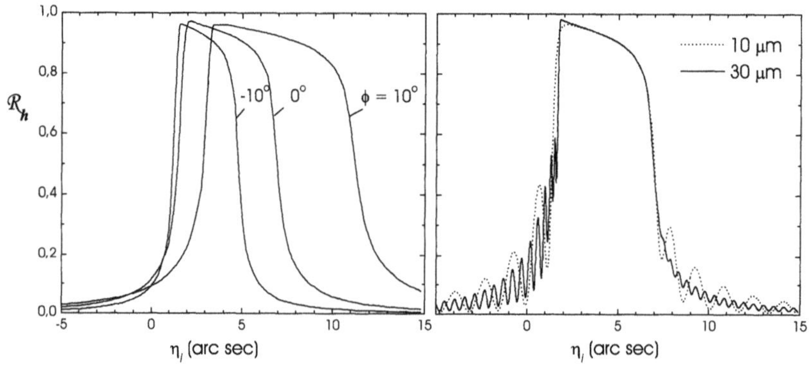

Fig. 6.7. Diffraction curves of a semi-infinite Si crystal, diffraction 004, $\lambda = 1.54$ Å, calculated for various asymmetry angles ϕ (left) and symmetrical diffraction curves of a Si crystal with two different thicknesses.

to the imaginary part of χ_h. The tails of the diffraction curve depend on $Q_z - h_z$ as $(Q_z - h_z)^{-2}$. This plateau is analogous to the total reflection region in the reflection curve shown in Fig. 6.3. The diffraction curve of a crystal with finite thickness exhibits oscillations; for very thin crystals, their period equals $\delta Q_z = 2\pi/T$ as in the kinematical case. For very thin crystals, the dynamical diffraction curve is identical to the kinematical one. For thick crystals, the intensity oscillations are too rapid and a smooth curve can be measured, corresponding to a semi-infinite sample.

In a coplanar geometry with strong asymmetry ($|\phi| \to \Theta_B$) the general formulae (6.15–6.17) and (6.21) must be used (*general dynamical diffraction*). As an example, in Fig. 6.8 the reflection and diffraction curves in strongly asymmetric diffraction geometry are depicted for both small incidence angles (left) and small exit angles (right).

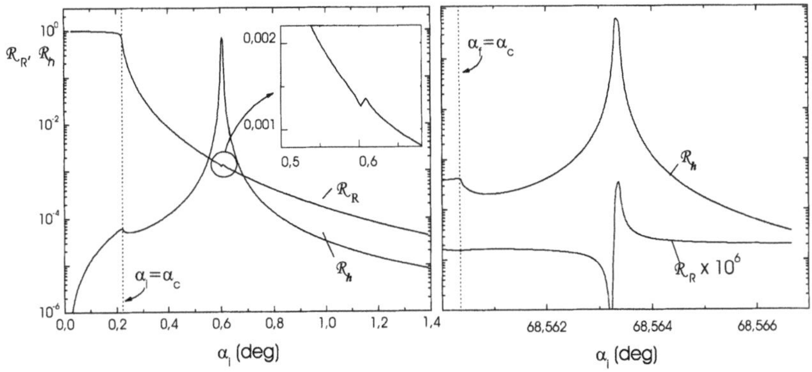

Fig. 6.8. Strongly asymmetric diffraction from a semi-infinite Si crystal, CuKα_1 radiation, S-polarization, for small (*left*) and large (*right*) angles of incidence. The asymmetry angle was $\phi = -34$ deg and $\phi = +34$ deg in the left and right panels, respectively. Both the reflection and diffraction curves are plotted as functions of the incidence angle.

In the case of very small incidence angles, a tiny S-shaped feature on the reflection curve can be seen – this effect is caused by the interaction of the reflected and the diffracted beams. Similarly, a subsidiary maximum occurs in the diffraction curve at the position of the critical angle α_c (the *Yoneda wing*). If the exit angle is very small (the right panel in Fig. 6.8), the reflected intensity is very weak. On the diffraction curve, a subsidiary maximum occurs if the exit angle α_h equals the critical angle α_c.

Another example shows the diffractivity in the grazing-incidence geometry of a semi-infinite Si substrate (Fig. 6.9) as a function of the take-off angle α_h for different incidence angles α_i, assuming a large azimuthal divergence of the primary beam (see Sect. 4.3). The diffraction maximum is close to the critical angle α_c. The shape of the maximum depends sensitively on α_i. If $\alpha_i \approx \alpha_c$ the maximum is broad and rounded; for other values of α_i, the diffraction maximum is sharp.

The calculation of the diffracted intensity was based on the dispersion surface (6.17) and the boundary conditions (6.24). We assumed a sufficiently large azimuthal divergence of the primary beam that the diffraction condition was fulfilled for every value of α_h (see Eq. 4.34).

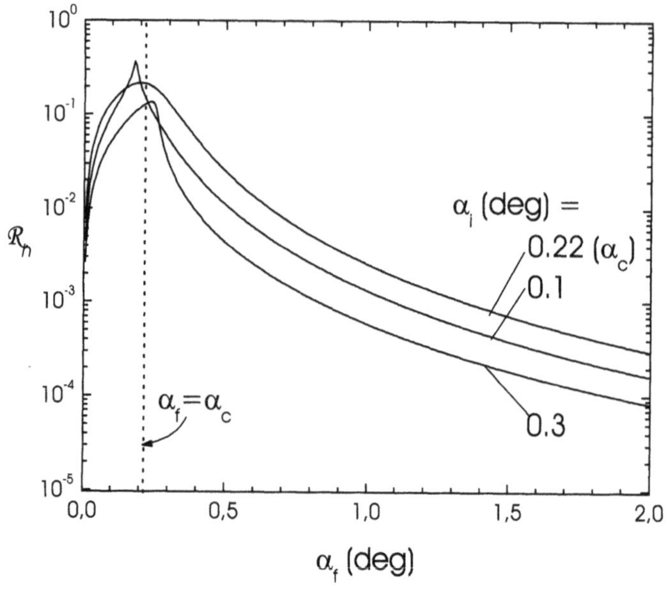

Fig. 6.9. Grazing-incidence diffraction from a semiinfinite Si crystal, the diffraction vector 004 was parallel to the surface, S polarization.

6.5 Layered Samples

In the previous section we found explicit formulae for the boundary conditions for the internal wavefield at an interface in the sample. These formulae were written in such a way that they can be used in all scattering geometries (coplanar Bragg and Laue cases, and grazing-incidence case) and both in the one-beam case (x-ray reflection) and in the two-beam diffraction case. They make it possible to express the internal and emitted wavefields in the case of a layered sample.

In this section, we describe dynamical x-ray scattering in a layered system using a matrix formalism. We find formulae that can be used both for x-ray reflection and for x-ray diffraction; in the latter case these formulae are applicable in the coplanar case (symmetrical as well as strongly asymmetrical) and in grazing-incidence geometry.

The matrix formalism describing the wavefield in a layered system in strongly asymmetrical diffraction has been developed in [98, 349, 352]. In grazing-incidence diffraction, the theoretical description depends strongly on the lateral mismatch in the layers. If the multilayer structure is pseudomorph, a matrix formalism similar to that in the strongly asymmetric coplanar diffraction case can be used [350]. In the case of relaxed layers, the theoretical description is complicated and was published for the first time in [380].

In the following, we restrict ourselves to a pseudomorph multilayer only. We assume the same layered structure described in Sect. 5.4 (see figure 5.6). The sample consists of N layers and $N+1$ flat interfaces. The layer $N+1$ is the semi-infinite substrate. Each layer is characterized by its thickness T_j, its polarizability coefficients $\chi^{(j)}_{0,\boldsymbol{h},-\boldsymbol{h}}$ and the vertical component $h_z^{(j)}$ of the diffraction vector. We assume a pseudomorph structure of the layers; i.e., the in-plane components of the diffraction vectors in all the layers are the same and they are equal to that in the substrate. In each layer, the positions of the tie-points on the particular dispersion surface are calculated. These positions depend on $h_z^{(j)}$; this is demonstrated in Fig. 6.10.

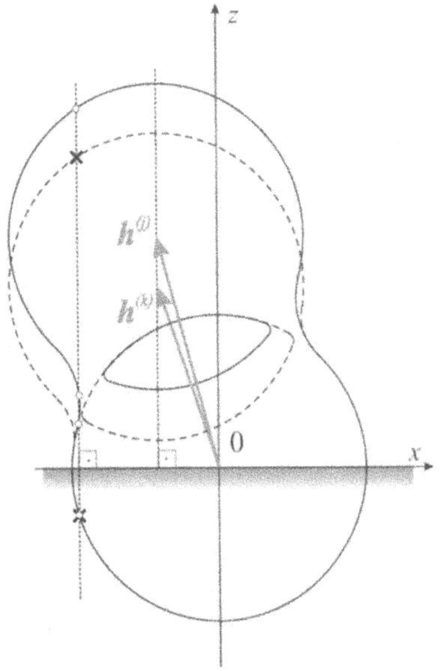

Fig. 6.10. The dispersion surfaces for the layers j (solid line) and k (dashed line) in a pseudomorph multilayer. Due to different values of h_z, the positions of the tie-points in these two layers are different; the tie-points of a layer j are denoted by circles, those of layer k by crosses. Notice that two tie-points for layer k are imaginary and cannot be presented in this sketch.

The position of the tie point is a nonlinear function of h_z. If the difference between the values of h_z in the multilayer is small (i.e., if the layers have small vertical mismatch), the difference between h_z in layer j and in a reference structure (substrate, for instance) is approximately equivalent to the change in the incidence angle of the primary beam by

$$\delta\alpha_i^{(j)} = \delta_z^{(j)} \cos^2\phi(\tan\phi + \tan\Theta_B),$$

where $\delta_z^{(j)} = (a_z^{(j)} - a)/a$ is the vertical mismatch of the layer j with respect to the reference structure. The angles ϕ and Θ_B are calculated for the reference structure. This approximative method, however, has to be tested in any particular case by comparing the results with exact calculations.

For S-polarization, we introduce the column vector

$$\boldsymbol{E}_j = \begin{pmatrix} E_{01}^{(j)} \\ E_{02}^{(j)} \\ E_{03}^{(j)} \\ E_{04}^{(j)} \end{pmatrix} . \tag{6.30}$$

$E_{0n}^{(j)} \equiv E_{0n}(z = z_j)$ denotes the nth transmitted wave in the top of the layer j just below the interface j, including the phase term $\mathrm{e}^{\mathrm{i}k_{0nz}z}$. The in-plane phase terms $\mathrm{e}^{\mathrm{i}\boldsymbol{k}_\parallel \cdot \boldsymbol{r}_\parallel}$ vanish in the boundary conditions since the in-plane components of all the wave vectors are same. In P-polarization, the column vector $\boldsymbol{E}_j$ contains the amplitude components $E_{0\beta n}^{(j)}$, $n = 1, \ldots, 4$ in the top of layer j below the interface j.

The boundary conditions at the interface j can be written in the following uniform way in all geometries and both polarization states:

$$\hat{\mathbf{C}}_{j-1}\hat{\Phi}_{j-1}\boldsymbol{E}_{j-1} = \hat{\mathbf{C}}_j\boldsymbol{E}_j,\ j = 1, \ldots, N+1, \tag{6.31}$$

where, in S-polarization,

$$\hat{\mathbf{C}}_j = \begin{pmatrix} 1 & 1 & 1 & 1 \\ k_{01z}^{(j)} & k_{02z}^{(j)} & k_{03z}^{(j)} & k_{04z}^{(j)} \\ c_1^{(j)} & c_2^{(j)} & c_3^{(j)} & c_4^{(j)} \\ k_{\boldsymbol{h}z1}^{(j)}c_1^{(j)} & k_{\boldsymbol{h}z2}^{(j)}c_2^{(j)} & k_{\boldsymbol{h}z3}^{(j)}c_3^{(j)} & k_{\boldsymbol{h}z4}^{(j)}c_4^{(j)} \end{pmatrix} , \tag{6.32}$$

and in P-polarization the n-th column of the matrix $\hat{\mathbf{C}}_j$ is

$$\begin{pmatrix} \frac{1}{K}(k_{0\alpha n}^{(j)} - k_{0\beta n}^{(j)}p_n^{(j)}) \\ \frac{K}{h^{(j)}}(-h_z^{(j)}p_n^{(j)} - h_x^{(j)}) \\ \frac{1}{K}(k_{0\alpha n}^{(j)}d_n^{(j)} - (k_{0\beta n}^{(j)} + h^{(j)})c_n^{(j)}) \\ \frac{K}{h^{(j)}}(-h_z^{(j)}c_n^{(j)} - h_x^{(j)}d_n^{(j)}) \end{pmatrix} . \tag{6.33}$$

The matrix Φ_j is the same for both polarizations:

$$\hat{\Phi}_j = \begin{pmatrix} \mathrm{e}^{-\mathrm{i}k_{01z}^{(j)}T_j} & 0 & 0 & 0 \\ 0 & \mathrm{e}^{-\mathrm{i}k_{02z}^{(j)}T_j} & 0 & 0 \\ 0 & 0 & \mathrm{e}^{-\mathrm{i}k_{03z}^{(j)}T_j} & 0 \\ 0 & 0 & 0 & \mathrm{e}^{-\mathrm{i}k_{04z}^{(j)}T_j} \end{pmatrix} . \tag{6.34}$$

In these expressions, the superscript $^{(j)}$ denotes the index of the layer and the subscript $_n$ the index of the tie-point.

The expressions for $\hat{\mathbf{C}}_j$ and $\hat{\Phi}_j$ are valid for general two-beam diffraction; in other cases (namely, conventional diffraction and reflection), the corresponding submatrices of the matrices for S-polarization must be taken. In conventional coplanar diffraction, the submatrices are created from the first and the third rows and columns of the matrices in (6.32–6.34); for x-ray reflection, the first and the second rows and columns must be taken.

The matrix $\hat{\Phi}_j$ expresses the phase shifts of the waves between the top and the bottom of layer j. In deriving this matrix, we have taken into account the condition $\exp(-ik_{\boldsymbol{h}nz}^{(j)}T_j) = \exp(-ik_{0nz}^{(j)}T_j)$; i.e., we have assumed that the thickness of layer j is an integer multiple of its lattice parameter.

Equation (6.31) represents a recurrence formula enabling the calculation of the amplitudes of the emitted waves. From this equation we obtain

$$\begin{aligned} \boldsymbol{E}_0 &= \hat{\Phi}_0^{-1}\hat{\mathbf{C}}_0^{-1}\hat{\mathbf{C}}_1\hat{\Phi}_1^{-1}\hat{\mathbf{C}}_1^{-1}\hat{\mathbf{C}}_2\hat{\Phi}_2^{-1}\hat{\mathbf{C}}_2^{-1}\cdots\hat{\Phi}_N^{-1}\hat{\mathbf{C}}_N^{-1}\hat{\mathbf{C}}_{N+1}\hat{\Phi}_{N+1}^{-1}\boldsymbol{E}_{\mathrm{sub}} \\ &\equiv \hat{\mathbf{M}}\boldsymbol{E}_{\mathrm{sub}}. \end{aligned} \tag{6.35}$$

Here $\hat{\Phi}_0$ and $\hat{\Phi}_{N+1}$ represent the phase shifts of the wavefields above the crystal surface between the surface and the detector, and of the wavefields in the substrate between the last (the $N+1$th) interface and some fictive interface in the substrate, respectively. These shifts correspond to the thicknesses of (fictive) layers T_0 and T_{N+1}, respectively. The intensity of the emitted waves does not depend on these thicknesses; however, $\hat{\Phi}_0$ and $\hat{\Phi}_{N+1}$ are important if we use this formalism for rough surfaces. We define

$$\hat{\mathbf{C}}_0 = \begin{pmatrix} 1 & 1 & 0 & 0 \\ K_{0z} & K_{\mathrm{R}z} & 0 & 0 \\ 0 & 0 & 1 & 1 \\ 0 & 0 & K_{\boldsymbol{h}z} & K_{\boldsymbol{h}z} \end{pmatrix}, \quad \boldsymbol{E}_0 = \begin{pmatrix} E_i \\ E_{\mathrm{R}} \\ E_{\boldsymbol{h}} \\ 0 \end{pmatrix}, \tag{6.36}$$

where E_i is the amplitude of the incident field, and $E_{\boldsymbol{h}}$ and E_{R} are the amplitudes of the diffracted and reflected waves, respectively. The z components of their wave vectors are K_{0z}, $K_{\boldsymbol{h}z}$, and $K_{\mathrm{R}z}$, respectively. For conventional coplanar diffraction or for x-ray reflection, the appropriate submatrices should be taken.

The column vector $\boldsymbol{E}_{\mathrm{sub}}$ contains the amplitudes in the substrate. Since the substrate is assumed to be semi-infinite, only two of the four tie-points (or one of two tie-points in conventional diffraction or reflection) are physically relevant. Therefore, only two (or one) terms of $\boldsymbol{E}_{\mathrm{sub}}$ are not zero.

The procedure for the calculation of the intensity of the wave diffracted (and/or reflected) from a layered sample consists of the following steps.

1. In each layer, we find the intersection points (the tie-points) of the appropriate dispersion surface ((6.10) for reflection, (6.17) for diffraction, and (6.26) for conventional diffraction) with the common interface normal. The relative position of the normal to the dispersion surface is determined by the direction of the primary beam. The positions of the tie-points are

defined by the z components $k_{hzn}^{(j)}$ of the wave vectors of the internal wavefields. In the case of the substrate, we find only the physically relevant tie-points.

2. From the known tie-points we find the amplitude ratios $c_n^{(j)}$ in each layer in S-polarization and $c_n^{(j)}$, $d_n^{(j)}$, $p_n^{(j)}$ in P-polarization (Eqs. (6.19) and (6.20)). For a conventional diffraction, only the $c_n^{(j)}$ coefficients are used; these are calculated for two values of the linear polarization factor C. This step is omitted in the case of reflection.
3. We construct the matrices $\hat{\mathbf{C}}_j^{-1}$ (6.32) or (6.33) and $\hat{\Phi}_j^{-1}$ (6.34) and their product $\hat{\mathbf{M}}$ (6.35).
4. We solve the system (6.35) for four linear equations. In the general case, the unknowns are E_{R}, $E_{\boldsymbol{h}}$, and the substrate amplitudes $E_{\mathrm{sub},1}$, $E_{\mathrm{sub},2}$. In the cases of reflection and conventional diffraction, the system contains only two equations with the unknowns E_{R} (or $E_{\boldsymbol{h}}$ for conventional diffraction) and E_{sub}.
5. From the known amplitudes of the emitted waves we calculate the diffractivity

$$\mathcal{R}_{\boldsymbol{h}} = \frac{I_{\boldsymbol{h}} A_{\boldsymbol{h}}}{I_i A_i} \equiv \left| \frac{E_{\boldsymbol{h}}}{E_i} \right|^2 \frac{|\gamma_{\boldsymbol{h}}|}{\gamma_0}$$

or the reflectivity

$$\mathcal{R}_{\mathrm{R}} = \left| \frac{E_{\mathrm{R}}}{E_i} \right|^2 .$$

6.5.1 Multilayers: X-Ray Reflection

Taking only the first and second rows and columns of the matrices $\hat{\mathbf{C}}_j$, $\hat{\Phi}_j$ into account, from (6.35) we can obtain the amplitude of the wave reflected from a layered sample. In this case, however, it is useful to rewrite this formula in a slightly different form,

$$\boldsymbol{E}_0 = \hat{\Phi}_0^{-1} \hat{\mathbf{R}}_1 \hat{\Phi}_1^{-1} \hat{\mathbf{R}}_2 \hat{\Phi}_2^{-1} \cdots \hat{\Phi}_N^{-1} \hat{\mathbf{R}}_{N+1} \hat{\Phi}_{N+1}^{-1} \boldsymbol{E}_{\mathrm{sub}} \equiv \hat{\mathbf{M}} \boldsymbol{E}_{\mathrm{sub}}, \tag{6.37}$$

where

$$\hat{\mathbf{R}}_j = \hat{\mathbf{C}}_{j-1}^{-1} \hat{\mathbf{C}}_j = \frac{1}{t_j} \begin{pmatrix} 1 & r_j \\ r_j & 1 \end{pmatrix}, \; j = 1, \ldots, N+1,$$

$$\hat{\Phi}_j = \begin{pmatrix} \mathrm{e}^{-\mathrm{i}k_z^{(j)} T_j} & 0 \\ 0 & \mathrm{e}^{\mathrm{i}k_z^{(j)} T_j} \end{pmatrix}, \; j = 0, \ldots, N+1$$

and we have denoted by t_j and r_j the Fresnel complex transmittivity and reflectivity, respectively, of the interface j (see Eq. (6.14)); $k_z^{(j)}$ is the vertical component of the wave vector of the transmitted wave in the jth layer. The index $N+1$ denotes the substrate.

The wavefield on the top of the jth layer is represented by the two-dimensional column vector

$$\boldsymbol{E}_j = \begin{pmatrix} E_{\mathrm{T}}^{(j)} \\ E_{\mathrm{R}}^{(j)} \end{pmatrix}$$

containing the amplitudes of the transmitted and the reflected beams, including the appropriate phase terms. The reflectivity of the multilayer stack is simply given by

$$\mathcal{R}_{\mathrm{R}} = \left| \frac{M_{21}}{M_{11}} \right|^2 . \tag{6.38}$$

This expression represents the exact dynamical formula for the x-ray reflectivity [259].

If we denote $\Re_j = E_{\mathrm{R}}^{(j)}/E_{\mathrm{T}}^{(j)}$, the ratio of the amplitudes of the reflected and transmitted waves in layer j, the following recursive formula can be found

$$\Re_j = \mathrm{e}^{-2\mathrm{i}k_z^{(j)}T_j} \frac{r_{j+1} + \Re_{j+1}}{1 + r_{j+1}\Re_{j+1}}, j = 0, \ldots, N. \tag{6.39}$$

This formula can also be used for the numerical calculation, starting from the condition $\mathcal{R}_{\mathrm{sub}} \equiv \Re_{N+1} = 0$, since there is no reflected wave in the semi-infinite substrate. The resulting reflectivity is

$$\mathcal{R}_{\mathrm{R}} = |\Re_0|^2.$$

In Fig. 6.11 we compare the reflection curve calculated dynamically with the results of kinematical calculations based on Eqs. (5.30), where we take only one term $\boldsymbol{g} = 0$ into account. For larger incidence angles, both curves are very similar; however, the dynamical curve is shifted by refraction at the free surface. In this region, the kinematical approach is applicable only if one includes "ad hoc" the refraction condition in the calculation (see Chap. 7). For the incidence angle comparable or smaller than the critical angle α_{c} the kinematical approach fails.

6.5.2 Multilayers: Conventional X-Ray Diffraction

The theoretical description of conventional coplanar diffraction from a multilayer is based on the general formalism derived above in (6.30)–(6.36). The formula (6.35) can be written in the form (6.37), similar to the case of x-ray reflection, but the meaning of the matrices $\hat{\mathbf{R}}_j$ and $\hat{\Phi}_j$ is different:

$$\hat{\mathbf{R}}_j = \frac{1}{c_2^{(j-1)} - c_1^{(j-1)}} \begin{pmatrix} c_2^{(j-1)} - c_1^{(j)} & c_2^{(j-1)} - c_2^{(j)} \\ -c_1^{(j-1)} + c_1^{(j)} & -c_1^{(j-1)} + c_2^{(j)} \end{pmatrix}, \; j = 2, \ldots, N+1,$$

$$\hat{\Phi}_j = \begin{pmatrix} \mathrm{e}^{-\mathrm{i}k_{01z}^{(j)}T_j} & 0 \\ 0 & \mathrm{e}^{-\mathrm{i}k_{02z}^{(j)}T_j} \end{pmatrix}, j = 1, \ldots, N, \tag{6.40}$$

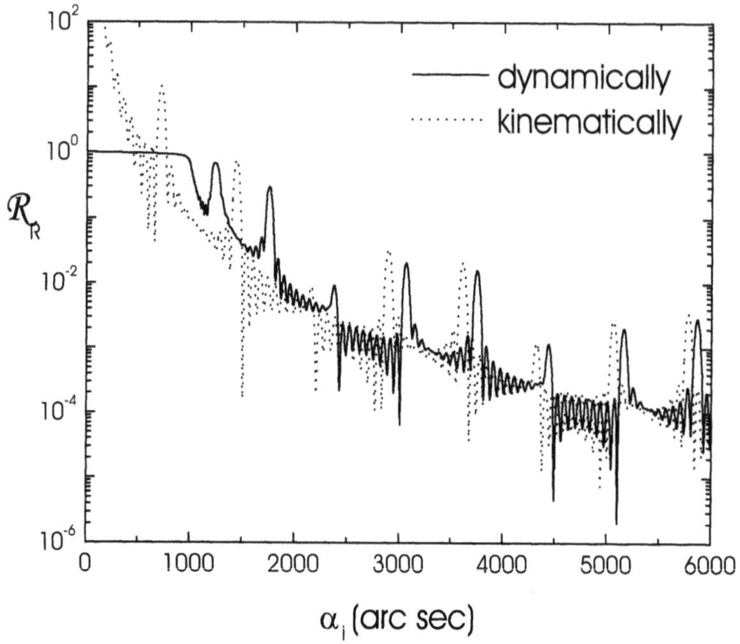

Fig. 6.11. The reflection curves of a periodic multilayer consisting of 10 periods of 7 nm GaAs and 15 nm AlAs with flat interfaces, wavelength $\lambda = 1.54$ Å. Full dynamical and kinematical calculations were performed.

where $c_n^{(j)}$ are the amplitude ratios in the layer j defined in Eq. (6.19). The matrix $\hat{\mathbf{R}}_1$ has a different form as well:

$$\hat{\mathbf{R}}_1 = \begin{pmatrix} 1 & 1 \\ c_1^{(1)} & c_2^{(1)} \end{pmatrix} .$$

The diffractivity of the multilayer is given by the general definition (6.29):

$$\mathcal{R}_{\boldsymbol{h}} = \left| \frac{M_{21}}{M_{11}} \right|^2 \left| \frac{\gamma_{\boldsymbol{h}}}{\gamma_0} \right|^2 .$$

Although the above expressions are quite general, we can derive other expressions for conventional diffraction from a layered sample that are, in many cases, more convenient for practical use. Let us define the complex diffractivity in the jth layer as the ratio of the total diffracted amplitude and the total transmitted amplitude just below the interface j:

$$\Re_j = \frac{c_1^{(j)} E_{01}^{(j)} + c_2^{(j)} E_{02}^{(j)}}{E_{01}^{(j)} + E_{02}^{(j)}} \equiv \frac{A_2}{A_1} ,$$

where

$$\boldsymbol{A} = \begin{pmatrix} A_1 \\ A_2 \end{pmatrix} = \hat{\mathbf{C}}_j \boldsymbol{E}_j .$$

Then, from (6.31), the following recurrence formula can be derived (see also [28]):

$$\Re_j = \frac{\Re_{j+1}(c_1^{(j)} - M_j c_2^{(j)}) + c_1^{(j)} c_2^{(j)} (M_j - 1)}{\Re_{j+1}(1 - M_j) + c_1^{(j)} M_j - c_2^{(j)}},$$

$$M_j = \mathrm{e}^{-\mathrm{i}(k_{0z1}^{(j)} - k_{0z2}^{(j)})T_j}, \; j = 0, \ldots, N. \tag{6.41}$$

The complex diffractivity of the substrate is simply

$$\Re_{\mathrm{sub}} \equiv \Re_{N+1} = c_s^{(N+1)}, \tag{6.42}$$

where $s = 1$ or 2 is the index of the physically relevant tie-point for the substrate.

The diffractivity of the multilayer is

$$\mathcal{R} = |\Re_0|^2 \frac{|\gamma_{\boldsymbol{h}}|}{\gamma_0}. \tag{6.43}$$

Numerical examples showing the application of these formulae can be found in Chapters 8 and 9, along with the measured diffraction and reflection curves.

6.6 A Comment on the Three-Beam Diffraction

Recently, many beam diffraction effects have been widely studied since they yield the possibility of a direct determination of the phase of the structure factor [72, 173, 387]. In this section, we do not discuss this aspect, we use the many-beam approach for the calculation of a diffraction curve of a pseudomorph multilayer. We choose the simplest case, namely, a three-beam diffraction, where the wave in a crystal is a superposition of three plane waves:

$$\boldsymbol{E}(\boldsymbol{r}) = \boldsymbol{E}_0 \mathrm{e}^{\mathrm{i}\boldsymbol{k}_0 . \boldsymbol{r}} + \boldsymbol{E}_{\boldsymbol{h}} \mathrm{e}^{\mathrm{i}\boldsymbol{k}_{\boldsymbol{h}} . \boldsymbol{r}} + \boldsymbol{E}_{\boldsymbol{g}} \mathrm{e}^{\mathrm{i}\boldsymbol{k}_{\boldsymbol{g}} . \boldsymbol{r}}; \tag{6.44}$$

the in-plane components of both diffraction vectors $\boldsymbol{g}_{\parallel}$ and $\boldsymbol{h}_{\parallel}$ are assumed to be same. Further, we assume that all three wave vectors $\boldsymbol{k}_0$, $\boldsymbol{k}_{\boldsymbol{g}} = \boldsymbol{k}_0 + \boldsymbol{g}$, $\boldsymbol{k}_{\boldsymbol{h}} = \boldsymbol{k}_0 + \boldsymbol{h}$ lie in the same scattering plane (*three-beam coplanar case*). In the S-polarization, the amplitude equations are rather simple:

$$\begin{aligned}
&(k_0^2 - k^2)E_0 - K^2\chi_{-\boldsymbol{h}}E_{\boldsymbol{h}} - K^2\chi_{-\boldsymbol{g}}E_{\boldsymbol{g}} = 0 \\
&-K^2\chi_{\boldsymbol{h}}E_0 + (k_{\boldsymbol{h}}^2 - k^2)E_{\boldsymbol{h}} - K^2\chi_{\boldsymbol{h}-\boldsymbol{g}}E_{\boldsymbol{g}} = 0 \\
&-K^2\chi_{\boldsymbol{g}}E_0 - K^2\chi_{\boldsymbol{g}-\boldsymbol{h}}E_{\boldsymbol{h}} + (k_{\boldsymbol{g}}^2 - k^2)E_{\boldsymbol{g}} = 0.
\end{aligned} \tag{6.45}$$

This equation system is homogeneous and it has a non-zero solution, if the dispersion equation is fulfilled:

$$(k_0^2 - k^2)(k_{\boldsymbol{h}}^2 - k^2)(k_{\boldsymbol{g}}^2 - k^2) - (k_0^2 - k^2)K^4\chi_{\boldsymbol{g}-\boldsymbol{h}}\chi_{\boldsymbol{h}-\boldsymbol{g}}$$

$$-(k_{\boldsymbol{h}}^2 - k^2)K^4\chi_{\boldsymbol{g}}\chi_{-\boldsymbol{g}} + (k_{\boldsymbol{g}}^2 - k^2)K^4\chi_{\boldsymbol{h}}\chi_{-\boldsymbol{h}}$$

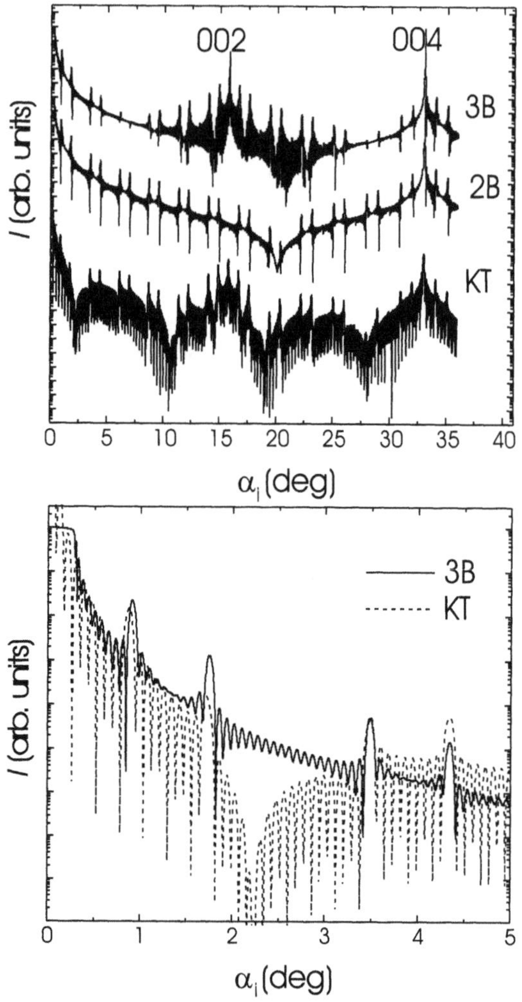

Fig. 6.12. The diffraction curves of a periodic multilayer calculated by different methods; see text for details. In the upper panel, the whole curves are depicted, and the curves are shifted vertically for clarity. The lower panel presents the details of the 000 maximum.

$$-K^6\chi_{\boldsymbol{h}}\chi_{-\boldsymbol{g}}\chi_{\boldsymbol{g}-\boldsymbol{h}} - K^6\chi_{\boldsymbol{g}}\chi_{-\boldsymbol{h}}\chi_{\boldsymbol{h}-\boldsymbol{g}} = 0. \qquad (6.46)$$

As in the two-beam case, this equation represents a dispersion surface in reciprocal space; this surface has the order six, i.e., for a given direction of primary beam we obtain 6 tie-points and 18 waves in the crystal:

$$E_{0n}\mathrm{e}^{\mathrm{i}\boldsymbol{k}_{0n}\cdot\boldsymbol{r}},\ E_{\boldsymbol{h}n}\mathrm{e}^{\mathrm{i}\boldsymbol{k}_{\boldsymbol{h}n}\cdot\boldsymbol{r}},\ E_{\boldsymbol{g}n}\mathrm{e}^{\mathrm{i}\boldsymbol{k}_{\boldsymbol{g}n}\cdot\boldsymbol{r}},\ n = 1,\ldots,6.$$

In P-polarization, the amplitude equations and the dispersion surface are rather complicated, since the waves in the layers are not transversal.

The boundary conditions at the sample surface and at an interface in a multilayer have the same form as in the two-beam case and we can construct analogous matrix formulas; the matrices have the rank 6. Instead of presenting huge formulas here, we present a numerical example showing the differences between the two-beam and three-beam diffraction curves of a multilayer. In the example, we assume the same multilayer structure as in Sect. 5.4 – namely, a pseudomorph superlattice with 10 periods, each period consisting of an GaAs layer (thickness $3a_{GaAs}$) and an AlAs layer (thickness $6a_{AlAs}$). The wavelength is $\lambda = 1.54$ Å and we assume S-polarization. The two-beam diffraction curves have been calculated for $\boldsymbol{h} = (004)$; in the calculation of the three-beam curve we took $\boldsymbol{h} = (004)$ and $\boldsymbol{g} = (002)$.

In Fig. 6.12 we compare the diffraction curves calculated by the general two-beam case method (denoted as 2B), using Eqs. (6.30) – (6.36), using the three-beam diffraction (3B) (6.44) – (6.46), and finally using the many-beam approach in kinematical theory (KT) according to Eqs. (5.30) – (5.40). The 3B and KT curves exhibit both 002 and 004 main diffraction maxima; all the curves also show total reflection maxima for $\alpha_i \to 0$ (maximum 000). This maximum is not correct in the KT curve since the kinematical approximation is not valid for such small angles; see the right panel in Fig. 6.12. The main difference between the curves lies in their shapes between the maxima. In the calculation of the KT curve the dependence of the structure factor on Q was considered. This dependence is responsible for the slow oscillations of the intensity envelope between the main maxima. These oscillations cannot be obtained by the dynamical approach, since the n-beam dynamical theory uses the values of the structure factor in n reciprocal lattice points $\boldsymbol{g}$.

7 Semikinematical Theory

In the previous two chapters we dealt with two limiting approaches for the theoretical description of x-ray scattering, namely, kinematical theory, where we have neglected multiple scattering, and dynamical theory, where the multiple scattering processes have been treated exactly. In most cases, the kinematical approach is sufficient for scattering in thin layers; however, it fails in some cases, especially in a surface-sensitive arrangement, where the interaction of the x-rays with the sample is rather strong. The application of the dynamical theory is limited to the simplest trivial cases (a perfect crystal or a perfectly pseudomorph layered system); in many practical cases the dynamical formulas are too complicated and not practicable. In this chapter we describe a *semikinematical* approach that can be used in these cases.

7.1 Basic Formulas

Let us divide the scattering potential $\hat{\mathrm{V}}(\boldsymbol{r})$ defined in (4.12) into two parts:

$$\hat{\mathrm{V}}(\boldsymbol{r}) = \hat{\mathrm{V}}_A(\boldsymbol{r}) + \hat{\mathrm{V}}_B(\boldsymbol{r}). \tag{7.1}$$

The first part describes the scattering from a non-disturbed system; the second corresponds to the disturbance. The non-disturbed system has to be chosen to be so simple that the corresponding wave equation

$$(\Delta + K^2)\boldsymbol{E}^{(A)}(\boldsymbol{r}) = \hat{\mathrm{V}}_A(\boldsymbol{r})\boldsymbol{E}^{(A)}(\boldsymbol{r}) \tag{7.2}$$

can be solved *exactly*. We find two independent solutions of this equation denoted in the Dirac notation as $|\boldsymbol{E}_1^{(A)}\rangle$ and $|\boldsymbol{E}_2^{(-A)}\rangle$. We choose the solution $|\boldsymbol{E}_1^{(A)}\rangle$ so that the corresponding incident wave is the actual incident plane wave $|\boldsymbol{K}_i\rangle$ defined in (4.2). Therefore, $|\boldsymbol{E}_1^{(A)}\rangle$ describes the wavefield inside and outside the non-disturbed sample excited by $|\boldsymbol{K}_i\rangle$. The incident wave corresponding to $|\boldsymbol{E}_2^{(-A)}\rangle$ is the wave $|\boldsymbol{K}_s\rangle$ emitted by the sample that is measured in the experiment. Therefore, the solution $|\boldsymbol{E}_2^{(-A)}\rangle$ must be time-inverted; this time-inversion is denoted by the minus sign in the superscript.

The *exact* (dynamical) expression for the differential cross section of the scattering from the non-disturbed system is (4.21)

$$\left(\frac{\mathrm{d}\sigma}{\mathrm{d}\Omega}\right)_A = \frac{1}{16\pi^2}\left|\langle \boldsymbol{K}_s|\hat{\mathbf{T}}_A|\boldsymbol{K}_i\rangle\right|^2 \equiv \frac{1}{16\pi^2}\left|\langle \boldsymbol{E}_2^{(-A)}|\hat{\mathbf{V}}_A|\boldsymbol{K}_i\rangle\right|^2 . \tag{7.3}$$

The scattering process from the disturbance is assumed *kinematical* and we neglect multiple scattering due to $\hat{\mathbf{V}}_B$.

This approximation is usually called *semikinematical approximation*, or equivalently *distorted-wave Born approximation* (DWBA). The method of DWBA is described in more detail in textbooks on the scattering theory; see [92, 288], for instance. The application of the DWBA approach for x-ray scattering is described thoroughly in [95, 332, 384].

The differential cross section in the DWBA approach is a coherent sum of the contribution of the non-disturbed system (7.3) and the contribution of $\hat{\mathbf{V}}_B$:

$$\left(\frac{\mathrm{d}\sigma}{\mathrm{d}\Omega}\right)_{DWBA} = \frac{1}{16\pi^2}\left|\langle \boldsymbol{E}_2^{(-A)}|\hat{\mathbf{V}}_A|\boldsymbol{K}_i\rangle + \langle \boldsymbol{E}_2^{(-A)}|\hat{\mathbf{V}}_B|\boldsymbol{E}_1^{(A)}\rangle\right|^2 . \tag{7.4}$$

The contribution of $\hat{\mathbf{V}}_B$ is therefore proportional to the matrix element of $\hat{\mathbf{V}}_B$ expressed in the basis of the non-disturbed solutions $|\boldsymbol{E}_{1,2}^{(\pm A)}\rangle$. Since these solutions can be expressed as sums of plane waves (see the examples below), this matrix element equals a sum of Fourier transformations of $\hat{\mathbf{V}}_B(\boldsymbol{r})$.

The whole scattering can be described as a sequence of the following processes (see also Fig. 7.1):

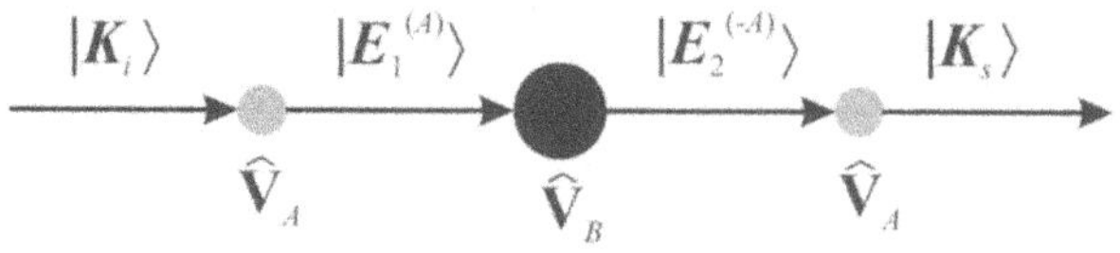

Fig. 7.1. Schematic illustration of the calculation procedure in the DWBA approximation.

1. The primary wave $|\boldsymbol{K}_i\rangle$ is scattered dynamically in the non-disturbed system and the wavefield $|\boldsymbol{E}_1^{(A)}\rangle$ is created. This wavefield is a solution of the non-disturbed wave equation (7.2).
2. The wavefield $|\boldsymbol{E}_1^{(A)}\rangle$ is scattered kinematically by the disturbance $\hat{\mathbf{V}}_B(\boldsymbol{r})$ resulting in the wavefield $|\boldsymbol{E}_2^{(-A)}\rangle$; this wavefield is a solution of (7.2) as well.
3. The wavefield $|\boldsymbol{E}_2^{(-A)}\rangle$ is scattered by the non-disturbed system, and the vacuum wave $|\boldsymbol{K}_s\rangle$ is created. This scattering process is time-inverted; if one irradiates the non-disturbed sample with the wave $|-\boldsymbol{K}_s\rangle$, i.e., with the wave time-inverted to the true emitted wave $\boldsymbol{K}_s$, the wavefield $|\boldsymbol{E}_2^{(A)}\rangle$ will result.

The DWBA formula for the differential cross-section (7.4) can be improved if we take into account double scattering due to $\hat{\mathbf{V}}_B$; i.e., we consider the processes in Fig. 7.2. The corresponding formulas are rather complicated;

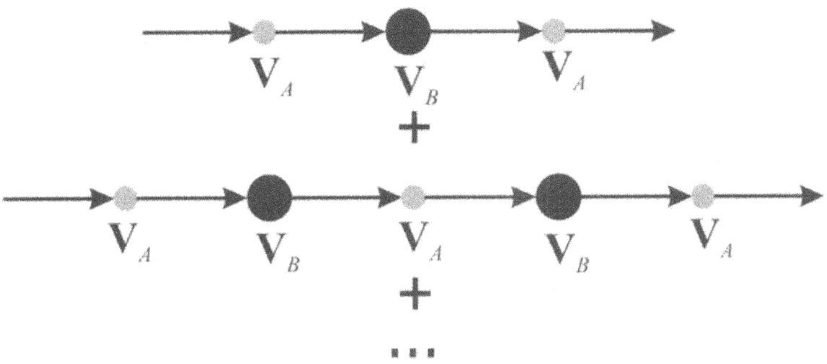

Fig. 7.2. Higher iteration steps in the DWBA.

however, the analysis performed in [94] showed that the second-order DWBA must be considered for the description of diffuse scattering from a rough surface with a large lateral correlation length (see Chap. 11).

7.2 Examples

The DWBA method is widely used for the simulation of a variety of scattering problems. We will show its application in the following chapters of this book; in this chapter we present three typical examples showing characteristic features of the method.

7.2.1 Small-Angle Scattering from Empty Holes in a Semi-infinite Matrix

Let us consider a semi-infinite substrate containing randomly distributed spherical empty holes. Their density is so small that we can neglect double scattering from the holes, and the DWBA method may be used. The positions of the holes are statistically non-correlated. The problem has been solved in [277] for the first time. We choose a semi-infinite homogeneous continuum as the non-disturbed system. Since we deal with small angle scattering, where the relevant Fourier coefficient χ_0 of the polarizability and the refractive index n do not depend on the crystal structure, we can assume an amorphous substrate. The solution of the wave equation (7.2) of this system is simple: above the surface ($z > 0$) the incident wave and the reflected wave exist; below the sample surface $z < 0$ the transmitted wave propagates. The amplitudes of the reflected and transmitted waves are determined by the Fresnel coefficients r and t depending on the incidence angle α_i. We choose two independent solutions of (7.2) in the following way (we assume S-polarization only):

$$E_1^{(A)}(\boldsymbol{r}) \equiv \langle \boldsymbol{r}|E_1^{(A)}\rangle = \begin{cases} \mathrm{e}^{\mathrm{i}\boldsymbol{K}_{i1}.\boldsymbol{r}} + r_1 \mathrm{e}^{\mathrm{i}\boldsymbol{K}_{R1}.\boldsymbol{r}} & \text{for } z > 0 \\ t_1 \mathrm{e}^{\mathrm{i}\boldsymbol{k}_{T2}.\boldsymbol{r}} & \text{for } z < 0 \end{cases} \tag{7.5}$$

and

$$E_2^{(-A)}(\boldsymbol{r}) \equiv \langle \boldsymbol{r}|E_2^{(-A)}\rangle = \begin{cases} \mathrm{e}^{\mathrm{i}\boldsymbol{K}_{i2}.\boldsymbol{r}} + r_2^* \mathrm{e}^{\mathrm{i}\boldsymbol{K}_{R2}.\boldsymbol{r}} & \text{for } z > 0 \\ t_2^* \mathrm{e}^{\mathrm{i}\boldsymbol{k}_{T2}^*.\boldsymbol{r}} & \text{for } z < 0 \end{cases} . \tag{7.6}$$

The wave vectors of these solutions are sketched in Fig. 7.3.

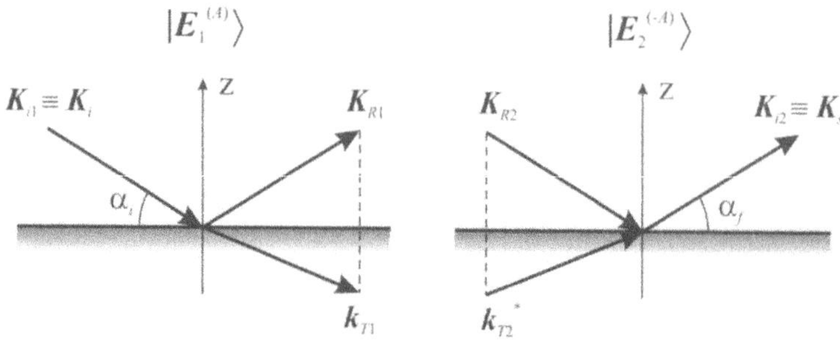

Fig. 7.3. The solutions of the non-disturbed wave equation, a semi-infinite amorphous substrate.

Here $\boldsymbol{K}_{i1} \equiv \boldsymbol{K}_i$ is the true incident wave and $\boldsymbol{K}_{i2} \equiv \boldsymbol{K}_s$ is the true scattered wave, $\boldsymbol{K}_{R1,2}$ are the wave vectors of the reflected waves and $\boldsymbol{k}_{T1,2}$ are the wave vectors of the (complex) transmitted waves in the substrate. The time inversion in (7.6) converts all complex quantities in their conjugates.

The disturbance $\hat{\mathbf{V}}_B(\boldsymbol{r})$ comprises the holes; the difference between the polarizability of the holes and their neighborhood is $-\chi_0$ and then

$$\hat{\mathbf{V}}_B(\boldsymbol{r}) = K^2 \chi_0 \sum_n \Omega_n(\boldsymbol{r} - \boldsymbol{R}_n) \tag{7.7}$$

holds, where $\boldsymbol{R}_n$ is the position of the n-th hole and $\Omega_n(\boldsymbol{r})$ is its shape function (unity in the hole and zero outside). This scattering potential is random and the total differential cross section can be divided into the coherent and incoherent (diffuse) parts

$$\left(\frac{\mathrm{d}\sigma}{\mathrm{d}\Omega}\right)_{\mathrm{coh}} = \frac{1}{16\pi^2} \left| \langle \boldsymbol{E}_2^{(-A)}|\hat{\mathbf{V}}_A|\boldsymbol{K}_i\rangle + \langle \boldsymbol{E}_2^{(-A)}|\langle \hat{\mathbf{V}}_B\rangle|\boldsymbol{E}_1^{(A)}\rangle \right|^2 \tag{7.8}$$

and

$$\left(\frac{\mathrm{d}\sigma}{\mathrm{d}\Omega}\right)_{\mathrm{incoh}} = \frac{1}{16\pi^2} \mathrm{Cov}\left(\langle \boldsymbol{E}_2^{(-A)}|\hat{\mathbf{V}}_B|\boldsymbol{E}_1^{(A)}\rangle, \langle \boldsymbol{E}_2^{(-A)}|\hat{\mathbf{V}}_B|\boldsymbol{E}_1^{(A)}\rangle \right) . \tag{7.9}$$

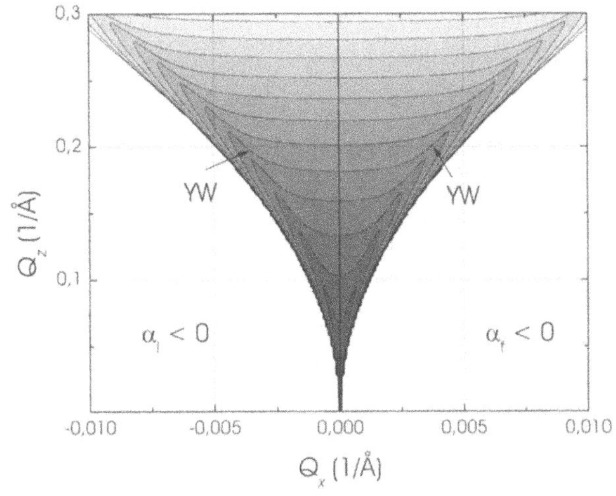

Fig. 7.4. Reciprocal-space distribution of the incoherent intensity scattered from a system of randomly placed holes. The average hole radius was $R = 1$ nm, and an exponential distribution of the hole sizes was assumed. YW denotes the Yoneda wings; the regions $\alpha_i < 0$ and $\alpha_f < 0$ are not accessible in the coplanar geometry.

Since the positions of the holes are random and statistically not correlated, the incoherent differential cross section can be expressed by

$$\left(\frac{\mathrm{d}\sigma}{\mathrm{d}\Omega}\right)_{\text{incoh}} = \frac{N}{16\pi^2}|K^2\chi_0 t_1 t_2|^2 \left\langle \left|\Omega^{\text{FT}}(\boldsymbol{Q}_T)\right|^2 \right\rangle_{\text{size}}, \tag{7.10}$$

where $\boldsymbol{Q}_T = \boldsymbol{k}_{T2} - \boldsymbol{k}_{T1}$ is the complex scattering vector *in the substrate*, i.e., corrected to refraction and absorption; $\Omega^{\text{FT}}(\boldsymbol{Q}_T)$ is the Fourier transformation of the shape function of a hole; N is the number of the holes in the irradiated volume; and $\langle\ \rangle_{\text{size}}$ is the averaging over the hole sizes.

In comparison with the kinematical formula

$$\left(\frac{\mathrm{d}\sigma}{\mathrm{d}\Omega}\right)_{\text{incoh,kin}} = \frac{N}{16\pi^2}|K^2\chi_0|^2 \left\langle \left|\Omega^{\text{FT}}(\boldsymbol{Q})\right|^2 \right\rangle_{\text{size}}, \tag{7.11}$$

the DWBA expression (7.10) differs in the factor $|t_1 t_2|^2$ that stems from the amplitudes of the non-disturbed states. The Fresnel transmittivities $t_{1,2}$ exhibit a maximum for $\alpha_{i,f} = \alpha_c$; therefore, the intensity distribution in reciprocal space exhibits maxima along the lines $\alpha_i = \alpha_c$ and $\alpha_f = \alpha_c$. These maxima are called the *Yoneda wings* [332, 395]. In Fig. 7.4 they are denoted by YW.

Another difference between the DWBA and kinematical formulas (7.10) and (7.11) lies in the argument of Ω^{FT}; in the DBWA, the refraction corrected value is used. This deforms the intensity distribution for small Q's; this deformation must be taken into account in an analysis of measured data.

7.2.2 Small-Angle Scattering from Pyramidal Islands Randomly Placed on a Flat Surface

As in the previous example, we use the non-disturbed wavefields (7.5) and (7.6). However, the islands are placed *on the sample surface* so that the parts of these wavefields for $z > 0$ occur in the matrix elements $\langle \boldsymbol{E}_2^{(-A)} | \langle \hat{\mathbf{V}}_B \rangle | \boldsymbol{E}_1^{(A)} \rangle$. Since the non-disturbed wavefield consists of two plane waves for $z > 0$ (the incident and the reflected waves), this matrix element is a coherent superposition of four terms describing four scattering processes depicted in Fig. 7.5. The first scattering process is the scattering of the incident wave; in the second process the incident wave is reflected by the surface and the reflected wave is scattered by the island, etc. (see [278] for details). In the kinematical approximation, only the first scattering process is included.

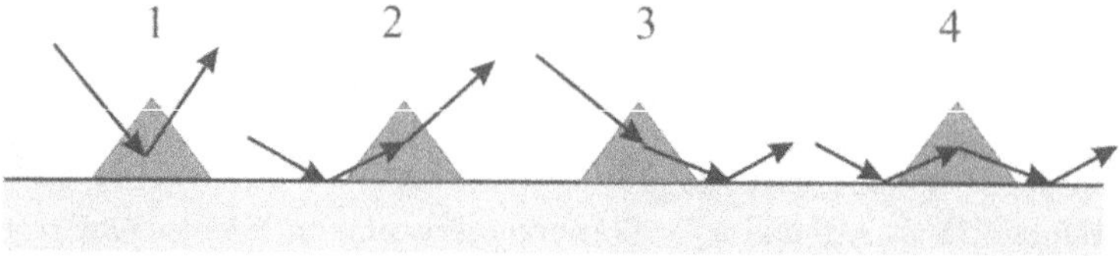

Fig. 7.5. Four scattering processes considered in the semikinematical calculation of diffuse scattering from free-standing islands.

If we assume that the islands have identical shapes and they are randomly placed on the surface, the differential cross section of diffuse scattering is

$$\left(\frac{\mathrm{d}\sigma}{\mathrm{d}\Omega}\right)_{\text{incoh}} = \frac{N}{16\pi^2}|K^2\chi_0|^2 \left|\sum_{p=1}^{4} A_p \Omega^{\text{FT}}(\boldsymbol{Q}_p)\right|^2 . \tag{7.12}$$

Here we have denoted A_p, the amplitudes of the scattering processes:

$$A_1 = 1,\ A_2 = r_1,\ A_3 = r_2,\ A_4 = r_1 r_2 \tag{7.13}$$

and $\boldsymbol{Q}_p$ are the corresponding scattering vectors (see Fig. 7.3),

$$\begin{aligned} \boldsymbol{Q}_1 = \boldsymbol{K}_{i2} - \boldsymbol{K}_{i1} \equiv \boldsymbol{Q},\ \boldsymbol{Q}_2 = \boldsymbol{K}_{i2} - \boldsymbol{K}_{R1}, \\ \boldsymbol{Q}_3 = \boldsymbol{K}_{R2} - \boldsymbol{K}_{i1},\ \boldsymbol{Q}_4 = \boldsymbol{K}_{R2} - \boldsymbol{K}_{R1}. \end{aligned} \tag{7.14}$$

We have used this theoretical approach for the simulation of GISAXS from PbSe quantum dots on PbTe(111). These pyramidal dots have a threefold symmetry (the pyramid base is an equilateral triangle) with the side facets {100}. Figure 7.6 shows the intensities of individual scattering processes as functions of the in-plane scattering angle ψ. The definition of this angle and the orientation of the pyramid with respect to the primary beam is explained in the figure. The calculation has been carried out for the constant incidence angle $\alpha_i = 0.2°$ and the constant exit angle $\alpha_f = 0.5°$. The amplitude A_p

of the p-th process is determined by the Fresnel reflectivity coefficients $r_{i,f}$; therefore, the processes 2–4 can be neglected for larger angles $\alpha_{i,f}$, where the corresponding reflectivity coefficients are negligible.

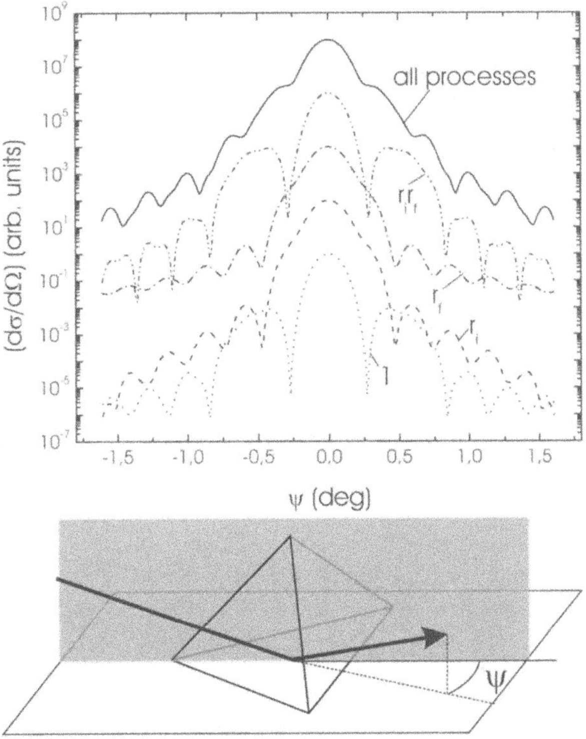

Fig. 7.6. The GISAXS scans calculated for PbSe pyramids on PbTe, the intensities of particular scattering processes are plotted (left). The right picture shows the azimuthal orientation of the incident beam and the definition of the in-plane scattering angle ψ.

7.2.3 Diffuse Scattering in Diffraction from Empty Holes in a Crystal

In this example, we assume a random set of empty holes in a semi-infinite crystal and we calculate the incoherent component of the *diffracted* wave. In this case, we choose a semi-infinite *crystalline* substrate as the non-disturbed system. The solution of the non-disturbed wave equation (7.2) can be find using dynamical x-ray diffraction explained in Chapter 6.

In this example we restrict ourselves to the conventional two-beam diffraction in the coplanar case, S-polarization. Then, a primary plane wave excites in the crystal one transmitted wave $E_0 e^{\mathrm{i}\boldsymbol{k}_0.\boldsymbol{r}}$ and one diffracted wave $E_{\boldsymbol{h}} e^{\mathrm{i}\boldsymbol{k}_{\boldsymbol{h}}.\boldsymbol{r}}$. From the boundary conditions at the sample surface (6.28) we obtain $E_0 = E_i, E_{\boldsymbol{h}} = cE_i$, where c is the amplitude ratio (6.19) for the physically relevant tie-point (see Sect. 6.4 for details). The amplitude E_i of the incident wave equals unity in the calculation. A suitable choice of the two independent solutions of Eq. (7.2) is shown in Fig. 7.7. In the solution $|E_1^{(A)}\rangle$,

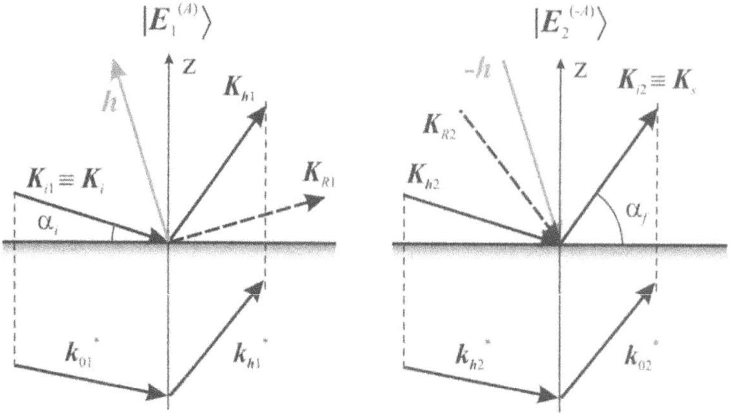

Fig. 7.7. Two independent solutions of the non-disturbed wave equation, a semi-infinite crystalline matrix.

the vacuum wave with the wave vector $\boldsymbol{K}_{i1}$ is the actual incident wave, the waves in the substrate have the wave vectors $\boldsymbol{k}_{0,\boldsymbol{h}1}$. In the other solution $|E_2^{(-A)}\rangle$, the incident (time inverted) wave is the true diffracted wave; therefore, this solution corresponds to the diffraction with the *opposite* diffraction vector $-\boldsymbol{h}$. The wave vectors of the waves in the substrate are $\boldsymbol{k}_{0,\boldsymbol{h}2}$; the direction of the wave vector $\boldsymbol{k}_{02}$ is close to the direction of $\boldsymbol{k}_{\boldsymbol{h}1}$. The specularly reflected waves $\boldsymbol{K}_{R1,2}$ are not included in the conventional dynamical theory. The amplitudes of the $\boldsymbol{k}_{\boldsymbol{h}1,2}$-waves are c_1 and c_2^*, respectively, where $c_{1,2}$ are the amplitude ratios (6.19) for the relevant tie-point in the solutions $|E_1^{(A)}\rangle$, and $|E_2^{(-A)}\rangle$.

Since the holes are empty, their local polarizability is zero and the disturbance $\hat{\mathbf{V}}_B(\boldsymbol{r})$ is similar to the small-angle case (7.7):

$$\hat{\mathbf{V}}_B(\boldsymbol{r}) = K^2 \sum_n \Omega_n(\boldsymbol{r} - \boldsymbol{R}_n) \left(\chi_0 + \chi_{\boldsymbol{h}} e^{\mathrm{i}\boldsymbol{h}.\boldsymbol{r}} + \chi_{-\boldsymbol{h}} e^{-\mathrm{i}\boldsymbol{h}.\boldsymbol{r}}\right) . \tag{7.15}$$

Using the expression (7.9) for the incoherent cross section of the scattering from a random sample, we obtain after some algebra

$$\left(\frac{\mathrm{d}\sigma}{\mathrm{d}\Omega}\right)_{\mathrm{incoh}} = \frac{N}{16\pi^2} \langle |\chi_{\boldsymbol{h}} \Omega^{\mathrm{FT}}(\boldsymbol{k}_{02} - \boldsymbol{k}_{01} - \boldsymbol{h}) + \chi_0 c_1 \Omega^{\mathrm{FT}}(\boldsymbol{k}_{02} - \boldsymbol{k}_{\boldsymbol{h}1}) +$$

$$+\chi_0 c_2 \Omega^{\mathrm{FT}}(\boldsymbol{k}_{\boldsymbol{h}2} - \boldsymbol{k}_{01}) + \chi_{-\boldsymbol{h}} c_1 c_2 \Omega^{\mathrm{FT}}(\boldsymbol{k}_{\boldsymbol{h}2} - \boldsymbol{k}_{\boldsymbol{h}1} + \boldsymbol{h})\Big|^2\Big\rangle_{\mathrm{size}}. \quad (7.16)$$

The scattered intensity is therefore a coherent superposition of the contributions of four processes (Fig. 7.8). The first process (i.e., scattering from

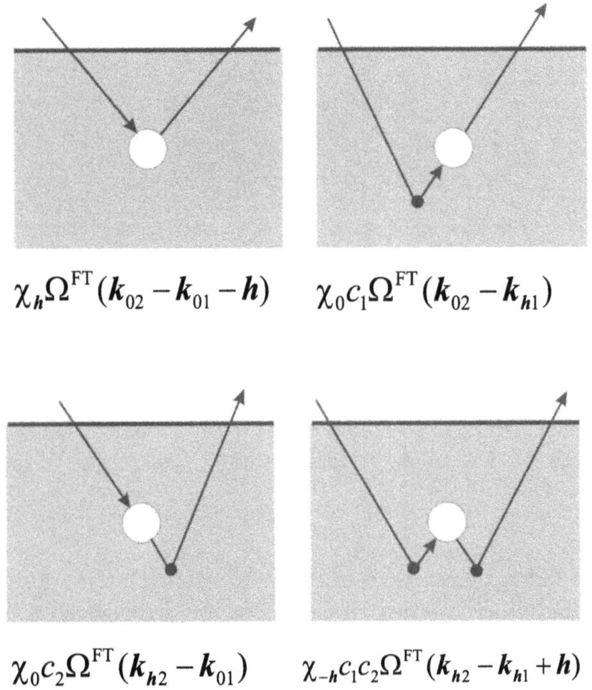

Fig. 7.8. Scattering processes included in Eq. (7.16). The black dots denote the dynamical diffraction in the substrate, the empty circles represent the kinematical scattering by the holes.

$\boldsymbol{k}_{01}$ to $\boldsymbol{k}_{02}$ by the $\chi_{\boldsymbol{h}}$-term in $\hat{\mathbf{V}}_B$) is analogous to the kinematical scattering, with the wave vectors $\boldsymbol{K}_i$ and $\boldsymbol{K}_s$ of the primary and scattered waves; the other processes are dynamical, since they involve (dynamical) diffraction in the substrate.

In Fig. 7.9 we show the reciprocal-space map of the incoherently scattered intensity calculated by Eq. (7.16) for randomly distributed empty holes in GaAs with uniform sizes (radius 50 nm) assuming symmetrical diffraction 004, $\lambda = 1.54$ Å. The map is plotted in the relative coordinates q_x, q_z, $\boldsymbol{q} = \boldsymbol{Q} - \boldsymbol{h}$. The dynamical effects can be observed along the lines $\alpha_i \approx \Theta_B$ and $\alpha_f \approx \Theta_B$ where the diffraction in the substrate occurs. This diffraction produces streaks that deform the intensity distribution. The maximum of the intensity is shifted from the reciprocal lattice point to larger q_z due to refraction. From the kinematical approximation, a circularly symmetric intensity

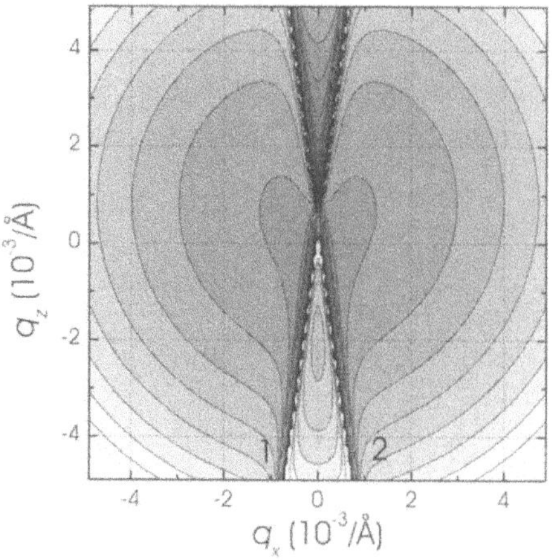

Fig. 7.9. Reciprocal-space distribution of diffusely scattered intensity around the 004 reciprocal lattice point, calculated for a GaAs crystal with randomly distributed empty holes. The oblique lines 1 and 2 correspond to the points in reciprocal space where the incident and the scattered waves obey the diffraction condition in the crystal, respectively. The whole pattern is shifted upward due to the refraction in the crystal.

distribution follows, being centered in the reciprocal lattice point $\boldsymbol{q} = 0$, since both refraction and diffraction in the substrate are neglected. From the figure it is also obvious that the dynamical effects are hard to measure, since they are located exactly in the positions of the monochromator and analyzer streaks (see Chap. 3).

7.2.4 Diffraction from a Thin Layer on a Semi-infinite Substrate

Diffraction from a perfect thin layer on a semi-infinite perfect substrate can be easily calculated using the general or conventional dynamical theory (Chap. 6). In some practical applications, however, it is sometimes suitable to replace the full dynamical formula by a semikinematical expression, where the diffraction in the substrate is dynamical and the layer is so thin that the diffraction in the layer can be assumed to be kinematical. This is the idea of the *semikinematical approximation* [210]. In this section we show how to derive the semikinematical formula using the DWBA method. We choose the same non-disturbed system as in the previous section, of the semikinematical, a semi-infinite crystalline ideal substrate. The two independent solutions of the non-disturbed wave equation (7.2) are shown again in Fig. 7.7. The disturbance comprises the layer above the substrate:

$$\hat{\mathbf{V}}_B(\boldsymbol{r}) = -K^2 \left(\chi_{\boldsymbol{h}L} \mathrm{e}^{\mathrm{i}\boldsymbol{h}_L \cdot \boldsymbol{r}} + \chi_{-\boldsymbol{h}L} \mathrm{e}^{-\mathrm{i}\boldsymbol{h}_L \cdot \boldsymbol{r}}\right) \Omega_{\text{layer}}(z), \tag{7.17}$$

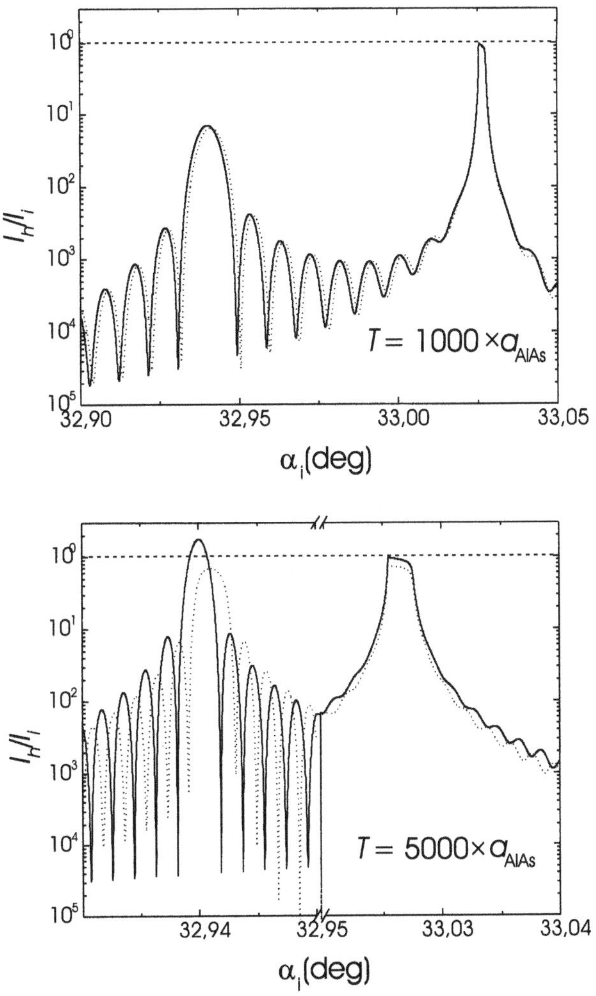

Fig. 7.10. The comparison of the semikinematical diffraction curves (full) with two-beam dynamical theory (dotted); a single AlAs layer on a semiinfinite GaAs substrate, conventional two-beam case, S-polarization, symmetrical 004 diffraction $\lambda = 1.54$ Å. Two different layer thicknesses T were considered.

where $\Omega_{\text{layer}}(z) = 1$ for $z \in [0, T]$ and zero elsewhere, and T is the layer thickness. The subscript $_L$ denotes the quantities characterizing the layer. In contrast to the previous examples, the disturbance $\hat{\mathbf{V}}_B$ is not random and the crystal emits only the coherent wave. Using the expression (7.8) for the coherent scattering cross section and the formula (4.24) for the calculation of the diffracted intensity, we find that the diffracted intensity is a coherent superposition of the contribution of the non-disturbed substrate and the disturbance

$$I_h = \left| E_{h\mathrm{sub}} - \mathrm{i}E_i \frac{K^2}{2K_{iz}} \left[\chi_{hL} S_{\mathrm{layer}}(-2K_{iz} - h_{zL}) + \right.\right.$$

$$\left.\left. + \chi_{-hL} c^2 S_{\mathrm{layer}}(2K_{iz} + h_{zL}) \right] \right|^2 , \qquad (7.18)$$

where $c = E_h/E_0$ is the ratio of the amplitudes in the non-disturbed substrate,

$$S_{\mathrm{layer}}(Q_z) = \int_0^T \mathrm{d}z \mathrm{e}^{-\mathrm{i}Q_z z}$$

is the geometrical factor of the layer, K_{iz} is the vertical component of the wave vector of the primary beam, and $E_{h\mathrm{sub}}$ is the amplitude of the wave diffracted from the substrate, calculated dynamically. In the *ansatz* (7.17) we took into account the diffraction in the layer with the diffraction vectors $\boldsymbol{h}$ and $-\boldsymbol{h}$, but we neglected the χ_0-term in the polarizability of the layer; therefore we neglected the absorption in the layer.

The physical meaning of individual terms on the right-hand side of Eq. (7.18) is explained in Fig. 7.11. The first term describes the dynamical diffraction from the substrate (process (a)), the second term represents the kinematical scattering from the layer (b). The third term corresponds to the following process: the primary wave is dynamically diffracted in the substrate, the diffracted wave is (kinematically) diffracted in the layer with the diffraction vector $-\boldsymbol{h}$, and the resulting wave is diffracted in the substrate again (process (c) in the figure).

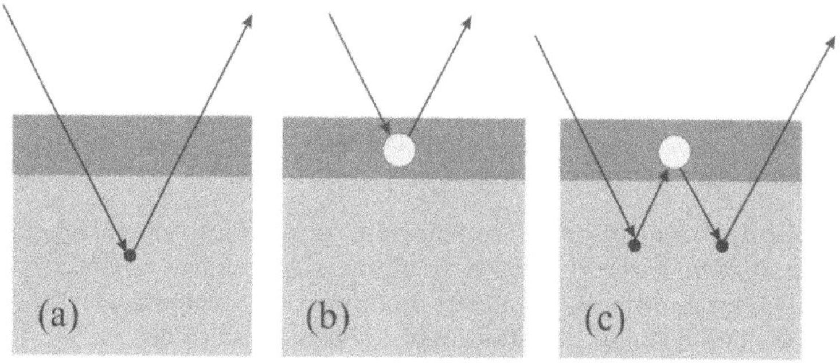

Fig. 7.11. Scattering processes included in Eq. (7.18). The black dots denote the dynamical diffraction in the substrate; the empty circles represent the kinematical diffraction in the layer.

In Fig. 7.10 we show a comparison of the full dynamical calculation and semikinematical calculation of the diffraction curve of a thin AlAs layer on a semi-infinite GaAs substrate, symmetrical 004 diffraction, conventional two-beam diffraction, and S-polarization. For a thinner layer ($T = 1{,}000 \times a_{\mathrm{AlAs}}$)

both curves are very similar; the layer maximum in the semikinematical curve is slightly shifted, since the refraction in the layer is neglected. Here the semikinematical approach is fully applicable, if one includes an empirical refraction correction. This can be done by replacing the arguments of S_{layer} by the refraction-corrected values of K_{iz}. In the case of a thicker layer ($T = 5,000 \times a_{\text{AlAs}}$), the semikinematical approach completely fails; for these layer thicknesses, the full dynamical approach must be performed.

Due to dynamical diffraction effects, the intensity of the transmitted beam is reduced due to the diffraction. This reduction is described by the characteristic length called *extinction length*:

$$\tau = \frac{\lambda}{C}\sqrt{\frac{\gamma_0 \gamma_{\boldsymbol{h}}}{\chi_{\boldsymbol{h}} \chi_{-\boldsymbol{h}}}}. \tag{7.19}$$

A detailed analysis [155] has shown that the diffraction process in a thin layer can be treated kinematically, if the extinction length of the crystal is much larger than the *absorption length*, describing usual photoelectric absorption of the radiation along its entire beam path in the diffracting layer

$$L_{\text{abs}} = \frac{1}{2\text{Im}(q_{zT})},$$

where q_{zT} is the vertical component of the reduced scattering vector corrected to absorption. This condition is fulfilled if the Fourier polarizability coefficient $\chi_{\pm \boldsymbol{h}}$ is small or if the absorption length is extremely large. The former case occurs for disturbed crystals, where the static Debye-Waller factor is small (Chap. 10), the latter case corresponds to grazing-incidence or grazing-exit geometries (see Sect. 9.1).

Part III

Solution of Experimental Problems

An understanding of the physical behavior of crystalline thin films and multilayers and their variation with changes of the layer growth conditions requires knowledge of their structure properties. A detailed structure characterization supports the optimization of the growth conditions and the design of new devices. X-ray scattering methods are non-destructive, with no need for special sample preparation or a special sample environment. Presuming lateral homogeneity, an investigation of any particular sample area provides structure information which is representative of the whole sample structure. All of this makes x-ray methods complementary to locally probing microscopy methods such as transmission electron microscopy, scanning electron microscopy, scanning tunneling microscopy, or atomic force microscopy.

X-ray scattering methods have been successfully employed for the structure characterization of thin crystalline layers, multilayers, and superlattices. Nowadays, they have also been applied to characterize laterally periodic arrays such as surface gratings, quantum wires, and quantum dots. The determination of geometrical dimensions such as the thickness of single layers, the vertical period in superlattices and the lateral period of gratings, and the mean lattice parameters of a layer are today routine tasks in any x-ray laboratory associated to an epitaxy growth group. But much more sophisticated structure characterization has been performed in the last decade, including real structure effects and the study of their statistical properties.

The structure parameters of interest include :

1. the form and perfection of the mesoscopic superstructure, i.e., layer thickness, superlattice period, vertical compositional profile, miscut angle, miscut direction, grating or dot array periodicity, shape of gratings or dots, correlation of statistical dot assemblies,
2. the quality of the surface and interfaces, i.e., graduated heterointerface transition and interdiffusion, surface- and interface roughness, their lateral correlation properties and vertical replication,
3. the crystalline properties, i.e., the elastic lattice distortion and strain relaxation, porosity, and structure defects such as misfit dislocations or clusters of point defects.

In this book we are focused on the investigation of structural properties of vertically layered multilayers and laterally periodic nanostructures.

Part II of this book outlined how the surface and interfaces of a sample break the translational periodicity of the crystal lattice in vertical direction. This has the consequence that the scattered intensity in reciprocal space becomes distributed in the form of crystal truncation rods passing through all the reciprocal lattice points. In pseudomorphic layered systems without any structure defects, the lateral translational symmetry is maintained, and the crystal truncation rods of different layers coincide. In these systems, no intensity between the truncation rods can be found. The concentration of the scattered intensity into the truncation rods expresses the exactly fulfilled lateral (in-plane) diffraction condition.

The intensity distribution along a particular truncation rod reflects the vertical setup of the sample. It is expressed by the product of structure factor $F(\boldsymbol{Q})$ and geometrical factor $S(Q_z - g_z)$ defined in Chapter 5. This distribution is also influenced by multiple scattering effects, especially by x-ray refraction.

In the case of perfect pseudomorph systems, the intensity distribution along a truncation rod can be measured by the use of a double crystal diffractometer or reflectometer with open detector slits, since the direction of (coherently) scattered waves is unambiguously determined by the wave vector of the incident wave (see Chaps. 2 and 3).

The determination of the depth profile of the lattice constant or the polarizability coefficient (or refraction index in the reflectivity case) from the intensity distribution along a truncation rod is not straightforward since the intensity of the scattered radiation is the only accessible quantity and not its phase. The phase problem is a crucial point of the analysis of this kind.

The simplest method for solving this problem is based on a numerical fitting procedure of the measured data to the theoretical diffraction curve calculated dynamically or semikinematically as introduced in Chapters 6 and 7. The actual layer is decomposed into a sequence of thin, homogeneous sublayers, and their parameters (thickness, vertical lattice constant, and the polarizability in the diffraction case, or thickness and refractive index in the reflection case) are determined by a suitable numerical fitting procedure (see [244, 397], for instance).

Another method (*the direct method*) is based on an integral transform analogical to the Kramers–Kronig analysis, which is well-known in conventional optics. The variant of this method suitable for diffraction curves was published in a series of papers [249]. In the reflectivity case, the phase could be retrieved using the Marchenko integral equation [219].

Laterally random or non-random modulations of the layer properties give rise to diffuse x-ray scattering away from the truncation rods. The entire distribution of the scattered intensity in reciprocal space (consisting of coherent and diffuse components) can be measured by a diffractometer (reflectometer) equipped with a directionally sensitive detector, i.e., by a narrow detector slit or an analyzing crystal (a triple axis diffractometer, see Chaps. 2 and 3). Coplanar triple-axis diffractometry resolves the scattered intensity as a function of the wavelength and of the scattering vector $\mathbf{Q}$ in the scattering plane. The accessible region of a coplanar scattering method in the reflection geometry is limited by the conditions $\alpha_i > 0$, $\alpha_f > 0$. The non-coplanar reflection geometry permits the access also in those regions that are forbidden in the coplanar case.

Various x-ray scattering techniques measure the scattered intensities in different regions of reciprocal space, and they use different scattering geometries in real space. Therefore, they give complementary structure information.

Specular x-ray reflection measures the intensity distribution along the truncation rod close to the origin of reciprocal space. Thus, it determines the depth profile of the *mean electron density*.

The *diffraction* methods study the truncation rods crossing reciprocal lattice points far from the origin, yielding information about *crystalline* properties of the sample. *Symmetrical x-ray diffraction* with the diffracting planes parallel to the sample surface is sensitive to the *vertical strain* and the thickness of the crystalline part of the layers. *Asymmetrical diffraction* measures the intensity diffracted by inclined crystallographic planes; it is sensitive both to the lateral and vertical strain components. *X-ray grazing-incidence diffraction* uses a strongly non-coplanar reflection geometry. Therefore, it probes the in-plane diffractions, employing the diffraction planes perpendicular to the surface, and allows one to study the in-plane lattice strain separately. In principle, the same information can be obtained by a symmetrical diffraction in the transmission geometry. This arrangement, however, is less common for thin layer analysis since the diffracted wave penetrates the substrate and it suffers strong absorption.

In all scattering geometries mentioned above, the reciprocal space maps of diffusely scattered intensity can also be measured, which allows us to detect random static fluctuations of the structure properties. In particular, all the scattering geometries using very small incidence and/or exit angles (x-ray reflection, strongly asymmetrical coplanar diffraction, grazing incidence diffraction) are highly sensitive with respect to geometric and chemical fluctuations of the interfaces.

A particular arrangement is the grazing-incidence small-angle scattering (GISAXS) method. It combines x-ray reflectivity at shallow incidence and diffuse scattering in a direction perpendicular to the incidence plane. It is well suited to measuring the correlation properties of statistically distributed surface nanostructures, such as quantum dots. This method is the most effective probe since a 2D detector is used.

In Part III, we discuss the application of various methods to the structure characterization of layered samples. First, we deal with the question of the determination of the mean structure parameters, such as layer thicknesses, mean strains in single layers, and multilayers (Chaps. 8 and 9). In Chapter 10 we discuss the scattering from disturbed layers containing volume defects and interface roughness. Finally, in Chapter 11 we discuss problems related to the investigation of surface and interface roughness, considering the phenomenon of correlated roughness.

8 Determination of Layer Thicknesses of Single Layers and Multilayers

The measurement of layer thickness is a basic problem, and can be solved both by x-ray reflection and x-ray diffraction (see [121] for a review). In both methods, the thickness of a thin layer can be determined from the angular positions of the subsidiary maxima on the reflection (or diffraction) curves.

In a reflectivity curve, these maxima are caused by the interference of the waves reflected from the upper and lower interfaces of the layers. This phenomenon is equivalent to the interference fringes that can be observed with visible light, known as Pohls interference pattern [128]. The visibility of this interference effect depends substantially on the reflectivities of both boundaries, i.e., on the differences in the x-ray refraction indices above and below the boundaries and on the interface roughnesses. In the x-ray region, the latter factor is especially important since, as we show later, even a very fine roughness on the nanometer scale gives rise to a considerable decrease in interface reflectivity.

The range of the layer thicknesses that can be measured by x-ray reflectometry depends on the intensity and divergence of the primary beam, on the angular resolution, and on the total angular range of the goniometer used, as well as on the wavelength λ (see Chap. 2).

As we show later, in the case of a single layer of the thickness T, the distance between the adjacent interference maxima is given by

$$\delta\eta_i = G\frac{\lambda}{2T}, \tag{8.1}$$

where G is a geometry factor, which is unity for x-ray reflectivity. Therefore, the primary-beam divergence and/or the angular resolution of the diffractometer determines the *upper* limit of the measurable thickness T. If, for instance, the divergence of the primary beam is $0.01°$ and $\lambda = 0.15405$ nm ($\mathrm{CuK}\alpha_1$ line), the maximum measurable layer thickness is smaller than about 0.43 μm. The lower limit for thickness analysis is given by the accessible angular range, i.e., in fact, by the maximum incidence angle α_i that yields a measurable reflectivity. Therefore, the minimum layer thickness which can be determined, depends on how many decades of intensity are accessible by the experiment. For instance, the determination of a layer thickness of 1.5 nm requires measurements up to $\alpha_i = 3°$ at least.

The subsidiary maxima on the diffraction curve of a layered sample can be explained as a result of interference of the beam diffracted by the layer (or layers) with the beam diffracted by the substrate. The distance of the adjacent maxima depends on the layer thickness according to a formula similar to Eq. (8.1), where the value of the geometrical factor G can differ from unity depending on the diffraction asymmetry. The scattering contrast of the interference maxima depends mainly on the difference between the polarizability coefficients χ_h of the layer and the substrate and on the lattice mismatch between layer and substrate. For the latter case the thickness determination is not straightforward and requires computer simulation.

In this chapter we will describe the possibilities for determining the layer thickness in single-layer and multilayer structures by x-ray reflectometry and diffraction measurements. On the basis of the general theory formulated in Sect. 6.5, we will demonstrate the dependence of the positions of the intensity maxima on the reflection (diffraction) curves on the layer thicknesses and we will discuss the influence of the inhomogeneities of the layer thickness on these curves.

8.1 X-Ray Reflection by Single Layers

From general dynamical formulae (6.14), (6.39) we can derive the following expression for the reflectivity of a single layer deposited on a semi-infinite substrate:

$$\mathcal{R} = \left| \frac{r_1 + r_2 \mathrm{e}^{-2\mathrm{i}k_{0z}T}}{1 + r_1 r_2 \mathrm{e}^{-2\mathrm{i}k_{0z}T}} \right|^2 , \tag{8.2}$$

where $r_{1,2}$ are the Fresnel reflectivity coefficients of the free surface and the substrate interface, respectively, k_{0z} is the vertical component of the wave vector of the beam transmitted through the layer, and T is the layer thickness. From this formula it follows that in an angle-dispersive experiment the intensity maxima appear whenever $\exp(-2\mathrm{i}k_{0z}T) = 1$, this means at angle positions α_{im}. This condition can be expressed by

$$2T\sqrt{\sin^2 \alpha_{im} - \sin^2 \alpha_c} = m\lambda, \tag{8.3}$$

where m is an integer, $\sin \alpha_c = \sqrt{2(1-n)}$ and α_c is the critical angle of total external reflection of the layer and n is the layer refractive index. Eq. (8.3) is analogous to the Bragg equation but modified by the influence of refraction. The appearing thickness fringes are called *Kiessig fringes*, in honor of their discoverer [193].

Since, in most cases, the incidence angle α_i is sufficiently small, Eq. (8.3) has the following approximative form:

$$\alpha_{im}^2 - \alpha_c^2 = m^2 \left(\frac{\lambda}{2T} \right)^2 . \tag{8.4}$$

This relation shows a simple method to determine the layer thickness from the measured reflectivity curve. One plots the squares of the angular positions of the intensity maxima versus the squares of the Kiessig fringe order. In the range of validity of Eq. (8.4) it gives a straight line with the layer thickness T as parameter. From the intersection point of this straight line with the ordinate one obtains the critical angle α_c of the layer material, and, consequently its refractive index.

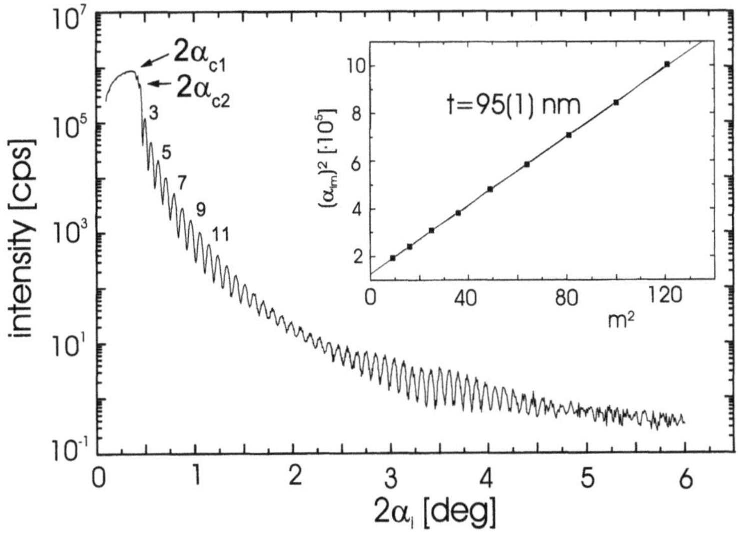

Fig. 8.1. X-ray reflectivity curve of BN coated onto silicon substrate recorded as a function of the detector angle $2\alpha_i$. α_{c_1} is the critical angle of the layer; α_{c_2} that of the substrate. The numbers denote the fringe order m. The inset shows the plot α_i^2 versus m^2, which gives a layer thickness of $T = 95 \pm 1$ nm.

Figure 8.1 shows a reflectivity curve of a BN layer deposited on a silicon substrate. It was measured by means of a powder x-ray diffractometer introduced in Chapter 2.4 using $\lambda = 0.154$ nm. The reflected intensity is recorded over six orders of magnitude. This corresponds to $2\alpha_i \leq 6.0°$.

For intensity reasons and to improve the angular resolution, the low angle region between $0 < 2\alpha_i < 2\alpha_c$ was measured with the highest angular resolution possible, what is determined by a step width of $\delta\alpha_i = 0.001°$ and a width of the incident beam of 0.05 mm. A counting time of 2 seconds per angular step was sufficient for good counting statistics. For larger α_i the slit width and the counting time were increased to 0.5 mm and 30 to 60 seconds, respectively. As is visible in Fig. 8.1 the intensity increases slightly for $\alpha_i < \alpha_c$ and drops very rapidly if α_i exceeds α_c. The first dependence is governed by the *illumination correction* (see later).

Beyond α_c the reflected intensity is proportional to α_i^{-4} as follows from the kinematical formula (5.17). This drop is modulated by the interference of the x-ray beam reflected at the upper and lower boundaries of the layer. Furthermore, there are two different frequencies of oscillations. The high frequency is a measure of the thickness of the sputtered BN layer, and the low frequency is that of the native SiO_2 covering the silicon substrate. The layer thickness T of the BN is obtained from the angular distance between the oscillation maxima according to (8.4). This is demonstrated in the inset of Fig. 8.1 using the third to eleventh Kiessig maximum of the reflection curve. Its graphical evaluation gives $T = 95 \pm 1$ nm. The extrapolation to $m = 0$ gives $\alpha_{c1}^2 \approx 10^{-5}$, which represents a rough estimate of the average electron density of the layer. α_{c2} corresponds to the density of the silicon substrate (see below). Extracted from the long-range beating of the reflectivity curve, the thickness of the SiO_2 layer amounts to 3.4 ± 0.4 nm. Note that this layer becomes visible only if the reflectivity curve has been recorded over more than five orders of magnitude.

Expressed in reciprocal space, Eq. (8.3) looks much simpler:

$$T = \frac{2\pi}{\Delta Q_{zT}}. \tag{8.5}$$

That means T can be measured from a difference of the scattering vectors *inside* the crystal (i.e., corrected for refraction).

The accuracy of the thickness determination depends on the smallest angular step $\delta\alpha_i$ of the goniometer and on the layer thickness T. Neglecting refraction the accuracy can be estimated from

$$\frac{\Delta T}{T} = \frac{\Delta\alpha_i}{\alpha_i} \approx \frac{\alpha_i}{m_{\max}}. \tag{8.6}$$

This accuracy is of the order of 1% if the oscillation maximum measured at $\alpha_i = 1°$ is determined with an accuracy better than $\Delta\alpha_i = 0.01°$. Eq. (8.6) can be expressed also in terms of the largest fringe order $m_{\max}$ that is detected in the reflectivity curve with an accuracy of one-half of a fringe period. In the example shown in Fig. 8.1 one finds $m_{\max} = 45$ at $2\alpha_i \approx 5.0°$. In this case the layer thickness t is determined with a relative error of $\frac{\Delta T}{T} \approx 2\%$.

The accuracy of the layer thickness can be preserved as long as a sufficient number of fringe maxima appear within the detectable angular interval, i.e., if T is sufficient large. Owing to the α_i^{-4} dependence, the reflectivity of the silicon substrate decreases to $\mathcal{R} = 2 \times 10^{-4}$ at $2\alpha_i = 2°$ and to $\mathcal{R} = 5 \times 10^{-5}$ at $2\alpha_i = 3.0°$. Considering the low counting statistics at large angles, the thickness cannot be estimated with an accuracy better than 1% in practical cases. A dynamical range of up to ten orders of magnitude is required in order to detect one single fringe period corresponding to the thickness of a single atomic layer ($T \approx 0.3$ nm). Such dynamical range cannot be realized under common laboratory conditions, it requires synchrotron radiation . Nowadays, a dynamic range of seven to eight orders of magnitude is available using modern home laboratory equipment (see Sect. 2.1).

However, using a low-power x-ray source, a rough estimate of the layer thickness of a very thin layer can be determined exploiting the small-angle part of the reflectivity curve, in particular the angular position of the first oscillation minimum [382].

The electron density of the material can be determined by measuring the critical angle of total external reflection α_c. From theory (see Chap. 6) one would expect the reflecting intensity to remain constant between $0 < \alpha_i < \alpha_c$. That is not the case in experiments: as seen in Fig. 8.1, the intensity increases within this angular range. For a given beam width, b_{beam}, and a very small α_i, the projection of the incoming beam onto the sample surface, A_{beam}, can exceed the sample size, A_{sample} (see Fig. 8.2) . Under this condition the recorded intensity depends on the ratio $b_{\text{sample}}/b_{\text{beam}}$ and has to be corrected by

$$\begin{aligned} I &= I_{\text{meas}} \cdot \sin(\alpha_i) \quad \text{for } \tfrac{A_{\text{sample}}}{A_{\text{beam}}} < 1 \\ &\text{and} \\ I &= I_{\text{meas}} \qquad\qquad\quad \text{for } \tfrac{A_{\text{sample}}}{A_{\text{beam}}} \geq 1 \,. \end{aligned} \tag{8.7}$$

The particular angle α_i, where $A_{sample}/A_{beam} = 1$, depends on the sample size and the slit width b_{beam} defining the beam in front of the sample. Both parameters have to be defined for each sample under investigation. A correct

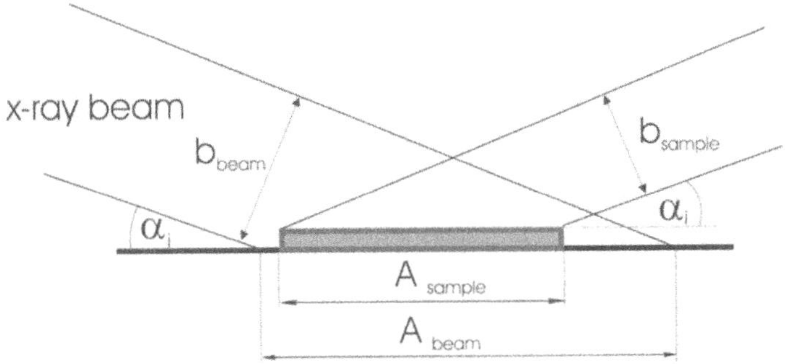

Fig. 8.2. Illumination of a terminated sample area while scanning the reflectivity at very small α_i.

determination of α_c is not straightforward. As long as absorption is negligible and the sample is infinitely large, α_c is that value of α_i where the reflecting intensity I_c is decreased to 50% compared of the maximum intensity $I_{\text{max}} = 1$. In this case I_{max} corresponds to the incident beam intensity I_0 measured at $\alpha_i = 0$. For finite-sized samples and highly absorbing materials I_{max} is always smaller than unity and α_c appears at an intensity smaller than 50% (see Eq.

(8.7)). This problem becomes significant if the electron density of the layer is lower than that of the substrate and if the layer is thin. Then two critical angles may appear: one belongs to the layer and a second one, at slightly larger α_i, corresponds to the substrate. This has already been illustrated in Fig. 8.1.

Generally the average electron density ϱ_{el} can be determined using the relation

$$\alpha_c = \sqrt{-\chi_0}, \tag{8.8}$$

which results in

$$\varrho_{\mathrm{el}} = \frac{\pi \alpha_c^2}{\lambda^2 r_{\mathrm{el}}}. \tag{8.9}$$

Instead of ϱ_{el} the mass density ϱ_{m} is often of interest. These two densities are connected by

$$\varrho_{\mathrm{m}} = \frac{\varrho_{\mathrm{el}} A}{N_A Z}, \tag{8.10}$$

where r_{el} is the electron radius defined in Sect. 5.1, Z is the atomic number, A is the mass number and N_A is the Avogadro constant. Figure 8.3 shows three

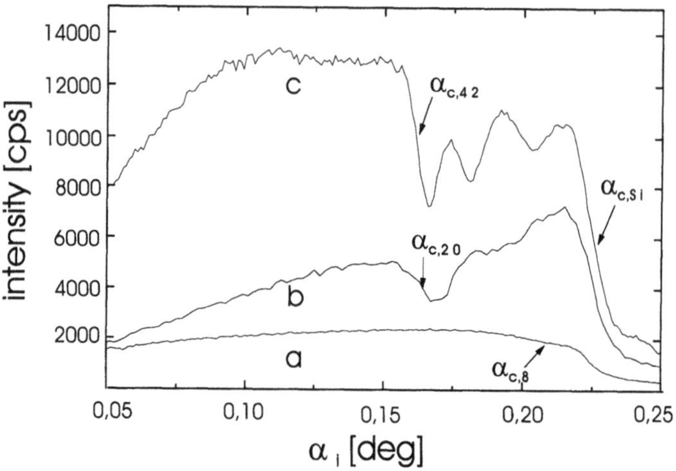

Fig. 8.3. The angular range of total external reflection, recorded for three different organic films made of from fatty acid salt molecules coated onto a silicon support by means of the Langmuir-Blodgett technique. As can be seen here, the critical angle of film decreases as the number of layers increases due to the increased number of structural defects within the film.

reflectivity curves of organic films made of different numbers of monolayers coated onto a crystalline silicon support. Besides the critical angle of silicon at 0.22°, there is a second α_c which belongs to the organic film. This smaller critical angle decreases with an increase in the number of monolayers, due to the increasing number of defects within the layer. For the 20–monolayer sample, for example, $\alpha_c = 0.18°$ corresponds to an electron density of $\varrho_{\mathrm{el}} = 4.6 \times 10^{23}\ \mathrm{cm}^{-3}$, i.e., a mass density of $\varrho_{\mathrm{m}} = 1.54\ \mathrm{gcm}^{-3}$. The density values of the silicon substrate are $6.99 \times 10^{23}\ \mathrm{cm}^{-3}$ and $2.32\ \mathrm{gcm}^{-3}$, respectively. A density determination by eye is not possible if the layer density is close to that of the substrate or if the layer is very thin. The latter reason is evident in the bottom curve of Fig. 8.3. Here, the layer density can only be extracted using computer simulation. In that particular example, the decreasing density of the layer is caused by the incomplete layer coverage on the substrate which decreases with the number of transferred layers [353].

For an approximate determination of ϱ_{el}, we recommend measuring the reflectivity curve in the angle range $0 \leq \alpha_i \leq 1.5 \times \alpha_c$ using the smallest possible step width of the goniometer $\delta\alpha_i$ and find α_c at the angle position where $I(\alpha_i) = I_{\mathrm{max}}/2$. Using $\delta\alpha_i \sim 0.001°$, the accuracy of the density determination may be estimated as

$$\Delta\varrho/\varrho = 2\frac{\delta\alpha_i}{\alpha_c} \sim 0.01\ , \tag{8.11}$$

which is sufficiently precise for many technological applications. This procedure works well if the rotational axis of the sample circle is aligned exactly at the sample surface (see Sect. 2.1).

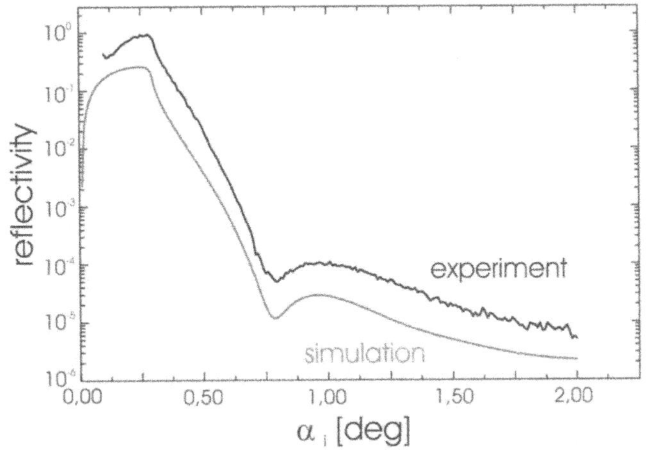

Fig. 8.4. Experimental and fitted x-ray reflectivity curves of a thin antimony layer on GaAs (110) substrate. The experiment cannot be explained assuming a single-layer model.

The following examples will illustrate some problems one may encounter while studying extremely thin layers. Figure 8.4 shows the reflectivity curve of a thin antimony layer grown epitaxially on GaAs (110) . This curve was recorded using a reflectometer with low angular resolution. It demonstrates the limit of layer thickness estimation made by eye. The electron density of antimony is about 15% larger than that of GaAs. Therefore the critical angle of the layer is larger than that of the substrate and it is not visible. At larger angles a single fringe minimum and maximum are visible above the background, the shape of the oscillation is asymmetric. A complete fit of the reflectivity curve which considers the experimental resolution function gives $T_{Sb} = (4.0 \pm 0.5)$ nm, an interface roughness of $\sigma = 0.5$ nm and a refraction index $n \approx 1 - \delta$ with $\delta_{Sb} = 1.65\ 10^{-5}$. The interface roughness was treated according to Sect. 11.3. Additionally one has to consider a second layer with slightly reduced density ($\delta = 1.05\ 10^{-5}$) on the top of the antimony. Its thickness is about $T = (2.8 \pm 0.5)$ nm, and it corresponds to microcrystalline aggregates caused by the transition of the two-dimensional into the three-dimensional growing mode during preparation.

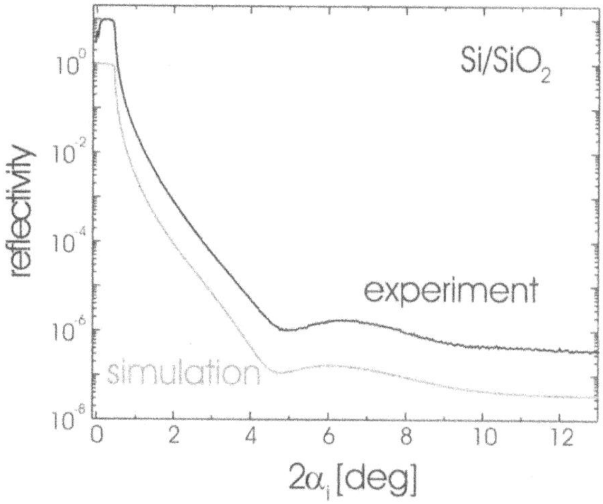

Fig. 8.5. Experimental simulated reflectivity curves of a silicon surface measured with the use of a home reflectometer similar to that shown in Fig. 2.1. The enhanced dynamical range of the experiment enables us to determine a thin top layer of native oxide. The thickness of the native oxide is 1 nm with a surface roughness of about 0.35 nm. The fit is possible only considering a gradual increase of the electron density from oxide to the pure silicon.

This exact data evaluation is in contrast to a rough estimate by eye. Here one can suppose a single-layer model. Using the fringe minimum at $\alpha_i = 0.76°$, one get a thickness of $T = (9 \pm 1)$ nm, which is larger than the sum of both layers determined above. Naturally this model does not reproduce the observed fringe asymmetry [144].

The lower limit for the determination of a thin surface layer can be estimated measuring the native oxide of a silicon wafer. Figure 8.5 shows the reflectivity curve of a *clean* silicon surface measured with a home reflectometer similar to Fig. 2.1 using $\lambda = 0.154$ nm. The experimental curve is quite similar to that one which can be measured with synchrotron radiation [167, 356]. Only the large dynamical range of about eight orders of magnitude makes it possible to identify the native oxide. The measured angular position of α_c corresponds to the silicon mass density of $\varrho_{\mathrm{m}} = 2.32$ g/cm^3. At higher α_i the intensity decrease is modulated due to the existence of a very thin surface layer. At the angular position of the destructive interference the reflecting intensity is about 10^{-7}. Under circumstances of a limited dynamical range the reflection curve would probably have been misinterpreted by a clean surface only. Here one clearly can identify the existence of the native oxide. The minimum at $2\alpha_i \approx 4.8°$ corresponds to a thickness of $T_{\mathrm{top}} = 1.0$ nm. The full fit of the reflectivity curve supplies additional parameters, i.e., the mass density of the top layer($\varrho_{\mathrm{m}} = 1.7$ g/cm^3) and the interface roughnesses of the SiO_2 surface and the SiO_2-Si interface, which are $\sigma_{\mathrm{Si}} = 0.15$ nm and $\sigma_{\mathrm{SiO_2}} = 0.35$ nm, respectively. Furthermore the fit requires consideration of a gradual change of the density from the top layer down to the pure silicon. This reflects the property of SiO_2 to protect the silicon against further oxidation.

After the substrate has been characterized, the layers on top of it can be investigated. This can be a thin organic film, as shown in Fig. 8.6. The layer consists of lipids(l-1,2-dipalmitoylphosphatidic acid – DPPA) attached to polyelectrolyte molecules (poly-diallyldimethylammonium chloride – PDADMAC). Both have been transferred onto a silicon substrate by means of the Langmuir-Blodgett technique. The main problem here is the low density difference between the molecular sub-units. Both lipids and polyelectrolytes consist of carbon and hydrogen atoms. The only difference is the molecular arrangement which is laterally ordered in the case of the lipids but rather random for the polyelectrolyte molecules. The reflectivity curve has to be recorded over eight orders of magnitude to yield sufficient structure information (Fig. 8.6). As shown in the inset, the data evaluation does not result in a unique electron density distribution. Assuming either a two-layer or a four-layer model, one cannot decide whether the polyelectrolytes built the sub-layer with larger or smaller thickness compared to the lipid layers [271]. This ambiguity is a consequence of the *phase problem* of crystallography.

Similar information can be obtained using the energy-dispersive set-up (see Sect. 2.1). Instead of the angular coordinate the intensity varies as a

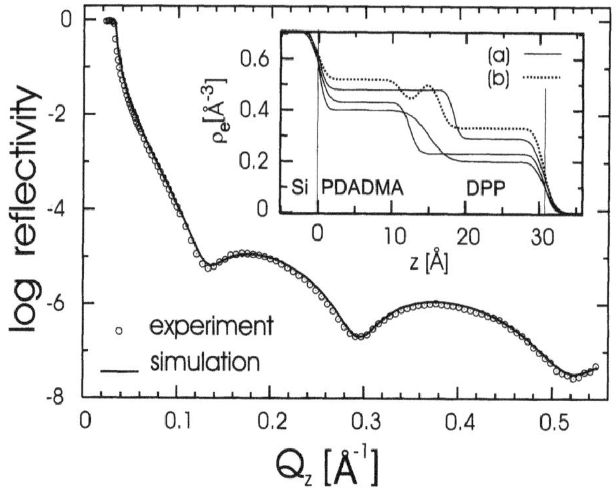

Fig. 8.6. X-ray reflectivity curve of a lipid monolayer attached to a polyelectrolyte molecule. Both are covered onto silicon support. Although the fit to the experimental curve is perfect there is an ambiguity with respect to the correct electron density distribution. This problem cannot be solved without additional structural information. The inset displays the fitted vertical density profile, (a) two box model, (b) four box model.

function of energy at fixed α_i. Figure 8.7 shows reflectivity spectra taken from a lipid monolayer of DPPA spread onto a water surface. The experiment has been performed onto a Langmuir through installed at the sample position shown in Fig. 2.3 at an energy-dispersive beamline. The incoming beam is reflected first at two super-mirrors (see Fig. 1.12) using an incident angle of $\alpha_i = 0.25°$ in each case. This provides an incidence angle of 2° with respect to the water surface. The reflectivity spectra mainly reflect the reflectivity of the super-mirror, which gives an almost uniform intensity up to about 16 keV, multiplied with the incident spectrum of the storage ring. There is a distinct difference in reflectivity between the spectra taken from the pure water surface and from the film on water. After division of the film spectrum by the water spectrum one can clearly identify one maximum and one minimum between $4 \leq E \leq 13$ keV. Both change as a function of the applied lateral surface pressure π. The thickness of $T = 3.2$ nm at $\pi = 40$ and 29 mN/m corresponds to a phase where the molecules stand upright with respect to the water surface. At $\pi = 11$ mN/m, the minimum and maximum shift toward higher energies. The respective thickness of $T = 2.9$ nm corresponds to a phase of tilted molecules. Each spectrum was recorded for 300 seconds. This time is sufficient to observe in-situ phase transitions of various amphiphilic molecules on the water surface as a function the applied pressure.

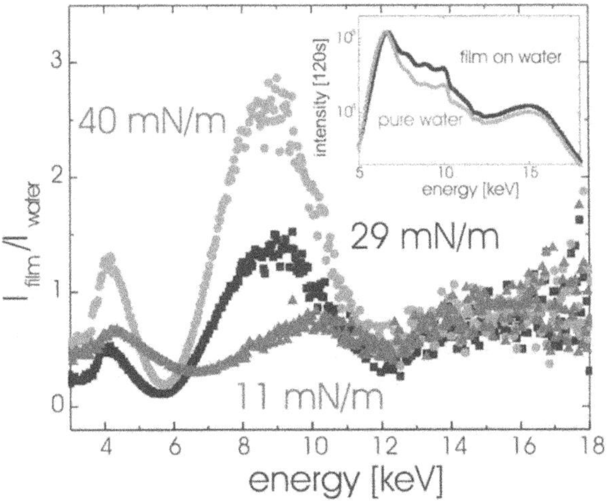

Fig. 8.7. Energy-dispersive reflectivity spectra taken from a lipid monolayer onto the water surface. The experiment is performed on a Langmuir trough when a lateral pressure π can be applied to the molecules spread onto the water surface. The reflectivity spectrum (see inset) mainly corresponds to that of the super-mirror shown in Fig. 1.12. The main figure displays the normalized spectra taken at three different values of lateral pressure. The first minimum and second maximum of the monolayer reflectivity is visible. Their positions at $\pi = 40$ and 29 mN/m correspond to a phase where the molecules stand upright with respect to the surface; $\pi = 11$ mN/m reflects a phase of tilted molecules.

8.2 X-Ray Reflection by Periodical Multilayers

Up to now, we have dealt with systems containing one or two layers. We have demonstrated that the parameters of the system consisting of a single layer on a substrate can be estimated from the measured reflectivity curve by eye. A simple analysis of the experimental reflectivity curve is possible if the sample consists of a periodical stack of layers (a periodical multilayer).

The x-ray reflectivity of a periodical multilayer can be calculated using the kinematical theory, or, more exactly, using Eq. 6.37 in the dynamical theory presented in Chap. 6. In many cases, the single-reflection approach (SRA) is quite sufficient. In this approach we neglect multiple reflections from different interfaces within the multilayer, and for the reflectivities of the interfaces we use the exact dynamical expressions (Fresnel coefficients – see Eq. (6.14)). In the following, we analyze the SRA formula in order to discuss some characteristic features of the reflectivity curve of a periodical multilayer.

Let us assume a multilayer being created by N periods, each consisting of a layer A with a thickness T_A and the refraction index $n_A = 1 - \delta_A$ and the layer B (T_B, $n_B = 1 - \delta_B$); the multilayer period is $D = T_A + T_B$. We denote the appropriate phase factors of layers A and B by

$$\Phi_S = \mathrm{e}^{-ik_z^S T_S}, \quad S = A, B,$$

where k_z^S is the z-component of the wave vector of the transmitted wave in the layer of type S. For the Fresnel reflection coefficients of the A–B and B–A interfaces, the relation

$$r_{BA} = -r_{AB}$$

holds, i.e., the amplitude of the reflection originating from the interface A–B is opposite that of the interface B–A. Using the matrix expression (6.37) and neglecting all the terms containing the second and higher powers of the Fresnel reflectivities, the reflectivity of the periodical multilayer is

$$\mathcal{R} = \left| r_{0A} + r_{AB}\left[\Phi_A^2 - \Phi_A^2\Phi_B^2 + \Phi_A^2\Phi_B^2\Phi_A^2 - \cdots \right.\right.$$
$$\left.\left. \cdots + (\Phi_A^2\Phi_B^2)^{N-1}\Phi_A^2\right] + r_{BS}(\Phi_A^2\Phi_B^2)^N\right|^2, \tag{8.12}$$

where r_{0A} and r_{BS} are the Fresnel reflection coefficients of the free sample surface (interface between the vacuum and layer A) and the substrate surface (interface between layer B and the substrate). The sum (in the square brackets) can be evaluated, and we obtain

$$\mathcal{R} = \left| r_{0A} + \frac{r_{AB}\Phi_A^2\Phi_B^2(\Phi_A^2-1)(\Phi_A^2\Phi_B^2)^{N-1}+\Phi_B^2-1}{(\Phi_A\Phi_B)^2-1} + \right.$$
$$\left. + r_{BS}(\Phi_A^2\Phi_B^2)^N\right|^2. \tag{8.13}$$

Using the SRA it is straightforward to derive parameters which characterize the multilayer structure. Several of these parameters can simply be extracted from the experimental reflection curves and can be used as an input for the fitting of the experimental reflection curves by means of full dynamical theory according to Eq. (6.37).

First, let us consider the second term on the right-hand side of formula (8.13). A maximum of this term occurs if

$$(\Phi_A\Phi_B)^2 = 1,$$

i.e., for

$$k_z^A T_A + k_z^B T_B = \pi m,$$

where m is an arbitrary integer. Now we introduce the averaged z-component of the wave vector:

$$\langle k_z \rangle = \frac{k_z^A T_A + k_z^B T_B}{D},$$

making an angle $\langle\alpha_t\rangle$ with the internal surface normal. The condition for a reflectivity maximum is

$$2D\langle n\rangle \sin\langle\alpha_t\rangle = m\lambda, \tag{8.14}$$

where $\langle n\rangle$ is the average refractive index of the multilayer or, using the angle of incidence,

$$2D\sqrt{\sin^2\alpha_i - \sin^2\langle\alpha_c\rangle} = m\lambda. \tag{8.15}$$

This formula is equivalent to Eq. (8.3) for a single layer; but in (8.15) the critical angle of total external reflection $\langle\alpha_c\rangle$ depends on the refraction index averaged over the multilayer period.

As in the case of a single layer, the modified Bragg law can be simplified if the angles are sufficiently small:

$$\alpha_{im}^2 - \langle\alpha_c\rangle^2 = m^2\left(\frac{\lambda}{2D}\right)^2. \tag{8.16}$$

Formulas (8.14) and (8.15) represent the modified Bragg law; and, consequently, optical reflection from a periodical multilayer can be interpreted as a *diffraction* from a one-dimensional crystal. The Bragg equation (8.15) is corrected by the refraction of x-rays in an averaged medium that replaces the actual multilayer structure. The reflectivity maxima can be considered as satellite maxima close to the reciprocal lattice point 000.

If one neglects the refraction, the distance of the satellite maxima can be approximated to

$$\Delta\alpha \approx \frac{\lambda}{2D},$$

which similar to (8.1).

The intensity of the satellite maxima are influenced by the thicknesses T_A and T_B of the layers in the period. The envelope curve of these maxima is described by the structure factor of the one-dimensional crystal, i.e., the multilayer period that, in the case of reflection, has the form

$$F_{\mathrm{period}}(G) = \int_{-D}^{0} \mathrm{d}z\chi_0(z)\mathrm{e}^{-\mathrm{i}Gz} = \frac{\mathrm{i}}{G}(\chi_{0B} - \chi_{0A})\left(\mathrm{e}^{-\mathrm{i}GT_A} - 1\right), \tag{8.17}$$

where $G = \frac{2\pi m}{D}$ is the value of Q_z in the m-th satellite. Like the diffraction case already explained in Chapter 5, the m-th satellite peak vanishes, if the layer thicknesses $T_{A,B}$ obey the following relation:

$$m = p\left(\frac{T_A}{T_B} + 1\right), \tag{8.18}$$

where p is an integer. For instance, every fourth satellite maximum vanishes if $T_A/T_B = 3$.

Now, let us investigate the first and the third terms in Eq. (8.13). These terms provide a maximum of the reflectivity if $(\Phi_A \Phi_B)^2 = 1$; i.e., a maximum occurs for the angles $\langle \alpha_t \rangle$ given by the relation

$$2ND\langle n \rangle \sin\langle \alpha_t \rangle = p\lambda, \tag{8.19}$$

where p is an integer. Neglecting refraction, the angular spacing between these maxima,

$$\Delta\alpha \approx \frac{\lambda}{2ND},$$

is inversely proportional to the *total thickness* $T = ND$ of the multilayer stack. The nature of these maxima (Kiessig fringes) is obvious. They are caused by the interference of the waves reflected at the free surface and at the substrate interface. Simple consideration shows that $N-2$ Kiessig fringes occur between two neighboring satellite maxima. Often the Kiessig fringes are not visible due to lateral sample inhomogeneities.

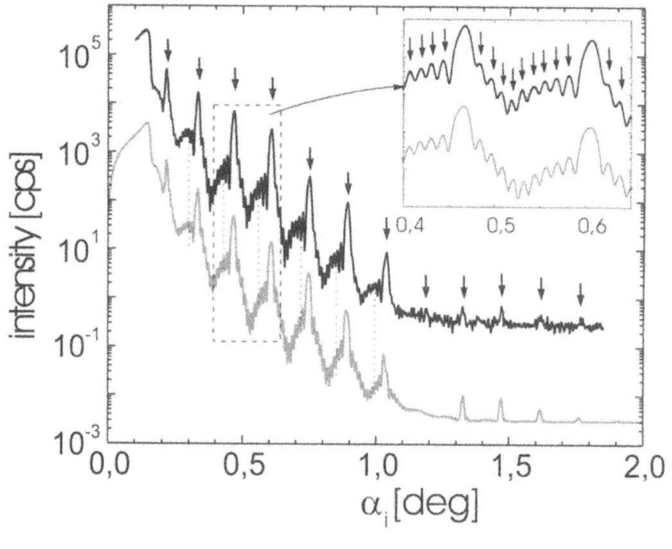

Fig. 8.8. Reflection curve of a SiGe/Si multilayer covered by 210-Å-thick cap layer, CuKα radiation. The satellite maxima are denoted by *vertical arrows*, the maxima stemming form the capping layer are denoted by *vertical dotted lines*. In the inset, the Kiessig fringes corresponding to the total thickness of the multilayer are denoted by *arrows*.

As an example, we show the measured reflection curve of a SiGe/Si multilayer (Fig. 8.8) covered by a Si capping layer with thickness T_C. On the experimental curve, three types of maxima can be resolved:

1. Satellite maxima (indicated by vertical arrows in the main part of Fig. 8.8, whose angular spacing depends on the multilayer thickness D according to (8.15).
2. Kiessig fringes (indicated by the vertical arrows in the inset). Their period depends on the total multilayer thickness $T = ND + T_C$ according to (8.19).
3. Maxima indicated by vertically dotted lines correspond to the thickness T_C of the capping layer.

Knowing the positions of the maxima of these types, we can estimate the corresponding thicknesses using the modified Bragg law in Eqs. (8.15) and (8.19).

Similar to the treatment shown in Fig. 8.1, we have plotted the square of the angular positions of the respective maxima versus the m^2 and obtained the thicknesses $D = (20.5 \pm 0.3)$ nm, $T = ND + T_C = (232 \pm 5)$ nm, and $T_C = (21 \pm 2)$ nm.

These values can serve as starting estimates for the numerical fitting of the whole measured curve using the dynamical theory presented in Sect. 6.4. The result of the fit procedure also is shown in Fig. 8.8. In order to obtain a good correspondence between the measured and calculated curves, we had to assume an oxide layer on top of the multilayer stack (having the thickness T_{ox}). From the fit we obtained the thicknesses of the individual layers as well as the average root mean square roughness σ of their interfaces. The fitting procedure was almost insensitive to the Ge concentration x in the SiGe layers. The fit yielded the following values: $T_{ox} = (3 \pm 1)$ nm, $T_C = (21 \pm 0.5)$ nm, $D = (20.6 \pm 0.2)$ nm, $T_A/T_B = 7.0 \pm 0.2$, $x = 0.35 \pm 0.15$, and $\sigma = (0.7 \pm 0.1)$ nm. The interface roughnesses were considered using the formalism presented in Sect. 11.2.

We can see that the estimates of the layer thicknesses from the positions of the reflectivity maxima nearly coincide with the more reliable values obtained by the numerical fit to the whole curve. The thickness of the additional oxide layer, however, could be estimated with an relative error of only about 30%, because no respective intensity maxima could be identified within the angular range of the measurement.

Figure 8.9 displays the reflectivity curve of a vanadium/mica multilayer sputtered onto a sapphire substrate measured at a wavelength of $\lambda = 0.139$ nm. Due to the huge difference of the electron densities between both constituents the reflectivity at the first-order Bragg peak is close to unity. Thus the multilayer can be used as broad band monochromator for synchrotron radiation use. The accepted band pass depends on the peak width, i.e., the number of coated double layers. In the present case there are 40 periods, which can be verified by the 38 Kiessig oscillations measured between two neighboring Bragg peaks (see inset of Fig. 8.9). The multilayer period amounts to 3.5 nm. The reflectivity curve could be recorded over nine orders of magnitude. The 7th-order Bragg peak appears at $\alpha_i \approx 9°$. Using

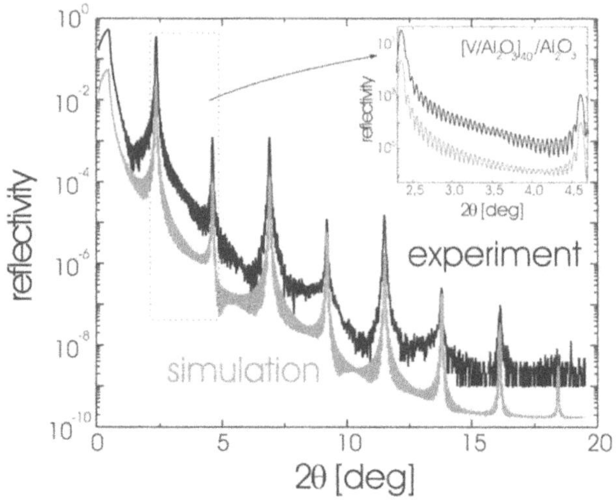

Fig. 8.9. Reflectivity curve of a V/Al_2O_3 multilayer film coated on mica. The experiment has been performed at $\lambda = 0.139$ nm using synchrotron radiation. The reflectivity could be recorded over nine orders of magnitude. The inset shows the angular range between the 1st- and 2nd-order Bragg peak [247].

Eq. (8.6) this corresponds to a relative error of about 2%. The thicknesses of the vanadium and mica layers have been determined by curve simulation and amount to 1.61 nm and 1.87 nm, respectively. The interface roughnesses were determined to be 0.24 nm and 0.18 nm. The substrate roughness amounts to 0.17 nm.

The following example shows a typical reflectivity curve of an organic film. It consists of 28 cadmium–behenate monolayers transferred onto silicon support by means of the Langmuir–Blodgett technique. The behenic acid molecules are amphiphilic in nature. They consist of a hydrophilic COO^- head and a $(CH_2)_nCH_3$ tail. One Cd^{2+} ion is attached to two molecular head groups. This is the reason that a single period of the multilayer always consists of two monolayers of upright standing molecules where the head groups are coupled via the Cd^{2+} ion. Figure 8.10 shows the respective reflectivity curve measured by a powder diffractometer and Cu$K\alpha$ radiation. There are two types of periodic maxima: the main satellites measure the period thickness. Due to the large resonant diffuse scattering (see Chap. 11) which appears in addition to the coherent scattering, these small-angle Bragg peaks are visible over a large angular range. In the present case they appear up to the 14th-order. The multilayer period could be determined as $D = 6.020 \pm 0.006$ nm. The inset of the figure shows the evaluated electron density profile. The peculiarity of the multilayer consists in the large density difference between the

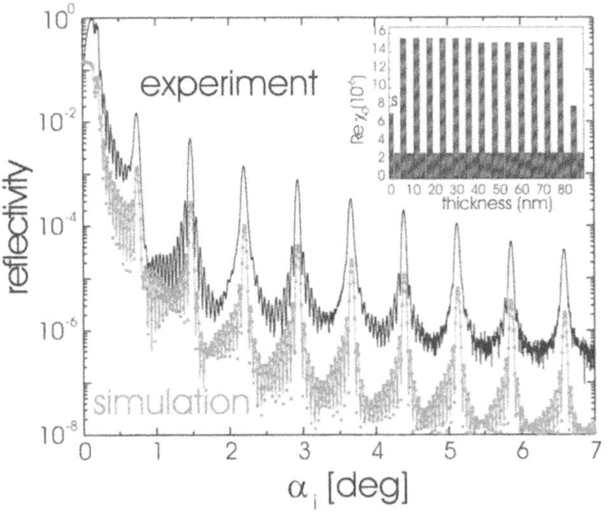

Fig. 8.10. Reflection curve of a 29-monolayer Cd-behenate film deposited on silicon support, measured with CuKα radiation. There are two distinguished series of satellite maxima: the main satellite maxima, measuring the multilayer spacing which consists of two monolayers with opposite molecular orientation, and the Kiessig fringes, measuring the total thickness of the film. Note the large difference in intensity of Kiessig peak maxima left and right with respect to the first satellite maximum which is created by the odd number of monolayers within the film. Curve simulation reveals that there is a fluctuation in the density of the individual sublayers.

chains and the head groups. The head groups are about 0.2 nm thick, but they have a density which is twice as large as that of the silicon substrate. On the other hand, the hydrocarbonic chains have a density of less than one-half of that of the silicon.

Kiessig maxima are clearly visible between the main satellites. Their number is $N = 12$; i.e., the total thickness should correspond to 14 double layers. The evaluation of the angular spacing between the Kiessig maxima results in a thickness of $T = 90.15 \pm 0.35$ nm which corresponds to 15 double layers. This discrepancy has two causes. Due to the hydrophilic nature of the silicon surface the molecules of the first monolayer are deposited with the head straight down to the substrate. Therefore, this layer is not a part of a double layer and the film consists of 29 monolayers. This non-centrosymmetry becomes visible as the strong asymmetry in the Kiessig intensities on left and right with respect to the first main satellite [283]. A second reason for the larger total thickness is the molecular pile-up effect; a few molecules leave the molecular layers and jump on top of the film, creating islands. Because this process is already associated with a very small activation energy at room temperature, a sufficient number of molecules alter their vertical positions within the multilayer film, increasing the total film thickness as a function of time [108].

Finally one can determine the average density of the organic film. As in Fig. 8.3, the critical angle of the films is smaller than that of the silicon substrate. From $\langle \alpha_{\mathrm{c,film}} \rangle = 0.175°$, one obtains an average density of $\varrho_{\mathrm{m}} = 1.5\ \mathrm{gcm}^{-3}$. Note there are Kiessig maxima in the angular range between the critical angles of film and substrate. This effect is similar to that already shown in Fig. 8.3. Figure 8.11 shows a similar organic multilayer, a cadmium

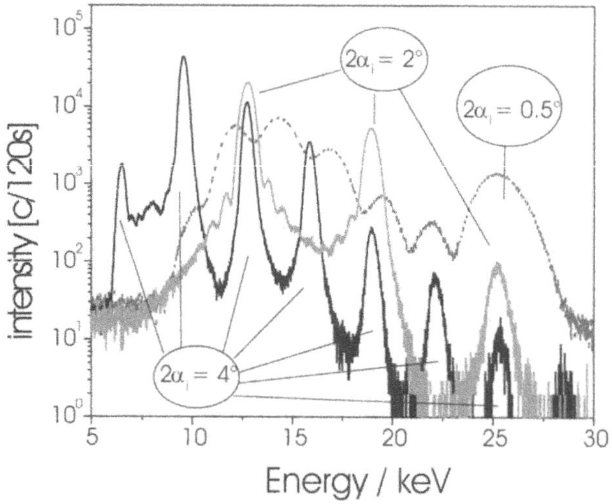

Fig. 8.11. Energy-dispersive reflection curves of a cadmium-behenate multilayer film measured at different incidence angles. The counting time per spectrum was 120 seconds each. The number of Bragg peaks increases with increasing α_i. Due to the limited detector resolution the Kiessig fringes disappear at large α_i.

behenate multilayer film covering a silicon substrate. Here, the experiment has been performed at the energy-dispersive beamline at BESSY II. As introduced in Sec. 3.2, the accessible range of the reflectivity curve depends on the incidence angle. For $\alpha_i = 0.5°$, the range between the first and second-order Bragg peak is probed. Several Kiessig fringes and the second-order Bragg peak are visible. The first-order Bragg peak is attenuated due to the large absorber thickness used to protect the detector. More Bragg peaks become visible, increasing the incidence angle. Eight Bragg peaks appear for $\alpha_i = 4°$, but no Kiessig fringes. This is due to the fact that the peak width now is determined by the limited detector resolution. Due to the different absorber thicknesses used, the onset of the reflectivity differs between the spectra measured at different α_i.

The layer thickness and total thickness can be determined from the peak distance at the energy scale. In energy-dispersive reflectometry the Kiessig peak maxima and Bragg peaks appear at different energies, changing α_i due to the relation $q_z E$. Rewriting Eq. (8.3) in terms of energy, the energy spacing ΔE between two neighboring intensity maxima decreases for increasing α_i. The layer thickness T follows from

$$\delta E = \frac{hc}{2T \sin \alpha_i} \approx \frac{6.2}{T\alpha_i}, \tag{8.20}$$

where h and c are the Planck constant and the velocity of light, respectively. Refraction is neglected in Eq. (8.20) and $\sin \alpha \approx \alpha$. The accuracy of the thickness determination depends on the energy resolution of the detector ΔE:

$$\frac{\Delta T}{T} = \frac{\Delta E}{E}. \tag{8.21}$$

For a germanium or Si:Li detector ΔE is about 180 eV. This results in a relative accuracy of $\frac{\Delta T}{T} \sim 1\%$ for peaks measured at $E = 10$ keV. The upper limit for evaluating a layer thickness depends on the minimum separation which can be resolved between two peaks. Assuming $\Delta E = 0.5$ keV and $\alpha_i = 0.25°$, the limit amounts to about 300 nm. The limited energy band pass of the experiment determines the lower limit of the thickness determination. Using $\alpha_{i,\max} = 4°$ and a band pass of about 15 keV, the lower limit is on the order of 1 nm. This limit has been determined by measuring the thermal expansion coefficient of polymer films with thicknesses of about 100 nm [48].

The evaluation of spectra shown in Fig. 8.11 gives a multilayer period of $D_{LB} = 5.65 \pm 0.05$ nm and a total thickness of $T_{\text{tot}} = 56$ nm, which verifies preparation conditions. The comparison of the various spectra manifests the validity of Eq. (8.20). As seen, the number of the Bragg peaks is doubled, increasing α_i by a factor of two.

In comparison with the angle-dispersive set-up, the accuracy of the absolute thickness determination is lower. Nevertheless each spectrum shown in Fig. 8.11 was collected in 120 seconds which is a small fraction of the time necessary for recording the analogous angle-dispersive curves.

8.3 Coplanar X-Ray Diffraction by Single Layers

X-ray reflection is sensitive to the gradient of the electron density normal to the air-sample interface; that means the layer thickness can be determined independent of crystal perfection. In contrast to this, coplanar x-ray diffraction measures the lattice spacing of the layer as well, presuming crystalline perfection. Therefore, it is advantageous to combine reflection and diffraction measurements in order to obtain complete information on the investigated sample. An example of a successful combination of both methods for the determination of the layer thicknesses in a multilayer can be found in [55].

To measure the layer thickness, let us investigate the coplanar diffraction from a single layer pseudomorphically grown on a semi-infinite substrate. Applying the recurrence formula (6.41), we obtain its diffractivity

$$\mathcal{R} = \left| \frac{\Re_{\text{sub}}(c_1 - Mc_2) + c_1 c_2 (M-1)}{\Re_{\text{sub}}(1-M) + c_1 M - c_2} \right|^2, \quad M = E^{-\mathrm{i}(k_{0z1} - k_{0z2})T}, \tag{8.22}$$

where the thickness T, the amplitude ratios $c_{1,2}$, and the z-components of the wave vectors k_{0zn} refer to the layer. M is a factor containing the phase difference of the waves corresponding to different tie-points in the layer; $\Re_{\text{sub}}$ is the (dynamical) diffractivity of the semiinfinite substrate (6.42). In the semikinematical approach (7.18), the reflectivity of the single layer on the substrate is a coherent superposition of the scattering amplitudes of the layer and substrate:

$$\mathcal{R} = \left| \Re_{sub} \mathrm{e}^{-\mathrm{i}\kappa T} + \frac{CK\chi_{\boldsymbol{h}}}{2\gamma_{\boldsymbol{h}}\kappa} \left(1 - \mathrm{e}^{-\mathrm{i}\kappa T}\right) \right|^2, \tag{8.23}$$

where we have neglected the third process in Fig. 7.11. κ is the modified deviation of the primary beam from the diffraction position in the layer corrected to refraction

$$\kappa = Q_z - h_{z\text{layer}} + \frac{K\chi_0}{2\gamma_{\boldsymbol{h}}} \left(1 - \frac{\gamma_{\boldsymbol{h}}}{\gamma_0}\right). \tag{8.24}$$

This formula can be used for the analysis of the diffraction curve.

Besides the Bragg peak, the contribution of the layer to the diffraction curve provides a series of maxima, whose distance follows from the condition

$$\Delta\kappa = \Delta Q_z = \frac{2\pi}{T}.$$

These maxima create a periodical sequence in reciprocal space along the Q_z-axis and can be understood as multiple scattering of the diffracted wave between the upper and lower boundary of the layer. This behavior is fully analogous to x-ray reflection. If we restrict ourselves to small angular deviations from the Bragg diffraction maximum, we find that this series of maxima is periodical also in the angular scale, and the angular spacing of the *thickness oscillation* maxima is

$$\Delta\eta_i = \lambda \frac{\gamma_{\boldsymbol{h}}}{\sin(2\Theta_B)T}. \tag{8.25}$$

For symmetrical reflections it is reduced to

$$\Delta\eta_i = \frac{\lambda}{2T\cos\Theta_B}, \tag{8.26}$$

which is similar to the derivative of the Bragg law and Eq. (8.1).

In general, both Bragg peaks of layer and substrate are separated by an angular distance $\Delta\eta_i$ (see Chap. 9). Due to the superposition of both

Bragg peaks, the intensity of thickness oscillations is increased in between them compared to the region outside. Since in the kinematical approximation the amplitude of the wave diffracted by the layer is proportional to $\chi_{\boldsymbol{h},\text{layer}}$ the phase difference of the substrate and layer waves depends on difference $\chi_{\boldsymbol{h},\text{layer}} - \chi_{\boldsymbol{h},\text{substrate}}$. Therefore, this difference affects the positions of maxima and minima at the diffraction curve.

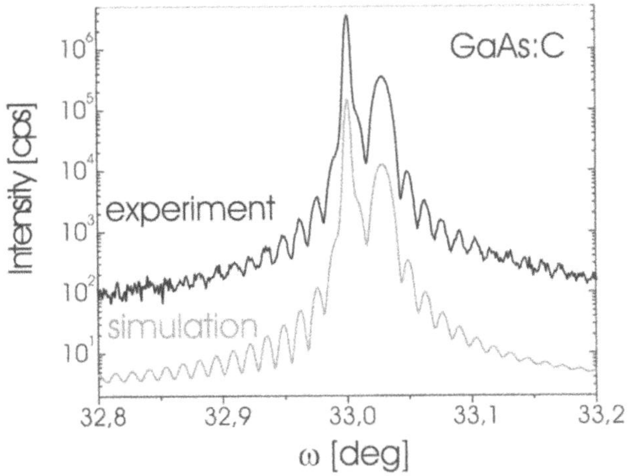

Fig. 8.12. 004 X-ray diffraction curve of a 300 nm thick GaAs:C covered on GaAs (001). The spacing between the Bragg peaks corresponds to a carbon content of about 5 $\times 10^{19} cm^{-3}$.

Figure 8.12 shows a sample where a thick GaAs layer doped with carbon was epitaxially grown on a GaAs substrate. Measuring the Bragg diffraction in vicinity of the GaAs (004) Bragg reflection, one can clearly distinguish between the substrate and layer Bragg peaks. The peak separation corresponds to a lattice mismatch of $\frac{\Delta a}{a} = -390$ ppm (see Sect. 9.1) which corresponds to a carbon content within the layer of about 5×10^{19} cm^{-3}. The layer thickness can simply be evaluated from thickness oscillations and yields $T = 392 \pm 32$ nm.

Figure 8.13 shows the example of a $In_{0.06}Ga_{0.94}As$/GaAs/ $In_{0.06}Ga_{0.94}As$ heterostructure. Here a 25-nm-thick GaAs is embedded by two InGaAs layers and covered by a thick GaAs top layer. The angular separation between the fringes corresponds to a layer thickness of $T_1 = (300 \pm 3)$ nm. A second modulation appears in addition; it is an overmodulation of the former ones with much larger angular separation. From the four oscillations, the thickness of the GaAs layers was determined to be $T_2 = (33 \pm 5)$ nm. The active GaAs layer in between is not visible. The accuracy of thickness determination is estimated using Eq. (8.6) from the maximum fringe order $m_{\max}$ detectable at the diffraction curve, counted from the layer Bragg peak at $\Theta_{B,\text{layer}}$. In the present case up to $m_{\max} = 40$ periods of the fast oscillation are visible at the low-angle side which determines the relative accuracy of the InGaAs layer thickness of 2.5%. On the other hand, the four oscillations appearing for the second layer, give an accuracy on the order of 20% only.

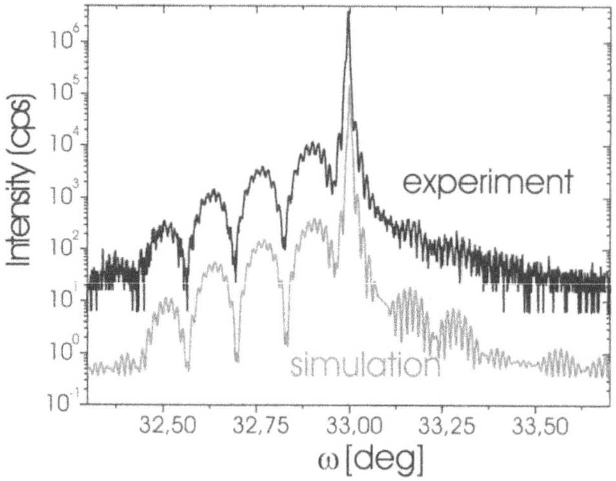

Fig. 8.13. 004 x-ray diffraction curve of $In_{0.06}Ga_{0.94}As/GaAs/\ In_{0.06}Ga_{0.94}As$ heterostructure on GaAs (001) buried under a thick GaAs top layer. There are two distinguished thickness oscillations: the first oscillations measure a thickness of the heterostructure of about 33 nm; the second one is 300 nm and measures the thickness of the GaAs top layer.

The thickness determination becomes complicated if both the lattice mismatch between layer and substrate and the layer thickness are small. Then the center of the layer Bragg peak cannot be definitely identified and no thickness oscillations may appear. Under this condition both quantities cannot be determined separately (see Sect. 9.2). This problem has been discussed extensively in our previous book [167].

The appearance of an electron density gradient normal to the interface gives rise to an inaccurate thickness determination. Equation (8.25) determines the thickness of a layer where the layer is a rectangular box of constant electron density. If the density varies gradually between two neighboring layers, one is unable to distinguish between a box of constant density and a slightly larger one of gradually changing density so long as

$$\int \varrho \mathrm{d}z = \text{const.} \tag{8.27}$$

From the intensity profile along the truncation rod, one cannot draw any conclusion on the origin of the grading, whether interdiffusion between neighbored layers or graduated variation of the composition, while layer growth, or an enhanced interface roughness may occur [393]. By computer simula-

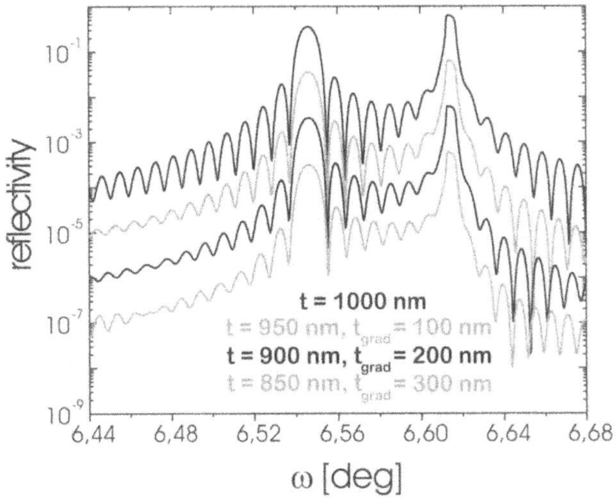

Fig. 8.14. Calculated variation of the diffraction curve of a $Ga_{0.6}Al_{0.4}As/GaAs$ heterostructure close to the 422 reciprocal lattice point assuming linear compositional grading close to the GaAs–$Ga_{0.6}Al_{0.4}As$ interface. The relative thickness of the grading layer amounts to $\frac{\Delta T}{T} = 0, 0.1, 0.2,$ and 0.3 (top to bottom).

tions one is able to estimate the thickness of a sublayer with a compositional grading close to the interface. The mean compositional grading reduces the scattering amplitude of thickness oscillations progressively with increasing angular distance from layer Bragg peak. Assuming a composition gradient at the interface as shown in Fig. 8.14, the thickness ratio between the graded layer T_{grad} and the layer of constant density T can be estimated from

$$\frac{T_{\text{grad}}}{T} = \frac{1}{m_{\max}} \tag{8.28}$$

by counting the number of thickness fringes $m_{\max}$ which can be measured at a particular diffraction curve beginning at the layer Bragg peak [29, 47]. If $m_{\max} = 7$, as shown in the bottom curve of Fig. 8.14, then 1/7 of the total thickness belongs to the graded part of the layer where a linear grading was supposed for calculation. Such an estimation is possible so long as a sufficient number of thickness oscillations appear above the background. Note that the conclusion from Eq. (8.28) is similar to that given for x-ray reflection according to Eq. (8.6). In both cases the scattering experiment probes the electron density in one particular direction, i.e., normal to the air–sample interface. Any lateral inhomogeneity of the electron density acts like a density gradient along the surface normal and disturbs the appearance of multiple scattering within the layer. At a certain reflection order m, the partial waves are out of phase and are extinguished.

In summary we can conclude that the lower limit for determining a single layer onto substrate amounts to 5–10 nm using home laboratory sources and triple-axis equipment (see Chaps. 2 and 3). Very thin single layers show very broad peaks of very low intensity. Here a precise thickness determination requires a highly intense x-ray beam in order to record the rocking curve over a very large angular range. The sensitivity for thickness determination increases if the thin layer is embedded between much thicker cover layers. Then the thickness determination is achieved via phase contrast. The application of this particular case is described in Sect. 8.6.

8.4 Coplanar X-Ray Diffraction by Periodical Superlattices

The diffraction curve of a periodical superlattice that has a perfect crystal structure can be calculated kinematically using Eq. (5.37). This approach is limited to the region far from the substrate Bragg maximum, since then the diffraction in the substrate must be treated dynamically. Therefore, full dynamical (6.41) or semikinematical (7.18) methods are more desirable.

Here, we use the semikinematical formula (7.18) in order to show some common features of the diffraction curves of the periodic, perfect superlattices. The superlattice consists of N periods of thickness D, and each contains two layers A and B with the thicknesses T_A and T_B, respectively ($T_A + T_B = D$). Their vertical lattice parameters are $a_{\perp A,B}$. In case of phseudomorphic growth, their lateral lattice parameters $a_{\|A,B}$ are the same as the parameter a of the substrate. In Chapter 9 we define the vertical lattice misfit in the layer M (M=A or B) as

$$\delta_M = \frac{a_{M\perp} - a_S}{a_S}, M = A, B,$$

and its mean value in the superlattice as

$$\langle\delta\rangle = \frac{\delta_A T_A + \delta_B T_B}{T_A + T_B}.$$

From the semikinematical equation (7.18) we obtain

$$\Re_0 = \Re_{sub}(\phi_A\phi_B)^N + (P_A + P_B\phi_A)\frac{(\phi_A\phi_B)^N - 1}{\phi_A\phi_B - 1}, \tag{8.29}$$

where

$$\phi_S = \exp\left(\frac{-\mathrm{i}K\beta_S}{2\gamma_h}T_S\right), \quad P_S = \frac{C\chi_{hS}}{\beta_S}(1-\phi_S),$$

$$\beta_S = 2\eta_i \sin(2\Theta_B) + \chi_{0S}\left(1 - \frac{\gamma_h}{\gamma_0}\right) + \delta_S h_z \frac{2\gamma_h}{K}, \quad S = A, B.$$

The first term on the right-hand side of (8.29) represents the wave diffracted by the substrate, the factor $(\phi_A\phi_B)^N$ is the phase shift of the wave due to the transmission through the multilayer stack. The second term in Eq. (8.29) is the wave diffracted by the multilayer itself. This term has a maximum value if $\phi_A\phi_B = 1$, where the deviation η_m from the kinematical Bragg position yields

$$\eta_m = m\frac{\lambda}{D}\frac{\gamma_h}{\sin(2\Theta_B)} - \left(1 - \frac{\gamma_h}{\gamma_0}\right)\frac{\langle\chi_0\rangle}{2\sin(2\Theta_B)} + \frac{\gamma_h(\gamma_0 - \gamma_h)}{\sin(2\Theta_B)}\langle\delta\rangle. \quad (8.30)$$

Here m is an arbitrary integer and

$$\langle\chi_0\rangle = \frac{1}{D}(\chi_{0A}T_A + \chi_{0B}T_B)$$

is the mean value of the zero-th polarizability coefficient.

Like the reflection curve, the diffraction curve of a periodical superlattice exhibits a periodic sequence of the satellite maxima. The distance between the maxima is

$$\Delta\eta = \frac{\lambda}{D}\frac{\gamma_h}{\sin(2\Theta_B)}. \quad (8.31)$$

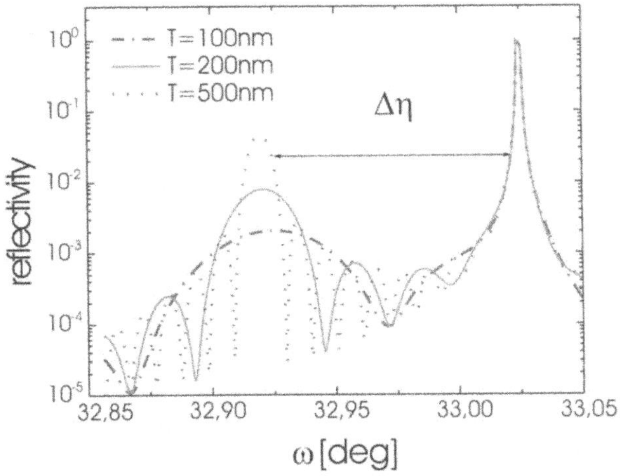

Fig. 8.15. The angular distance $\Delta\eta$ between the layer and substrate peaks depends on the layer thickness. The diffraction curves (004 symmetrical Bragg case diffraction, $CuK\alpha_1$) were calculated for an AlAs layer on GaAs substrate.

Since we denote as η_{sub} the relative angular position of the substrate peak with respect to the kinematical Bragg angle,

$$\eta_{\text{sub}} = -\frac{\chi_{0\text{sub}}}{2\sin(2\Theta_B)}\left(1-\frac{\gamma_{\boldsymbol{h}}}{\gamma_0}\right),$$

the zero-th satellite maximum is shifted with respect to the substrate by

$$\begin{aligned}\delta\eta_0 &= \eta_0 - \eta_{sub} \\ &= -\frac{1}{2\sin(2\Theta_B)}\left(1-\frac{\gamma_{\boldsymbol{h}}}{\gamma_0}\right)(\langle\chi_0\rangle - \chi_{0\text{sub}}) + \frac{\gamma_{\boldsymbol{h}}(\gamma_0-\gamma_{\boldsymbol{h}})}{\sin(2\Theta_B)}\langle\delta\rangle .\end{aligned}$$

The shift is given by a superposition of two terms. The first term is the result of the difference of χ_0 (i.e., of the refractive index) of the substrate and the superlattice and can be neglected in most cases. The second term depends on the mean vertical lattice misfit $\langle\delta\rangle$ of the superlattice. In practice, the last formula must be used with some care. Numerical simulations based on the semikinematical formula (8.29) have shown that in the case of a single very thin layer deposited on the substrate the distance between the layer peak and the substrate maximum may slightly decrease with decreasing layer thickness (see Fig. 8.15). Therefore, the above formula can be used only if the layer Bragg peak is not influenced by the substrate (see also Sect. 9.1). A preliminary numerical simulation is always important. The intensities of

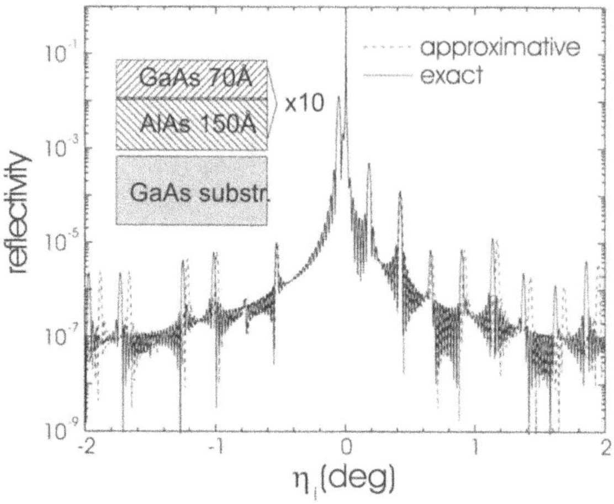

Fig. 8.16. The diffraction curves of an AlAs/GaAs superlattice (symmetrical 004 Bragg case) calculated by means of the general and simplified theories. In the simplified theory, the dispersion spherical surface is replaced by a tangential plane that leads to a shift of the satellites with greater orders.

the satellite peaks are modulated by the envelope function $|P_A + P_B\phi_A|^2$, which depends on the thicknesses T_A and T_B and the lattice mismatch of the layers. Analogously to the reflection case, this function is proportional to $|F_{\boldsymbol{h}}^{(p)}|^2$, where $F_{\boldsymbol{h}}^{(p)}$ is the structure factor of the multilayer defined in Eq. (5.38).

If the mismatch $\delta_{A,B}$ is not too large, the simplified expression (5.40) for the structure factor can be used. In this case, the formula (8.18) is valid and the thickness ratio T_A/T_B can be deduced.

From Eq. (8.29) another type of maxima follows. The phase $(\phi_A \phi_B)^N$ of the substrate contribution $\Re_{\text{sub}}$ to the total complex diffractivity depends on the total thickness $N(T_A + T_B)$ of the multilayer and on the angular deviation from Bragg peak maximum η. This phase term creates thickness fringes which are similar to the Kiessig maxima in reflectivity. The angular spacing between them is

$$\Delta\eta = \frac{\lambda}{ND} \frac{\gamma_{\boldsymbol{h}}}{\sin(2\Theta_B)}. \tag{8.32}$$

It is necessary to note that the formulas in this section are valid only within the conventional coplanar diffraction, i.e., for small angular deviations η and for small asymmetry angles ϕ. The general formulas can be derived using the approach given in Chapter 6.

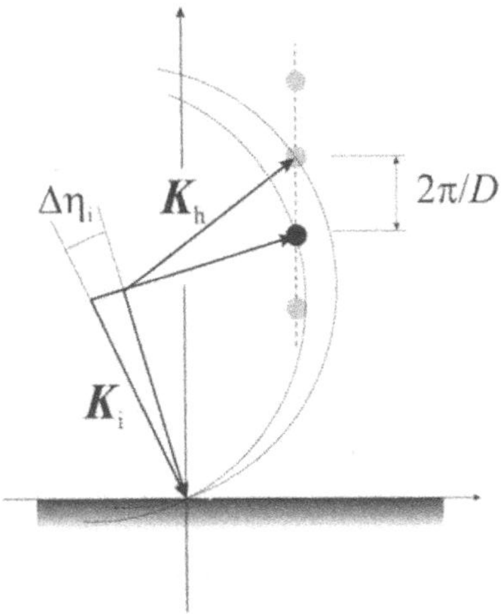

Fig. 8.17. The Ewald construction of a periodical superlattice.

Figure 8.16 shows a comparison of the diffraction curves of a short-period AlAs/GaAs superlattice calculated by means of the conventional diffraction theory and the general diffraction theory (Eqs. 6.36 – 6.40) and the general diffraction theory is shown in Chapter 6 (Eqs. 6.30 – 6.42). It is obvious that in the symmetrical Bragg case and small $|\eta|$ both of the theories yield the same curves; for larger $|\eta|$ the satellite maxima of the approximate *conventional* curve are slightly shifted. The satellite maxima on the curve, calculated by means of the general theory, are not equidistant. This fact can easily be understood from the Ewald construction of the satellite positions (see Fig. 8.17). Therefore, for large $|\eta|$, formula (8.31) must be extended to the following form:

$$\Delta\eta = \frac{\lambda}{D}\frac{\sqrt{1-(\cos\alpha_i - 2\sin\phi\sin\Theta_B)^2}}{\sin\alpha_i(\cos\alpha_i - 2\sin\phi\sin\Theta_B)}, \quad \alpha_i = \Theta_B - \phi + \eta. \tag{8.33}$$

In the symmetrical Bragg case ($\phi = 0$) we obtain a simple relation

$$\Delta\eta = \frac{\lambda}{D}\frac{1}{\cos\alpha_i}.$$

In some cases the true diffraction curve can also be described if one introduces high-order terms to the conventional description of the semikinematic theory approximating the true curvature of the dispersion surface by asymptotes of hyperbolic sheets [364, 398].

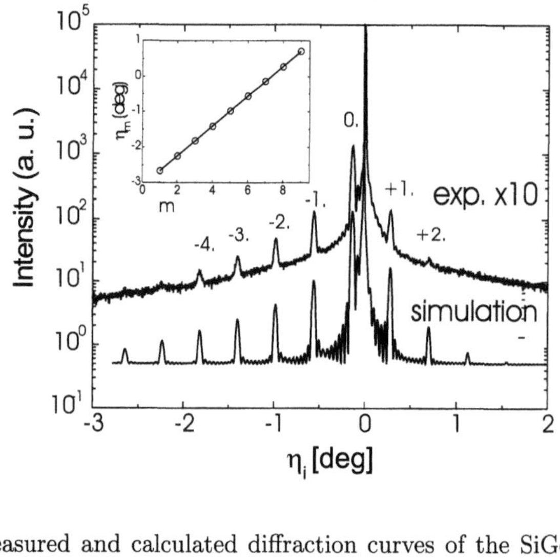

Fig. 8.18. Measured and calculated diffraction curves of the SiGe/Si multilayer (224 asymmetrical Bragg case, CuKα_1). The inset shows the satellite positions m as a function of the satellite order, measured from substrate peak. The zero-order satellite peak measures the average lattice misfit of the superlattice with respect to the substrate. The simulation was performed using Eq. 6.41.

An example of measured and simulated diffraction curves in the case of asymmetrical 224 diffraction (low incidence angle) of a SiGe/Si multilayer (the same sample as in Fig. 8.8) is plotted in Fig. 8.18. The positions and heights of the measured satellite maxima agree well with the theory; however, the tiny thickness fringes between the satellite maxima are not visible in the experimental curve. These maxima are very sensitive to the presence of defects; even a very weak diffuse scattering from the defects smears out the thickness fringes.

As in the reflection case, the superlattice period D can be determined from the positions of the satellite peaks (indicated by arrows). In the inset of Fig. 8.18 we compared the positions of these peaks with (8.28) and obtained $D = (20.6 \pm 0.7)$ nm. This value served as a starting estimate for the fitting of the whole diffraction curve. Additional parameters of fitting were T_A/T_B and the concentration x of Ge in the GeSi layers. These parameters were determined to be $D = (20.5 \pm 0.1)$ nm, $T_A/T_B = 7 \pm 0.5$, and $x = 0.36 \pm 0.02$. Comparing these values with those determined by x-ray reflection (previous subsection), we find that the values of D coincide. Using x-ray reflection, one can achieve a better accuracy of the thicknesses T_A, T_B of the layers in the multilayer period, but x-ray diffraction is more suitable for determining x. On the other hand, x-ray diffraction was completely insensitive to the oxide layer on the surface (it does not diffract) and nearly insensitive to the thickness of the capping Si layer.

8.5 X-Ray Grazing Incidence Diffraction

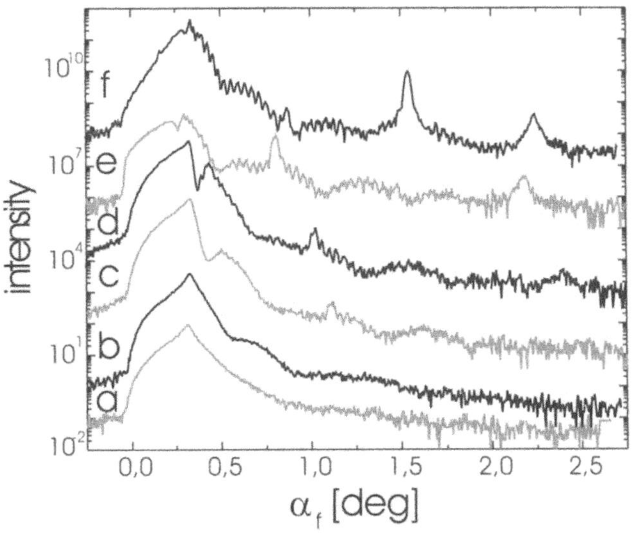

Fig. 8.19. Selected α_f scans at different α_i recorded in GID geometry at the 200 in-plane diffraction of a complex stacked multilayer structure grown on GaAs (001) substrate. The curves correspond to incidence angles of 0.2° (a), 0.3° (b), 0.35° (c), 0.4° (d), 0.5° (e), and 1.0° (f).

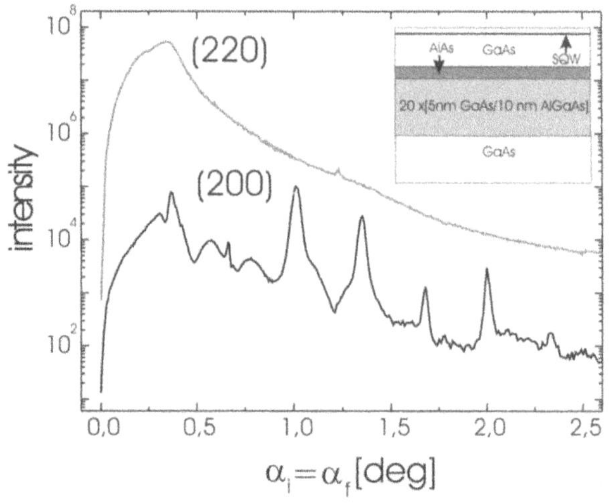

Fig. 8.20. α_f scans recorded at different in-plane Bragg positions. The 220 reflection of GaAs is strong in intensity but does not provide out-of plane structure information. The 200 reflection is weak, because it depends of the difference of the electron density of both fcc sublattices of the zinc-blende structure. The inset displays the layer stacking.

The GID technique is very sensitive to a change of the electron density close to the surface. Due to its opportunity to reduce the penetration depth of the probing x-ray close to the sample surface, this technique is well suited for thin film analysis. To apply it one has to inspect the intensity variation along the crystal truncation rod close to the angular position of an in-plane Bragg peak excited at $\Theta_{B\|}$ (see Sect. 4.3) using a fixed angle of incidence $\alpha_i > \alpha_c$. Under relaxed resolution with respect to the in-plane angle $\Delta\theta_{i,f}$, the vertical momentum transfer is solely varied by changing α_f. Consider this peculiarity, the thickness T can be determined using Eq. (5.39). In angular space and for $\alpha_f \ll \alpha_c$, it yields

$$T = \frac{2\pi}{\Delta q_z} \approx \frac{\lambda}{\alpha_{f,m+1} - \alpha_{f,m}}. \tag{8.34}$$

This is demonstrated in Fig. 8.19. It shows the intensity distribution of the 200 in-plane diffraction of an (001) oriented multilayer structure grown on GaAs substrate. It consists of a GaAs/$Al_{0.2}Ga_{0.8}As$ multilayer covered by a 30-nm-thick $Al_{0.2}Ga_{0.8}As$ layer followed by a 200-nm-thick GaAs layer. On top of this structure there is a 5-nm-thick $In_{0.07}Ga_{0.93}As$ single quantum well covered by a 20 nm GaAs top layer (see inset of Fig. 8.20). This complicated sequence of layers can be discovered by means of α_f resolved scans. The curve is smooth as long as $\alpha_i \leq \alpha_c$ (bottom curve in Fig. 8.19). The GaAs top layer becomes visible if α_i equals or is slightly larger than the critical angle ($\alpha_c = 0.31°$). Bragg peaks from the buried multilayer starts to appear at $\alpha_i \approx 0.5°$ (curve e). This depth resolution simplifies the structure analysis of this complicated multilayer. Figure 8.20 shows two α_i scans of the same sample recorded at the 220 and the 200 in-plane Bragg position. Although the intensity at the 220 position is about three orders of magnitude larger compared to the 200 one, the first one does not provide out-of plane structure information. Because the Fourier components of the sublayers χ_h for strong reflections differ by less than 5%, there is no measurable scattering contrast. This equals the situation of x-ray reflectivity where the difference of χ_0 is essential for the appearance of Bragg peaks of thickness oscillations. The situation differs using weak Bragg reflections. Here the difference of χ_h is large and provides strong variations in the angular dependence of reflectivity.

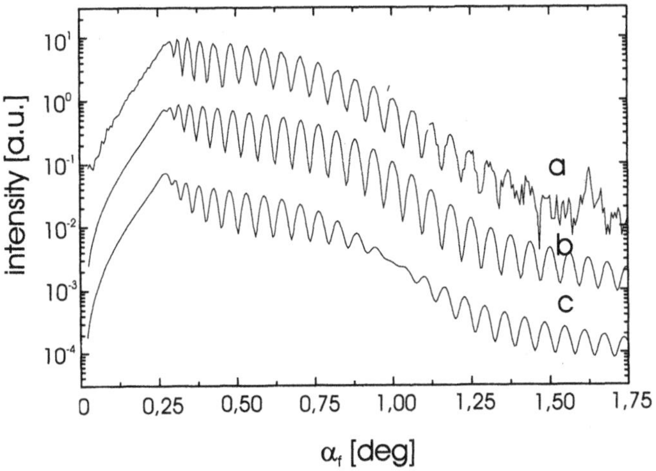

Fig. 8.21. Experimental and simulated q_z scans of a GaAs/$Ga_{0.87}In_{0.13}As$/GaAs (001) double heterostructure recorded under GID geometry close to the 200 in-plane diffraction of GaAs. The two simulations stand for the expected rod structure, assuming abrupt interfaces and a compositional grading close to a single interface.

Another example, shown in Fig. 8.21, displays α_f scans of a double heterostructure GaAs/$Ga_{0.87}In_{0.13}As$/GaAs. The task here was to verify the appearance of a compositional grading close to the GaAs-GaInAs interface which had been indicated by photoluminescence measurements [399]. The use of the 200 in-plane diffraction was preferred for this purpose because it gives a large contrast between the scattering amplitudes $\chi_{h\mathrm{GaInAs}}$ and $\chi_{h\mathrm{GaAs}}$ which can not be expected using 400. The angular spacing between the oscillation provides the top layer thickness $T = 112$ nm. Numerical simulation reveals that for sharp interfaces the truncation rod structure should show a periodicity $\Delta\alpha_f = \frac{\lambda}{T_{\mathrm{GaInAs}}}$. This is not visible in the experiment. The measured curves can indeed be explained by taking into account the expected compositional grading . Unfortunately we are not able to tell whether the grading is close to the bottom or the upper intrinsic interface. However, it amounts to one-fifth of the expected homogeneous thickness of the GaInAs layer [291].

8.6 Buried Layers

In previous sections we have demonstrated that the thickness of a single layer can be determined down to a few 10 nm by measuring the angular separation of thickness fringes (see Sect. 8.1). This behavior can be explained in terms of intensities, i.e., the square of the scattering amplitude. Thus the appearance of thickness fringes is denoted as *amplitude contrast* . In this section we will show that the sensitivity for layer thickness determination increases to a single monolayer if this layer is embedded between much thicker layers of a different material. Now the thickness of the buried layer becomes available via the change of scattering phases of confinement layers. This mechanism is denoted as *phase contrast* .

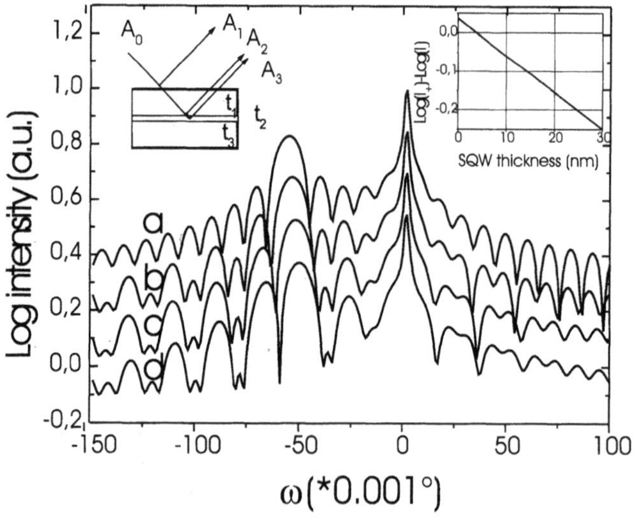

Fig. 8.22. Calculated diffraction curves close to the 400 reciprocal lattice point of a $Ga_{0.6}Al_{0.4}As/GaAs/Ga_{0.6}Al_{0.4}As$ double heterostructure on GaAs[001] with varied thickness of the GaAs quantum well varying between 0 and 30 nm. The inset shows the intensity ratio (solid line) between the subsidiary maxima I_{+1}/I_{-1} close to the AlGaAs Bragg peak.

It should be demonstrated for the case of a $Ga_{0.6}Al_{0.4}As/GaAs/Ga_{0.6}Al_{0.4}As$ double heterostructure grown on GaAs[001] by rocking-curve simulation [30]. In this model, the fringe periodicity of the heterostructure corresponds to its total thickness $T_{tot} = T_1 + T_2 + T_3$, where both confinement layers have equal thicknesses $T_1 = T_3 = 400$ nm. The results of calculation are shown in Fig. 8.22.

The buried GaAs has a thickness T_2 in the range $0 < T_2 < 30$ nm. This is the situation of a buried single quantum well (SQW). The scattering amplitude of the SQW is too low and its peak width too diffuse to contribute significantly to the diffraction curve. But it creates a phase shift between the scattering amplitudes A_1 and A_2 of the embedding layers. This causes a modulation of the fringe maxima. The modulation period depends on the thickness ratio of the confinement layers T_1/T_3.

If $T_1 = T_3$, as shown in Fig. 8.22, the phase shift induces an alternate increase and decrease of fringe maxima. If the thickness ratio is $T_1/T_3 = 2$ or 3, the intensity of the third or fourth fringe period, respectively, would be reduced. The modulation amplitudes itself depends on T_2. In Fig. 8.22 the SQW thickness is modeled for $T_2 = 0$, 10, 20 and 30 nm (curves *a*–*d*). The oscillations are uniform for $T_2 = 0$. At $T_2 = 40$ nm, every second fringe maximum vanishes almost completely. As mentioned above, the scattering amplitude corresponding to T_2 is not visible. The intensity ratio between the

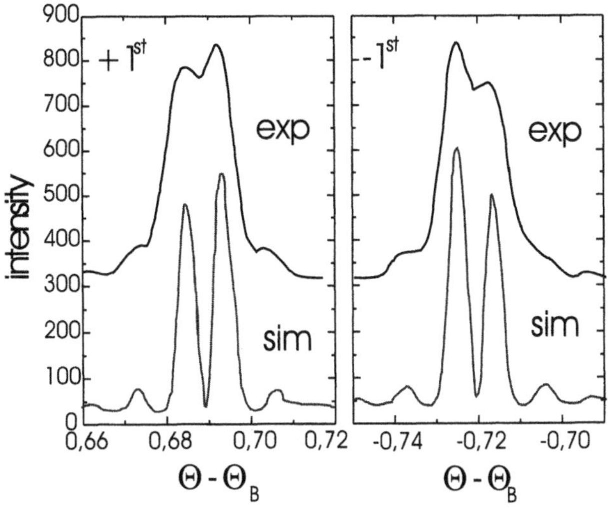

Fig. 8.23. Oscillation behavior close to the +1st- and -1st-order satellites of a $[GaAlAs/GaAs]_{49}$ on GaAs[001] multilayer with a single enlarged GaAs quantum well; top curves – measured(exp), bottom curves – simulated (sim).

first two subsidiary maxima close to the AlGaAs Bragg peak I_{+1}/I_{-1} can be used as a measure for T_2. Its ratio depends on sign and amount of the lattice mismatch between confinement and the active layers (see Sect. 9.2). In the present case the mismatch of the SQW (GaAs) with respect to AlAs (see Sec. 9.1) is $\frac{\Delta a}{a} < 0$, so that I_{-1} is larger but I_{+1} smaller compared to the case $T_2 = 0$. As shown in inset of Fig. 8.22 the ratio I_{+1}/I_{-1} decreases approximately exponentially for increasing T_2. Taking into account an uncertainty of the measured fringe ratio of about 5%, T_2 can be determined with an accuracy of about 1 nm.

This behavior can be understood by the phase contrast between the amplitudes A_1 and A_3 scattered at both AlGaAs confinement layers. For $T_2 \ll (T_1 + T_3)$ the total scattering amplitude A is given approximately by

$$A = A_1 + A_3 \mathrm{e}^{-\mathrm{i}\frac{2\pi}{\lambda}\frac{2T_2}{\sin\Theta_B}} . \tag{8.35}$$

The amplitude of A_3 experiences a phase shift during its passing across the SQW; it changes in magnitude and sign relative to A_1. Thus the determination of T_2 is not unique. Multiples of the argument in the exponential of Eq. (8.32) gives similar diffraction pattern.

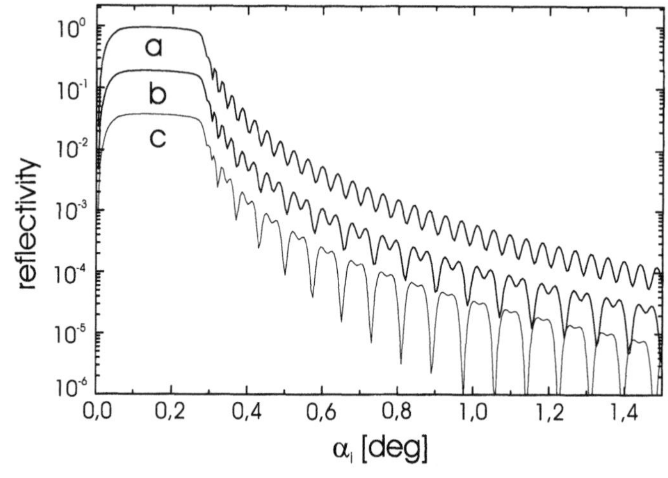

Fig. 8.24. Simulated x-ray reflectivity curves of 100 nm AlAs layer on GaAs (a), and for double heterostructure in which 50 nm AlAs layers each embed one (b) or two monolayers of GaAs (curve c).

This *fingerprint* of a heterostructure can be found using different x-ray scattering techniques and multiply stacked multilayers. The structure shown in Fig. 8.13 consists of two such heterostructure systems: two $In_{0.06}Ga_{0.94}As$ layers enclose the GaAs SQW. The GaAs top layer and the GaAs substrate built a confinement for the InGaAs/GaAs/InGaAs layer system. One can clearly see the modulation of the thickness fringe peak intensities. Every oscillation related to the InGaAs layers carries nine faster oscillations related to the thick GaAs top layer. Following from the ratio of periods, the thickness ratio amounts to 1:9.

The same behavior can be found at a multilayer structure where an enlarged single quantum well (ESQW) of GaAs is embedded between two superlattices [31]. Two 49-period GaAs/GaAlAs superlattices provide the optical and energetic confinement of the semiconductor laser structure. Similar to Eq. (8.35), the scattering phases of both superlattices are shifted against each other because of the enlarged optical path length across the ESQW. The induced phase shift is mainly visible close to the angular position of the $+1^{st}$ and -1st superlattice peaks (see Fig. 8.23). This phase shift induces a double peak shape. If the thickness of both superlattices equals $T_{SL1} = T_{SL2}$, the ESQW thickness T_{ESQW} can be expressed in terms of a single superlattice period D_{SL} as

$$T_{ESQW} = mD_{SL}.$$

Computer simulation reveals that each second thickness oscillation disappears if $2m$ is an even number ($c = 1, 2, \ldots$), whereas the fringe behavior is almost unchanged whenever $2c$ is odd ($c = 1.5, 2.5, \ldots$).

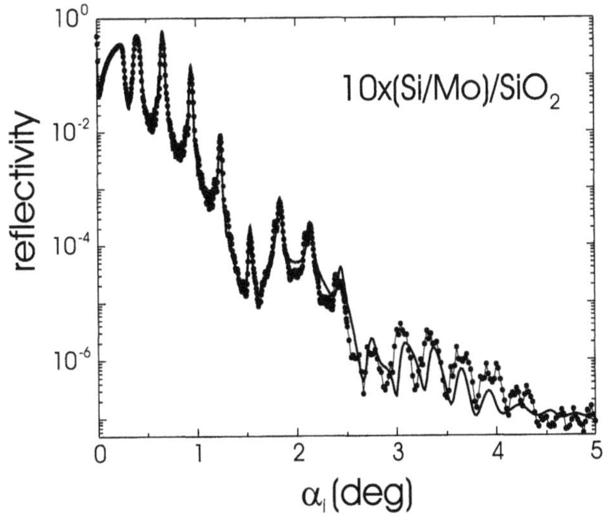

Fig. 8.25. Measured and calculated reflectivity of a 10-period silicon/molybdenum multilayer grown on quartz substrate. The thickness period amounts to $T_{Si} + T_{Mo} =$ 146 Å. The thickness ratio $T_{Si}/T_{Mo} = 3.64$; therefore, each fifth Kiessig peak is reduced in intensity, which provides a long-range beating in intensity.

The complete fit of the experimental data, including higher-order superlattice side maxima gives $m \approx 1$, which explains the observed double peak behavior in Fig. 8.23 as well. D_{SL} is evaluated to be 6.50 ± 0.03 nm and therefore, T_{ESQW} is about 6 nm.

The discussed phase contrast can also be observed in x-ray reflectivity experiments. This is shown in Fig. 8.24 for a hypothetical AlAs/GaAs/AlAs heterostructure. The intensities of thickness fringes become significantly modified if two equally thick confinement layers are separated by a 2-monolayer-thick GaAs. Each second fringe is nearly damped out. But this behavior does not significantly change when the quantum well thickness increases. This prevents its accurate determination via reflectivity measurements.

An experimental example of this peculiarity is shown in Fig. 8.25. Here the thickness ratio between the silicon and molybdenum layers is close to 4:1. Subsequently each fifth Kiessig oscillation is reduced, which provides a long-range beating of the peak intensities. This effect is also seen in Fig. 8.9, where the thicknesses of the vanadium and mica layers are in the ratio 4:3. One can see clearly that the Bragg peak intensies display a long-range modulation: the second and fourth peaks are lower than the first and third.

Finally the splitting of Bragg peaks in case of a buried quantum well can also be found by using grazing-incidence diffraction. In [167] we presented an example of a superlattice structure containing an enlarged GaInAs quantum well embedded between two equal [InP/GaInAs]×15 period superlattices. The measured rod scans show the superlattice peaks, which appear as double peaks due to the reasons mentioned before [239].

9 Lattice Parameters and Strains in Epitaxial Layers and Multilayers

In this chapter we describe the method of strain analysis in expitaxial layer systems. Epitaxial growth onto a crystalline substrate always is associated with the appearance of strain. Since the free lattice parameters of the layer system differs from that of the substrate, the lattice matching at the interface will result in lattice deformation perpendicular to it to minimize the strain energy of the layer. Often the strain energy of the substrate is neglected, as long as the substrate can be assumed as infinitely thick. For cubic material and epitaxy onto (001) the applied strain in the layer is bi-axial and an initially cubic material becomes tetragonally deformed. For epitaxy onto higher indexed planes the strain relaxation will result in a lattice of much lower symmetry. X-ray high-resolution diffraction is a nondestructive tool to analyze the strain state of epitaxial layers. Here the accuracy is high as long as one measures the lattice mismatch between the (non-deformed) substrate and the deformed layer lattice on the relative scale. In a completely lattice-matched system it is often sufficient to measure the out-of-plane lattice mismatch, partially relaxed systems require two independent measurements, one parallel and another perpendicular to the surface normal. This can be performed by measuring one Bragg diffraction with symmetric and the other with asymmetric scattering geometry. Since the layer is very thin one has to use strong-asymmetric diffraction in order to enhance the scattering signal of the layer. In this case the in-plane lattice mismatch can be measured uniquely by x-ray grazing-incidence diffraction.

The measurement of the lattice mismatch in epitaxial systems is routine work for any laboratory dealing with epitaxial crystal growth, and there are a lot of papers in the literature investigating one or another epitaxial system. We will refer these papers to [15, 74, 116, 184, 211, 344, 355, 379] when we are dealing with different epitaxial systems of semiconductor or metallic material. Reviews of the method of x-ray high-resolution diffraction for strain analysis can be found in the books by Bowen and Tanner [61] and Fewster [121].

9.1 Conventional Coplanar Diffraction

For micro- and optoelectronic applications thin semiconductor layers are mainly grown onto (001) or other low indexed basic planes of a cubic substrate. Whereas the lattice parameters of layer a_L and of the substrate a_S may coincide at crystal growth temperatures, they become different when they cool down to room temperature. The resulting lattice mismatch between both *cubic* lattice parameters is

$$\left(\frac{\Delta a}{a}\right)_{\infty} \equiv \delta_{L\infty} = \frac{a_{L\infty} - a_S}{a_S} \tag{9.1}$$

and depends on the chemical composition of layer. If the substrate thickness is greater than several times the layer thickness, the lattice mismatch induces lattice strain solely within the layer. Under conditions of biaxial strain, that means for epitaxy on (001), for example, the non-vanishing strain components are

$$\varepsilon_{xx} = \varepsilon_{yy} = \varepsilon_{\parallel} = \frac{a_{L\parallel} - a_{L\infty}}{a_{L\infty}}, \; \varepsilon_{zz} = \varepsilon_{\perp} = \frac{a_{L\perp} - a_{L\infty}}{a_{L\infty}} \tag{9.2}$$

parallel ($\varepsilon_{\parallel}$) and perpendicular ($\varepsilon_{\perp}$) to the interface, where $a_{L\perp}$ and $a_{L\parallel}$ are the lattice parameters of the strained layer. From elasticity theory, a simple connection between $\varepsilon_{\perp}$ and $\varepsilon_{\parallel}$ follows

$$\varepsilon_{\perp} = -2\frac{c_{12}}{c_{11}}\varepsilon_{\parallel}, \tag{9.3}$$

where c_{11} and c_{12} are the corresponding elastic constants.

Instead of the strain components, it is convenient to express the lattice distortion by means of the parallel and vertical lattice mismatches (*misfit*) of the layer with respect to the lattice parameter of substrate,

$$\delta_{L\parallel} = \frac{a_{L\parallel} - a_S}{a_S}, \; \delta_{L\perp} = \frac{a_{L\perp} - a_S}{a_S} \tag{9.4}$$

(see Fig. 9.1). Now the cubic lattice mismatch follows from

$$\delta_{L\infty} = \frac{\delta_{L\perp} + 2\delta_{L\parallel}c_{12}/c_{11}}{1 + 2c_{12}/c_{11}} \tag{9.5}$$

for epitaxy on (001). For other substrate orientations Eq. (9.5) contains other ratios of the elastic constants, as published in [28].

If a_S is known precisely, a diffraction measurement using double-crystal equipment (see Chap. 2) provides the misfits $\delta_{L\parallel,\perp}$ on the relative angular scale with high accuracy. In a diffraction curve of a heterostructure, the lattice misfit between layer and substrate causes an angular separation between the layer and the substrate Bragg peaks of $\Delta\eta_i$.

Generally, the angular separation contains three components:

$$\Delta\eta_i = \Delta\Theta_0 + \Delta\Theta_B + \Delta\phi. \tag{9.6}$$

Θ_0 is the angular separation due to the different amount of the refraction of the X-ray beam at the air-layer and layer-substrate interfaces (see Sect.5.3), $\Delta\Theta_B$ is the kinematical Bragg angle difference, for symmetric diffractions it is given by

$$\Delta\Theta_B = -\delta_\perp \tan\Theta_B, \tag{9.7}$$

which follows from the derivative of the Bragg law.

For non-symmetrical diffraction geometries, the third term in Eq. (9.6) has to be considered. This describes the different inclination angles ϕ_L and ϕ_S of the diffracting lattice planes within the layer and the substrate with respect to the sample surface, as a result of the layer lattice distortion (see Fig. 9.1).

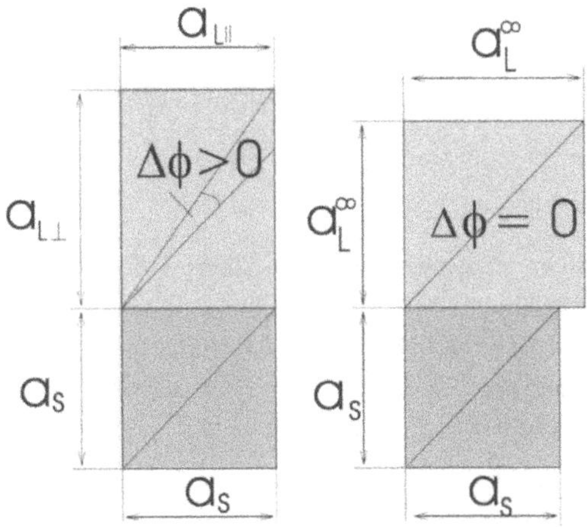

Fig. 9.1. Schematic sketch of the lattice parameters in pseudomorphic (*left*) and partially relaxed (*right*) single-layer heterostructures.

If refraction can be neglected, $\Delta\Theta_B$ and $\Delta\phi$ can be separated by measuring the angular separation $\Delta\eta_i$ in two complementary diffraction geometries. Figure 9.2 shows the ϕ_+ and the ϕ_- set-ups . This figure refer to the high-incidence and low-incidence asymmetry shown in Fig. 4.3. In both we assume $\phi > 0$. In the ϕ_- set-up the incidence angle with respect to the surface, $\alpha_{Bi} = \Theta_B - \phi$ is smaller than the exit angle $\alpha_{Bf} = \Theta_B + \phi$, which gives the angular separation

$$\Delta\eta_+ = \Delta\Theta_B + \Delta\phi. \tag{9.8}$$

The reverse geometry referred to as ϕ_+ set-up, gives the separation

$$\Delta\eta_- = \Delta\Theta_B - \Delta\phi. \tag{9.9}$$

$\Delta\Theta_B$ and $\Delta\phi$ can be separated by means of

$$\Delta\Theta_B = \frac{1}{2}(\Delta\eta_+ + \Delta\eta_-) \tag{9.10}$$

$$\Delta\phi = \frac{1}{2}(\Delta\eta_+ - \Delta\eta_-). \tag{9.11}$$

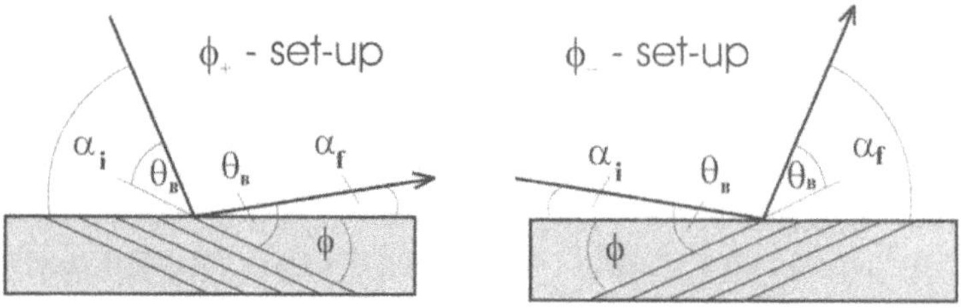

Fig. 9.2. Schematic view of both possible set-ups for asymmetric Bragg diffraction, ϕ_+ set-up (left), ϕ_- set-up (right).

Now one is able to determine the two components of lattice mismatch from

$$\delta_{L\perp} = \Delta\phi \tan\phi - \Delta\Theta_B \cot\Theta_B, \tag{9.12}$$

$$\delta_{L\|} = -\Delta\phi \ \cot\phi - \Delta\Theta_B \cot\Theta_B. \tag{9.13}$$

So long as the layer thickness does not exceed the critical value necessary for creating misfit dislocations (the critical thickness, T_c) , the layer growth is pseudomorphic on the substrate. That means its lateral lattice parameter $a_{L\|}$ equals a_S. Using Eqs. (9.8) and (9.9), the validity of this assumption may be verified by the identity

$$\delta_{L\perp} = \frac{\Delta\Theta_+ + \Delta\Theta_-}{2\cos^2\phi\tan\Theta_B} = \frac{\Delta\Theta_+ - \Delta\Theta_-}{2\cos\phi\sin\phi}. \tag{9.14}$$

If the identity (9.14) is not valid, the layer lattice is partially or fully relaxed. The degree of relaxation R may be defined in terms of the measured in-plane lattice parameter:

$$R = \frac{a_{L\|} - a_S}{a_{L\infty} - a_S}. \tag{9.15}$$

R is zero if $\delta_{L\|} = 0$ and unity for $a_\| = a_{L\infty}$ (see Fig. 9.1). If $R = 1$, then $\Delta\phi = 0$ and $a_{L\|} = a_{L\perp} = a_{L\infty}$ (see Eq. (9.4)).

The cubic lattice parameter of a ternary solid solutions such as $Al_xGa_{1-x}As$ is an approximately linear function of the composition x (Vegard's law) [227]. The mismatch scales as

$$\delta_{L\infty} = \delta_{\text{bin}}\, x. \tag{9.16}$$

δ_{bin} is the mismatch between the binary end components. For quaternary systems such as $In_xGa_{1-x}As_yP_{1-y}$, Vegard's law is valid if $x = \delta_{\text{bin}}(y)$. Thus x can be determined from the knowledge of $\delta_{L\|}$ and $\delta_{L\perp}$ in (9.5). For pseudomorph layers ($R = 0$), x can be obtained by

$$x = \frac{c_{11}}{\delta_{\text{bin}}(c_{11} + 2c_{12})}\delta_{L\perp} \equiv G\,\delta_{L\perp}. \tag{9.17}$$

For the $Ga_{1-x}Al_xAs/GaAs$ (001) $\delta_{\rm bin} = 1.4 \times 10^{-3}$ and following $G = 392$. For pseudomorphic $Ga_{1-x}In_xAs_yP_{1-y}$ layers on InP (001) ($x \approx 0.47y$) follows $G = 225$ ([2]). The respective parameter for $In_{0.16}Ga_{0.84}As/GaAs$ yields $G = 55$.

Epitaxy on step-faced substrate planes, such as (113), induces a non-

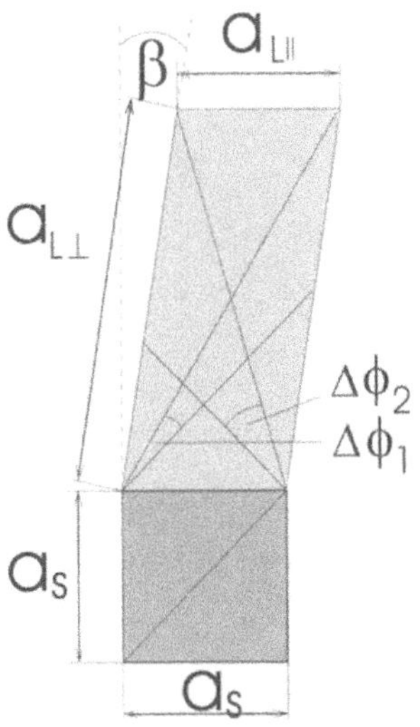

Fig. 9.3. Schematic set-up of lattice parameters for monoclinic relaxation.

tetragonal deformation status of the layer. The analysis of this status requires measurements of a particular Bragg diffraction at different azimuthal orientations with respect to the surface normal. If the symmetry of the growing face is rectangular (2D space group: 2mm), the deformation of the layer is monoclinic. Then the tilt angle β and the tilt direction with respect to the surface normal of the surface can be determined by recording the (hkl), (-h,k,l), (h,-k,l) and (-h,-k,l) diffraction curves, i.e., the hkl Bragg diffraction at the four different azimuthal angles ψ with respect to the surface normal. This problem has been analyzed in [26].

If $\Delta\phi_1$ is the inclination angle difference between layer and substrate at the diffracting lattice planes (hkl) measured at the azimuth $\psi = 0°$ (see Fig. 9.3) and $\Delta\phi_2$ is the corresponding difference at the lattice planes (-h,-k,l), i.e., at $\psi = 180°$, the z-axis of layer lattice is tilted by

$$\beta = \frac{\Delta\phi_1 - \Delta\phi_2}{\sin\phi} - \frac{1}{2} \tag{9.18}$$

relative to the (h,k,l) plane. Similar investigations are necessary for the (-h,k,l) and (h,-k,l) Bragg diffractions, i.e. for $\psi = 90°$ and $\psi = 270°$.

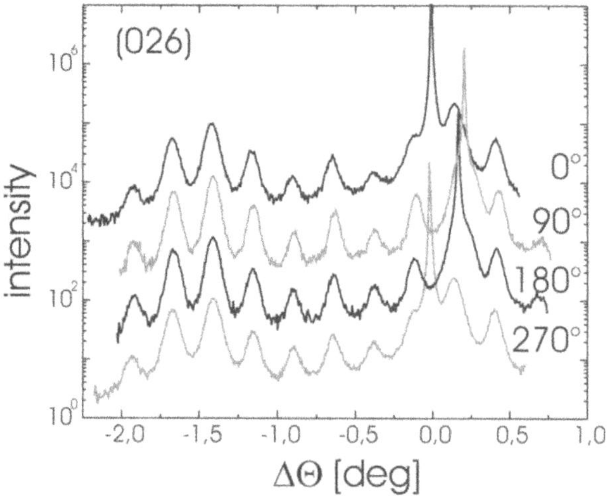

Fig. 9.4. (026) rocking curves of a [GaInAs/GaAs]×38 multilayer grown on GaAs(001) measured for four different azimuthal orientations.

This is demonstrated for an epitaxial [$Ga_{0.76}In_{0.24}As$/GaAs]×38 multilayer grown on (001) GaAs, for example, as shown in Fig. 9.4 [176]. The 026 diffraction curves were measured at four azimuthal orientations. When the curves are normalized to equal angular positions of the superlattice satellite peaks, the respective substrate peaks do not coincide. Applying Eq. (9.18), the layer lattice is tilted approximately 1000 arc seconds toward the [110] and 300 arc seconds toward [$1\bar{1}0$] direction. After the determination of the direction of deformation the average lateral and vertical lattice misfits of the superlattice can be determined to be $\frac{\Delta a_\parallel}{a} = 2.2\times10^{-3}$ and $\frac{\Delta a_\perp}{a} = 1.11\times10^{-2}$, which corresponds to an average degree of relaxation of about 15%.

Figure 9.5 shows SiGe monolayer grown on Si (001) substrate. Its thickness has been determined by x-ray reflectometry and amounts to 66 nm. The lattice misfit was probed at the asymmetric 115 diffraction in ϕ_+ and ϕ_- geometry. The separation between the layer and substrate Bragg peaks differ. For reasons of comparison the ϕ_+ and ϕ_- curves were shifted to equal substrate Bragg angles and normalized to equal substrate intensities. The data evaluation gives the values $\frac{\Delta a_\perp}{a} = 6.3 \times 10^{-3}$ and $\frac{\Delta a_\parallel}{a} = 4 \times 10^{-6}$. Because the value of $\frac{\Delta a_\parallel}{a}$ is negligibly small one can conclude that the layer is pseudomorphically grown and uniaxially expanded toward the surface normal.

The angular position of the two layer Bragg peaks in Fig. 9.5 corresponds to the maximum peak separation and represents the pseudomorphic status of strain ($R = 0$). The opposite limit, the status of total relaxation, i.e., $R = 1$,

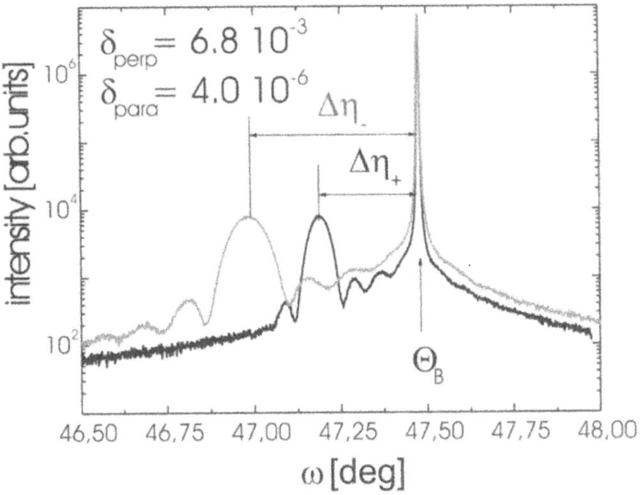

Fig. 9.5. The 115 diffraction curves of a $Si_{1-x}Ge_x$ monolayer grown onto Si (001) recorded in ϕ_+ and ϕ_- set-up. For better clarity the substrate peaks were plotted on each other and normalized at equal substrate intensity.

would be indicated by the coincidence of the two layer Bragg peaks of the ϕ_+ and ϕ_- diffraction curves.

Figure 9.6 shows two 004 diffraction curves of a $Ga_{1-x}Al_xAs/GaAs(001)$ heterostructure. The layer and substrate peaks are well separated from each other. The measured angular difference corresponds to an aluminum content of about 21% in the lower curve and about 45% in the upper one. By asymmetric diffractions it was verified that the layer was grown pseudomorphically (i.e., $\frac{\Delta a_{L\parallel}}{a} \approx 0$) [360]. Because of the very small difference of atomic radii of gallium and aluminum, pseudomorphic crystal growth can be expected for any compositions of AlAs and GaAs on GaAs substrate. Indeed, pseudomorphic crystal growth has been observed for a layer thickness of about 10 μm.

In the following examples we will show how one can evaluate the structure of a multilayer which consists of sublayers with different compositions and layer thicknesses. First we show a rocking curve of an InGaAs single quantum well grown on GaAs (001) and buried under a GaAs layer. This is the basic layer sequence of a semiconductor laser structure (Fig. 9.7). Lattice mismatch determination from the diffraction curves by eye is possible so long as the layer Bragg peak appears separated from that of the substrate. This becomes difficult if the layer is very thin and the lattice mismatch is small. If the layer and substrate peaks overlap, the layer thickness and its composition cannot be determined independent from each other.

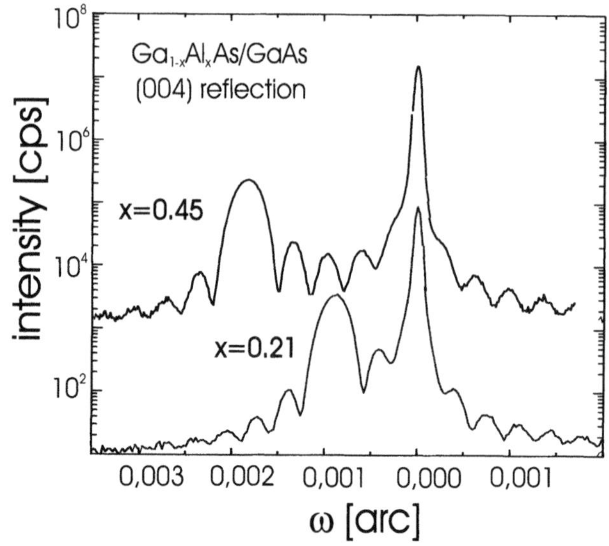

Fig. 9.6. X-ray double crystal diffraction curves of thin $Ga_{1-x}Al_xAs$ single layers covered epitaxially on GaAs(001) substrate.

The GaInAs thickness amounts to a few nanometers only and appears as a broad peak to the left of the substrate peak. The rapid oscillations measure

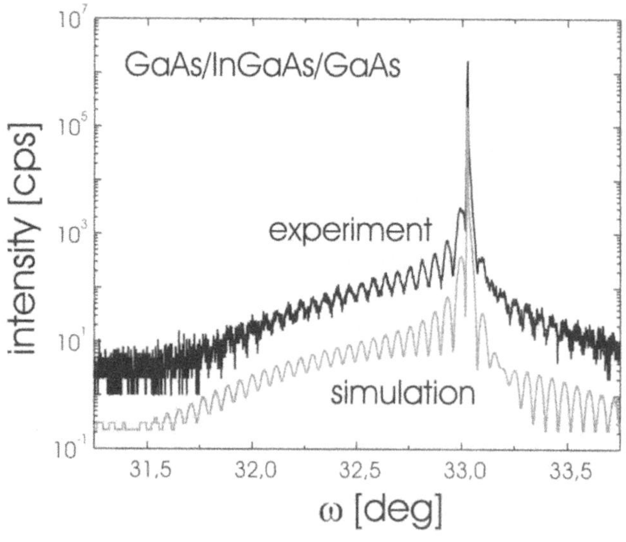

Fig. 9.7. 004 diffraction curves of two $GaAs/Ga_{1-x}In_xAs/GaAs[001]$ single quantum well structures with small peak separation as well as small layer thicknesses.

the top layer thickness with $T = 108$ nm. A computer simulation of the diffraction curves always provides the product xT only. This means means the accurate thickness can only be determined using additional information about the indium content within the layer or the layer thickness. A least-square-fit procedure gave a reasonable fit within the parameter intervals 6.25 nm< T <6.65 nm and $0.13 < x < 0.15$ [399].

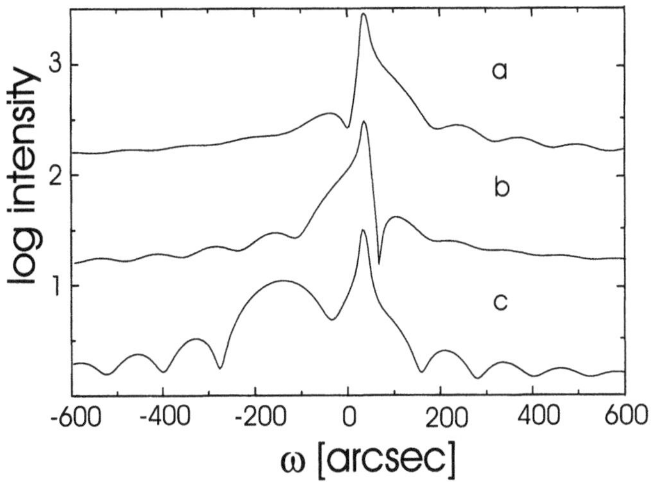

Fig. 9.8. Calculated X-ray rocking curves of the asymmetric 224 diffraction of a 250-nm-thick $Ga_{0.72}In_{0.28}As_{0.4}P_{0.6}$ layer on InP(001). The lattice misfits are zero (curve a), 2.5×10^{-4} (b), and 1×10^{-3} (c).

For other compositions of layers, in particular if the layer mismatch of the layer is close to zero, the interaction between the substrate and layer Bragg peaks becomes essential if the lattice mismatch is small. Even close to the substrate the interference term is large and can modify the rocking curve. This effect is important, if the electron density of the layer and substrate differs. In this case layer and substrate experience a different Bragg shift, $\Delta\theta_0$, as shown in (9.6), and the diffraction curve may display a small peak separation even for $\frac{\Delta a}{a} = 0$. The effect becomes significant when using asymmetric diffraction geometries . This is shown in Fig. 9.8, which illustrates the effect for a single layer of InGaAsP grown on InP. Although the electron density of the layer is about 10% larger than that of substrate and $\frac{\Delta a}{a} = 0$, the layer peak may appear at the high-angle side of the rocking curve (Fewster-Curling effect). As shown, layer and substrate peaks are well separated in curve c only [120].

The next example shows the 004 rocking curve of one typical example of a laser structure (Fig. 9.9) . It consists of a stacking of eight layers (see bottom graph of Fig. 9.9). The active layer is an $In_{0.35}Ga_{0.65}As$ layer with a thickness of about 10 nm. Due to the large indium content the lattice mismatch

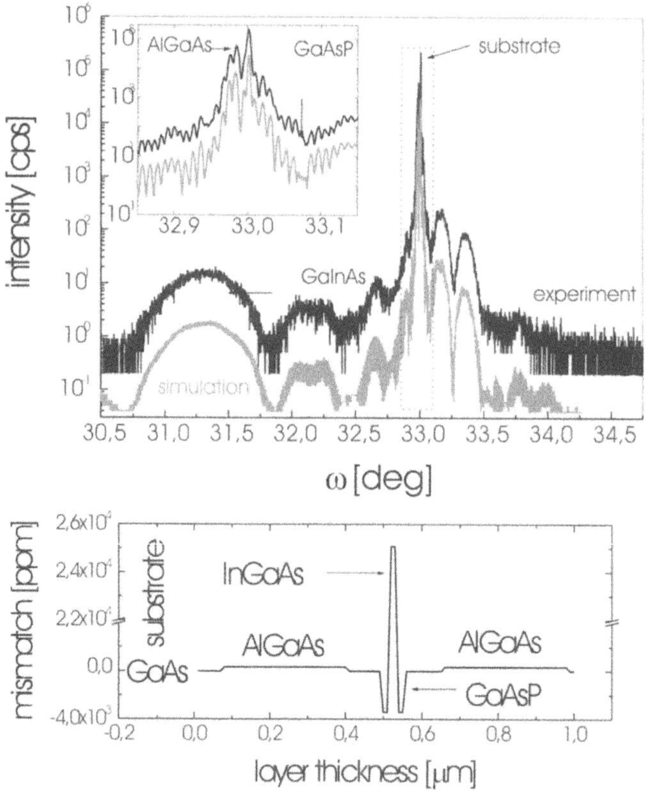

Fig. 9.9. 004 diffraction curves of a semiconductor laser structure consisting of eight layers of different compositions and thicknesses. The bottom graph shows the stacking of layers. The inset of the top figure is an enlargement of the rocking curve close to the substrate region, black curves – experiment, gray curves – simulation.

is large, which gives a broad peak, well separated from the substrate. For electronic confinement the active layer is embedded within two $GaAs_{0.9}P_{0.1}$ layers. Their thicknesses are similar to the active layer, but the mismatch is negative, which supplies a broad peak on the high angle side with respect to the substrate. Because the angular separation between the GaAsP and the substrate Bragg peaks is smaller than the width of the GaAsP peak, the shape of the GaAsP Bragg peak is strongly affected by the interference with the scattering amplitudes of the GaAs and AlGaAs layers. In fact, all the features ranging from $32.5° < \omega < 33.5°$ belong to the GaAsP Bragg peak. The oscillations in this angular range are caused by the GaAlAs layers embedding the GaAsP/InGaAs/GaAsP layer system because of optical confinement. The oscillation period measures a thickness of the embedding layer system. The thickness of the AlGaAs confinement layers are visible in the left inset of Fig. 9.9, showing an enlargement of the rocking curve close to the substrate Bragg peak. The AlGaAs Bragg peak is found at the small angle side of the substrate Bragg peak. It is split into two peaks due to the phase shift induced by the active layer system (see Sect. 8.6). The spacing between the oscillation maxima measures the AlGaAs layer thickness of about 330 nm.

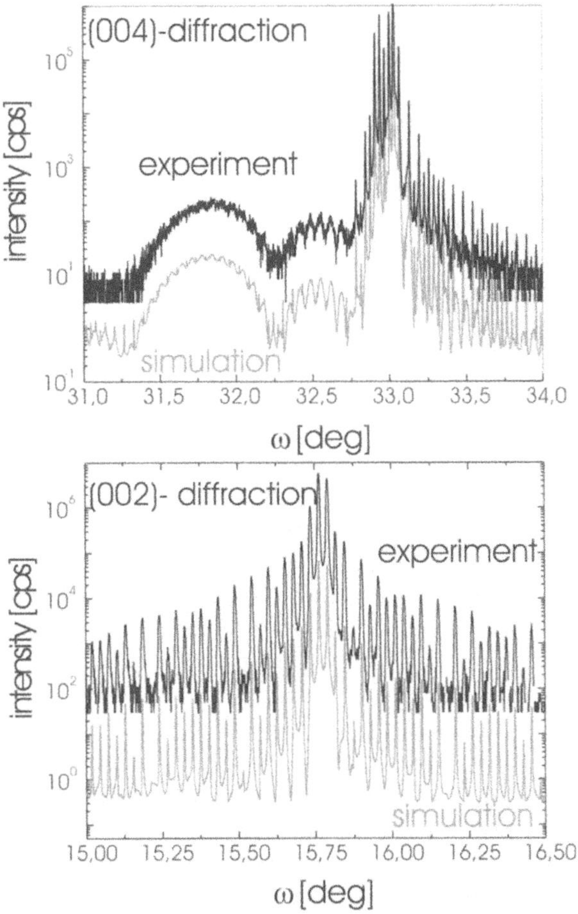

Fig. 9.10. 004 and 200 rocking curves of a semiconductor laser structure containing a InGaAs single quantum well embedded within GaAs confinement layers and grown on an AlAs/GaAs superlattice. black curves – experiment, gray curves – simulation.

Finally we show that the measurement of diffraction curves with different contrast can help to reveal the layer structure of a multilayer. Figure 9.10

shows a rocking curve of another semiconductor laser structure measured in the vicinity of the GaAs 004 Bragg diffraction. (top curve) In contrast to Fig. 9.9 the $In_{0.25}Ga_{0.75}As$ single quantum well is embedded in GaAs layers and the whole structure is grown out of a superlattice, consisting of 20 periods of AlAs/GaAs with $T_{GaAs} = 72.5$ nm and $T_{AlAs} = 76.0$ nm. The average Al content of the superlattice can be determined after the identification of the zero-th-order satellite peak of the superlattice at the small-angle side of the GaAs substrate peak. The superlattice period is extracted from the angular difference between the satellite peaks. The indium content and the quantum well thickness is obtained from the broad peak on the left, which is well separated from the main peaks of the AlAs/GaAs layers. The bottom graph of Fig. 9.10 shows the rocking curve of the same structure, measured in vicinity of the GaAs 002 Bragg diffraction. In contrast to to the 004 Bragg diffraction, the scattering contrast is larger because it depends on the difference between the ordinary numbers of the concerned atoms of the sublayers. Because this difference is particularly small for GaAs but large for AlAs, the satellite peaks of the superlattice are large in intensity. For this reason one can easily evaluate the thickness ratio T_{GaAs}/T_{AlAs} using Eq. (8.18). Note that almost every even-numbered satellite is smaller in intensity compared to the odd-numbered ones. The even-numbered satellites alter in intensity with a period of about 11 periods of the superlattice, which is caused by the phase shift with respect to the GaAs top layers (see Sect.8.6).

9.2 Reciprocal-Space Mapping

With triple-axis diffractometers the resolution element is small enough to resolve closely adjacent features in reciprocal space. This makes it possibile to characterize heterostructures and multilayer systems by reciprocal-space mapping (see Chap. 3). Instead of running a line scan along Q_z, the mapping of the scattering intensity distribution of a reciprocal area Q_x, Q_z has the advantage that the lattice misfits and the degree of relaxation can be obtained independent of the chosen scanning ratio between ω and 2θ and independent of the miscut of the diffracting lattice plane with respect to the surface. Additionally, reciprocal-space mapping records the the diffuse scattering distribution in the vicinity of the Bragg peaks (see Chap. 10).

The vertical lattice misfit appears as a peak separation between layer and substrate reciprocal lattice points along Q_z, whereas the lateral misfit is measured along Q_x. A non-zero value of the lateral lattice misfit is a direct indication of the appearance of lattice relaxation, i.e., $R > 0$ (see Eq. (9.15)).

As shown in Fig. 9.11, the layer and substrate reciprocal lattices are separated along the Q_z-axis if $R = 0$. In contrast, both peaks appear on the connection line between the origin of reciprocal space and the substrate reciprocal lattice point if $R = 1$. The connection line between the points for

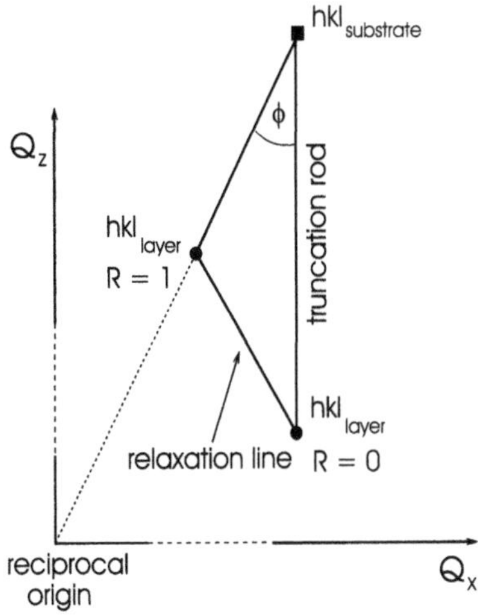

Fig. 9.11. The positions of the layer reciprocal lattice point with respect to that of the substrate for fully strained and completely relaxed multilayer systems.

$R = 1$ and $R = 0$ is called, the *relaxation line.* The layer peaks of samples with $0 < R < 1$ always appear on this line.

The technique of reciprocal-space mapping was first applied to the system ZnSe/GaAs by Heinke et al. [151]. Figure 9.12 shows reciprocal-space maps

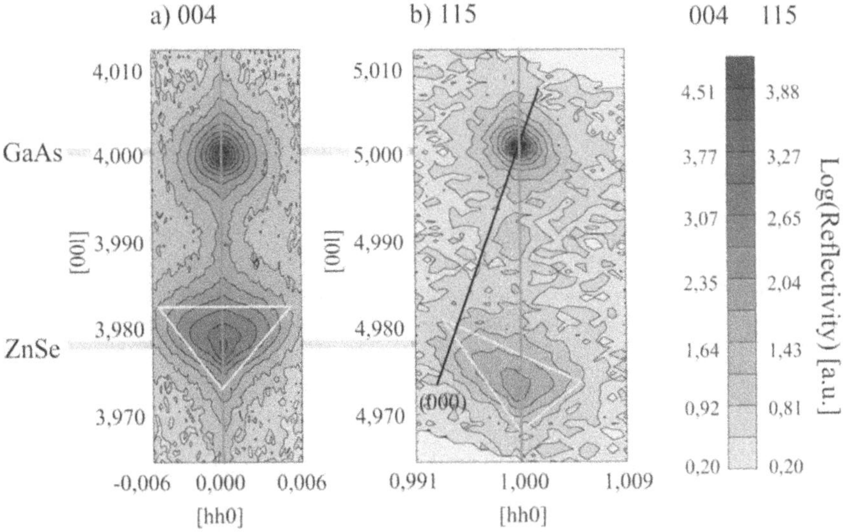

Fig. 9.12. Reciprocal-space map of ZnSe/GaAs single heterostructure measured close to the 004 and close to the asymmetric 115 diffraction of the substrate [151].

of a 310-nm-thick ZnSe layer epitaxially grown on GaAs [001] substrate. Because the 004 is insensitive for the $\Delta a_{\|}$, the layer peak appears on the Q_z-axis at $h = Q_z \times \frac{a}{2\pi} = 3.878$. The spacing between the layer and substrate peak at $h = 4.0$ provides the vertical lattice misfit. A mapping of the vicinity of the 115 reflection, measured in (ϕ_-) geometry, shows that the ZnSe layer is partially relaxed. The maximum intensity of the layer Bragg peak is not centered onto the Q_z-axis. Its intensity distribution shows a shoulder toward the relaxation line. This is an indication for the onset of relaxation. The degree of relaxation amounts to about $R \approx 10\%$. Similar measurements at asymmetric reflections with larger asymmetry (see Sect. 9.3) reveal that R decreases with increasing depth toward the substrate [127].

The method of reciprocal-space mapping is suitable for the determination

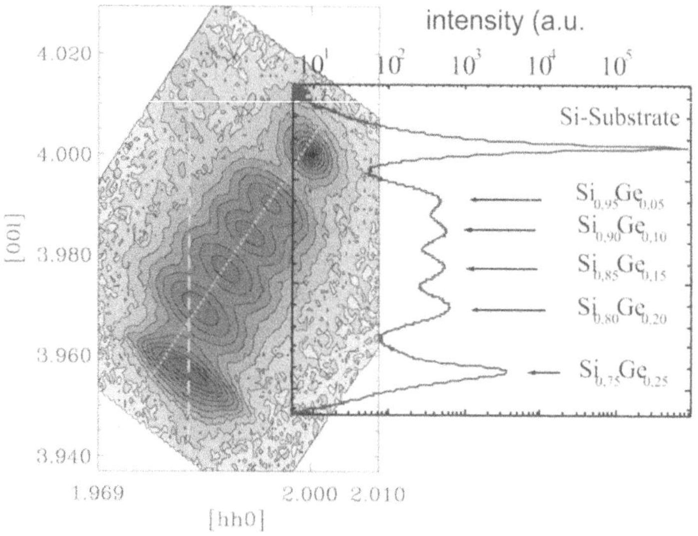

Fig. 9.13. Reciprocal-space mapping in vicinity of the 224 reflection of multilayer structure consisting in five SiGe layers of different composition grown on Si substrate. The thicknesses of the the first four layers are 500 nm, the top layer is 1,000 nm thick. The figure on right shows a Q_z line scan with open detector window [127].

of the lattice relaxation in multilayers as well. This is demonstrated in Fig. 9.13, where the asymmetrical 224 reciprocal-space map of a SiGe multilayer is plotted. The multilayer consists of five layers with Ge content x, varying from $x = 5\%$ to $x = 25\%$. The layer with ($x = 25\%$) is covered by a thin Si capping layer. The layer thicknesses are $T = 500$ nm, except the fifth layer with $T = 100$ nm. In the reciprocal-space map the substrate peak and the peaks from individual layers are well resolved. The peaks are elliptically shaped, which gives evidence for a mosaic structure of the layers (see Chap. 10). The layer peaks (1–4) lie on the relaxation line; they are completely relaxed ($R = 1$). The intensity maximum of the fifth layer is slightly strained ($R \approx 0.9$) due to the influence of the thin silicon based layer on top of it. From the position of the intensity maxima one can determine the lattice parameters $a_{L\perp}$, $a_{L\parallel}$ of each layer: using Eqs. (8.4) and (8.5), one obtains the cubic lattice mismatch $\delta_{L\infty}$ and, subsequently, on the basis of the Vegard's law (8.12) the Ge content x of each layer.

9.3 Coplanar Extremely Asymmetric Diffraction

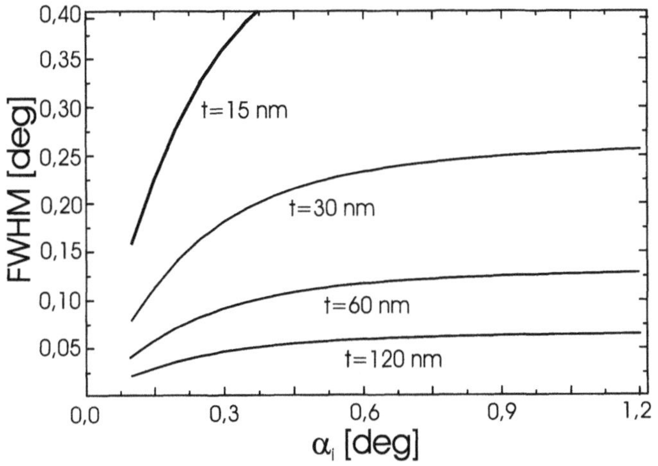

Fig. 9.14. Functional behavior of the layer peak width with varying angle of incidence, 422 diffraction, and $CuK\alpha$-radiation.

As shown in Fig. 9.7, the full width of half-maximum (FWHM) of the layer peak becomes very broad if the layer thickness decreases. Additionally the peak intensity is very small. The diffractivity of a layer follows from Eq. (5.9) and increases proportionally with T^2. Using the symmetrical 004 reflection and $\lambda = 0.154$ nm, the reflectivity of a GaAs layer is about 5×10^{-5} and 5×10^{-3} at $T = 10$ nm and $T = 100$ nm, respectively. The use of an asymmetric reflection geometry modifies the diffractivity proportional to the factor $\frac{|\gamma_h|}{\gamma_0}$ (see Eq. (6.29)); i.e., the diffractivity of an asymmetric diffraction in ϕ_- geometry is larger compared to the symmetric scattering geometry.

The angular width $\Delta\eta$ of the layer Bragg peak follows a relation similar to that derived for the angular spacing between thickness oscillations (see Sect. 5.1). That means $\Delta\eta$ is proportional to T^{-1}. For the 004 reflection of GaAs and CuK_α radiation the half-width of a 100-nm- and 10-nm-thick layer amounts to 0.44° and 0.044°. Due to the proportionality to $\gamma_{\boldsymbol{h}}$, the peak width again is larger when the ϕ_- geometry is used but smaller when the ϕ_+ geometry is used compared with the symmetric case. The ϕ_+ set-up provides smaller peak width but lower intensity. For the 335 diffraction, $\alpha_i \approx 0.54°$ using $CuK\alpha$ radiation and (001) oriented substrate. FWHM and peak intensity vary by a factor 85 between the ϕ_+ and ϕ_- set-ups. The use of the 113_+ diffraction narrows the layer peak by a factor 20 compared with 400 [202]. This demonstrates that the possibility of identification of a thin layer improves when measuring diffraction curves at extremely asymmetric diffraction geometries.

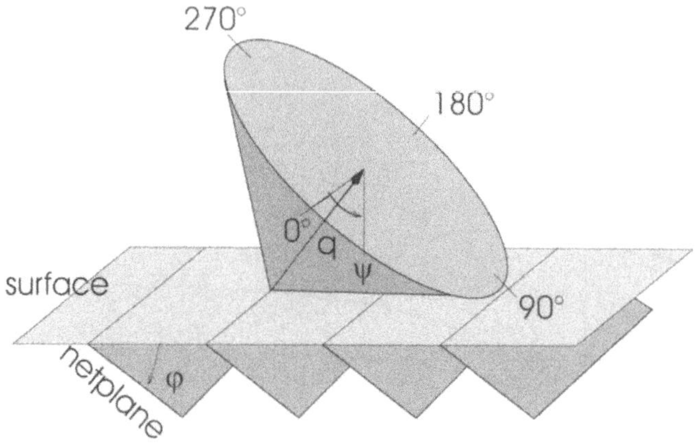

Fig. 9.15. The ψ-cone of extreme asymmetric diffraction geometry. The angle of incidence is changed by rotation of the sample around the normal of the diffracting lattice plane. Shallow incidence can be tuned if $\theta_B \approx \phi$.

The layer to substrate peak intensity ratio changes significantly if α_i approaches the critical angle of total external reflection. Under this condition the effect of refraction of the x-ray beam at the air–sample interface becomes essential. That has the consequence that the peak width becomes narrow when the layer Bragg peak appears very close to α_c. That is due to the fact that the small-angle part of the diffraction curve cannot be larger than α_c.

The effect of the extreme asymmetry on the peak width is shown in Fig. 9.14 for four different layer thicknesses of GaAs. The peak width decreases by a factor of 2 when the Bragg peak maximum appears at $\alpha_i \approx \alpha_c$ compared with $\alpha_i = 2\alpha_c$. Additionally the refraction shift of the layer peak differs from that of the substrate (see Eq. (9.6)). The peak separation is no longer a direct measure of the lattice misfit [67]. Because the refraction shift

$\Delta\theta_0$ increases with increasing asymmetry, two materials can be separated at extremely small α_i even if the lattice mismatch is zero. This was done for the system $Ca_{0.43}Sr_{0.57}F_2$ covered epitaxially on GaAs (001) [167, 264].

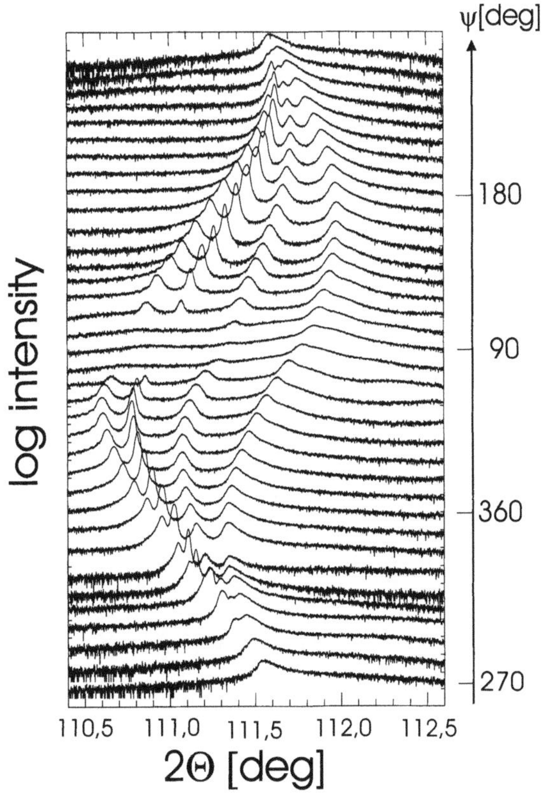

Fig. 9.16. Collection of 36 scans at the 444 reflection of a multilayer containing four CdCaTe layers of different composition. The scans are taken at different azimuthal angles ψ. The minimum penetration depth is reached close to $\psi = 90°$ [280].

Extremely asymmetric diffraction geometry can be realized by either changing incidence angles or changing wavelength. In the first case one has to choose a net plane where the inclination angle with respect to the surface $\phi \geq \theta_B$. An azimuthal rotation of the sample around the net plane normal can change $\alpha_{i,\min} = (\theta_B - \phi) \approx 0$ to $\alpha_{i,\max} \approx \theta_B + \phi$. The geometry was realized first by Borchard et al. [264] and improved by Ress et al. [280]. A general scheme of the set–up is shown in Fig. 9.15.

Under conditions of extreme asymmetric diffraction, the absorption length is always smaller than the extinction length τ (see Eq. (7.19)) and controls the effective penetration depth of the probing x-ray beam within the sample. This effect makes it possible to measure extremely thin layers. In addition one can investigate the depth variation of the strain status of a multilayer by recording rocking curves at different α_is of the same Bragg reflection. This is shown in Fig. 9.16 using the example of a multilayer system containing four different $Cd_{1-x}Ca_xTe$ layers on CdTe (001), where the Ca concentration increases from $x = 4\%$ to $x = 100\%$ from the substrate to the top. The layer thicknesses are about 300 nm each. The measurements have been performed at the 444 reflection with $\theta_B = 55.4°$ and $\phi = 54.73°$. The different Ca concentrations cause different peak shifts $\Delta\theta_0$ which differently vary with changing α_i as well. Whereas at $\psi \approx 270°$, i.e., $\alpha_{i,\max} \approx \theta_B + \phi$, the peaks nearly coincide, they are well separated at $\psi \approx 0$. Note that at $\psi = 0$ the effective incidence angle is smaller than α_c where the Bragg diffraction disappears [280].

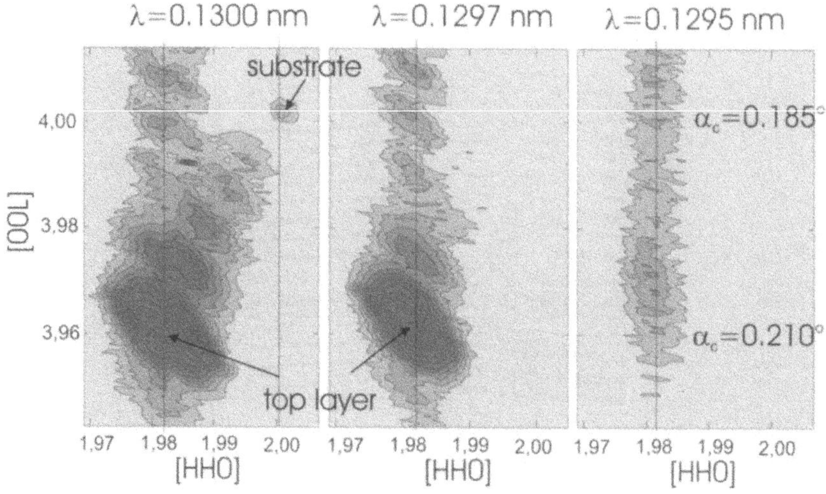

Fig. 9.17. Three reciprocal-space maps of a SiGe multilayer on Si substrate recorded with different wavelengths at the 224 Bragg reflection. The asymmetry increases with decreasing wavelength. The two values of the critical angle corresponds to those of the silicon and the $Si_{0.75}Ge_{0.25}$ top layer. The same sample was already shown in Fig. 9.13 measured at a much larger wavelength [127].

The second way to create extremely asymmetric Bragg diffraction is by varying wavelength for fixed α_i. This scheme was first devised by Brühl et al. [68, 167]. Here we show the result of the reciprocal space-mapping of the SiGe multilayer which was already shown in Fig. 9.13. There the wavelength used was $\lambda \approx 0.14$ nm, and the penetration depth of the probing x-ray for all the materials within the film was larger than the total thickness of the multilayer. Figure 9.17 shows three additional area maps of the same sample recorded at smaller wavelengths. Here the penetration depth becomes a function of the wavelength used and the material. In the wavelength range of interest the critical angles of the silicon and the $Si_{0.75}Ge_{0.25}$ top layer are $\alpha_{c,\text{sub}} = 0.185°$ and $\alpha_{c,\text{layer}} = 0.21°$. For $\lambda = 0.13$ nm the penetration depth is about 0.8 μm for silicon but 0.33 μm for the top layer. This is the reason that all the five layers and the substrate are visible. For $\lambda = 0.1297$ nm the top layer and the following layer are visible, but crystal truncation rod of the top layer dominates the reciprocal space map. This is caused by the penetration depth of the top layer, which amounts to 0.24 μm only. Finally, for $\lambda = 0.1295$ nm, the incidence angle of the top layer is smaller than the $\alpha_{c,\text{layer}}$. Therefore, the top layer Bragg peak disappears but not its truncation rod, where the critical angle is smaller. Details of the data treatment can be found in [127].

9.4 Utilization of Anomalous Scattering Effects

One advantage of experiments at synchrotron facilities is the possibility to tune the energy. This is of particular interest if one has to investigate a sample with small electron density contrast. In this case one can perform the measurement at the K- , L-, or higher absorption edges of one particular element of the sample. Due to the large change of anomalous dispersion the real and imaginary parts of the atomic form factor become essentially changed, which modifies the scattering power of the particular element. *Anomalous scattering* is a well-known technique in x-ray structure analysis. For high-resolution reflectivity it was applied first by Klemradt et al. in GaAs/InGaAs/GaAs SQW structures close to the GaK edge [197]. Here we present another instructive example published by Schülli et al. [318]. The sample under investigation consists of 100 periods of a $(PbSe)_m/(EuSe)_n$ bilayer grown on PbTe (111) substrate, where m,n denote the number of monolayers of an individual layer, when m =123 and n =13, respectively. The aim of the measurements was to verify the growth parameter and to determine the interdiffusion between the layers. Because both constituents crystallize in sodium chloride structure the 111 Bragg diffraction is weak and subsequently most sensitive for the determination of composition. Unfortunately the thickness ratio between the sublayers $T_{PbSe}/T_{EuSe} \approx 10$ and the PbSe layers will dominate the rocking curve. Because the lattice parameter of PbSe is smaller compared to PbTe, $\Delta\theta_B$ is positive and one expect the zero-th satellite peak at the high-angle side of the PbTe substrate Bragg peak. In fact this is the case, when we measure close to the EuL_{III} edge at $E = 6.970$ keV, which suppresses the structure factor of the EuSe sublayers. The situation differs when changing the energy to $E = 2.496$ keV, where the scattering factor of the PbSe layers is reduced due to the reduction of the Pb atomic form factor at the PbM_V edge. Now one can clearly identify the EuSe. Its lattice parameter is larger compared to PbTe which gives negative $\Delta\theta_B$. The thickness ratio of the layer follows from the modulation period of the satellite maxima. Clearly one can observe that such a period consists of about 10 satellite of the superlattice.

The numerical evaluation of both experimental curves helps to determine

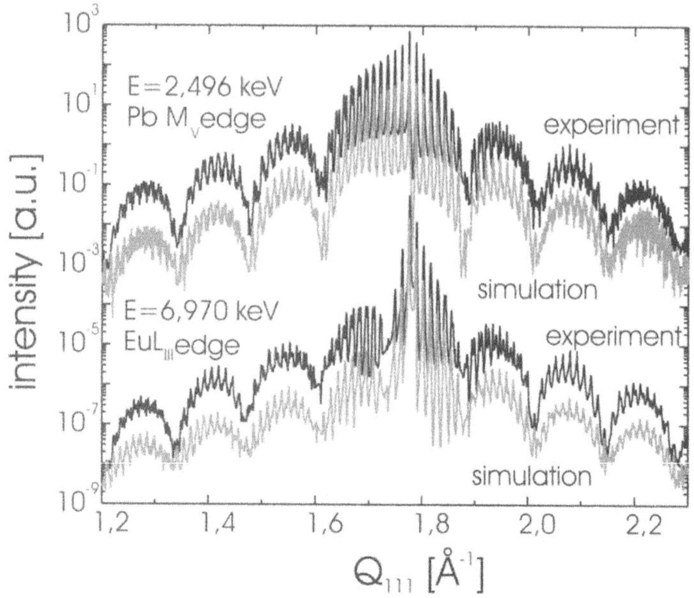

Fig. 9.18. 111 X-ray diffraction curves of a PbSe/EuSe multilayer measured at E = 2.496 keV, the PbM_V edge (top curves), and at E =6.97 keV, which is the EuL_{III} edge. The strong reduction of a single atomic scattering factor at the respective absorption edge changes the scattering power of one of the sublayers and subsequently increases the scattering contrast.

the thickness ratio and the particular lattice mismatch of both sublayers. The simulated rocking curves shown in Fig. 9.18 were calculated using a double layer thickness $D = T_{\text{PbSe}} + T_{\text{EuSe}} = 48$ nm, which corresponds to (135.8 ± 0.4) monolayers where the EuSe layers contribute with (12.45 ± 0.1) monolayers. The out-of-plane lattice parameters were determined to be $a_{\perp,\text{PbSe}} = 0.612$ nm. The interdiffusion length is less than 0.1 nm.

9.5 Grazing-Incidence Diffraction

In contrast to the coplanar diffraction, the grazing-incidence diffraction scheme is suited for measurement of the in-plane lattice parameter even for very thin layers. This was the very first application of this scheme published by Marra and Eisenberger in 1979 [234]. For semiconductor multilayers the GID scheme can be used to determine the degree relaxation R of short period

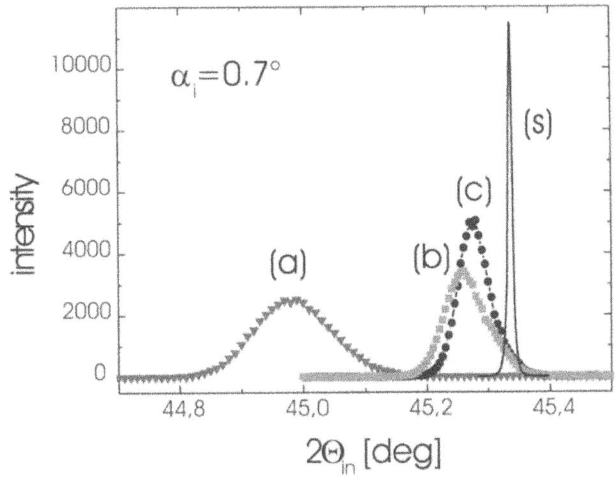

Fig. 9.19. 200 in-plane rocking curves of various samples of $In_{0.8}Ga_{0.2}As/GaAs$ multilayers, with $T_{InGaAs} = 18$ nm but varying thickness of the GaAs spacer layer. The measurements have been performed with home goniometers using a Göbel mirror and a three-bounce analyzer. The $a_{\|}$ amounts to 0.56651 nm (a), 0.56616 nm (b) and 0.56601 nm (c). The lattice parameter of the substrate was determined to be 0.5653 nm.

multilayers. This has been done at several GaInAs/GaAs multilayers [292], where the thickness and composition of the InGaAs sublayers was close to the critical thickness t_c. The samples have equal total thickness ($T_{tot} = 800$ nm) and equal thickness of the $Ga_{0.8}In_{0.2}As$ layer ($T_{GaInAs} = 18$ nm) but they differ in the thickness of the GaAs spacers. In fact, different angular positions of the 200 in-plane Bragg diffraction corresponds to different average lateral lattice parameters of the multilayer, $a_{\|}$ (see Fig. 9.19). Except for the sample with $t_{GaAs} = 30$ nm (not shown), the measured diffraction maximum of samples decreases for decreasing spacer thicknesses, indicating the onset of relaxation.

The evaluated degree of relaxation R is verified by the out-of-plane lattice parameter measured by HRXRD (see Eq. (9.17)). As shown in Fig. 9.20 the multilayer system behaves approximately pseudomorphicaly for thick spacer layers ($T_{GaAs} = 30$ nm), but it shows almost complete relaxation for $T <$ 2 nm.

The main advantage of the GID arrangement lies in the capability to measure the in-plane lattice parameter as a function of depth below the sample surface. If both α_i and α_f are smaller than α_c, the penetration depth L of the x-ray is extremely reduced. For GaAs L is about 5 nm. In this condition, only the top region of the sample is under investigation. L becomes

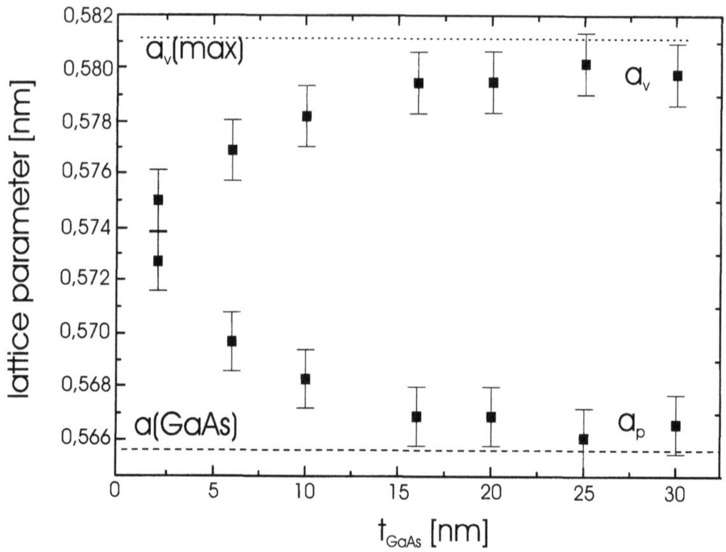

Fig. 9.20. In-plane and out-of-plane lattice parameters of an 800-nm-thick $In_{0.8}Ga_{0.2}As/GaAs$ multilayer system, with $T_{InGaAs} = 18$ nm but varying thicknesses of the GaAs spacer layer. The vertical lattice parameters $a_\perp$ were measured at the 004 diffraction, the in-plane parameters $a_\|$, at the 200 diffraction in GID geometry.

larger when one of both angles is increased. When the x-ray absorption of material is taken into account, L does not increase instantaneously if one of the angles exceeds α_c. For GaAs and $\lambda \approx 0.150$ nm, it increases between 10 nm$< L <$400 nm changing α_i between $\alpha_i < \alpha_c < 2\ldots3\alpha_i$ [265].

Figure 9.21 shows an example of such depth-resolved measurements. With a sample similar to that mentioned above but covered by a 100-nm-thick GaAs top layer, the in-plane rocking curve was recorded at $\alpha_i \approx 2\alpha_c$. The measurements have been performed using a position sensitive detector (see Sect. 2.3) aligned perpendicularly to the sample surface. At a bending magnet beamline the in-plane resolution is relaxed so that the whole intensity distribution along Q_z can be recorded by a single α_f scan at a certain in-plane angle $\theta - \theta_B$. The top curve *integrated* in Fig. 9.21 shows the intensity distribution recorded over the whole α_f range accessible by the detector. The other two curves show the corresponding intensities from the same scan but from selected channels of the PSD; the curves *surface peak* and *SL-peaks* were integrated between $\alpha_f \leq 2.5\alpha_c$ and $\alpha_f > 2.5\alpha_c$, respectively. These two curves show maximum intensity at different $\theta - \theta_B$ positions indicating a different in-plane lattice parameter in the top layer and within the multilayer. Whereas the value $a_\|$ of the GaAs top layer measured from the intensity close to the surface peak of the α_f scans coincides with the value of the substrate,

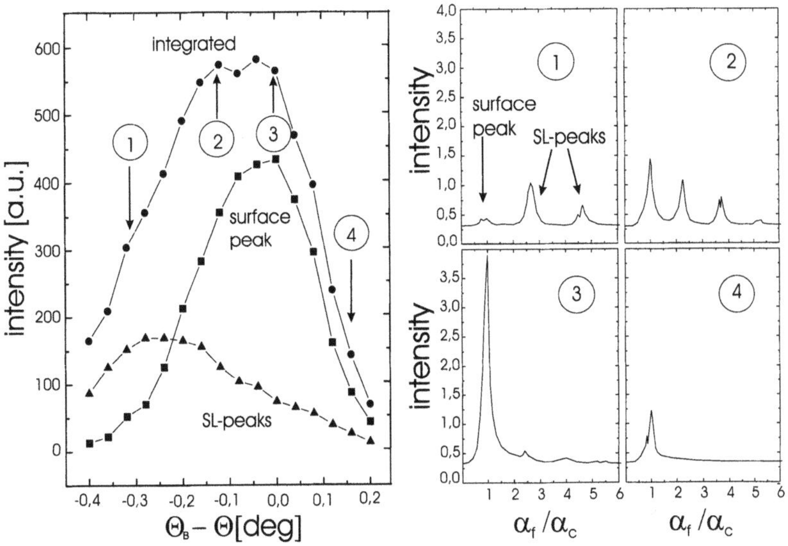

Fig. 9.21. Depth resolved evaluation of in-plane lattice parameters of a GaInAs/GaAs multilayer system covered by a 100 nm thick GaAs top layer via recording of trucation rods at different in-plane positions $\theta - \theta_B$. The angular positions of rod scans shown in the right graph are indicated by numbers 1 - 4. The rocking curves *surface peak* and *SL-peaks* were obtained from the ranges $\alpha_f \leq 2.5\alpha_c$ and $\alpha_f > 2.5\alpha_c$, respectively. The curve *integrated* was obtained from the whole α_f range.

the $a_{||}$ of the multilayer beneath is is larger; i.e., the multilayer is relaxed. The detailed relaxation profile of the sample could be estimated via computer simulation from the measured GID curves [266].

The realization of grazing-incidence diffraction is not limited to synchrotron radiation users. Home equipment can also provide sufficient intensity to measure in-plane rocking curves. One example is shown in Fig. 9.22, where the $10\bar{1}0$ in-plane rocking curve of a 1μm-thick hexagonally GaN layer has been recorded for different angles of incidence. To gain the intensity of a sealed tube, the authors used the line focus and collimated the incident beam in the in-plane direction. The incident angle was varied by moving the sample vertically through the divergent incident beam, a certain vertical position of the sample corresponds to a definite α_i [56].

The MOCVD growth of GaN suffers from the large lattice mismatch between layer and substrate. For a sapphire substrate the mismatch amounts to 16.1%. Thus, a 20-nm AlN layer was grown prior to the GaN. Nevertheless, the GaN grows in different columns slightly twisted against each other. This results in a large angle distribution perpendicular to the growth direction. The development of this in-plane angular distribution was measured by GID. For the example shown in Fig. 9.22, the *full width at half-maximum* de-

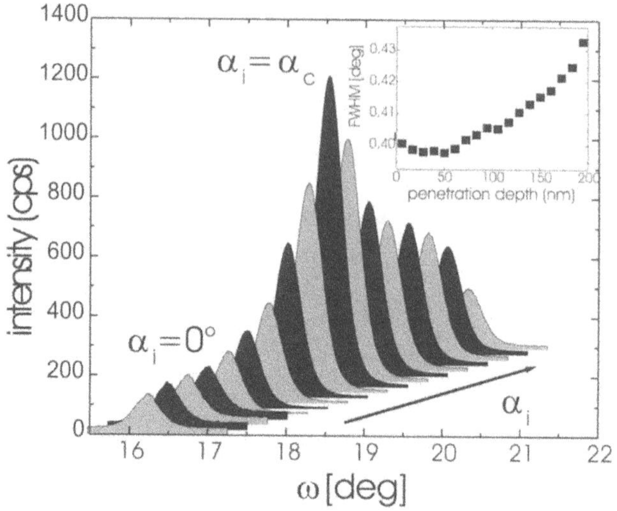

Fig. 9.22. In-plane rocking curves of a GaN layer epitaxially grown on sapphire substrate by MOCVD. The measurement has been performed in a home laboratory using the line focus of a sealed tube. The inset displays the development of the in-plane curve width as a function of the penetration depth controlled by α_i. For better clarity the rocking curves are displayed with an increasing offset in ω.

creases with decreasing penetration depth of the probing x-ray; that means the crystal quality increases with crystal growth.

GID enables an accurate determination of the in-plane lattice mismatch. This is realized by recording α_f scans instead of in-plane diffraction curves. If the thin top layer and layer beneath differ in the in-plane lattice parameter, both reciprocal lattice points create separate crystal truncation rods . They can be measured as a function of α_f and θ_i (Fig. 3.9). As shown in Fig. 3.10 a PSD scan corresponds to an oblique scan across the reciprocal space. If the detector is fixed at origin of the substrate truncation rod (CTR_0), i.e., at the Bragg position $\sin\Theta_{i1} = \frac{\lambda}{2a_1}, \alpha_f \approx 0$ with $a_1 < a_2$ the PSD cuts, the second truncation rod (CTR_1), corresponding to a_2, at the position $\sin\Theta_{i2} = \frac{\lambda}{2a_2}$ and $\alpha_f > 0$ [239].

From Eq. (4.33), this is situation can be expressed as

$$\Delta\alpha_f^2 = \alpha_i^2 - 2(\Theta_{i1} - \Theta_{i2})\sin 2\Theta_B. \tag{9.19}$$

If both α_i and α_f are small, a small angular separation $\Theta_{i1} - \Theta_{i2}$ becomes expanded toward the α_f- scale. Using $\alpha_i \approx \alpha_c$ and $\theta_B \approx 15°$, the stretching factor is in the order of 100.

Figure 9.23 shows several α_f scans of a thin silicon layer sputtered on sapphire , recorded at different in-plane positions Θ_i. Both the substrate and an interface peak can be clearly distinguished. The peak left always marks the

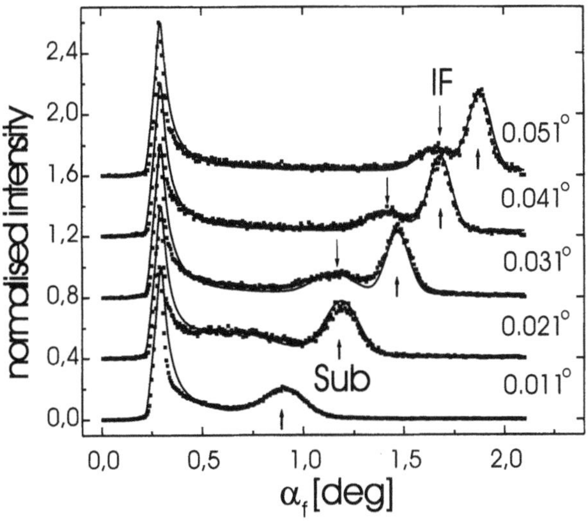

Fig. 9.23. α_f resolved Bragg intensity of a silicon layer on sapphire. The PSD scans cut two truncation rods, one from the sapphire substrate (Sub) and a second from a silicide interface layer (IF). The corresponding peaks change along α_f, changing the angle deviation with respect to the Bragg angle of the sapphire substrate θ_i.

critical angle α_c of the sapphire substrate. The silicon layer peak is not visible. The two peaks at larger α_f change their angular position when the in-plane angle θ_i is changed. The corresponding lattice parameters are obtained by extrapolation of the respective α_f positions of the two peaks towards $\alpha_f = 0$ and are located at $\theta_{\text{sapphire}} = 18.634°$ and $\theta_{IF} = 18.648°$. Both peaks are separated by $\delta_{L\|} = 8\ 10^{-4}$. The presence of the interface peak is explained by a very thin aluminum silicate layer which is formed during the growth of the silicon layer by interdiffusion of aluminum out of the sapphire substrate.

The sensitivity of this particular method has been predicted theoretically to be about $\delta_{L\|} \approx 10^{-7}$ [380].

10 Diffuse Scattering From Volume Defects in Thin Layers

In this chapter we present a detailed theoretical description of diffuse x-ray scattering from structure defects randomly placed in the volume of thin layers. We deal with so called weak defects (point defects and their clusters, small precipitates of another phase in the crystal lattice) as well as with strong defects (dislocations and their pile-ups). The description is based on the theory explained in the monograph [207]. In contrast to this work, we focus our description on defects in thin layers, where the relaxation at the interfaces affects the symmetry of the deformation field and consequently influences the reciprocal space distribution of scattered intensity.

Diffuse scattering from defects in thin layers will be described within the kinematical approach (Chap. 5). This method is suitable for the simulation of diffuse scattering in the conventional coplanar arrangement, where the dynamical effects can be neglected. In grazing-incidence diffraction, the kinematical method cannot be used, since x-ray reflection and refraction at the substrate surface cannot be neglected. In this case, the DWBA method (Chap. 7) is applicable, where a semi-infinite amorphous continuum is chosen as the non-disturbed system. In special cases, x-ray dynamical diffraction in the crystal surrounding the defects cannot be neglected. Then, a perfect crystal is used as the non-disturbed system taking into account dynamical diffraction in the crystal matrix but kinematical scattering from the defects.

10.1 Weak and Strong Defects

According to the classical monograph by Krivoglaz [207], we divide the defects into two classes according to the value of the static Debye-Waller factor defined in (5.48). We derive the expression for the static Debye-Waller factor for a crystal lattice containing randomly distributed defects. We neglect the correlation of the positions of the defects and the influence of the free surface on the displacement field around the defect. We characterize the defect distribution by random occupation numbers c_j^α; the value of c_j^α is unity if the lattice site $\boldsymbol{r}_j$ is occupied by a defect of type α, otherwise it is zero. The mean value of c_j^α is

$$\langle c_j^\alpha \rangle = c^\alpha = n^\alpha V_{\mathrm{cell}}, \tag{10.1}$$

where n^α is the volume density of the defects of type α. In a statistically homogeneous sample, this density is constant, and, in most cases, we can assume $c^\alpha \ll 1$.

Under the assumption above, the total displacement in the lattice point $\boldsymbol{r}_j$ is the superposition of the contribution of all defects

$$\boldsymbol{u}(\boldsymbol{r}_j) \equiv \boldsymbol{u}_j = \sum_k \sum_\alpha c_k^\alpha \boldsymbol{v}_{jk}^\alpha, \tag{10.2}$$

where $\boldsymbol{v}_{jk}^\alpha \equiv \boldsymbol{v}^\alpha(\boldsymbol{r}_j - \boldsymbol{r}_k)$ is the displacement of the lattice point $\boldsymbol{r}_j$ due to a defect of type α in point $\boldsymbol{r}_k$.

The mean deformation defining the averaged lattice is

$$\langle \boldsymbol{u}_j \rangle = \sum_\alpha c^\alpha \sum_k \boldsymbol{v}_{jk}^\alpha. \tag{10.3}$$

From this averaged lattice we construct the reciprocal lattice; the lattice point H of the reciprocal lattice (OH $\equiv \boldsymbol{h}^{\mathrm{def}}$) is the origin of the reduced scattering vector $\boldsymbol{q}$.

The static Debye-Waller factor is given by the expression

$$\mathcal{D} = \langle \mathrm{e}^{-\mathrm{i}\boldsymbol{h}.\delta\boldsymbol{u}} \rangle = \prod_{\alpha,k} \langle \mathrm{e}^{-\mathrm{i}(c_k^\alpha - c^\alpha)\boldsymbol{h}.\boldsymbol{v}_{jk}^\alpha} \rangle, \tag{10.4}$$

where we assume that the positions of individual defects are statistically independent. If we assume $c^\alpha \ll 1$, we obtain after some algebra

$$\mathcal{D} \approx \exp\left[-\sum_\alpha c^\alpha \sum_k \left(1 - \mathrm{i}\boldsymbol{h}.\boldsymbol{v}_{jk}^\alpha - \mathrm{e}^{-\mathrm{i}\boldsymbol{h}.\boldsymbol{v}_{jk}^\alpha}\right)\right]. \tag{10.5}$$

The main contribution to the value of the sum $\sum_k$ stems from the points $\boldsymbol{r}_k$ far away from $\boldsymbol{r}_j$. At these points, the function $\exp(-\mathrm{i}\boldsymbol{h}.\boldsymbol{v}_{jk}^\alpha)$ oscillates slowly so that we can replace this sum by the integral

$$\mathcal{D} \approx \exp\left[-\sum_\alpha n^\alpha \int \mathrm{d}^3\boldsymbol{r} \left(1 - \mathrm{i}\boldsymbol{h}.\boldsymbol{v}(\boldsymbol{r})^\alpha - \mathrm{e}^{-\mathrm{i}\boldsymbol{h}.\boldsymbol{v}(\boldsymbol{r})^\alpha}\right)\right]. \tag{10.6}$$

The defects can be divided into two groups according to the behavior of the sum in Eq. (10.5). In the case of *weak defects*, this sum converges to a finite value and hence the static Debye-Waller factor $\mathcal{D}$ is not zero. In this case, the diffracted wave contains both coherent and incoherent (diffuse) components. For *strong defects*, the sum in (10.5) diverges and $\mathcal{D} = 0$. In this case, the diffracted wave consists of the diffuse component only.

The classification of defects into these two types is rather simple for an infinite crystal. As shown in [207], the decisive factor is the asymptotic behavior of the displacement field $\boldsymbol{v}^\alpha(\boldsymbol{r})$. If the size of the defects is finite in all directions (dislocation loops, small inclusions, etc.), the asymptotic displacement field decreases with $|\boldsymbol{r}|$ as $|\boldsymbol{r}|^{-2}$. In this case, the sum in Eq. (10.5) converges and the defect is weak. If the size of the defect is infinite in one direction (a straight dislocation, for instance), the displacement field behaves as $|\boldsymbol{r}|^{-1}$ and the sum diverges. Such a defect is strong.

In the case of a finite crystal (or a thin layer), the finiteness of the sum depends not only on the displacement field $\boldsymbol{v}^{\alpha}(\boldsymbol{r})$, but also on the crystal size. In this case, a distinct classification of the defect types does not exist [182].

10.2 Diffuse Scattering From Weak Defects

As stated above, finite defects in an infinite crystal matrix are weak. Such defects consist of two parts:

1. the defect core, where the structure differs appreciably from the crystal structure of the surrounding matrix;
2. the deformed area of the surrounding crystal matrix.

The polarizability coefficients $\chi_{\boldsymbol{g}}$ of the defect core are different from those of the non-deformed crystal; in the deformed area around the core the Takagi approximation can be assumed, i.e. the polarization coefficients are not affected by the deformation and they equal the coefficients of a perfect crystal matrix.

Let us calculate the correlation function of the crystal deformation defined in Eq. (5.43) assuming non-correlated positions of the defects. In order to obtain the coherent part of this correlation function, we evaluate the average

$$\Phi(\boldsymbol{r}) \equiv \left\langle \chi_{\boldsymbol{h}}^{\mathrm{def}}(\boldsymbol{r}) \mathrm{e}^{-\mathrm{i}\boldsymbol{h}.\delta\boldsymbol{u}(\boldsymbol{r})} \right\rangle .$$

We define $\Omega^{\alpha}(\boldsymbol{r}_j - \boldsymbol{r}_k) \equiv \Omega^{\alpha}_{jk}$ the shape function of a defect of type α; this function is unity, if both lattice points $\boldsymbol{r}_{j,k}$ lie in the defect core of the same defect. We assume, for the sake of simplicity, that all defects of the same type have the same shape, and we denote as $\Delta\chi_{\boldsymbol{h}}$ the difference of the values of $\chi_{\boldsymbol{h}}$ between the defect core and the medium outside. Assuming $c^{\alpha} \ll 1$, we obtain

$$\Phi(\boldsymbol{r}_j) \equiv \Phi_j = \chi_{\boldsymbol{h}} + \sum_{\alpha} \sum_{m} c^{\alpha}(m) \left[\chi_{\boldsymbol{h}} \left(\mathrm{e}^{-\mathrm{i}\boldsymbol{h}.\boldsymbol{v}^{\alpha}_{jm}} + \mathrm{i}\boldsymbol{h}.\boldsymbol{v}^{\alpha}_{jm} - 1 \right) + \right.$$
$$\left. + \Delta\chi_{\boldsymbol{h}} \Omega^{\alpha}_{jm} \mathrm{e}^{-\mathrm{i}\boldsymbol{h}.\boldsymbol{v}^{\alpha}_{jm}} \right] . \qquad (10.7)$$

Deriving this equation, we *did not assume* the statistical homogeneity of the sample; i.e., the defect density depends on the position in the sample, and $\langle c^{\alpha}_m \rangle \equiv c^{\alpha}(m)$. The coherent part of the correlation function of the deformation is then $\Phi_j \Phi^*_k$. The incoherent part of the correlation function, defined in Eq., (5.45) is [100, 207]

$$M_{jk} \equiv M(\boldsymbol{r}_j, \boldsymbol{r}_k) = \sum_{\alpha} \sum_{m} c^{\alpha}(m) \Psi^{\alpha}_{jm} \Psi^{\alpha *}_{km}, \qquad (10.8)$$

where we have denoted

$$\Psi^{\alpha}_{jm} = \chi_{\boldsymbol{h}}\left(1 - \mathrm{e}^{-\mathrm{i}\boldsymbol{h}.\boldsymbol{v}^{\alpha}_{jm}}\right) + \Delta\chi_{\boldsymbol{h}}\Omega^{\alpha}_{jm}\mathrm{e}^{-\mathrm{i}\boldsymbol{h}.\boldsymbol{v}^{\alpha}_{jm}}. \tag{10.9}$$

Replacing the sum $\sum_m$ in Eq. (10.8) by an integral, we obtain the final formula for the incoherent part of the correlation function

$$M(\boldsymbol{r},\boldsymbol{r}') = \int \mathrm{d}^3\boldsymbol{r}'' \sum_{\alpha} n^{\alpha}(\boldsymbol{r}'')\Psi^{\alpha}(\boldsymbol{r}-\boldsymbol{r}'')\Psi^{\alpha*}(\boldsymbol{r}'-\boldsymbol{r}''), \tag{10.10}$$

and

$$\Psi^{\alpha}(\boldsymbol{r}) = \chi_{\boldsymbol{h}}\left(1 - \mathrm{e}^{-\mathrm{i}\boldsymbol{h}.\boldsymbol{v}^{\alpha}(\boldsymbol{r})}\right) + \Delta\chi_{\boldsymbol{h}}\Omega^{\alpha}(\boldsymbol{r})\mathrm{e}^{-\mathrm{i}\boldsymbol{h}.\boldsymbol{v}^{\alpha}(\boldsymbol{r})}. \tag{10.11}$$

For a statistically homogeneous crystal, the density n^{α} of the defects of type α is constant, and consequently, the the covariance $M(\boldsymbol{r},\boldsymbol{r}')$ depends only on the difference $\boldsymbol{r}-\boldsymbol{r}'$.

Knowing the covariance, we calculate the intensity distribution $J_{\text{incoh}}(\boldsymbol{Q})$ of the incoherent part of the scattered intensity, using Eqs. (5.49) and (5.51) and assuming a statistical homogeneity of the sample. From Eqs. (10.10, 10.11) and (5.49) we obtain immediately

$$J_{\text{incoh}}(\boldsymbol{Q}) = I_i \frac{K^6T^2}{32\pi^3} \sum_{\alpha} n^{\alpha} \times$$

$$\times \int \mathrm{d}\kappa_z \left| \Psi^{\alpha\mathrm{FT}}(\boldsymbol{q}_{\|}, \kappa_z) \mathrm{sinc}\left(\frac{T}{2}(q_z - \kappa_z)\right)\right|^2, \tag{10.12}$$

where $\boldsymbol{q} \equiv \boldsymbol{Q} - \boldsymbol{h}^{\text{def}}$ is the reduced scattering vector and $\mathrm{sinc}(x) \equiv \sin(x)/x$. The term $\mathrm{sinc}(T(q_z - \kappa_z)/2)$ is the geometrical factor of the layer, i.e., the Fourier transformation of its one-dimensional shape function (see Sect. 5.1). The intensity distribution of the scattered wave is therefore proportional to the convolution of the square of Ψ^{FT} with the square of this shape function.

If a defect (i.e., the area where $M^{\alpha}(\boldsymbol{r}-\boldsymbol{r}')$ essentially differs from zero) is much smaller than the layer thickness, the shape function is very narrow. Then, we use the approximate expression (5.51) for the scattered intensity and we find that the reciprocal-space distribution of the diffusely scattered intensity is proportional to the square of the Fourier transformation of the function Ψ

$$J_{\text{incoh}}(\boldsymbol{Q}) = I_i \frac{K^6T}{16\pi^2} \sum_{\alpha} n^{\alpha} \left|\Psi^{\alpha\mathrm{FT}}(\boldsymbol{q})\right|^2. \tag{10.13}$$

Eqs. (10.12, 10.13) have been derived within the kinematic approximation. Similar formulas can be obtained using the DWBA method described in Chap. 7, where we take a semi-infinite amorphous continuum as the undisturbed system. The resulting formula for the intensity distribution $J_{\text{incoh}}(\boldsymbol{Q})$ is similar to (10.12). The function $\Psi^{\alpha\mathrm{FT}}(\boldsymbol{q})$ is replaced by $\Psi^{\alpha\mathrm{FT}}(\boldsymbol{q}_T)$, where

$\boldsymbol{q}_T$ is the reduced scattering vector corrected to refraction at the sample surface, and the intensity is multiplied by the Yoneda term $|t_1 t_2|^2$ containing the Fresnel transmittivities of the surface in the non-disturbed eigenstates $|E_1^{(A)}\rangle$ and $|E_2^{(-A)}\rangle$, see Eq. (7.11).

From Eqs. (10.12, 10.13) it follows that the diffusely scattered intensity is a coherent superposition of two contributions. The first contribution is caused by the diffuse scattering from the deformed area around the defect core; this part is represented by the Fourier transformation of the term

$$\Psi^{\alpha}_{\text{Huang}}(\boldsymbol{r}) = \chi_{\boldsymbol{h}}\left(1 - \mathrm{e}^{-\mathrm{i}\boldsymbol{h}.\boldsymbol{v}^{\alpha}(\boldsymbol{r})}\right)$$

in Eq. (10.11). If we assume that the displacements of the atoms around the defect core are always much smaller than the distance $2\pi/|\boldsymbol{h}|$ of diffracting crystal planes, we can simplify this expression to

$$\Psi^{\alpha}_{\text{Huang}}(\boldsymbol{r}) \approx \mathrm{i}\chi_{\boldsymbol{h}}\boldsymbol{h}.\boldsymbol{v}^{\alpha}(\boldsymbol{r}).$$

For very small defects, Eq. (10.13) is valid. Then, the intensity of this part of the scattered wave is

$$J_{\text{Huang}}(\boldsymbol{Q}) = I_i \frac{K^6 T}{16\pi^2}|\chi_{\boldsymbol{h}}|^2 \sum_{\alpha} n^{\alpha}\left|\boldsymbol{h}.\boldsymbol{v}^{\alpha\text{FT}}(\boldsymbol{q})\right|^2, \tag{10.14}$$

i.e. it is proportional to the square of the Fourier transformation of the scalar product $\boldsymbol{h}.\boldsymbol{v}(\boldsymbol{r})$. This scattering process is usually called *Huang scattering* [107, 172, 207].

The second term in Eq. (10.11)

$$\Psi^{\alpha}_{\text{core}}(\boldsymbol{r}) = \Delta\chi_{\boldsymbol{h}}\Omega^{\alpha}(\boldsymbol{r})\mathrm{e}^{-\mathrm{i}\boldsymbol{h}.\boldsymbol{v}^{\alpha}(\boldsymbol{r})}$$

describes the contribution of the defect core to the scattering. The corresponding intensity $J_{\text{core}}(\boldsymbol{Q})$ is proportional to the square of the Fourier transformation of the defect shape function $\Omega(\boldsymbol{r})$ slightly modified by the phase factor $\exp(-\mathrm{i}\boldsymbol{h}.\boldsymbol{v})$.

As a simple example, we present the intensity distributions calculated for a crystal with randomly distributed spherical inclusions with radius R. Due to the mismatch δ between the lattices of the inclusion and the surrounding crystal, the inclusions induce an elastic deformation of the crystal matrix. The deformation field $\boldsymbol{v}(\boldsymbol{r})$ can easily be calculated, if we neglect the surface relaxation of internal stresses and assume elastic isotropy of both lattices. Then, the displacement field has a Coulomb-like form [114]:

$$\boldsymbol{v}(\boldsymbol{r}) = \begin{cases} P\boldsymbol{r}/r^3 & \text{for } r > R \\ P\boldsymbol{r}/R^3 & \text{for } r \le R \end{cases}. \tag{10.15}$$

The constant P expresses the defect "strength"

$$P = \frac{\delta}{4\pi}\frac{1+\nu}{1-\nu}, \tag{10.16}$$

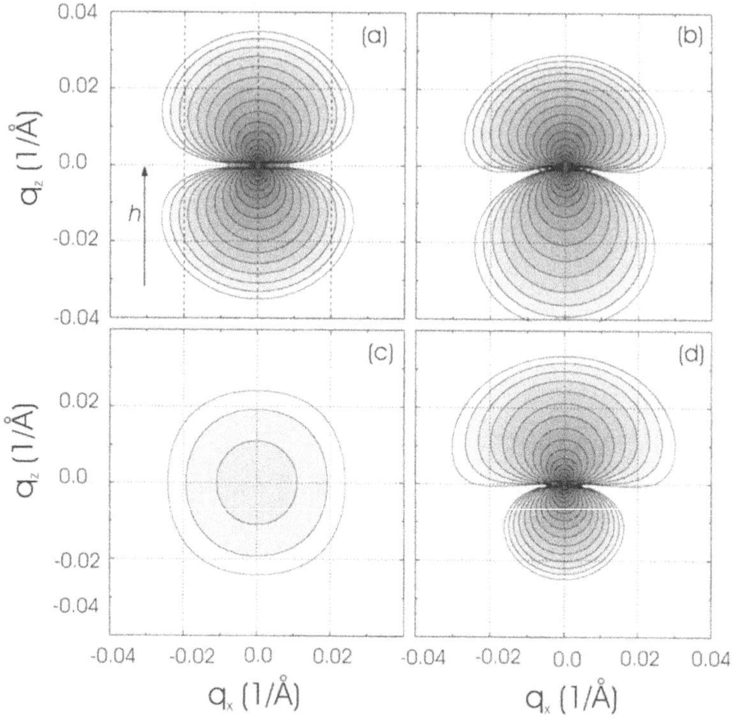

Fig. 10.1. Diffuse intensity distribution in reciprocal space calculated for spherical inclusions with radius 10 nm and the lattice mismatch 0.5%. Panel (a) shows the Huang scattering using the approximation $\exp(-i\boldsymbol{h}.\boldsymbol{v}) - 1 \approx -i\boldsymbol{h}.\boldsymbol{v}$; in (b) the Huang scattering is presented without this approximation. Panel (c) represents the core scattering; in (d) the total scattered intensity is depicted. The contour step is $10^{0.25}$.

where ν is the Poisson ratio. In the next section we present the calculation method for the displacement field of an inclusion embedded in a semi-infinite anisotropic crystal.

From the symmetry it follows that $\langle \boldsymbol{v} \rangle = 0$; thus, the average lattice constant of the crystal is not affected by the inclusions themselves if they are homogeneously distributed in the sample. The expression for the intensity of the Huang scattering (Eq. (10.14)) contains the term

$$\left|\boldsymbol{h}.\boldsymbol{v}^{\mathrm{FT}}(\boldsymbol{q})\right|^2 = \left|4\pi P \frac{\boldsymbol{h}.\boldsymbol{q}}{q^2}\right|^2 .$$

For $\boldsymbol{h}.\boldsymbol{q} = 0$, the intensity is zero; therefore, the reciprocal space distribution of the scattered intensity has a *nodal plane* perpendicular to $\boldsymbol{h}$. The existence of this plane follows from the symmetry of the displacement field. This plane may vanish if the material is elastically anisotropic [378], if the finite layer thickness is considered [24], or if the approximation $\boldsymbol{h}.\boldsymbol{v}(\boldsymbol{r}) \ll 1$ is not fulfilled.

In Fig. 10.1 we show numerical results for $R = 10$ nm; the mismatch was $\delta = 0.005$. The density of the defects is arbitrary, since it is a pre-factor that affects the absolute intensity and not the shape of the intensity contours, provided that $c \ll 1$. We calculated the diffuse scattering in symmetrical 004 diffraction and we assumed that the core of the inclusion does not scatter at all ($\Delta\chi_{\boldsymbol{h}} = -\chi_{\boldsymbol{h}}$). Then, the core scattering represents the scattering from the empty holes in the crystal matrix. Because of the symmetry of the displacement field (10.15), the intensity distribution of the Huang scattering using Eq. (10.14) is symmetric and it consists of two "lobes" along the diffraction vector. The symmetry is broken, if we use the exact formula without assuming $\exp(-\mathrm{i}\boldsymbol{h}.\boldsymbol{v}) - 1 \approx -\mathrm{i}\boldsymbol{h}.\boldsymbol{v}$.

The total scattered intensity is a coherent superposition of the Huang term and the core term. Since the sign of $\boldsymbol{h}.\boldsymbol{v}^{\mathrm{FT}}(\boldsymbol{q})$ equals the sign of $\boldsymbol{h}.\boldsymbol{q}$, the resulting intensity distribution is asymmetric. For $P > 0$ (interstitial-type defects), the Huang lobe with $\boldsymbol{h}.\boldsymbol{q} > 0$ is positive and therefore the intensity in the points with $\boldsymbol{h}.\boldsymbol{q} > 0$ is larger than the intensity for $\boldsymbol{h}.\boldsymbol{q} < 0$ (see Fig. 10.1(d)). For $P < 0$ (vacancy-type defects), the asymmetry of the intensity distribution is opposite in sign.

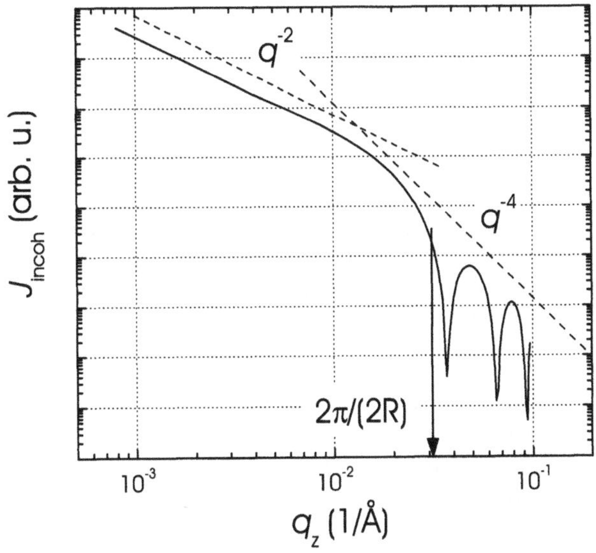

Fig. 10.2. The q_z dependence of the scattered intensity symmetrized with respect to the reciprocal lattice point, calculated from the same defect parameters used in Fig. 10.1. The slopes q^{-2} and q^{-4} are labeled.

In Fig. 10.2 we show the q_z-dependence of the diffusely scattered intensity symmetrized with respect to the reciprocal lattice point. For $|q_z| < 2\pi/(2R)$ the Huang scattering dominates and the intensity decreases as q_z^{-2}. For larger deviations from the reciprocal lattice point, the core scattering is more important and the intensity drops as q_z^{-4} [102, 110, 222]. This dependence can be used for a rough estimate of the mean size of defects from experimental data. The slopes of the Huang and core parts of the intensity distribution become different if the measured intensity is integrated over q_y, as is the case in a laboratory set-up with a large vertical divergence of the primary beam. Then the slopes are -1 and -3 in the Huang region and in the core scattering region, respectively.

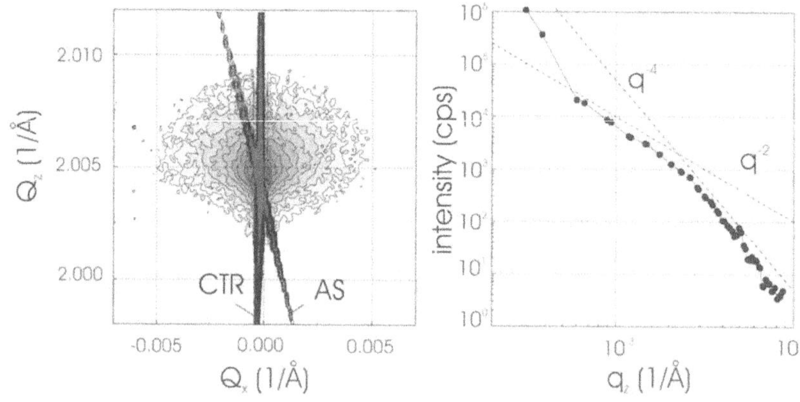

Fig. 10.3. The reciprocal space map of diffusely scattered intensity around the 111 reciprocal lattice point measured in an annealed Si wafer (left) and the q_z-distribution of the symmetrical part of the intensity (right). The maximum labeled AS is the analyzer streak explained in Section 3.2. CTR denotes the coherent crystal truncation rod. The contour step is $10^{0.25}$ in the left panel.

An example of measured data is given in Fig. 10.3. In this picture we plot the reciprocal space distribution of the intensity diffusely scattered from a (111)Si wafer after annealing at 750°C for 40 hours. During this annealing, clusters of point defects are created (mainly interstitial and oxygen atoms) that may act as centers for nucleation of SiO_2 precipitates. The reciprocal space map (the left panel) has been measured in symmetric 111 diffraction using the CuKα_1 radiation. From the asymmetry of the intensity distribution it follows that $P > 0$ and the defects have interstitial nature. From the measured intensity map we have calculated the symmetric part of the intensity. In the right panel, we have plotted this symmetric part along the line parallel to the q_z-axis just beside the coherent crystal truncation rod (CTR). In the loglog representation of the intensity, the slopes -2 and -4 can be resolved.

From the position of the intercept of both slopes we estimate the mean size of the defects to 300 nm.

In another example, we present diffuse scattering from a Si crystal containing randomly placed small circular stacking faults. The theory of diffuse scattering of stacking faults was investigated in several works [109, 212, 225, 250, 251]. In these works, the Huang scattering has been calculated both by a continuum elasticity approach and by atomistic simulations and the continuum approach was based on the elastic Green function $\boldsymbol{w}$ in Eq. (10.23) in the next section. Here we present more exact simulations of the Huang scattering based on an exact expression for the displacement field of a stacking fault. We assume that the faults lie in the planes $\{111\}$ and that their Burgers vectors are $\boldsymbol{b} = 1/3\langle 111\rangle$ perpendicular to the fault plane (the Frank stacking faults) [156, 169]. The displacement field $\boldsymbol{v}(\boldsymbol{r})$ of a single stacking fault can be calculated using the Bartels formula for the displacement around a dislocation loop [156, 174]:

$$\boldsymbol{v}(\boldsymbol{r}) = \frac{\boldsymbol{b}}{4\pi}\int_A \frac{\boldsymbol{R}.\mathrm{d}^2\boldsymbol{A}}{R^3} - \frac{1}{4\pi}\oint_C \frac{\boldsymbol{b}\times \mathrm{d}\boldsymbol{r}'}{R} +$$

$$+\frac{1}{8\pi(1-\nu)}\mathrm{grad}\oint_C \frac{(\boldsymbol{b}\times\boldsymbol{R}).\mathrm{d}\boldsymbol{r}'}{R}. \tag{10.17}$$

Here $\boldsymbol{r}$ denotes the position of the observation point; $\boldsymbol{r}'$ is an arbitrary point in the loop; $\boldsymbol{R} = \boldsymbol{r}' - \boldsymbol{r}$; and A is the loop area and C is the loop circumference (see Fig. 10.4).

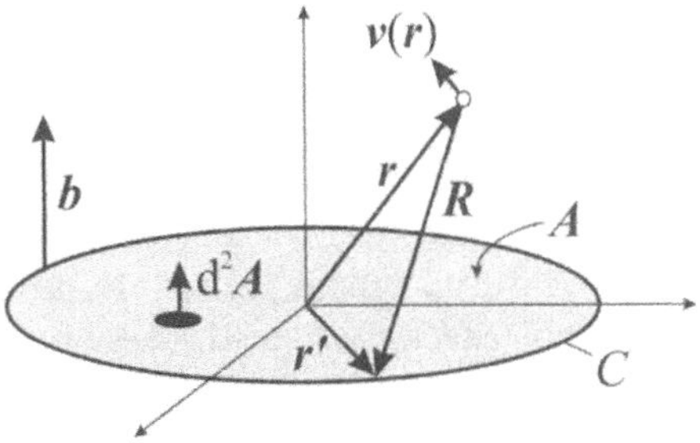

Fig. 10.4. Sketch of a dislocation loop.

Using Eq. (10.17), we calculate the displacement field around a stacking fault, the diffusely scattered intensity is obtained from Eq. (10.13) neglecting the contribution of the defect core. In the calculation we took the symmetric diffraction 111 and, for the sake of simplicity, we assumed that all the faults have the same Burgers vector $\boldsymbol{b} = \frac{1}{3}[11\bar{1}]$. The calculated displacement field and the resulting intensity distribution for loops with radius 1 μm are shown in Fig. 10.5. The scattered intensity is concentrated in a narrow stripe perpendicular to the loop plane. The width of the stripe is inversely proportional to the loop radius.

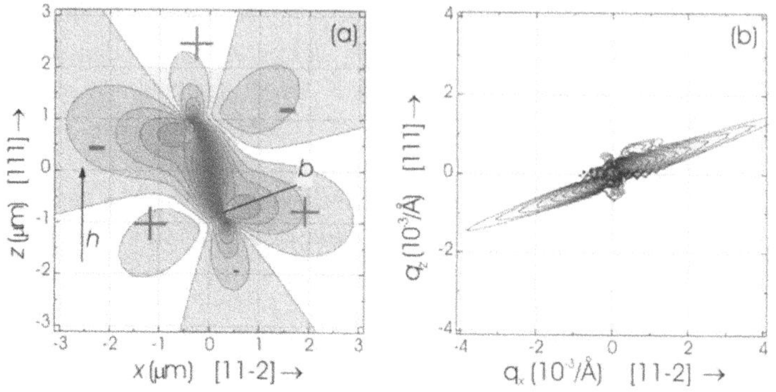

Fig. 10.5. The distribution of the function $\boldsymbol{h}.\boldsymbol{v}(\boldsymbol{r})$ calculated for a stacking fault loop with radius 1 μm (a) and the corresponding reciprocal space distribution of the scattered intensity (b). In (a), the directions of the Burgers vector $\boldsymbol{b}$ and the diffraction vector $\boldsymbol{h}$ as well as the sign of $\boldsymbol{h}.\boldsymbol{v}$ are denoted. The step of the contours is 0.1, the step of the intensity contours in the panel (b) is $10^{0.25}$.

In Fig. 10.6 we show an experimental intensity map of a Si wafer after a two-step annealing (750°C for 20 h and 1050°C for 20 h). During the heat treatment, stacking faults have been generated in {111} planes, producing characteristic intensity stripes in the intensity map. Due to the vertical divergence of the primary x-ray beam, we detect not only the stripe $[11\bar{1}]$ lying in the scattering plane $(\bar{1}10)$, but also the projections of the stripes along $[1\bar{1}1]$ and $[\bar{1}11]$ into the scattering plane. Since the azimuthal direction of the scattering plane was not exactly adjusted, these two projections do not occur exactly in same directions; this is why the $[\bar{1}11][1\bar{1}1]$ stripe seems to be broader than the stripe $[11\bar{1}]$. Similar studies have been published in [42, 225], where the diffuse scattering from microdefects in Si wafers implanted with B^+ ions has been measured in GID geometry after various stages of heat treatment. In Fig. 10.7 the intensity distributions are plotted in the plane $(q_x q_y)$ parallel to the sample surface, along with the linear scans extracted from these distributions. After a low-temperature annealing at 750°C for 15 min, the intensity map is nearly isotropic; thus the wafer contains approximately spherical defect clusters. During a rapid thermal annealing at 1060°C stacking faults are created, giving rise to an intensity lobe in the [110] direction, corresponding to the projection of the [111] intensity streak to the depicted reciprocal plane (001). The linear scans extracted from the maps along the

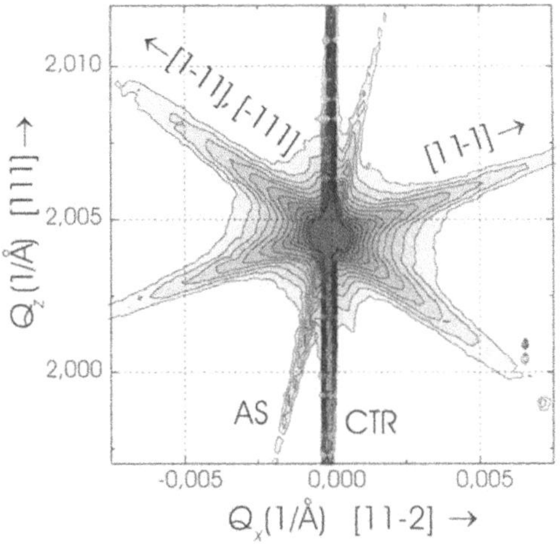

Fig. 10.6. The intensity distribution of diffuse scattering measured around the 111 reciprocal lattice point of a Si wafer after a high-temperature annealing (see text). The stripes stem from the stacking faults in {111} planes. "AS" is the analyzer streak (see Sec. 3.3); "CTR" denotes the coherent crystal truncation rod. The contour step is $10^{0.25}$.

dashed lines are plotted in Fig. 10.8. The characteristic slopes -2 and -4 are present in the intensity scan after the low-temperature annealing; from the position of the cross-over point the mean size of the clusters ($2R \approx 20$ nm) was determined.

In [205] GID geometry has been used to investigate the precipitation process of carbon in silicon. During annealing, small precipitates of SiC are nucleated, giving rise to diffuse scattering. From the measured intensity distribution, the size and density of the precipitates could be determined.

10.3 Weak Defects in a Subsurface Layer

In the previous section we have described diffuse x-ray scattering from weak defects randomly placed in an infinite, elastically isotropic crystal. The symmetry of the deformation field of such a defect is fully determined by the shape of the defect itself. In many cases, however, the deformation field is affected by a free surface of the sample or by an interface between the substrate and the layer. This fact is especially important for surface-sensitive x-ray methods (GID), probing the defects lying in a thin subsurface layer. In this chapter, we present a general method for the calculation of the deformation of a small inclusion lying below the surface of a semi-infinite crystal. The method takes into account the elastic anisotropy of the crystal; for the sake of simplicity we restrict ourselves to cubic crystals and to inclusion materials with cubic symmetry. The application of the method is limited for crystals with a flat surface. A similar method calculating the displacement field in a generally modulated thin layer has been published in [135].

Let us assume a single inclusion described by the shape function $\Omega(\boldsymbol{r})$ (unity in the inclusion volume, zero outside it) lying below the free surface $z = 0$ of a semi-infinite cubic crystal. We assume the same elastic constants c_{11}, c_{12} and c_{44} of the crystal and the inclusion material. The condition of the elastic equilibrium is [166, 207, 339]

$$\frac{\partial \sigma_{jk}}{\partial x_k} + f_j = 0, \ x_{j,k} = x, y, z, \tag{10.18}$$

where σ_{jk} is the elastic stress tensor and

$$f_k = -4\pi P c_{11} \frac{\partial \Omega(\boldsymbol{r})}{\partial x_k} \tag{10.19}$$

is the density of the force acting on a volume element of the crystal. P is the defect strength

$$P = \frac{\delta}{4\pi} \frac{c_{11} + 2c_{12}}{c_{11}}, \tag{10.20}$$

where δ is the mismatch of the inclusion material with respect to the surrounding crystal, and x, y, z are the crystallographic axes $\langle 100 \rangle$. If the material is elastically isotropic,indexelastic isotropy $c_{44} = (c_{11} - c_{12})/2$ holds and the defect strength can be expressed using the Poisson ratio after Eq.

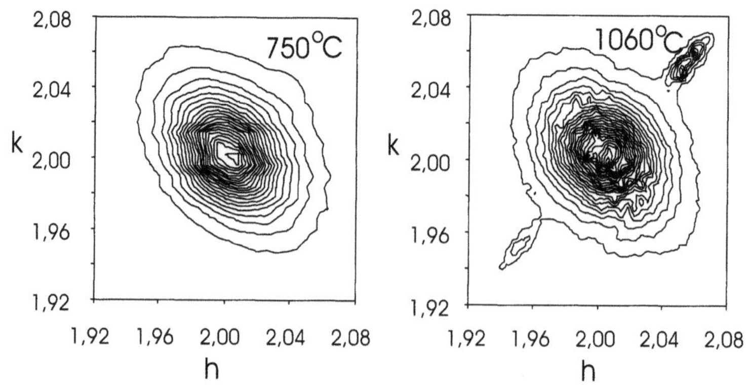

Fig. 10.7. The intensity distribution in the (001) plane parallel to the sample surface, measured in the GID geometry (diffraction 220) on a Si wafer after low- and high-temperature annealing [42].

(10.16). If the inclusion material has lower symmetry than cubic, the defect strength P has a tensor character.

The equations of equilibrium (10.18) are solved using the boundary condition

$$\sigma_{jz}\big|_{z=0} = 0, \; x_j = x, y, z, \tag{10.21}$$

expressing the fact that the surface is stress-free. We perform the two-dimensional Fourier transformation of the displacement field [207]:

$$\boldsymbol{v}^{\mathrm{FT}}(\boldsymbol{\kappa}; z) = \int_{-\infty}^{\infty}\int_{-\infty}^{\infty} \mathrm{d}x\mathrm{d}y \; \boldsymbol{v}(x, y, z)\mathrm{e}^{-\mathrm{i}(\kappa_x x + \kappa_y y)}. \tag{10.22}$$

Then, taking into account the Hook law connecting the elastic stress tensor σ_{jk} with the derivatives of $\boldsymbol{v}(\boldsymbol{r})$, we transform the equilibrium equation into an ordinary differential equation for $\boldsymbol{v}^{\mathrm{FT}}(\boldsymbol{\kappa}; z)$

$$\hat{\mathbf{A}}\left(\boldsymbol{v}^{\mathrm{FT}}\right)'' + \mathrm{i}\hat{\mathbf{B}}\left(\boldsymbol{v}^{\mathrm{FT}}\right)' - \hat{\mathbf{C}}\boldsymbol{v}^{\mathrm{FT}} = \boldsymbol{D},$$

where the primes denote the derivatives with respect to z, the matrices $\hat{\mathbf{A}}$, $\hat{\mathbf{B}}$ and $\hat{\mathbf{C}}$ contain the elastic constants and the components of $\boldsymbol{\kappa}$; the vector $\boldsymbol{D}$ contains $\boldsymbol{\kappa}$, the defect strength P and its shape function Ω. The boundary condition (10.21) is expressed in the form

$$\hat{\mathbf{A}}\left(\boldsymbol{v}^{\mathrm{FT}}\right)' + \mathrm{i}\hat{\mathbf{F}}\boldsymbol{v}^{\mathrm{FT}}\Big|_{z=0} = 0,$$

the matrix $\hat{\mathbf{F}}$ contains κ and the elastic constants. The differential equation is linear with constant coefficients and it can be solved by standard analytic

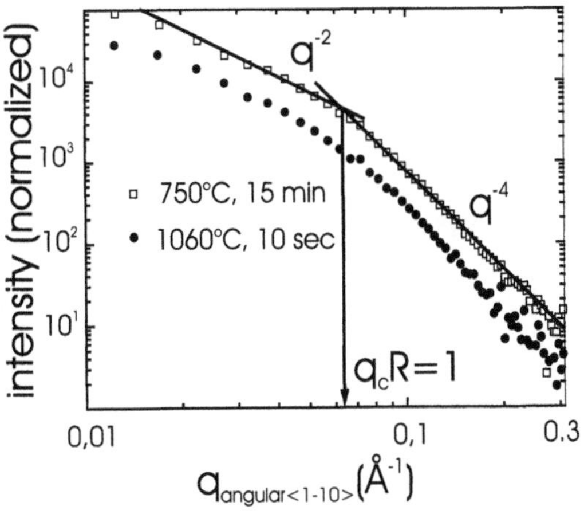

Fig. 10.8. The linear scans extracted from the intensity distributions in Fig. 10.7 along the dashed lines.

methods. The details of the calculation procedure can be found elsewhere [286].

The calculation procedure can be substantially simplified in two limiting cases, namely, for an infinite anisotropic crystal and for a semi-infinite isotropic continuum. In the former case, the displacement field of an inclusion can be expressed as a convolution of the shape function Ω of the inclusion with the elastic Green function of the displacement field $\boldsymbol{w}(\boldsymbol{r})$ [99, 101]:

$$\boldsymbol{v}(\boldsymbol{r}) = \int \mathrm{d}^3\boldsymbol{r}'\Omega(\boldsymbol{r}')\boldsymbol{w}(\boldsymbol{r}-\boldsymbol{r}'). \tag{10.23}$$

The elastic Green function $\boldsymbol{w}(\boldsymbol{r}-\boldsymbol{r}')$ represents the displacement in point $\boldsymbol{r}$ due to an elementary point-like defect in $\boldsymbol{r}'$. The Fourier transformation of $\boldsymbol{v}(\boldsymbol{r})$ is a *product* of the Fourier transformation $\Omega^{\mathrm{FT}}(\boldsymbol{q})$ of the shape function with the Fourier transformation of $\boldsymbol{w}$. This Fourier transformation can be explicitly obtained for a cubic crystal [101, 235]

$$w_j^{\mathrm{FT}}(\boldsymbol{q}) = -\mathrm{i}\frac{\delta}{q}(c_{11}+2c_{12})\frac{q_j^0}{c_{44}+A(q_j^0)^2}\times$$

$$\times\left[1+\sum_{m=x,y,z}\frac{c_{44}+c_{12}}{c_{44}+A(q_m^0)^2}(q_m^0)^2\right]^{-1}, \; j=x,y,z, \tag{10.24}$$

where $A = c_{11}-c_{12}-2c_{44}$ is the anisotropy factor (zero for an elastically isotropic continuum), and $\boldsymbol{q}^0 = \boldsymbol{q}/q$. Eq. (10.23) is suitable especially for the calculation of the Huang scattering under the assumption $\boldsymbol{h}.\boldsymbol{v}(\boldsymbol{r}) \ll 1$ (see Eq. (10.14)). In this case, the reciprocal space distribution of scattered intensity is proportional to the square of the product of $\boldsymbol{h}.\boldsymbol{w}^{\mathrm{FT}}$ with Ω^{FT}.

The latter simple case (an semi-infinite isotropic continuum) allows us to express the displacement $\boldsymbol{v}(\boldsymbol{r})$ using the integral expression [171]

$$\boldsymbol{v}(\boldsymbol{r}) = \int \mathrm{d}^3\boldsymbol{r}'\Omega(\boldsymbol{r}')\boldsymbol{w}^{(s)}(\boldsymbol{r}_\parallel-\boldsymbol{r}'_\parallel;z,z'). \tag{10.25}$$

In contrast to the infinite crystal, the elastic Green function $\boldsymbol{w}^{(s)}$ of a semi-infinite crystal depends separately on the coordinates z and z' perpendicular to the surface [171]:

$$\boldsymbol{w}^{(s)}(\boldsymbol{r}_\parallel-\boldsymbol{r}'_\parallel;z,z') = P\left[\frac{\boldsymbol{R}_1}{R_1^3}+(3-4\nu)\frac{\boldsymbol{R}_2}{R_2^3}-6z(z+z')\frac{\boldsymbol{R}_2}{R_2^5}-\right.$$

$$\left.-\frac{2\boldsymbol{n}}{R_2^3}((3-4\nu)(z+z')-z)\right], \tag{10.26}$$

where $\boldsymbol{R}_{1,2} = (x-x', y-y', z\mp z')$ are the position vectors of the observation point with respect to the point $\boldsymbol{r}'$ in the inclusion and in the "imaginary" inclusion reflected from the sample surface (see Fig. 10.9), and $\boldsymbol{n}$ is the unit vector of the outer normal to the surface. The first term in the square brackets

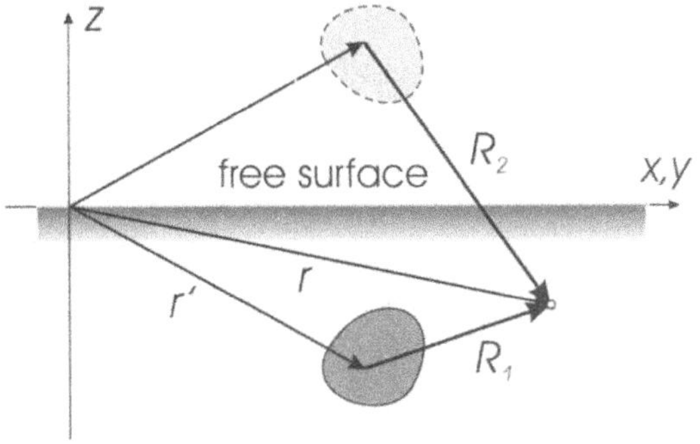

Fig. 10.9. Vectors introduced in Eq. (10.26).

corresponds to the case of an infinite crystal according to Eq. (10.15); the other terms account for the surface relaxation of internal stresses. In [24, 207] the Huang scattering has been calculated using Eqs. (10.14, 10.26) for an inhomogeneous depth distribution of small defects in a subsurface layer. For very small defects, we can replace the shape function $\Omega(\boldsymbol{r}')$ in Eq. (10.25) by a very narrow peak $\Omega(\boldsymbol{r}') \approx V\delta^{(3)}(\boldsymbol{r}' - \boldsymbol{r}'')$, where V is the defects volume and $\boldsymbol{r}''$ is its position vector. Then, from Eqs. (10.26) and (10.14), we obtain for the intensity distribution of the Huang scattering the following expression:

$$J(\boldsymbol{Q}) = \text{const.}(PV)^2 \int_{-\infty}^{0} \mathrm{d}z n(z) \left| \mathrm{e}^{-\mathrm{i}q_z z} \frac{\boldsymbol{h}.\boldsymbol{q}}{q^2} - \right.$$

$$\left. -2\mathrm{e}^{q_\parallel z} \frac{h_z q_\parallel - \mathrm{i}\boldsymbol{h}_\parallel . \boldsymbol{q}_\parallel}{q_\parallel (q_z + \mathrm{i}q_\parallel)} \left(\frac{q_z^2}{q^2} - 4\nu \right) \right|^2 , \tag{10.27}$$

where $\boldsymbol{q}_\parallel \equiv (q_x, q_y, 0)$ is the lateral component of the reduced scattering vector and $n(z)$ is the vertical profile of the defect density. The first term on the right-hand side corresponds to the Huang scattering from defects in an infinite matrix; the second term accounts both for the surface relaxation of the internal stresses and from the fact that the integral in Eq. (10.10) is performed only in the half space $z'' < 0$.

In Fig. 10.10 we show the displacement components v_x and v_z calculated for a spherical inclusion with the radius 10 nm in the depth 20 nm below the free surface. For the calculation we have used the exact formulas (10.18 – 10.21) considering the elastic anisotropy. The elastic constants of Si were used, the linear mismatch δ was 0.5%. If we neglect both the elastic anisotropy and the surface relaxation, the displacement field is highly symmetric, as follows from Eq. (10.15). The elastic anisotropy modifies the shape of the iso-displacement contours, keeping the mirror-like symmetry of the displacement field with respect to the horizontal plane going across the center of the inclusion. This symmetry is broken due to the surface relaxation. This relaxation *increases* the magnitude of the displacement close to the sample surface.

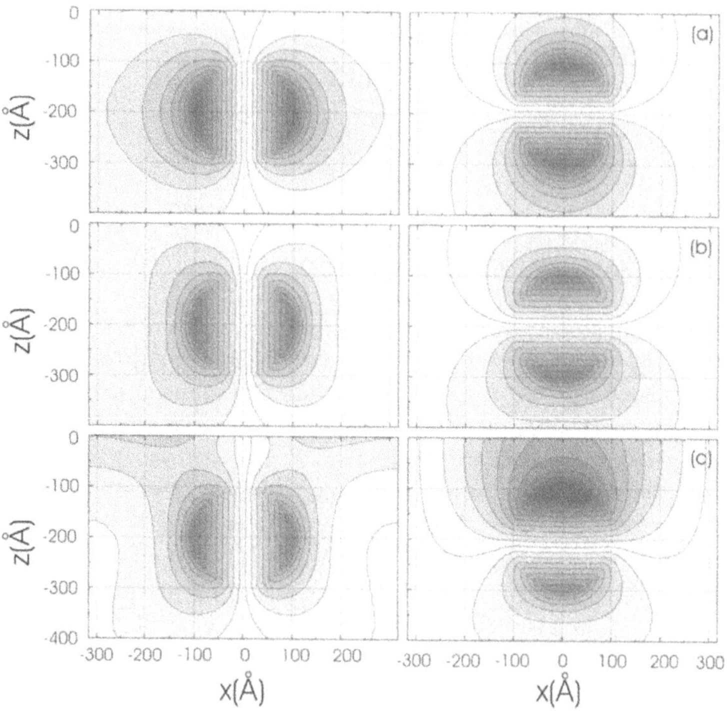

Fig. 10.10. The x (left) and z components (right) of the displacement vector $\boldsymbol{v}(\boldsymbol{r})$ of a spherical inclusion embedded in a semi-infinite Si matrix. The inclusion radius is 10 nm, the depth of the inclusion center below the surface is 20 nm, the lattice mismatch is $\delta = 0.5\%$. The panels (a) were calculated assuming elastic isotropy and without the surface relaxation; in panels (b) the elastic anisotropy was included. The panels (c) show the results of the full calculation with anisotropy and surface relaxation.

Using this displacement field, we have computed the reciprocal space distribution of diffuse scattering. Since we have assumed that both the inclusion and the sample lattice are made of Si; $\Delta\chi_h = 0$ and only the Huang scattering occurs. The results are plotted in Figs. 10.11 and 10.12. In the symmetric coplanar 004 diffraction (Fig. 10.11), the lobes of the Huang scattering similar to those in Fig. 10.1 can be seen. The presence of the free surface influences the volume integral in Eq. (5.49), giving rise to a narrow vertical maximum along the q_z-axis; this maximum, however, coincides with the CTR peak. In addition, the free surface affects the displacement field $\boldsymbol{v}(\boldsymbol{r})$ via surface relaxation; this changes the intensity distribution, too. In particular, the surface relaxation removes the intensity minimum in the plane perpendicular to the diffraction vector (the nodal plane); compare panels (b) and (c) in Fig. 10.11. In a rough estimate, the surface relaxation affects the intensity distribution in reciprocal space in the region $|q_x| \leq 2\pi/T$, where T is the depth of the defects below the surface.

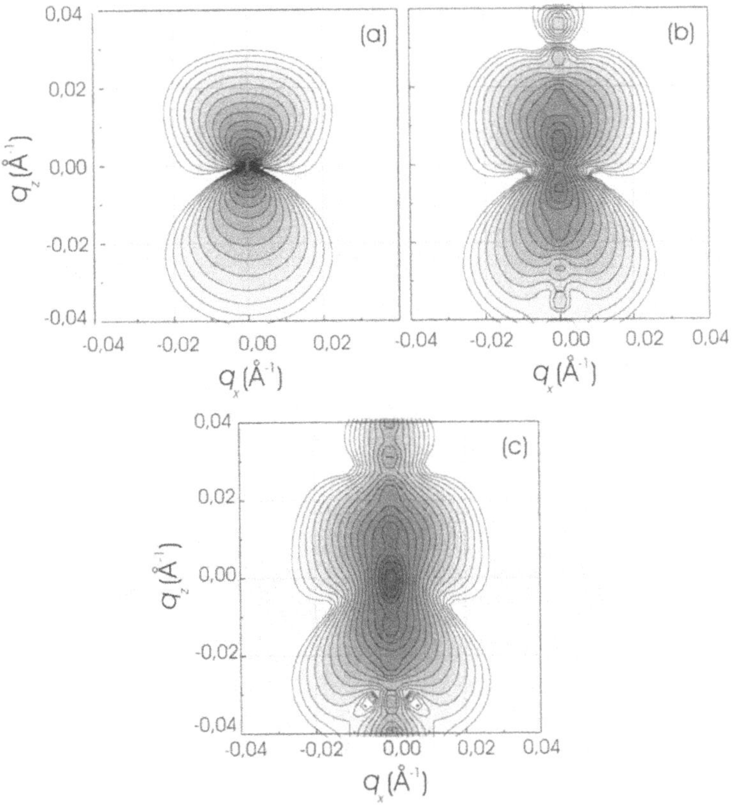

Fig. 10.11. The Huang diffuse scattering from spherical inclusions of the radius of 10 nm in the depth 20 nm below the surface, semi-infinite (001) Si crystal, symmetrical diffraction 004. The lattice mismatch between the inclusion and the surrounding lattice is $\delta = 0.5\%$. In (a) an infinite elastically anisotropic crystal was assumed. From the comparison with Fig. 10.1(a), calculated without elastic anisotropy, the role of the elastic anisotropy is obvious. In (b) we included the semi-infinite substrate in the intensity calculation but we took the displacement field of an inclusion in an infinite crystal. The panel (c) shows the complete calculation with the surface relaxation.

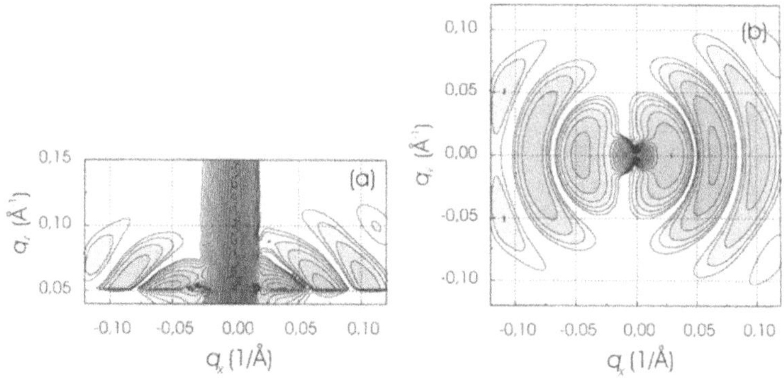

Fig. 10.12. The Huang diffuse scattering from spherical inclusions with the same parameters as in Fig. 10.11, semi-infinite (001) Si crystal, grazing-incidence diffraction 004. In (a), the intensity distribution in radial plane ($q_x q_z$) is plotted for the incidence angle $\alpha_i = 0.5°$. (b) shows the intensity distribution in the plane $q_z = 0.06$ Å^{-1}.

The intensity distribution depicted in Fig. 10.11 has been calculated within the kinematical approximation. In a surface-sensitive diffraction geometry (grazing-incidence diffraction – GID), the semikinematical approximation (DWBA) has to be taken into account, since the refraction of the primary wave at the sample surface cannot be neglected. Let us calculate the intensity distribution using the DWBA approach according to Chap. 7. We choose a semi-infinite *amorphous* substrate, so that we consider the refraction at the surface, but we neglect the dynamical diffraction in the substrate. The solutions of the non-disturbed wave equation (7.2) are in Eqs. (7.5, 7.6) and they are shown schematically in Fig. 7.3. The scattered intensity has been obtained from Eq. (5.49), where we have replaced the scattering vector $\boldsymbol{Q}$ by the scattering vector $\boldsymbol{Q}_T$ corrected to refraction, and we have multiplied the intensity by the Yoneda term $|t_1 t_2|^2$ containing the Fresnel transmittivities of the surface. For the covariance M we have used the formula (10.10). The calculations have been carried out for the in-plane diffraction 400; its diffraction vector was parallel to the sample surface. The results are plotted in Fig. 10.12. In panel (a), we show the intensity distribution in the vertical plane ($q_x q_z$) parallel to the diffraction vector (the radial plane); the angle of incidence was kept constant. The intensity distribution in a horizontal plane $q_z =$ const is depicted in panel (b). In (a), the Yoneda peak is visible as a horizontal maximum at $q_z = 0.05$ Å^{-1}. In these points, the exit angle α_f equals the critical angle α_c (see Chap. 7). The sequence of lateral maxima visible both in (a) and (b) are caused by the assumption that all the inclusions have the same size. A dispersion of the sizes would smear out these oscillations. The radial asymmetry of the pattern stems from the interstitial nature of the inclusions.

For the calculation of a displacement field $\boldsymbol{v}(\boldsymbol{r})$ of a defect, an atomistic approach can also be used [185, 218, 230, 251] based on the minimization of the potential energy of a crystal lattice containing the defect. The calculation is based on several models of empirical or semi-empirical potentials, and the calculation procedure must include the optimization of the positions of several millions of atoms. The method can be used also for defects in a shallow layer below the surface. Numerical tests [71, 90, 230] proved that the results of the atomistic and continuum approaches are fully comparable; the latter is more applicable for larger defects and for self-organized objects below the sample surface (see Sect. 14.3).

Experimental study of diffuse scattering from defects in thin subsurface layers is usually performed in GID geometry. In [139] this method has been used for the investigation of microdefects in a Si sample with a shallow As^+-implanted layer. The theoretical approach in [24] was used for the analysis of experimental data obtained in GID geometry of scattering of very thin surface layers [23]. The GID method has been also applied for the measurement of diffuse scattering from microdefects produced in superlattices by laterally modulated focused ion beam implantation [137].

10.4 Small-Angle Scattering From Small Defects in Thin Layers

In the previous sections we have investigated diffuse x-ray scattering from small defects in the layers; the scattered wave consisted of two parts, namely the wave scattered from the defect core (core scattering) and the wave scattered from the displacement field around the defect core. The latter contribution depends on the diffraction vector $\boldsymbol{h}$; the wave diffusely scattered with $\boldsymbol{h} = 0$ is sensitive only to the contrast $\Delta\chi_0$ and completely insensitive to the strain. This small-angle diffuse scattering is widely used for the investigation of dense clusters in thin layers. The description of small-angle diffuse scattering in thin layers is based on the traditional theory of small-angle scattering [141]. Since the scattering experiments are usually performed in the non-coplanar GISAXS geometry (see 4.3), in the theoretical description the refraction of the primary and scattered waves at the sample surface must be included. This can be done quite easily using the semikinematical DWBA approach according to Chap. 7, assuming a semi-infinite amorphous substrate as the non-disturbed system and the clusters as the disturbance (see Section 7.2.1). The resulting differential scattering cross section has the following form [277] (see also Eq. (7.10)):

$$\left(\frac{\mathrm{d}\sigma}{\mathrm{d}\Omega}\right)_{\mathrm{incoh}} = \frac{1}{16\pi^2}|K^2\Delta\chi_0 t_1 t_2|^2 \left\langle \left|\Omega^{\mathrm{FT}}(\boldsymbol{Q}_T)\right|^2 \right\rangle_{\mathrm{size}} G(\boldsymbol{Q}_T), \tag{10.28}$$

where $\boldsymbol{Q}_T$ is the scattering vector corrected to refraction, $t_{1,2}$ are the Fresnel transmittivites of the primary and the scattered waves at the sample surface,

$\Omega^{\rm FT}(\boldsymbol{Q}_T)$ is the Fourier transformation of the shape function of the cluster, and $G(\boldsymbol{Q})$ is the geometrical factor of the cluster positions (see Sec. 14.2).

As an example of the application, we present here the GISAXS measurement on Au nanoclusters embedded in a thin amorphous hydrogenated carbon film [21]. The measurement has been performed using a two-dimensional detector nearly perpendicular to the direction of the primary x-ray beam in GISAXS geometry, the resulting intensity distribution is shown in Fig. 10.13 in the left panel. The axis q_y is perpendicular to the primary beam and parallel to the sample surface; q_z is perpendicular to the surface. In the figure, the scattered intensity is concentrated in a broad arc, its radius is inversely proportional to the mean distance of the clusters. The region with small q_y containing the coherent crystal truncation rod was shaded off by a thin beam-stop. From the intensity distribution, linear scans have been extracted at various angles with the q_y-axis; an example of the scan is in Fig. 10.13, the right panel. Using Eq. (10.28) and assuming a suitable model for the cluster shapes and for the correlations of their positions, mean size and distance of the clusters were obtained.

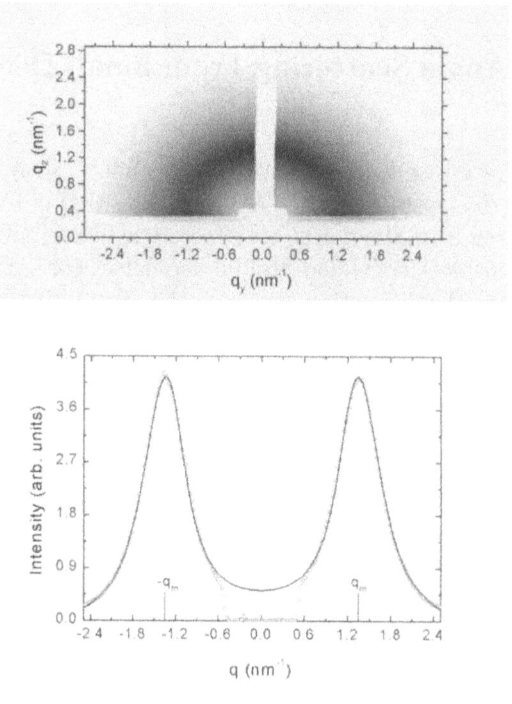

Fig. 10.13. The GISAXS intensity map of a thin carbon layer with Au nanoclusters (left) and a radial scan extracted from the map (points) along with its fit using Eq. (10.28) (right) [21].

10.5 Diffuse Scattering From an Array of Misfit Dislocations

As we stated above, crystal defects are divided into two classes according to the static Debye-Waller factor. Small defects such as inclusions and dislocation loops belong to the class of weak defects. The wave diffracted from a crystal containing these defects consists of a coherent and incoherent component. A typical example of a strong defect is a straight dislocation. A large crystal containing a random array of straight dislocations produces only the diffusely scattered wave, since the corresponding static Debye-Waller factor goes to zero.

Diffuse scattering from crystals with dislocations has been investigated for many years and the results can be found in several papers (see [17, 121, 126, 207, 386], among others). It has been found that the angular divergence of diffusely scattered wave is proportional to the square root of the dislocation density, and the intensity profile has a Gaussian shape. These studies, however refer to large samples, and they do not account for the influence of the free surface on the deformation field of the dislocations. In [182] a theory has been presented describing the diffuse scattering from straight dislocations parallel to the crystal surface lying in a finite depth below the surface. Typically, this approach is suitable for the description of diffuse scattering from an array of misfit dislocations at a substrate-layer interface.

In contrast to the previous approach, where we have defined the reduced scattering vector $\boldsymbol{q}$ with respect to the reciprocal lattice point $\boldsymbol{h}^{\mathrm{def}}$ of the average deformed lattice, for the description of diffuse scattering from dislocations, we define $\boldsymbol{q}$ with respect to the reciprocal lattice point $\boldsymbol{h}$ of the non-deformed lattice. Then, the correlation function of the crystal deformation is (compare to Eq. (5.43)):

$$C(\boldsymbol{r},\boldsymbol{r}') \equiv C(\boldsymbol{r}-\boldsymbol{r}') \equiv M(\boldsymbol{r}-\boldsymbol{r}') =$$
$$= \left\langle \chi_{\boldsymbol{h}}^{\mathrm{def}}(\boldsymbol{r})(\chi_{\boldsymbol{h}}^{\mathrm{def}}(\boldsymbol{r}'))^{*}\mathrm{e}^{-\mathrm{i}\boldsymbol{h}.(\boldsymbol{u}(\boldsymbol{r})-\boldsymbol{u}(\boldsymbol{r}'))}\right\rangle . \tag{10.29}$$

We repeat the procedure leading to Eqs. (10.8, 10.9). Since the dislocations are strong defects, we cannot replace the expression $\sum_m[1-\exp(-\mathrm{i}\boldsymbol{h}.(\boldsymbol{v}_{jm}-\boldsymbol{v}_{km}))]$ by $\sum_m \mathrm{i}\boldsymbol{h}.(\boldsymbol{v}_{jm}-\boldsymbol{v}_{km})$. Instead, we express the covariance M as

$$M(\boldsymbol{r}_j,\boldsymbol{r}_k) = \exp\left[-\sum_{\alpha} c^{\alpha}\sum_{m}\left(1-\mathrm{e}^{-\mathrm{i}\boldsymbol{h}.(\boldsymbol{v}_{jm}-\boldsymbol{v}_{km})}\right)\right]. \tag{10.30}$$

The sum $\sum_m$ runs over all possible positions of the dislocations. All dislocations lie in the depth T at the substrate-layer interface; the dislocations may differ in the directions of the dislocation lines and their Burgers vectors. In (001) Si wafers, for instance, the dislocation lines have two possible orientations $[1\bar{1}0]$ and $[\bar{1}10]$, and the Burgers vectors are $\frac{1}{2}\langle 110\rangle$. We denote x_α the coordinate perpendicular to the dislocation line of type α lying parallel to the crystal surface. Then, in the expression (10.30) we can replace the sum over m by the integral as follows:

$$M(\boldsymbol{r},\boldsymbol{r}') \equiv M(\boldsymbol{r}_{\|} - \boldsymbol{r}'_{\|}; z, z') = \exp\left(-\sum_{\alpha} n^{\alpha} \int_{-\infty}^{\infty} \mathrm{d}x''_{\alpha} \Psi\right), \tag{10.31}$$

using

$$\Psi = 1 - \mathrm{e}^{-\mathrm{i}\boldsymbol{h}.(\boldsymbol{v}^{\alpha}(x_{\alpha} - x'_{\alpha} + x''_{\alpha}, z+T) - \boldsymbol{v}(x''_{\alpha}, z'+T))}.$$

Here n^{α} denotes the one-dimensional density of the misfit dislocations of type α; $\boldsymbol{v}^{\alpha}(x_{\alpha}, z+T)$ is the displacement in the point $(x_{\alpha}, y_{\alpha}, z)$ caused by the dislocation crossing the point $(x_{\alpha} = 0, y_{\alpha}, -T)$. Due to the translational symmetry, this displacement does not depend on the coordinate y_{α} along the dislocation line. Eqs. (10.30) and (10.31) have been derived under the assumption that the positions of individual misfit dislocations are statistically independent. A correlation in the position of the dislocations modifies the intensity distribution. If the correlation distance of the dislocation positions is rather small, this correlation makes the intensity distribution narrower in the q_x-direction [182].

In [325] the formulas are presented for the displacement field of a general dislocation in a semi-infinite elastically isotropic continuum, taking the influence of the free surface into account. Using these expressions and Eqs. (5.49, 10.31), the reciprocal-space distribution of scattered intensity can be calculated directly. The calculation procedure is rather lengthy and time-consuming. If the mean distance between the dislocations is smaller than the depth T of the dislocation lines below the surface, the following simplification can be performed [182]:

$$\boldsymbol{h}.(\boldsymbol{v}^{\alpha}(x_{\alpha} - x'_{\alpha} + x''_{\alpha}, z+T) - \boldsymbol{v}(x''_{\alpha}, z'+T)) \approx$$

$$\approx \left. \frac{\partial \boldsymbol{h}.\boldsymbol{v}^{\alpha}}{\partial x_{\alpha}}(x_{\alpha} - x'_{\alpha}) + \frac{\partial \boldsymbol{h}.\boldsymbol{v}^{\alpha}}{\partial z}(z - z')\right|_{x''_{\alpha}, z'+T}. \tag{10.32}$$

Then, replacing the exponential function in (10.31) by the two first terms of its Taylor expansion, we obtain finally

$$M(\boldsymbol{r},\boldsymbol{r}') = \exp\Big[-\mathrm{i}\left(A_x(x-x') + A_z(z-z')\right) - \frac{1}{2}\big(B_{xx}(x-x')^2 +$$

$$+ B_{zz}(z-z')^2 + 2B_{xz}(x-x')(z-z')\big)\Big], \tag{10.33}$$

where

$$A_j = \sum_{\alpha} n^{\alpha} \int_{-\infty}^{\infty} \mathrm{d}x''_{\alpha} \left.\frac{\partial \boldsymbol{h}.\boldsymbol{v}^{\alpha}}{\partial x_j}\right|_{x''_{\alpha}, z'+T}, \tag{10.34}$$

and

$$B_{jk} = \sum_{\alpha} n^{\alpha} \int_{-\infty}^{\infty} \mathrm{d}x''_{\alpha} \left. \frac{\partial \boldsymbol{h}.\boldsymbol{v}^{\alpha}}{\partial x_j} \frac{\partial \boldsymbol{h}.\boldsymbol{v}^{\alpha}}{\partial x_k} \right|_{x''_{\alpha}, z'+T}, \tag{10.35}$$

where $x_{j,k} = x_{\alpha}, z$. These coefficients can be calculated numerically. The coefficients A_j are responsible for the shift of the maximum of the diffuse scattering; i.e., they describe the average deformation in the layer and the diffraction vector $\boldsymbol{h}^{\mathrm{def}}$:

$$h_j^{\mathrm{def}} = h_j - A_j. \tag{10.36}$$

The coefficients B_{jk} determine the shape of the intensity maximum. Since these coefficients occur in a quadratic form of the coordinates, the resulting intensity distribution in reciprocal space has a Gaussian shape; its width is proportional to the square root of the dislocation density.

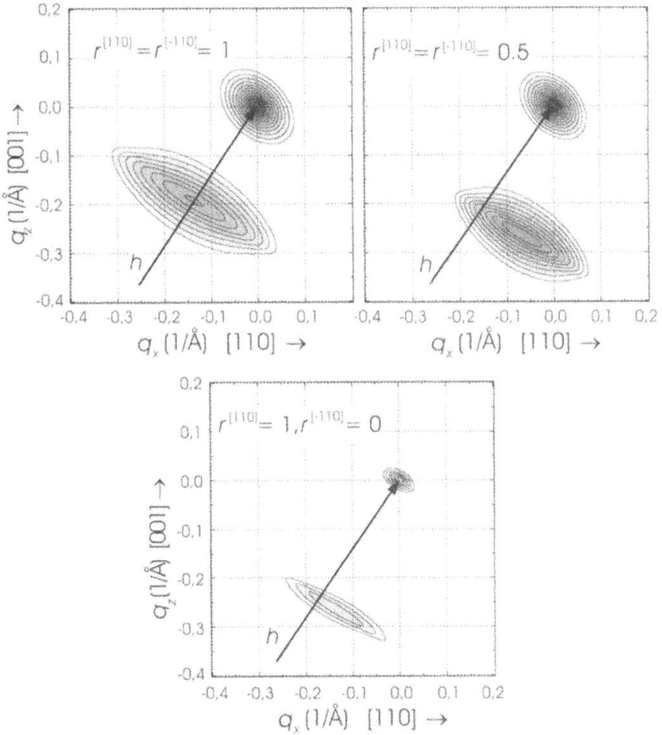

Fig. 10.14. Diffuse scattering from misfit dislocations calculated in the asymmetric 224 diffraction for various values of the relaxation degree R in two orthogonal directions [110] and [$\bar{1}$10]; a 50 nm Ge layer was deposited onto a (001) Si substrate. The step of the intensity contours is $10^{0.25}$, the direction of the diffraction vector is denoted by the arrow.

We demonstrate this theoretical approach in a simple example of a thin partially relaxed epitaxial layer on a Si semi-infinite substrate with surface (001). In this case, two families of misfit dislocations are present at the substrate-layer interface having the dislocation lines perpendicular to [110] and $[\bar{1}10]$. We assume that the degree R of plastic relaxation of the layer is different in these both directions (see Chap. 9), and the densities of both dislocation families are

$$n^{\alpha} = \delta \frac{R_{\alpha}}{b_{\parallel}}, \; \alpha = [110], [\bar{1}10],$$

where δ is the lattice mismatch between the layer and the substrate and $b_{\parallel}$ is the length of the component of the Burgers vector $\boldsymbol{b}$ lying in the interface perpendicular to the dislocation line. For the dislocations with the Burgers vectors $\frac{1}{2}\langle 110 \rangle$ $b_{\parallel} = a/(2\sqrt{2})$ holds, where a is the lattice constant.

We have calculated the diffuse scattering in an asymmetric coplanar diffraction 224, the results are shown in Fig. 10.14. In the figure, we can resolve the substrate maximum at $(q_x, q_z) = (0, 0)$ and the broader layer peak. The substrate peak is visible due to the diffuse scattering from the random deformation field propagating from the misfit dislocations at the substrate-layer interface into the substrate. If the layer is fully relaxed, the layer peak is shifted radially (i.e., parallel to the diffraction vector) from the substrate peak toward the origin of reciprocal space. In the case of a pseudomorph layer, the layer and the substrate peak have the same lateral positions (see also Chap. 9). Using this theoretical approach, the position and the width of the layer peak are not independent, since they follow from the relaxation degree R.

The position of the layer peak $\boldsymbol{q}_{\max} = \boldsymbol{h}^{\text{def}} - \boldsymbol{h}$ can also be calculated using simple elasticity theory,

$$q_j^{\max} = -h_j R^j \delta; \; q_z^{\max} = -h_z \delta \left(\frac{1+\nu}{1-\nu} - (R^{[110]} + R^{[\bar{1}10]}) \frac{\nu}{1-\nu} \right), \quad (10.37)$$

where the subscript j denotes the components parallel to the relaxation directions [110] and $[\bar{1}10]$. Using the approximation (10.32) and calculating the coefficients A_j, we find that Eqs. (10.36) and (10.37) yield exactly the same positions of the layer peak. However, if approximation (10.32) is not fulfilled, the actual position of the layer peak differs from that following from Eq. (10.37), and the simple elasticity theory *is not valid* [182]. Therefore, the formulas (10.37) are misleading for very thin layers and/or for a slight relaxation.

10.6 Diffuse Scattering From Mosaic Layers

In the previous section, we have described the calculation procedure of diffuse scattering from a random array of misfit dislocations at a substrate-layer interface. From this dislocation array, a random displacement field propagates both into the layer and into the substrate; this displacement field is the source of diffuse x-ray scattering. We have presented formulas for the intensity distribution in reciprocal space assuming statistical independence of the positions of the dislocations. This calculation method is based on a physically correct model of the structure of a relaxed layer; however, its application is limited only to dislocations parallel to the sample surface. An analogous calculation of diffuse scattering from dislocations crossing the epitaxial layer and ending at the free surface (threading dislocations) is rather complicated. For a routine characterization of relaxed layers a simpler model must be used.

The structure of many relaxed layers can be described as composed of densely packed randomly rotated mosaic blocks; the block boundaries are created by networks of threading dislocations. This mosaic model of relaxed layers is characterized by the mean radius R of the blocks and their root mean square misorientation $\varDelta$. We assume that crystal lattice of a block is randomly rotated and not strained, so that the symmetric part of the deformation tensor of the block lattice is zero, and $\langle \boldsymbol{u}(\boldsymbol{r})\rangle = 0$. The mosaic model is purely phenomenological and the connection of its parameters, the radius R, and the root mean square misorientation $\varDelta$ with "actual" parameters describing the layer structure (such as dislocation densities, their Burgers vectors, etc.) is only indirect. Because of its simplicity, the mosaic block model is widely used for the characterization of the structure of relaxed epitaxial layers.

It can be easily demonstrated that the static Debye-Waller factor of a mosaic structure $\mathcal{D} = \langle \exp(-\mathrm{i}\boldsymbol{h}.\boldsymbol{u})\rangle$ is negligibly small and the wave diffracted by such a structure contains only the diffusely scattered wave. In the calculation of the correlation function $C(\boldsymbol{r}-\boldsymbol{r}')$ of the crystal deformation we assume that the polarizability coefficients $\chi_{\boldsymbol{h}}$ of the blocks are the same and they equal to those of a perfect lattice. Then

$$C(\boldsymbol{r}-\boldsymbol{r}') = |\chi_{\boldsymbol{h}}|^2 \left\langle \mathrm{e}^{-\mathrm{i}\boldsymbol{h}.(\boldsymbol{u}(\boldsymbol{r})-\boldsymbol{u}(\boldsymbol{r}'))}\right\rangle = \\ = |\chi_{\boldsymbol{h}}|^2 P(\boldsymbol{r}-\boldsymbol{r}') \left\langle \mathrm{e}^{-\mathrm{i}\boldsymbol{h}.(\boldsymbol{u}(\boldsymbol{r})-\boldsymbol{u}(\boldsymbol{r}'))}\right\rangle_{\mathrm{rot}}, \quad (10.38)$$

where $P(\boldsymbol{r}-\boldsymbol{r}')$ is the probability that both points $\boldsymbol{r}$, $\boldsymbol{r}'$ lie in the same block and $\langle\ \rangle_{\mathrm{rot}}$ denotes the averaging over all possible orientations of the crystal lattice of a block.

The probability P is connected with the shape function $\Omega(\boldsymbol{r})$ of a block by

$$P(\boldsymbol{\varrho}) = \frac{\int \mathrm{d}^3\boldsymbol{r}''\Omega(\boldsymbol{r}'')\Omega(\boldsymbol{\varrho}+\boldsymbol{r}'')}{\int \mathrm{d}^3\boldsymbol{r}''\Omega(\boldsymbol{r}'')}, \quad \boldsymbol{\varrho} = \boldsymbol{r}-\boldsymbol{r}', \quad (10.39)$$

Therefore its Fourier transformation is proportional to $|\Omega^{\mathrm{FT}}|^2$. From Eq. (10.39) the probability P for simple block shapes can be calculated exactly. For spherical blocks with radius R,

$$P(\boldsymbol{\varrho}) = \begin{cases} 1 - \frac{3}{4}\frac{|\boldsymbol{\varrho}|}{R} + \frac{1}{16}\left(\frac{|\boldsymbol{\varrho}|}{R}\right)^3 & \text{for } |\boldsymbol{\varrho}| \le 2R \\ 0 & \text{for } |\boldsymbol{\varrho}| > 2R \end{cases}, \quad (10.40)$$

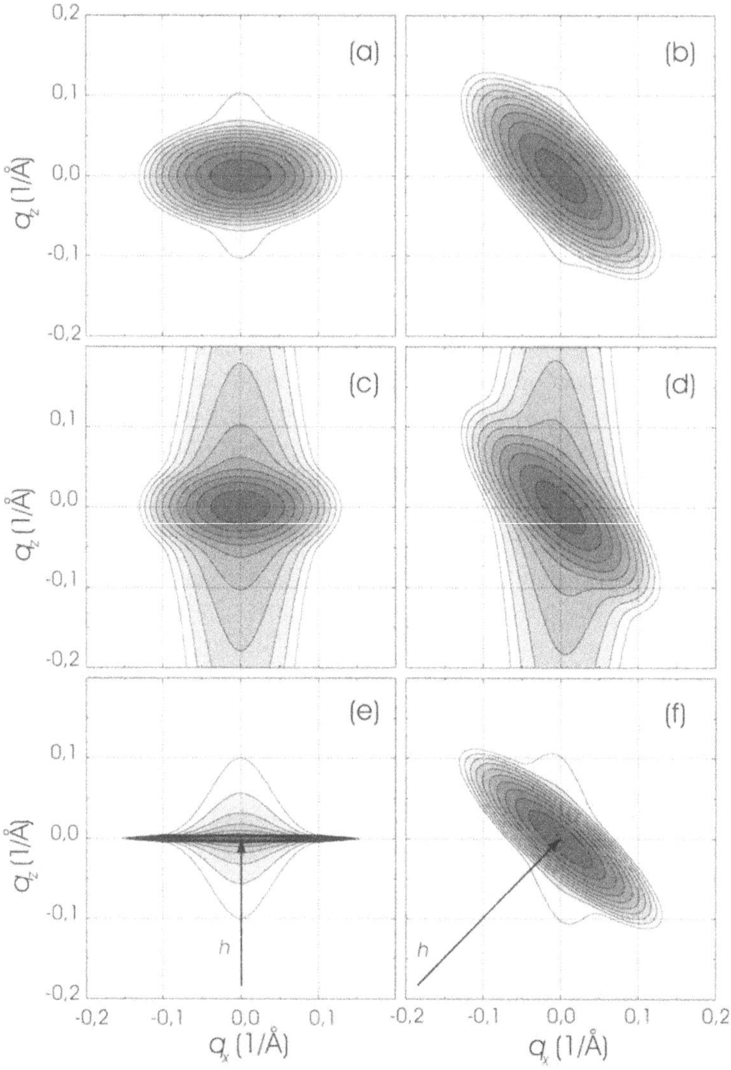

Fig. 10.15. The intensity maps of a mosaic layer calculated in symmetric 004 diffraction (a,c,e) and asymmetric 404 diffraction (b,d,f). The layer thickness was 10 μm (a,b,e,f) and 100 nm (c,d). In all panels, the root mean square misorientation of the blocks was 0.5°. In (a-d) the blocks were spherical with radius 10 nm; in (e,f) they were columnar with the horizontal radius 10 nm and height 2 μm. In (e) and (f) the directions of the diffraction vectors are denoted by arrows; the step of the intensity contours was $\sqrt{10}$.

holds [20, 142].

The result of the orientation averaging in Eq. (10.38) depends on the probability distribution of the block orientations. We introduce a vector $\boldsymbol{\phi}$ of a small random rotation of the lattice of a block with respect to the mean lattice orientation, and for simplicity we assume that these vectors are normally distributed with zero mean and root mean square deviation $\Delta \ll 1$. Then [157, 160, 203],

$$\left\langle \mathrm{e}^{-\mathrm{i}\boldsymbol{h}.(\boldsymbol{u}(\boldsymbol{r})-\boldsymbol{u}(\boldsymbol{r}'))} \right\rangle_{\mathrm{rot}} = \exp\left[-\frac{1}{6}\Delta^2|\boldsymbol{h}\times(\boldsymbol{r}-\boldsymbol{r}')|^2\right]. \tag{10.41}$$

Figure 10.15 shows several examples of intensity maps calculated for various parameters of the mosaic structure and various layer thicknesses. If the layer is much thicker than the mean block size and the mosaic blocks are spherical, the intensity contours have nearly elliptical shapes, and the main axes of the ellipses are perpendicular and parallel to the diffraction vector. The axis perpendicular to the diffraction vector is proportional to

$$\sqrt{1/R^2 + (h\Delta)^2/6}.$$

For $h\Delta R \gg 1$, i.e., if the blocks are large and/or strongly misoriented, the length of this axis is determined mainly by the misorientation Δ, and it is proportional to the length of the diffraction vector $\boldsymbol{h}$. The axis parallel to the diffraction vector is inversely proportional to R and it is not influenced by the misorientation. The determination of the parameters of the mosaic models is not straightforward if the layer thickness is comparable with the mean size of the blocks or if the block shape is anisotropic. In these cases, the shape of the intensity contours is not elliptical and no simple rule exists connecting this shape with the parameters. Figure 10.15(e,f) represents the intensity maps calculated for mosaic blocks having the form of narrow vertical columns. The intensity map in the symmetric diffraction (the diffraction vector is parallel with the column axis) exhibits a narrow horizontal stripe; the width of this stripe is inversely proportional to the column height. In an asymmetric diffraction, such a narrow stripe is not visible.

As an experimental example, we present here the intensity maps of a 300-nm-thick CdTe layer (orientation (111)) grown in a sandwich structure ZnSe/CdTe/ZnSe on a (001) GaAs substrate [299]. The maps have been measured in the symmetric 333 diffractions in two different azimuthal directions of the scattering plane (Fig. 10.16) and in the asymmetric diffraction 440 (Fig. 10.17). In the figures, we compare the experimental intensity maps with their best fit using the mosaic model above. We have fitted all the maps measured in a given azimuthal direction using the same set of free parameters. From the fit we have determined the mean radius of the mosaic blocks to $R = (29 \pm 4)$ nm in both azimuthal directions and the root mean square misorientation to $\Delta = (0.20 \pm 0.04)^\circ$ and $\Delta = (0.13 \pm 0.03)^\circ$ in the azimuthal directions $[11\bar{2}]$ and $[\bar{1}10]$, respectively. The agreement of the measured and fitted intensity distributions is quite good.

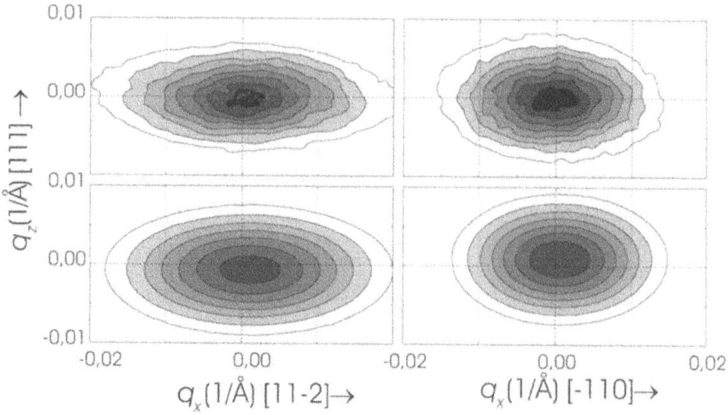

Fig. 10.16. The measured (upper row) and fitted (lower row) intensity maps of a CdTe mosaic layer, symmetrical diffraction 333, two different azimuthal directions of the scattering plane. The step of the intensity contours was 10% of the maximum intensity [299].

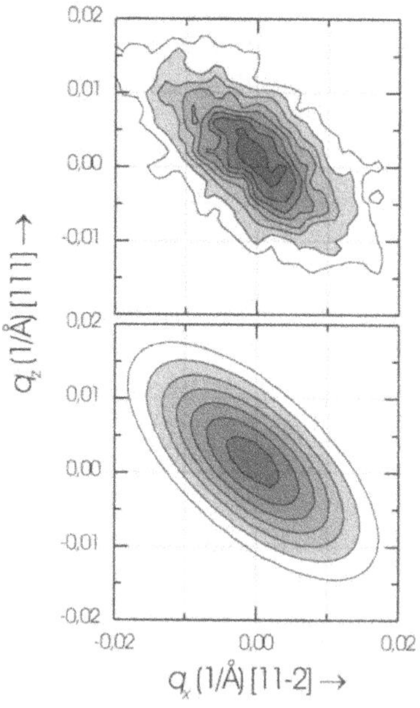

Fig. 10.17. The same situation as in Fig. 10.16, asymmetric diffraction 440.

The mosaic block model can be used for a rough characterization of the structure of relaxed layers. In some cases, however, this model fails, and no reliable fits of the measured intensity maps can be obtained. Then, the model can be refined, introducing random strains in the individual blocks, i.e., a random symmetric part of the local deformation tensor [160]. By increasing the number of the fitted parameters in this way, we can achieve a "nice" fit of the measured data; a physically sound interpretation of the fitted parameters is rather doubtful.

11 X-Ray Scattering by Rough Multilayers

This chapter deals with theoretical and experimental aspects of x-ray reflection and diffraction by thin layers and multilayers with rough interfaces illustrated by various examples. We describe the scattering potential of rough multilayers for x-ray reflection and later for x-ray diffraction; the multilayers are characterized by point properties and the correlation properties of their interfaces. General formulae are derived for the coherent part of the specularly scattered intensity by means of the coherent approach. The incoherent approach introduced in Chapter 5 is used for the calculation of the non-specularly scattered intensity. The end of the chapter discusses briefly the extension of the formalism to surface-sensitive diffraction methods.

X-ray methods investigate atomic-scale roughness in terms of its correlation properties on the mesoscopic (sub-micrometer) scale. By irradiating a macroscopic area of the sample, surface-sensitive x-ray scattering allows one to investigate the statistical behavior of the roughness profile.

Roughness of surfaces and interfaces has been investigated by different surface-sensitive x-ray scattering methods (x-ray reflection, grazing incidence diffraction, strongly asymmetrical x-ray diffraction) [22, 78, 79, 93, 159, 161, 164, 197, 204, 216, 246, 248, 261, 262, 301, 303, 307, 308, 310, 337, 338, 347, 351]. *Specular x-ray reflection* is the most frequently applied method (see, e.g., [22, 78, 93, 197, 248, 261, 337]). This method studies the depth profile of the mean electron density. The presence of diffuse x-ray scattering enables us to distinguish between interface roughness and graded interface profiles due to transition layers, interdiffusion or a graded heterotransition. Diffuse x-ray scattering in x-ray reflection has frequently been observed by the methods of *non-specular x-ray reflection* and *GISAXS (grazing-incidence small-angle x-ray scattering)* (e.g., [93, 159, 161, 164, 204, 246, 262, 301, 303, 307, 308, 310, 338, 347]). Measurements of diffuse x-ray scattering from interface roughness in diffraction modes have been reported for *grazing-incidence diffraction* and *strongly asymmetric coplanar diffraction* [34, 39, 351]. Diffuse x-ray scattering provides information about the lateral correlation properties of surfaces and interfaces. The occurrence of resonant diffuse scattering gives evidence of vertical roughness replication from interface to interface. Diffuse scattering allows us to examine the validity of layer growth models for different material systems and growth techniques.

11.1 Interface Roughness, Scattering Potential, and Statistical Properties

The form of the scattering potential $\hat{\mathbf{V}}$ occurring on the right-hand side of Eq. (4.11) depends on the scattering geometry. The general formula (4.12) for $\hat{\mathbf{V}}$ can be simplified for small-angle scattering, where the scattering vector $\boldsymbol{Q}$ is much shorter than any reciprocal lattice vector. In this case, the scattering is sensitive to the electron density averaged over the crystal unit cell (or over a sufficiently large volume in the case of an amorphous sample). Then, one of the Maxwell equations reads $0 = \mathrm{div}\boldsymbol{D} = \mathrm{div}(\epsilon\boldsymbol{E}) \approx \epsilon\mathrm{div}\boldsymbol{E}$ and the scattering potential is simply proportional to the zero-th Fourier coefficient of the crystal polarizability

$$\hat{\mathbf{V}} = -K^2\chi_0(\boldsymbol{r}). \tag{11.1}$$

The dependence of χ_0 on the coordinates reflects the inhomogeneity of the sample.

First, we consider a single rough surface, the profile of which can be described by a random function $U_0(x, y) \equiv U_0(\mathbf{r}_\parallel)$. We put $\langle U_0 \rangle = 0$ and the vertical coordinate of the mean surface as $z = z_0$. The position-dependent polarizability of a semi-infinite medium is expressed in form of a Heaviside function $H(x)$ ($H(x) = 1$ for $x > 0$ and $H(x) = 0$ for $x < 0$):

$$\chi(\boldsymbol{r}) = \chi_0 H(z_0 + U_0(\boldsymbol{r}_\parallel) - z). \tag{11.2}$$

For a multilayer with rough interfaces we express the polarizability by the sum of the individual layer contributions,

$$\chi(\boldsymbol{r}) = \sum_{m=1}^{N+1} \chi_0^{(m)} \Omega^{(m)}(\boldsymbol{r}), \tag{11.3}$$

where $\chi_0^{(m)}$ and $\Omega^{(m)}(\boldsymbol{r})$ are the polarizabilities and shape functions of layer m (see Fig. 5.6).

Within this chapter we consider vertically layered structures with a *random defect structure*, which we assume to be *laterally statistically homogeneous.* We concentrate on defects which vary *the layer shape and interface sharpness* $\Omega(\boldsymbol{r})$ (interdiffusion and roughness), in contrast to those influencing the *layer volume properties* $\chi_0^{(m)}$ (porosity, inclusions). Interdiffusion and graduated heterotransition between neighboring layers produce *vertically graded interfaces.* Then $\Omega(\boldsymbol{r})$ can have all values between 1 and 0. The layer is defined within the region $\Omega(\boldsymbol{r}) \neq 0$. We allow an intermixing of neighboring layers only, in order to keep the layer sequence. We define here by interface roughness the random profile of locally sharp interfaces. The vertical shift of the actual interface position with respect to its mean position is characterized by the displacement function U_m of each interface (see Fig. 11.1), modifying the actual layer size function

$$\Omega^{(m)}(\boldsymbol{r}) = H\left(z_m + U_m(\boldsymbol{r}) - z\right) - H\left(z_{m+1} + U_{m+1}(\boldsymbol{r}) - z\right), \tag{11.4}$$

and the actual layer thickness is $T_m(\boldsymbol{r}_\|) = T_m^{\mathrm{id}} + U_m(\boldsymbol{r}_\|) - U_{m+1}(\boldsymbol{r}_\|)$, where the *ideal* thickness equals $T_m^{\mathrm{id}} = z_m - z_{m+1}$. The multilayer consists of $N+1$ interfaces and N thin layers; layer $N+1$ is the substrate. Layer m lies between the interfaces m and $m+1$; interface 1 is the free surface. As before it is convenient to write the polarizability of the rough multilayer as the sum of the unperturbed contribution of a smooth multilayer and its disturbance by the roughness profile.

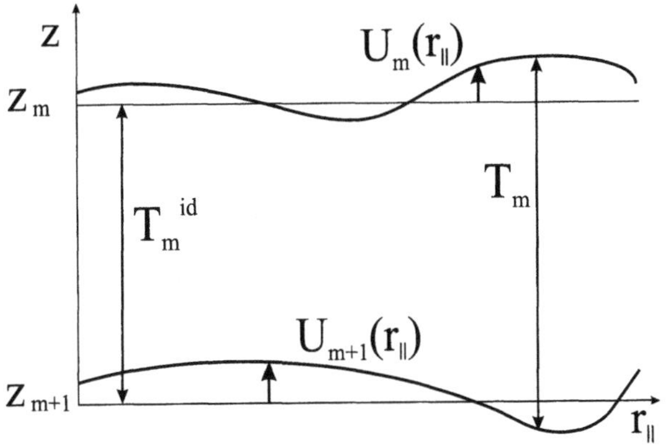

Fig. 11.1. Definition of the ideal and the actual layer thickness and of the interface displacement U_m.

The statistical behavior of a single interface is characterized by its point properties and its correlation properties. If the correlation length of the roughness profiles is small with respect to the irradiated sample area, we can assume that all possible configurations of the interfaces occur in the irradiated sample area. This assumption allows us to substitute the averaging over the statistical ensemble by the averaging over the sample surface.

The point properties of the interface named by the index m are described by use of the *probability density* of the displacement,

$$w_m(U) = \frac{1}{S}\int_S \mathrm{d}^2\boldsymbol{r}_\|\, \delta(U_m(\boldsymbol{r}_\|) - U), \tag{11.5}$$

where S is the area of integration, and by the corresponding *characteristic function*

$$\chi_{U_m}(Q) = \left\langle \mathrm{e}^{-\mathrm{i}QU_m} \right\rangle = \int_{-\infty}^{\infty} \mathrm{d}U\, w_m(U)\, \mathrm{e}^{-\mathrm{i}QU}\ . \tag{11.6}$$

Averaged over the interface, the relative displacement equals zero, $\langle U_m \rangle = 0$, with the square deviation

$$\sigma_m^2 \equiv \langle U_m^2(\boldsymbol{r}_\parallel)\rangle = \int_{-\infty}^{\infty} \mathrm{d}U\, w_m(U)\, U^2. \tag{11.7}$$

Here σ_m is the *root mean square (r.m.s.) roughness* of the interface. Since the interface profile is statistically homogeneous, σ_m is independent of the lateral position $\boldsymbol{r}_\parallel$.

The *mean coverage* of the upper (statistically homogeneous) interface at the distance $U = z - z_m$ from its mean position z_m is [80]

$$\Phi_m(z) = \int_{-\infty}^{z-z_m} \mathrm{d}U\, w_m(U). \tag{11.8}$$

The correlation properties between two points $\boldsymbol{r}_\parallel$ and $\boldsymbol{r}'_\parallel$ are characterized by the two-dimensional *probability density* (see, for example, [78, 159, 206, 256, 257, 310, 332])

$$w(U, U') = w(U(\boldsymbol{r}_\parallel), U(\boldsymbol{r}'_\parallel)). \tag{11.9}$$

The two-dimensional Fourier transformation of this two-point probability density is the *two-dimensional characteristic function*

$$\chi_{UU'}(Q, Q') = \left\langle \mathrm{e}^{-\mathrm{i}(QU - Q'U')} \right\rangle. \tag{11.10}$$

If the two-point probability density depends only on $\boldsymbol{r}_\parallel - \boldsymbol{r}'_\parallel$, the interface is statistically homogeneous. Then we define the *correlation function* of the interface by the expression

$$C(\boldsymbol{r}_\parallel - \boldsymbol{r}'_\parallel) = \left\langle U(\boldsymbol{r}_\parallel) U(\boldsymbol{r}'_\parallel) \right\rangle. \tag{11.11}$$

Now, we proceed to the description of the correlation properties of rough interfaces in a multilayer. Using Eq. (11.8) consequently the mean coverage of the layer m of a multilayer stack equals

$$\left\langle \Omega^{(m)}(z) \right\rangle \equiv \Phi_m(z - z_m) - \Phi_{m+1}(z - z_{m+1}). \tag{11.12}$$

In order to describe the correlation properties between different interfaces of the multilayer, we introduce the two-dimensional probability density of *two* interfaces

$$w(U_m, U'_n) = w(U_m(\boldsymbol{r}_\parallel), U_n(\boldsymbol{r}'_\parallel)), \tag{11.13}$$

the corresponding characteristic function $\chi_{U_m U'_n}$ and correlation function

$$C_{mn}(\boldsymbol{r}_\parallel - \boldsymbol{r}'_\parallel) = \left\langle U_m(\boldsymbol{r}_\parallel) U_n(\boldsymbol{r}'_\parallel) \right\rangle. \tag{11.14}$$

Here we have again assumed that the multilayer is laterally statistically homogeneous.

Usually the perfection of interfaces in multilayers is essentially influenced by the quality of the substrate or buffer surface. The surface defects can be replicated in the growth direction. Depending on the material system, the layer set-up and the growth conditions, different replication behaviors have been observed. The following replication model has been proposed in [338, 347]: 1) during the growth of the mth layer, the roughness profile $U_{m+1}(\boldsymbol{r}_\parallel)$ of the lower interface is *partially replicated*; and 2) an *intrinsic roughness* $\Delta_m(\boldsymbol{r}_\parallel)$ is induced by imperfections of the growth process. The resulting random profile $U_m(\boldsymbol{r}_\parallel)$ of the interface m is a sum of both contributions:

$$\begin{aligned} U_m(\boldsymbol{r}_\parallel) &= \int_S \mathrm{d}^2 \boldsymbol{r}'_\parallel \, U_{m+1}(\boldsymbol{r}'_\parallel) a_m(\boldsymbol{r}_\parallel - \boldsymbol{r}'_\parallel) + \Delta_m(\boldsymbol{r}_\parallel) \\ &= U_{m+1}(\boldsymbol{r}_\parallel) \otimes a_m(\boldsymbol{r}_\parallel) + \Delta_m(\boldsymbol{r}_\parallel). \end{aligned} \tag{11.15}$$

Here a non-random replication function $a_m(\boldsymbol{r}_\parallel)$ has been introduced. It determines the *degree of memory* of the top interface for the roughness profile of the corresponding bottom interface. If the replication function is zero, the upper interface of a layer forgets the interface profile at the layer bottom and its profile is entirely determined by the intrinsic roughness (*no replication*). *Identical profile replication* is achieved for zero intrinsic roughness and full replication ($a_m(\boldsymbol{r}_\parallel)$ equals the delta function). Other cases are discussed in detail in [338, 347] and will gain our interest when we proceed to the discussion of experimental results.

In later sections we will use the Fourier transformation of the interface correlation functions

$$\begin{aligned} C^{\mathrm{FT}}_{mn}(\boldsymbol{Q}_\parallel) &= \int \mathrm{d}^2(\boldsymbol{r}_\parallel - \boldsymbol{r}'_\parallel) \, C_{mn}(\boldsymbol{r}_\parallel - \boldsymbol{r}'_\parallel) \mathrm{e}^{-\mathrm{i}\boldsymbol{Q}_\parallel(\boldsymbol{r}_\parallel - \boldsymbol{r}'_\parallel)} \\ &= \langle U^{\mathrm{FT}}_m(\boldsymbol{Q}_\parallel) U^{\mathrm{FT}}_n(\boldsymbol{Q}_\parallel) \rangle \end{aligned} \tag{11.16}$$

with

$$U^{\mathrm{FT}}_m(\boldsymbol{Q}_\parallel) = \Delta^{\mathrm{FT}}_m(\boldsymbol{Q}_\parallel) + U^{\mathrm{FT}}_{m+1}(\boldsymbol{Q}_\parallel) a^{\mathrm{FT}}_n(\boldsymbol{Q}_\parallel). \tag{11.17}$$

In the following we neglect any statistical influence of the interface profile $U_{m+1}(\boldsymbol{r}_\parallel)$ on the intrinsic roughness $\Delta_m(\boldsymbol{r}_\parallel)$. Also the intrinsic roughness of different interfaces shall be statistically independent. Then we find the recursion formula for the Fourier transform of the correlation function,

$$C^{\mathrm{FT}}_{mn}(\boldsymbol{Q}_\parallel) = C^{\mathrm{FT}}_{m+1,n+1}(\boldsymbol{Q}_\parallel) a^{\mathrm{FT}}_m(\boldsymbol{Q}_\parallel) a^{\mathrm{FT}}_n(\boldsymbol{Q}_\parallel) + \delta_{mn} K^{\mathrm{FT}}_m(\boldsymbol{Q}_\parallel), \tag{11.18}$$

where $K^{\mathrm{FT}}_m(\boldsymbol{Q}_\parallel)$ is the Fourier transform of the correlation function of the intrinsic roughness:

$$K_m(\boldsymbol{r}_\parallel - \boldsymbol{r}'_\parallel) = \left\langle \Delta_m(\boldsymbol{r}_\parallel) \Delta_m(\boldsymbol{r}'_\parallel) \right\rangle. \tag{11.19}$$

If we assume for all layers the same replication function $a(\boldsymbol{r}_\parallel)$, and the same intrinsic roughness $\Delta(\boldsymbol{r}_\parallel)$ we obtain explicit expressions for the Fourier transforms of the *in-plane correlation function* [163]:

$$\begin{aligned} C^{\mathrm{FT}}_{mm}(\boldsymbol{Q}_\parallel) &= C^{\mathrm{FT}}_{NN}(\boldsymbol{Q}_\parallel) \left[a^{\mathrm{FT}}(\boldsymbol{Q}_\parallel) \right]^{2(N-m)} \\ &\quad + K^{\mathrm{FT}}_m(\boldsymbol{Q}_\parallel) \frac{\left[a^{\mathrm{FT}}(\boldsymbol{Q}_\parallel) \right]^{2(N-m-1)} - 1}{\left[a^{\mathrm{FT}}(\boldsymbol{Q}_\parallel) \right]^2 - 1}, \end{aligned} \tag{11.20}$$

where $C^{\mathrm{FT}}_{NN}(\boldsymbol{Q}_{\|})$ is the correlation function of the substrate, and of the *inter-plane correlation function*

$$C^{\mathrm{FT}}_{m>n}(\boldsymbol{Q}_{\|}) = C^{\mathrm{FT}}_{mm}(\boldsymbol{Q}_{\|}) \left[a^{\mathrm{FT}}(\boldsymbol{Q}_{\|}))\right]^{(m-n)} . \tag{11.21}$$

The physical meaning of the particular terms in (11.20) is obvious. The first term on the right-hand side represents the influence of the substrate surface modified by the replication function; the second term is due to the intrinsic roughness of the layers beneath the layer m. Knowing $C^{\mathrm{FT}}_{mm}(\boldsymbol{Q}_{\|})$, we can calculate the mean square roughness σ^2_m of the mth interface by putting (11.16) in (11.6):

$$\sigma^2_m = \int \mathrm{d}^2\boldsymbol{Q}_{\|}\, C^{\mathrm{FT}}_{mm}(\boldsymbol{Q}_{\|}). \tag{11.22}$$

In growth models of kinetic roughening , the interface roughness is related to the microscopic mechanism of the growth process. There, the time evolution of the surface displacement function during growth $U_{\mathrm{surface}}(\boldsymbol{r}_{\|}, t)$ plays a role and follows certain scaling laws (see, e.g., [25]). These scaling laws apply to the *height-height correlation function*, defined as

$$g(\boldsymbol{r}_{\|} - \boldsymbol{r}'_{\|}, t) = \left\langle [U_{\mathrm{surface}}(\boldsymbol{r}_{\|}, t) - U_{\mathrm{surface}}(\boldsymbol{r}'_{\|}, t)]^2 \right\rangle . \tag{11.23}$$

From this function, the correlation function of the interface roughness can be derived:

$$C(\boldsymbol{r}_{\|} - \boldsymbol{r}'_{\|}) = \langle U(\boldsymbol{r}_{\|}) U(\boldsymbol{r}'_{\|}) \rangle = \sigma^2 - \frac{1}{2} g(\boldsymbol{r}_{\|} - \boldsymbol{r}'_{\|}). \tag{11.24}$$

Growth processes far from the thermodynamical equilibrium create a good approximation self-affine surfaces [208]. The form of the function g can be postulated from the analogy of the stochastic process defined by the growth of a rough interface with the Brownian motion [337, 385]:

$$g(\boldsymbol{r}_{\|} - \boldsymbol{r}'_{\|}) \sim |\boldsymbol{r}_{\|} - \boldsymbol{r}'_{\|}|^{2h} \quad \text{for} \quad |\boldsymbol{r}_{\|} - \boldsymbol{r}'_{\|}| \to 0, \tag{11.25}$$

where h (the *fractal exponent*) is connected to the *fractal dimension* D_{f} of the interface by $D_{\mathrm{f}} = 3 - h$. This self-affine behavior of the interface is violated for larger distances $|\boldsymbol{r}_{\|} - \boldsymbol{r}'_{\|}|$, where the shifts $U(\boldsymbol{r}_{\|})$ and $U(\boldsymbol{r}'_{\|})$ can be assumed to be statistically independent, (i.e., $C(\boldsymbol{r}_{\|} - \boldsymbol{r}'_{\|}) \to 0$):

$$g\left(\boldsymbol{r}_{\|} - \boldsymbol{r}'_{\|}\right) \approx 2\sigma^2 \quad \text{for} \quad |\boldsymbol{r}_{\|} - \boldsymbol{r}'_{\|}| \to \infty . \tag{11.26}$$

In this way the self-affine surface should be characterized by the fractal exponent h and an upper cutoff, where the scaling law (11.25) is no longer valid; this is the lateral correlation length of the interface roughness. The atomic distance is considered as a lower cutoff.

A suitable form of C having the above asymptotic properties is

$$C(\boldsymbol{r}_{\|} - \boldsymbol{r}'_{\|}) = \sigma^2 \exp\left[-\left(\frac{|\boldsymbol{r}_{\|} - \boldsymbol{r}'_{\|}|}{\Lambda}\right)^{2h}\right], \tag{11.27}$$

where Λ is the correlation length of the interface roughness. The correlation function of the form of Eq. (11.27) has a serious disadvantage – its Fourier transformation can be calculated only numerically (except for $h = 0.5$ and $h = 1$) and, moreover, the Fourier integral of C diverges very slowly for $h \to 0$. Therefore, other forms of the correlation function are also used in the literature (see, e.g., [256, 302]), all fulfilling the scaling requirements of Eq. (11.25) for scaling coefficients $0 < h \leq 1$ that are better suited especially for the limiting case of $h \to 0$, where the correlation function has a logarithmic behavior.

For the replication of the roughness profile from interface to interface within a multilayer simplified models of vertical roughness correlation are introduced [246], where the inter-plane correlation function C_{mn} depends on the in-plane correlation function C_{jj} ($j = \max(m,n)$) of the lower interface according to [246]

$$C_{mn}(\boldsymbol{r}_\| - \boldsymbol{r}'_\|) = C_{jj}(\boldsymbol{r}_\| - \boldsymbol{r}'_\|)\mathrm{e}^{-|z_m - z_n|/\Lambda_\perp}; \tag{11.28}$$

here $\Lambda_\perp$ is the vertical correlation length. In contrast to (11.15), the simplified model of (11.28) does not explain effects of smoothing and roughening; however, it makes the calculations simpler.

A more realistic inter-plane correlation function was derived in [303],

$$C_{mn}(\boldsymbol{Q}_\|)^{\mathrm{FT}} = C_{jj}(\boldsymbol{Q}_\|)^{\mathrm{FT}} \exp\left[-\zeta|\boldsymbol{Q}_\||^n |z_m - z_n|\right], \tag{11.29}$$

where ζ is a constant and the exponent n depends on the dominant mechanism of the surface smoothing. If this smoothing is caused by desorption, then $n = 2$; if surface diffuse is important, $n = 4$. This formula follows from the stochastic Edwards–Wilkinson equation describing the growth kinetics [25].

11.2 Specular X-Ray Reflection

Let us now consider how surface and interface roughness influence x-ray reflection by single surfaces and multilayers. First, we incorporate the statistical properties of rough interfaces into the theoretical formalism of specular x-ray reflection by multilayers introduced in Sect. 6.5.

In Sect. 5.5 we have shown that the wave scattered by a randomly disturbed system consists of one coherent and one incoherent (diffuse) component. In many cases, the incoherent part of the specularly reflected wave can be neglected (this part corresponds to the portion of the diffusely scattered wave that has the same direction as the specularly reflected wave). Therefore, the specular reflectivity can be calculated using the coherent approach. In this approach, we have to solve the wave equation

$$(\triangle + K^2)\langle E(\boldsymbol{r})\rangle = \langle \hat{\mathbf{V}}(\boldsymbol{r})E(\boldsymbol{r})\rangle \equiv \hat{\mathbf{V}}_{\mathrm{eff}}(\boldsymbol{r})\langle E(\boldsymbol{r})\rangle. \tag{11.30}$$

Various treatments have been applied in the literature. One approach replaces the effective potential $\hat{\mathbf{V}}_{\text{eff}}(\boldsymbol{r})$ by the *averaged scattering potential* $\hat{\mathbf{V}}_{\text{eff}}(\boldsymbol{r}) \approx \langle \hat{\mathbf{V}}(\boldsymbol{r}) \rangle$ [161]. Averaging the potential over the irradiated sample surface leads to graded interfaces, expressed here by smoothly varying surface shape functions. In the case of a multilayer with rough interfaces, the individual shape functions of the layers are smooth without discontinuities (see Fig. 11.2). In this approach, the effective scattering potential is

$$\hat{\mathbf{V}}_{\text{eff}}(\boldsymbol{r}) \approx -K^2 \sum_{m=1}^{N+1} \chi_0^{(m)} \langle \Omega^{(m)}(\boldsymbol{r}) \rangle. \tag{11.31}$$

Since, in this approach, the effective scattering potential contains a sum of averaged contributions of particular layers, the possible correlation of roughness profiles of various interfaces does not affect the value of $\hat{\mathbf{V}}_{\text{eff}}$.

If the scattering sample is laterally statistically homogeneous, its effective scattering potential $\hat{\mathbf{V}}_{\text{eff}}$ depends only on the vertical coordinate z. Then, due to the lateral translational invariance of this potential, the reflection law holds for the coherently scattered wave and its intensity is concentrated along the truncation rod of the ideal smooth structure (see also Sec. 4.3). Therefore, if a randomly rough sample is irradiated by a plane wave, the coherent component of the scattered wave is plane, too, and its direction is the same as that for a smooth structure. However, the distribution of the scattered intensity along the rod may differ appreciably from that for a smooth structure.

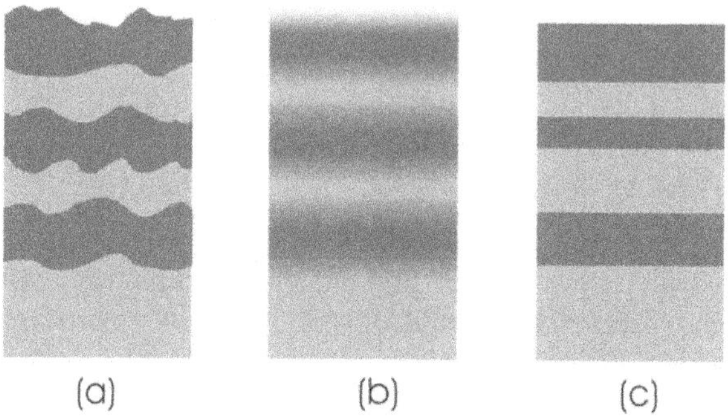

Fig. 11.2. Illustration of the coherent reflection treatment by rough interfaces: (*left*) real multilayer, (*center*) model of (11.31), (*right*) model of (11.32).

Using this approximation, the wave equation can be solved fully dynamically using the method described in Chapter 6, or within the semikinematical or kinematic approximations. Another method for the calculation of the coherently reflected wave is based on the concept of *averaging the reflected amplitudes* of samples with sharp, flat interfaces over the statistical ensemble of different vertical interface positions, first proposed in [78, 248]:

$$\langle E(\boldsymbol{r})\rangle = \int \mathrm{d}^{N+1}(U_1,\ldots,U_{N+1})E[\boldsymbol{r},(U_1,\ldots,U_{N+1})] \\ \times w(U_1,\ldots,U_{N+1}). \tag{11.32}$$

In both coherent treatments, any influence of the (finite) lateral correlation length of the roughness profile on the coherent reflectivity is neglected. Dynamical calculations of the specular x-ray reflectivity of rough interfaces based on the two models provide identical results within measurable precision. Here only the second procedure is discussed.

If the interfaces are rough, the layer thicknesses occurring in the propagation matrices (6.37) are random quantities $T_j(\boldsymbol{r}_\parallel) = T_j^{\mathrm{id}} + U_j(\boldsymbol{r}_\parallel) - U_{j+1}(\boldsymbol{r}_\parallel)$ (see Fig. 11.1). Applying the approximation (11.32), we average the scattered amplitudes of an ensemble of samples with flat interfaces. The actual inverse propagation matrix $\hat{\varPhi}_j^{-1}$ of layer j, defined in Sect. 6.5.1, can be written as a product of three matrices, $\hat{\varPhi}_j^{-1} = \hat{\mathbf{U}}_j \hat{\varPhi}_j^{\mathrm{id}-1} \widetilde{\mathbf{U}}_{j+1}$, where $\hat{\varPhi}_j^{\mathrm{id}-1}$ is the inverse of the propagation matrix of layer j, which has an averaged thickness $T_j^{\mathrm{id}} \equiv \langle T_j\rangle$. $\hat{\mathbf{U}}_j(\boldsymbol{r}_\parallel)$ and $\widetilde{\mathbf{U}}_{j+1}(\boldsymbol{r}_\parallel)$ are the propagation matrices of the thickness deviations

$$\hat{\mathbf{U}}_j = \begin{pmatrix} \mathrm{e}^{\mathrm{i}k_z^{(j)}U_j} & 0 \\ 0 & \mathrm{e}^{-\mathrm{i}k_z^{(j)}U_j} \end{pmatrix}, \quad \widetilde{\mathbf{U}}_{j+1} = \begin{pmatrix} \mathrm{e}^{-\mathrm{i}k_z^{(j)}U_{j+1}} & 0 \\ 0 & \mathrm{e}^{\mathrm{i}k_z^{(j)}U_{j+1}} \end{pmatrix}. \tag{11.33}$$

The wavefield at the detector window is then given by (compare with Eq. 6.37))

$$\boldsymbol{E}_0 = \hat{\varPhi}_0^{\mathrm{id}-1}\widetilde{\mathbf{U}}_1\hat{\mathbf{R}}_1\hat{\mathbf{U}}_1\hat{\varPhi}_1^{\mathrm{id}-1}\widetilde{\mathbf{U}}_2\hat{\mathbf{R}}_2\hat{\mathbf{U}}_2\hat{\varPhi}_2^{\mathrm{id}-1}\cdots\widetilde{\mathbf{U}}_{N+1}\hat{\mathbf{R}}_{N+1}\hat{\mathbf{U}}_{N+1}\hat{\varPhi}_{N+1}^{\mathrm{id}-1}E_{\mathrm{sub}} \\ \equiv \hat{M}\boldsymbol{E}_{\mathrm{sub}}, \tag{11.34}$$

where $\hat{\varPhi}_0^{\mathrm{id}-1}$ is the inverse propagation matrix between the averaged sample surface and the detector over the distance T_0. $\hat{\varPhi}_{N+1}^{\mathrm{id}-1}$ is the inverse propagation matrix between the averaged substrate surface and a fictitious interface lying at an arbitrary depth T_{N+1} below the substrate surface (see Fig. 11.3).

We assume that the wavefield $\boldsymbol{E}_{\mathrm{sub}}$ below this interface is not random, so that we can perform the averaging

$$\langle \boldsymbol{E}_0\rangle \approx \langle \hat{M}\rangle \boldsymbol{E}_{\mathrm{sub}}, \tag{11.35}$$

yielding the coherent reflectivity of the multilayer in the form

$$\mathcal{R}_{\mathrm{coh}} = \left|\frac{\langle M_{21}\rangle}{\langle M_{11}\rangle}\right|^2. \tag{11.36}$$

The averaging procedure additionally assumes the validity of the approximation

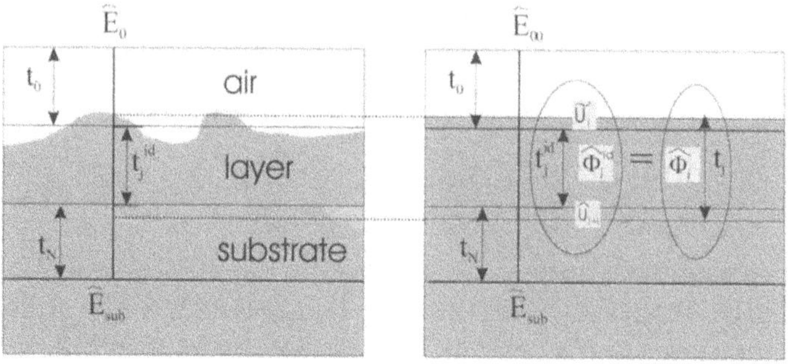

Fig. 11.3. Illustration of the coherent scattering approach discussed in the text: (*left*) real multilayer with rough interfaces, (*right*) the meaning of the matrices discussed in the text.

$$\left\langle \frac{M_{21}}{M_{11}} \right\rangle \approx \frac{\langle M_{12} \rangle}{\langle M_{11} \rangle}. \tag{11.37}$$

By use of numerical calculations it can be shown that this approximation is valid if the r.m.s. roughness is small compared to the layer thicknesses [167].

The averaging of the elements M_{12} and M_{11} of the matrix $\hat{\mathbf{M}}$ can easily be performed if we assume that the displacements U_j and U_k $(j \neq k)$ are statistically independent (no vertical correlation of the roughness profiles). Then, we obtain

$$\hat{\mathbf{R}}'_j = \langle \widetilde{\mathbf{U}}_j \hat{\mathbf{R}}_j \hat{\mathbf{U}}_j \rangle = \frac{1}{t'_j} \begin{pmatrix} 1 & r'_j \\ r'_j & 1, \end{pmatrix} \tag{11.38}$$

and the matrix $\hat{\mathbf{M}}$ can be expressed in the same way as in the case of a flat multilayer, as in (6.37) but replacing the interface matrices $\hat{\mathbf{R}}_j$ by $\hat{\mathbf{R}}'_j$. The effective Fresnel transmission and reflection coefficients are influenced by the interface roughness as follows:

$$t'_j = \frac{t_j}{\chi_{U_j}(k_z^{(j-1)} - k_z^{(j)})}, r'_j = r_j \frac{\chi_{U_j}(k_z^{(j-1)} + k_z^{(j)})}{\chi_{U_j}(k_z^{(j-1)} - k_z^{(j)})}. \tag{11.39}$$

If the roughness profiles of different interfaces are correlated (vertical replication of the roughness), the averaging of the matrix $\hat{\mathbf{M}}$ can still be performed, however, the resulting formulae are rather cumbersome. If all roughness profiles are identical, i.e., $U_j = U_k$, there are no thickness fluctuations in the layers ($T_j = T_j^{\mathrm{id}}$ except for the fictitious layers with thicknesses T_0 and T_{N+1}). The coherent reflectivity of the multilayer is then

$$\mathcal{R}_{\mathrm{coh}} = \mathcal{R}_{\mathrm{flat}} \frac{\chi_U(K_z + k_z^{\mathrm{sub}})}{\chi_U(K_z - k_z^{\mathrm{sub}})}, \tag{11.40}$$

which resembles the reflectivity of a single rough surface.

In the kinematical approximation we assume $t_j = 1$, and in the arguments of the characteristic functions χ_{U_j} we neglect refraction, putting $k_z^{(j)} = K_z$. This leads to the effective Fresnel coefficients,

$$t'_j = t_j = 1, \; r'_j = r_j^{\mathrm{kin}} \chi_{U_j}(2K_z), \tag{11.41}$$

where r_j^{kin} is the kinematical limit of the Fresnel reflectivity coefficient $r^{\mathrm{kin}} = (1 - n^2)/(4 \sin^2 \alpha_i)$.

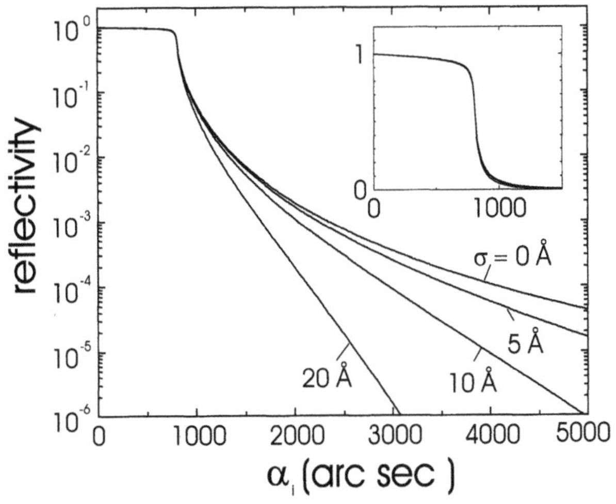

Fig. 11.4. The coherent reflectivity of a rough Si surface for different r.m.s. roughnesses, calculated with Eq. (11.44).

As the simplest example, let us deal with the coherent reflectivity of a single rough surface. We assume that the distribution function $w(U)$ is Gaussian:

$$w(U) = \frac{1}{\sigma\sqrt{2\pi}} \mathrm{e}^{-U^2/2\sigma^2}. \tag{11.42}$$

If we substitute its characteristic function

$$\chi_U(Q) = \int_{-\infty}^{\infty} \mathrm{d}U w(U) \mathrm{e}^{-\mathrm{i}QU} = \mathrm{e}^{-Q^2\sigma^2/2} \tag{11.43}$$

into (11.40), we obtain the coherent reflectivity of a single rough surface,

$$\mathcal{R}_{\mathrm{coh}} = \mathcal{R}_{\mathrm{flat}} \left| \mathrm{e}^{-Q_z Q_{z\mathrm{T}} \sigma^2/2} \right|^2, \tag{11.44}$$

where the reflectivity $\mathcal{R}_{\mathrm{flat}}$ is the reflectivity of a flat surface given by the square of the modulus of the appropriate Fresnel reflection coefficient r, and $Q_{z\mathrm{T}}$ is the z component of the scattering vector within the medium.

The coherent reflectivity of a rough surface in the kinematic approximation is

$$\mathcal{R}_{\mathrm{coh}}^{\mathrm{kin}} = \mathcal{R}_{\mathrm{flat}}^{\mathrm{kin}} e^{-Q_z^2\sigma^2}, \tag{11.45}$$

where the kinematic reflectivity of the flat surface is given by the square of the modulus of the kinematic Fresnel reflectivity (11.41).

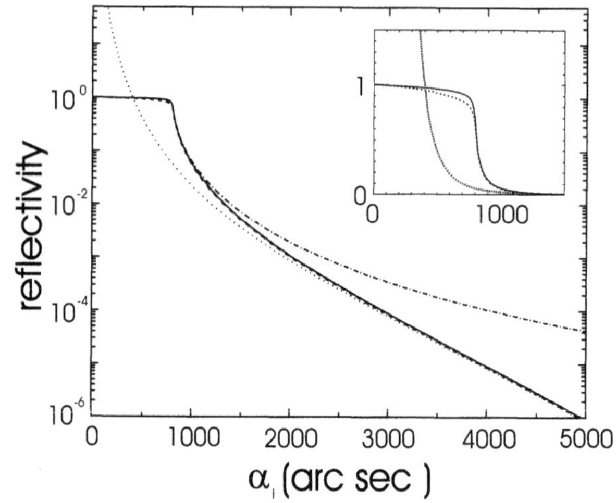

Fig. 11.5. Calculated coherent reflectivity of a rough Si surface. The reflectivity of a flat surface (dotted-dashed line) is compared to that for a roughness $\sigma = 1$ nm, calculated by the kinematical theory (11.45) (dotted line) and the dynamical theory (11.44) (full line). The kinematical reflectivity diverges for $\alpha_i \to \infty$. The dynamical curve coincides nearly with that of the flat surface below α_c. The dotted line represents the coherent reflectivity of a rough surface calculated with the *dynamical* Fresnel reflection coefficient and the *kinematical* diminution factor. This reflectivity is also damped by the roughness below α_c and coincides with the full calculation above it [39].

Both the kinematical and the dynamical reflectivities are multiplied by a diminution factor containing the r.m.s. roughness in the exponent. The kinematical diminution factor decreases with the square of the vacuum scattering vector Q_z, which is proportional to the incidence angle. The dynamical diminution factor contains the product of the scattering vectors in vacuum and in the medium, $Q_z Q_{z\mathrm{T}}$. The angular dependencies of these diminution factors differ substantially for angles below α_c. If absorption is neglected, the scattering vector in the sample, $Q_{z\mathrm{T}}$, becomes purely imaginary below α_c. Consequently, within the dynamical description, there is no influence of roughness on the coherently reflected intensity in this angular range. At large incidence angles both diminution factors coincide (see Figs. 11.4 and 11.5).

A more detailed discussion of the formulae (11.44) and (11.45) is given in [93]. There, the contribution of the incoherent scattering to the specular direction has been studied by means of a second-order DWBA, showing its

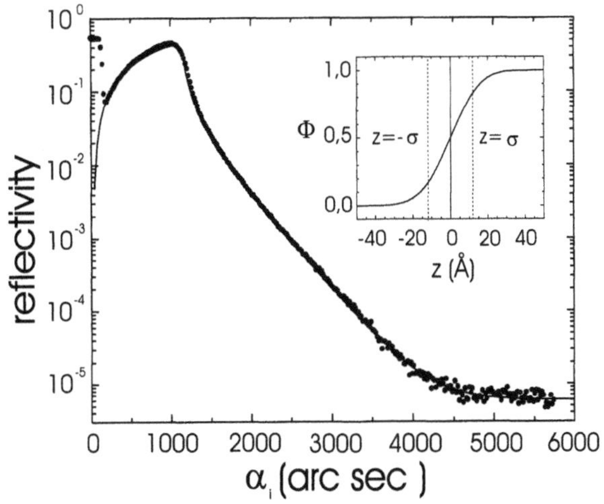

Fig. 11.6. Measured (points) and calculated (line) reflectivity curves of a GaAs substrate. In the inset, the mean depth profile of the surface, as extracted from the analysis, is plotted [167].

dependence on the lateral correlation length Λ. This study concluded that the coherently reflected intensity can be described by the dynamical equation (11.44) for short Λ ("fast" roughness), below 1 μm. For larger Λ ("slow"

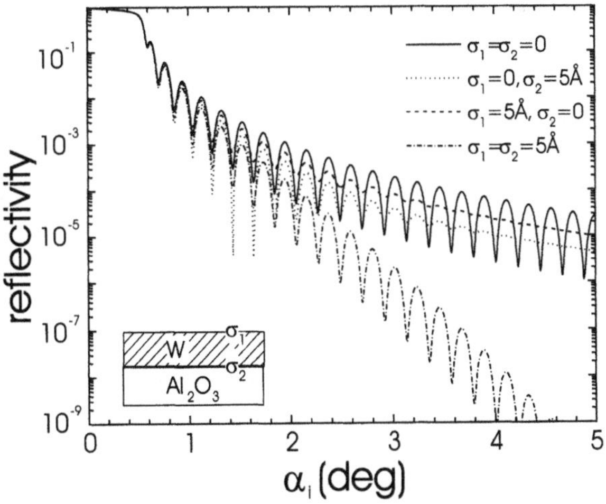

Fig. 11.7. Calculation of the specular reflectivity of a single layer (20 nm W) on sapphire for different r.m.s. roughnesses and dynamical diminution factors. *Upper* to *lower* curve: without roughness, interface roughness 0.5 nm, surface roughness 0.5 nm, surface and interface roughness both 0.5 nm. Surface roughness causes a faster decay of the reflectivity, while interface roughness damps the oscillations [39].

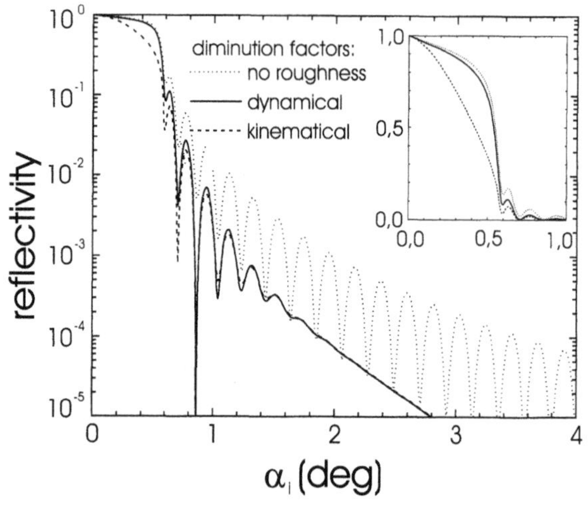

Fig. 11.8. Calculation of the specular reflectivity of a single layer (20 nm W) on sapphire for different diminution factors. Surface roughness of 1.2 nm and interface roughness of 0.3 nm. Kinematical slow roughness (lower curve, dashed), dynamical rapid roughness (middle curve, line) and without roughness (upper curve, dotted) [39].

roughness), the kinematical formula (11.45) becomes more appropriate (see Fig. 11.8).

The surface roughness of numerous samples of amorphous, polycrystalline and single-crystal material systems has been studied by specular x-ray reflection. In Fig. 11.6 we plot an experimental example, the reflectivity of a rough GaAs substrate. Conventional simulation and fit programs for the specular x-ray reflectivity of a rough multilayer are usually based on a multilayer model with independent roughness profiles of the interfaces, supposing a Gaussian probability density of the random shifts U_m. This leads, with (11.39) and (11.43), to the effective Fresnel reflection and transmission coefficients

$$r'_j = r_j \exp\left(-\frac{\sigma_j^2}{2} Q_z^j Q_z^{j-1}\right), \quad t'_j = t_j \exp\left[\frac{\sigma_j^2}{2}\left(\frac{Q_z^j - Q_z^{j-1}}{2}\right)^2\right] \qquad (11.46)$$

for each interface. The influence of the roughness on the transmission coefficient is rather small because of the small difference in the vertical scattering-vector components of the layers. However, the interface reflection is exponentially decreasing by the roughness, creating a strong change in the interference pattern.

The effect of interface roughness compared with surface roughness is shown in Fig. 11.7. The surface roughness decreases the specular intensity of the whole curve progressively with Q_z, whereas the interface roughness gives rise to a progressive damping of the interference fringes (thickness oscillations). However, the variation of the Fresnel coefficients which is due to

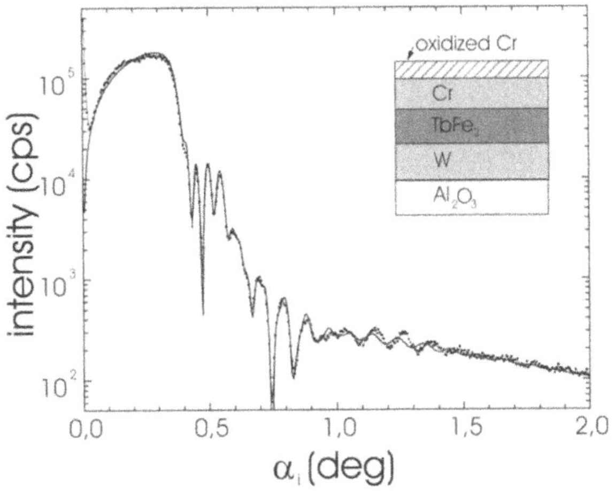

Fig. 11.9. Measurement (points) and fit (full line) of the specular reflectivity of a Cr/$TbFe_2$/W multilayer. The fitted thicknesses are W 34.6 nm, $TbFe_2$ 4.8 nm, Cr 50.5 nm and oxidized Cr 3 nm. The roughnesses are: sapphire substrate 0.2 nm, $TbFe_2$ 9.4 nm, and Cr 2.2 nm [39].

the roughness can locally cause more pronounced oscillations. In Fig. 11.9 we have plotted the experimental curve of a magnetic rare earth/transition metal multilayer (Cr/$TbFe_2$W on sapphire) grown by laser ablation deposition. This shows a quite complicated, irregular interference pattern. Good agreement with the simulation was obtained by assuming a thin oxide film at the sample surface and fitting the thicknesses and the r.m.s. roughnesses.

The main features of specular scans of *periodic* multilayers are the multilayer satellite peaks (superlattice maximum), which give evidence of the vertical periodicity (see Sect. 5.4). The intensity ratios of these peaks depend on the layer arrangement within the multilayer period. The differences in the electron densities determine the values of the Fresnel coefficients. The thickness ratios of the layers characterize the phase relations of the reflected waves from different interfaces. Interface roughness damps the intensity of the Bragg peaks progressively with Q_z (see Fig. 11.10); the surface roughness reduces the intensity of the whole curve. Usually, the perfection of the interfaces in multilayers is greatly influenced by the quality of the substrate or buffer surface. The surface defects can be replicated in the growth direction. Different replication behaviors have been observed, depending on the material systems, layer arrangement and growth conditions. The phenomenological replication model according to Eq. (11.15) was proposed in [338], assuming a Gaussian form of the replication function $a_m(\boldsymbol{r}_\parallel)$:

$$a(\boldsymbol{r}_\parallel - \boldsymbol{r}'_\parallel) = \frac{1}{2\pi L^2} \exp\left[-\left(\frac{|\boldsymbol{r}_\parallel - \boldsymbol{r}'_\parallel|^2}{2L^2}\right)\right]. \tag{11.47}$$

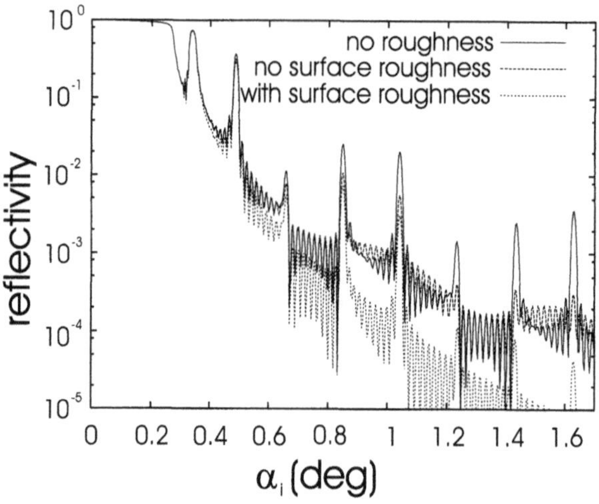

Fig. 11.10. Coherent reflectivity of a [GaAs(7 nm)/AlAs(15 nm)]×10 multilayer calculated by averaging the layer thickness. Without roughness (full line); with interface roughness but smooth surface (dashed line); with interface roughness and surface roughness (points).

The factor L determines the loss of memory from interface to interface. This choice arises from the aim of explaining the different limiting cases of roughness replication models by one class of functions. Let us further suppose a Gaussian roughness profile of the substrate surface and a single Gaussian correlation function for the intrinsic interface roughness of all layers, $K(\boldsymbol{r}_{\|} - \boldsymbol{r'}_{\|}) = \langle \Delta(\boldsymbol{r}_{\|})\Delta(\boldsymbol{r'}_{\|})\rangle$. If $a_m = 0$, K is also the correlation function of interface m: $C_m = K$.

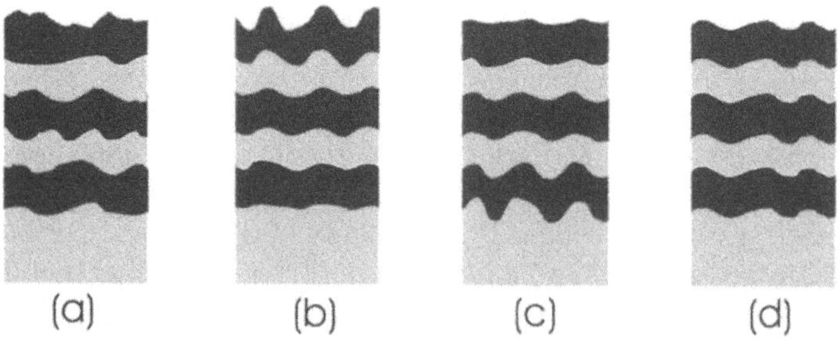

Fig. 11.11. Roughness models: (a) no vertical replication, (b) roughening, (c) smoothening, (d) identical replication.

The model allows us to describe all four limiting cases of roughness development illustrated in Fig. 11.11 as follows.

1. *No replication* occurs for $L \to \infty$, where a goes to zero. The roughness profiles of different interfaces are independent.
2. *Increasing roughness* toward the free surface is obtained by maximum replication ($L = 0$) and non-zero intrinsic roughness. One obtains
$$\sigma_j^2 = \sigma_N^2 + (\Delta\sigma)^2(N - j), \tag{11.48}$$
where $\Delta\sigma$ is the r.m.s. intrinsic roughness. Formula (11.48) describes the roughening during growth. By applying other growth models, other growth exponents α in the dependence $\sigma_j \sim (N - j)^\alpha$ can be obtained [25].
3. *Partial replication and no intrinsic roughness* ($L > 0$ and $\Delta\sigma = 0$) lead to decreasing r.m.s. roughness toward the free surface (smoothing of the multilayer during growth).
4. *Identical roughness* is achieved with maximum replication and no intrinsic roughness ($L = 0$ and $\Delta\sigma = 0$). Consequently, $\sigma_j = \sigma_N$, and all interfaces reproduce the profile of the substrate ($U_j(\boldsymbol{r}_\parallel) = U_N(\boldsymbol{r}_\parallel)$).

We examine here an experimental example of two periodic Si/Nb multilayers grown by sputtering. The multilayers were deposited on Si substrates, each with a thick SiO_2 layer and an Al buffer layer. The roughness of the buffer layer depends on its thickness and influences the quality of the interfaces. Two samples of different Al thicknesses were investigated and the results are shown in Fig. 11.12. The multilayer periodicity generates the superlattice Bragg peaks, which are damped by the interface roughness. The roughnesses of the substrate and of the buffer layers have less influence on the reflection pattern. Sample A can be fitted by a model of constant r.m.s. roughness for all interfaces. The first satellite peak is broadened by extinction due to dynamical multiple scattering. For all higher-order satellite peaks, we observe narrower (kinematical) peak widths.

The satellite maxima of sample B are rapidly damped, indicating a large interface roughness. In addition, the widths of the peaks increase with Q_z, which cannot be explained by a model of constant roughness. The satellite intensities and shapes can be successfully reproduced by supposing increasing roughness according to (11.48). Owing to their increased roughness, the upper layers near the surface contribute to the reflected wave with decreasing effective Fresnel coefficients. Within a satellite maximum, the contributions of all interfaces are in phase. However, slightly away from the maximum, the contributions of interfaces near the substrate do not completely cancel out the contributions of the interfaces near the sample surface, giving rise to the observed peak broadening.

Following the concept of terraces or small separated islands the surface morphology of crystalline samples can also be described by a discrete sur-

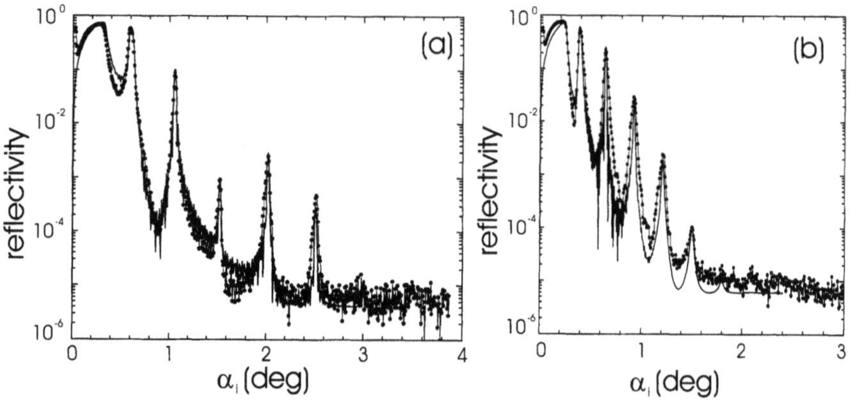

Fig. 11.12. Measurement (points) and simulation (full lines) of the specular reflectivity of a periodic Si/Nb multilayer with 10 periods. (a) Sample A, fitted by a model of constant roughness. (b) Sample B, fitted by a model of increasing roughness [39]

face probability distribution. An example of the simplest case consisting of randomly placed islands will be discussed in Chapter 14.

11.3 Non-Specular X-Ray Reflection

11.3.1 General Approach

Interface roughness also gives rise to scattering with an additional non-zero lateral scattering vector. This diffuse scattering occurs near all reciprocal-lattice points. Thus, it can be measured by x-ray reflection and diffraction methods. X-ray reflection is by far the most commonly used method to measure the diffuse x-ray scattering produced by rough interfaces, where the scattering intensity can be detected by simple off-specular scans. We develop here the basic theory for non-specular x-ray reflection and discuss the most common scattering features with reference to various experimental results.

The diffusely scattered intensity can be described in terms of the differential scattering cross section. We follow the formal incoherent approach introduced in Chapter 5. The kinematical treatment neglects the effect of refraction as well as all multiple-scattering processes. For incident angles near the critical angle of total reflection, both of these effects become important. In this case, the reflected and transmitted wave amplitudes are sufficiently strong that multiple scattering is expected. By use of the first-order distorted-wave Born approximation (DWBA) (see Chapt. 7), we take into account the specular reflection of the diffusely scattered wave, as well as the diffuse scattering of the specularly reflected wave. Multiple diffuse scattering (i.e., diffuse scattering of the diffusely scattered wave) can be described by DWBA methods of higher order [79].

In this book we restrict ourselves to the first-order DWBA described in Chapter 7. The scattering potential $\hat{\mathbf{V}}$, expressed by the polarizability, will be divided into the non-disturbed potential of the ideal plane multilayer $\hat{\mathbf{V}}_A$ and the disturbance $\hat{\mathbf{V}}_B$ due to the interface roughness. The incoherent (diffuse) differential cross section for the scattering can be calculated by means of Eq. (7.10).

Specular reflection by the undisturbed multilayer provides the wavefield, which consists of two non-disturbed plane waves (the reflected and transmitted waves) in each layer. In order to calculate the incoherent cross section we have to consider two independent solutions $E_1^{(A)}$ and $E_2^{(-A)}$ of the undisturbed wave equation (see Eqs. (7.5) and (7.6), for instance). In the case of non-specular reflection these solutions are

$$E_1^{(A)}(\boldsymbol{r}) = \sum_{m=0}^{N+1} \mathrm{e}^{\mathrm{i}\boldsymbol{k}_{1\|}\cdot\boldsymbol{r}_\|} \times \left[T_1^{(m)}\mathrm{e}^{\mathrm{i}k_{1z}^{(m)}(z-z_m)} + R_1^{(m)}\mathrm{e}^{-\mathrm{i}k_{1z}^{(m)}(z-z_m)}\right] \Omega_{\mathrm{id}}^{(m)}(z) \quad (11.49)$$

$$E_2^{(-A)}(\boldsymbol{r}) = \sum_{m=0}^{N+1} \mathrm{e}^{\mathrm{i}\boldsymbol{k}_{2\|}\cdot\boldsymbol{r}_\|} \times \left[T_2^{*(m)}\mathrm{e}^{\mathrm{i}k_{2z}^{*(m)}(z-z_m)} + R_2^{*(m)}\mathrm{e}^{-\mathrm{i}k_{2z}^{*(m)}(z-z_m)}\right] \Omega_{\mathrm{id}}^{(m)}(z). \quad (11.50)$$

The amplitudes of the transmitted and reflected waves in layer m belonging to the solutions $E_{1,2}^{(\pm A)}$ are denoted by $T_{1,2}^{(m)}$ and $R_{1,2}^{(m)}$; these amplitudes can be easily calculated by using, for example, the matrix approach described in Sect. 6.5.1. Since the phase terms contain only the differences $z - z_m$, these amplitudes include the phase shifts between the sample surface and the mth interface. The vertical components of the wave vectors of the transmitted waves in the layer m are $k_{1,2z}^{(m)}$. The complex conjugation in the second solution is due to the time inversion. The $m = 0$ terms in the sums express the vacuum waves above the sample surface. These waves correspond to the primary beam (the wave $T_1^{(0)}\mathrm{e}^{\mathrm{i}\boldsymbol{k}_1\cdot\boldsymbol{r}} \equiv E_i\mathrm{e}^{\mathrm{i}\boldsymbol{K}_i\cdot\boldsymbol{r}}$) and the actual scattered beam (the wave $T_2^{(0)}\mathrm{e}^{\mathrm{i}\boldsymbol{k}_2\cdot\boldsymbol{r}} \equiv E_s\mathrm{e}^{\mathrm{i}\boldsymbol{K}_2\cdot\boldsymbol{r}}$). The term $m = N+1$ corresponds to the (semi-infinite) substrate, where $R_{1,2}^{(N+1)} = 0$.

The disturbance $\hat{\mathbf{V}}_B = -K^2\Delta\chi(\boldsymbol{r})$ of the scattering potential can be expressed by means of the random shifts U_m (see (11.4)) as a sum of the contributions of the individual interfaces:

$$\hat{\mathbf{V}}_B = \sum_{m=1}^{N+1} \hat{\mathbf{v}}^{(m)},$$

where

$$\hat{\mathbf{v}}^{(m)} = -K^2 \Delta\chi_0^{(m)} [H(z_m + U_m - z) - H(z_m - z)] \tag{11.51}$$

and where $\Delta\chi_0^{(m)} = \chi_0^{(m)} - \chi_0^{(m-1)}$ and $\chi_0^{(0)} = 0$ is the polarizability of the vacuum above the sample surface.

For the calculation of the matrix element $\langle E_2^{(-A)} | \hat{\mathbf{V}}_B | E_1^{(A)} \rangle$ we use this sum and the non-disturbed solutions (11.49, 11.50). The matrix element equals a sum of volume integrals over the region $(z_m, z_m + U_m)$; in each integral the product of $\hat{\mathbf{v}}^{(m)}$ occurs with the non-disturbed wavefields. Therefore, in the mth term of this sum, the integral contains the non-disturbed wavefield of layer $m-1$ if $U_m > 0$ and that of layer m if $U_m < 0$. The resulting expression is rather cumbersome but can be simplified substantially using the following approximation. We assume that the non-disturbed wavefield does not change abruptly in the region $z \in (z_m, z_m + U_m)$. Then, this wavefield in the region where $\hat{v}^{(m)} \neq 0$ is always that of layer m below interface m, even if $U_m > 0$. Obviously, this simplification is valid if $\sigma q < 1$, where q is the vertical component of the scattering vector in the material. For larger values of σq this approximation cannot be used; in this case, however, the validity of the DWBA approach must be checked in any particular case. If this simplification is valid, the numerical results must be nearly identical if we use the non-disturbed wavefield also *above* the interface m. With this assumption,

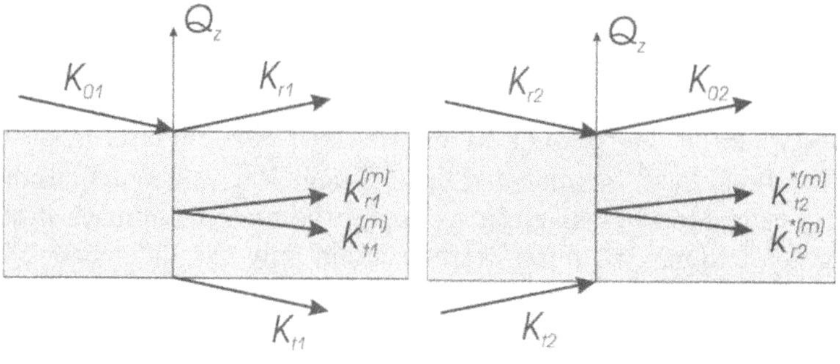

Fig. 11.13. Illustration of the wave vectors of the undisturbed states for the incident wave (index 1) (left) and for the scattered wave (index 2) (right). The indices t and r stand for transmitted and reflected wave, respectively.

the resulting matrix element is again a sum over the interfaces

$$\mathcal{T}_{12}^{(B)} \equiv \langle E_2^{(-A)} | \hat{\mathbf{V}}_B | E_1^{(A)} \rangle = \sum_{m=1}^{N+1} \tau_\Delta^{(m)} \equiv \sum_{m=1}^{N+1} \langle E_2^{(A)} | \hat{\mathbf{v}}^{(m)} | E_1^{(A)} \rangle ,$$

where

$$\begin{aligned}\tau_{\Delta}^{(m)} = -K^2 \Big[& T_1^{(m)} T_2^{(m)} \widetilde{S}_{\Delta}^{(m)}(\boldsymbol{k}_{2\|} - \boldsymbol{k}_{1\|}, q_1) \\ & + T_1^{(m)} R_1^{(m)} \widetilde{S}_{\Delta}^{(m)}(\boldsymbol{k}_{2\|} - \boldsymbol{k}_{1\|}, -q_2) \\ & + R_1^{(m)} T_2^{(m)} \widetilde{S}_{\Delta}^{(m)}(\boldsymbol{k}_{2\|} - \boldsymbol{k}_{1\|}, q_2) \\ & + R_1^{(m)} R_2^{(m)} \widetilde{S}_{\Delta}^{(m)}(\boldsymbol{k}_{2\|} - \boldsymbol{k}_{1\|}, -q_1) \Big] . \end{aligned} \tag{11.52}$$

Here we have denoted

$$\widetilde{S}_{\Delta}^{(m)}(\boldsymbol{k}_{2\|} - \boldsymbol{k}_{1\|}, q) = \Delta\chi_0^{(m)} \int \mathrm{d}^2\boldsymbol{r}_{\|} \mathrm{e}^{-\mathrm{i}(\boldsymbol{k}_{2\|} - \boldsymbol{k}_{1\|}).\boldsymbol{r}_{\|}} S_m(q; \boldsymbol{r}_{\|}) \tag{11.53}$$

with the one-dimensional geometrical factor of the interface disturbance

$$S_m(q; \boldsymbol{r}_{\|}) = \int_0^{U_m(\boldsymbol{r}_{\|})} \mathrm{d}z' \mathrm{e}^{-\mathrm{i}qz'} = \frac{\mathrm{i}}{q} \left(\mathrm{e}^{-\mathrm{i}qU_m(\boldsymbol{r}_{\|})} - 1 \right) , \tag{11.54}$$

and $q_{1,2}$ are the differences of the z-components of the wave vectors: $q_1 = k_{2z}^{(m)} - k_{1z}^{(m)}$, $q_2 = k_{2z}^{(m)} + k_{1z}^{(m)}$.

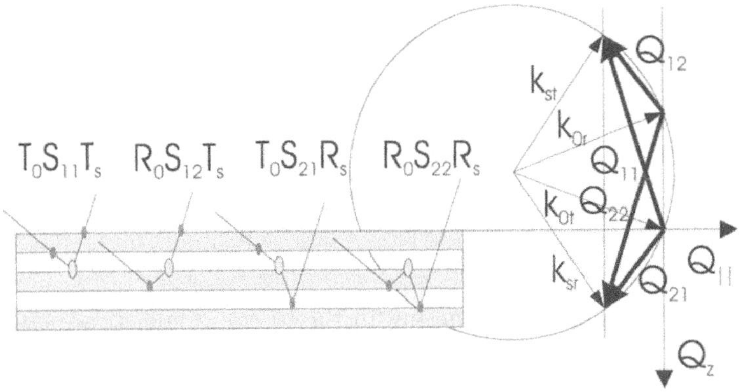

Fig. 11.14. Schematic representation of the four scattering processes in real space (a) and in reciprocal space (b), taken into account by the DWBA (Eq. (11.52)). The filled ellipses denote the dynamical reflection and transmission in the ideal multilayer, and the open ellipses indicate the diffuse scattering due to the interface roughness. The process with the indices "11" is the primary scattering process, $\boldsymbol{Q}_{11} = (\boldsymbol{k}_{2\|} - \boldsymbol{k}_{1\|}, q_1)$, described also by the kinematical approximation. The other three processes $\boldsymbol{Q}_{12} = (\boldsymbol{k}_{2\|} - \boldsymbol{k}_{1\|}, q_2)$, $\boldsymbol{Q}_{21} = (\boldsymbol{k}_{2\|} - \boldsymbol{k}_{1\|}, -q_2)$ and $\boldsymbol{Q}_{22} = (\boldsymbol{k}_{2\|} - \boldsymbol{k}_{1\|}, -q_1)$, are of purely dynamical nature.

The contribution of the disturbance of an individual interface to the whole scattering matrix element $\langle E_2^{(-A)} | \hat{\mathbf{V}}_B | E_1^{(A)} \rangle$ consists of four scattering processes described by the four terms in (11.52). The first term describes the following process. The primary wave $|E_i\rangle$ excites the wave $|E_1^{(A)}\rangle$ in the undisturbed sample. Its transmitted component $T_1^{(m)}$ in the mth layer is scattered by the disturbance potential $\hat{\mathbf{v}}^{(m)}$ and the scattered wave belongs to the transmitted component $T_2^{(m)}$ of the undisturbed wave $|E_2^{(-A)}\rangle$. Since this wave is time-inverted, it propagates toward the free surface. The corresponding vacuum wave $|E_s\rangle$ is the scattered wave emitted by the sample. This scattering process (*primary* scattering process) is also present within the kinematical description, where the transmitted wave amplitudes $T_{1,2}^{(m)}$ equal those of the vacuum primary and scattered waves.

The other three processes are secondary; they are of a purely dynamical nature and occur exclusively because of multiple scattering. So, for instance, the second process describes the diffuse scattering of the transmitted wave $T_1^{(m)}$; the scattered wave contributes to $R_2^{(m)}$ so that it is reflected dynamically by the ideal multilayer. All four processes are schematically depicted in Fig. 11.14. In order to obtain the differential cross section for diffuse scattering (see Chap. 5) the covariance

$$\mathrm{Cov}(\mathcal{T}_{12}^B, \mathcal{T}_{12}^B) = \sum_{m=1}^{N+1} \sum_{n=1}^{N+1} \mathrm{Cov}(\tau_\Delta^{(m)}, \tau_\Delta^{(n)}) \tag{11.55}$$

has to be calculated. This covariance contains the covariances belonging to particular pairs of interfaces and a particular pair of scattering processes (see Fig.11.15 for illustration)

$$\widetilde{Q}_{jk}^{(mn)} \equiv K^4 \Delta\chi_0^{(m)} (\Delta\chi_0^{(n)})^* \mathrm{Cov}[S_m(q_j; \boldsymbol{r}_\parallel), S_n(q_k; \boldsymbol{r}'_\parallel)], \tag{11.56}$$

with

$$q_{j,k} = \pm q_1, \pm q_2,$$

where the components $q_{j,k}$ belong to the layers m and n, respectively. For each pair (mn) of interfaces, 16 covariances $\widetilde{Q}_{jk}^{(mn)}$ exist describing 16 possible scattering processes.

Substituting from (11.6) and (11.10) yields a formal expression for the covariance function,

$$\widetilde{Q}_{jk}^{(mn)} = \frac{K^4 \Delta\chi_0^{(m)} (\Delta\chi_0^{(n)})^*}{q_j q_k^*} [\chi_{U_m,U_n}(q_j, q_k^*) - \chi_{U_m}(q_j)\chi_{U_n}(q_k^*)] \tag{11.57}$$

and for a Gaussian r.m.s. roughness we obtain finally

$$\widetilde{Q}_{jk}^{(mn)} = \frac{K^4 \Delta\chi_0^{(m)} (\Delta\chi_0^{(n)})^*}{q_j q_k^*} \mathrm{e}^{-\frac{1}{2}((\sigma_m q_j)^2 + (\sigma_n q_k^*)^2)}$$

$$\times \left[\mathrm{e}^{q_j q_k^* C_{mn}(\boldsymbol{r}_\parallel - \boldsymbol{r}'_\parallel)} - 1\right]. \tag{11.58}$$

A graphic representation of the covariance function is shown in Fig. 11.15. Since the rough interfaces are assumed to be statistically homogeneous, these covariances depend on $\boldsymbol{r}_\parallel - \boldsymbol{r}'_\parallel$ only. Performing the integrations over $\boldsymbol{r}_\parallel$ and

$\boldsymbol{r}'_{\parallel}$, we find that the differential cross section for the diffuse scattering is proportional to the area A of the irradiated sample surface.

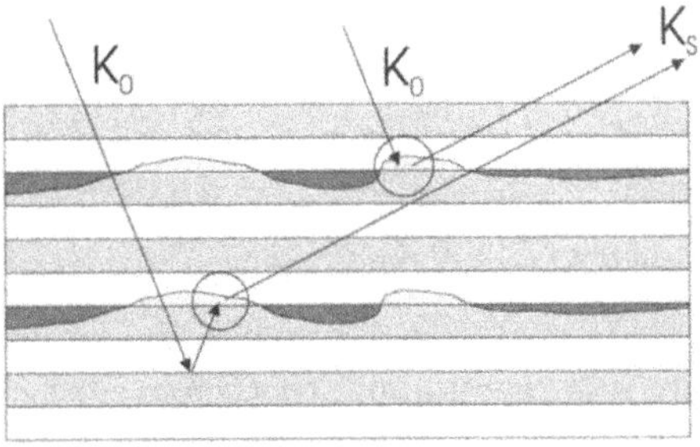

Fig. 11.15. Illustration of the covariance function $\widetilde{Q}_{jk}^{(mn)}$, which describes, for example, the correlation of the two drawn processes of scattering at the roughness profiles of different interfaces.

The complexity of the final expression can be reduced if we choose another non-disturbed system, namely, a semi-infinite averaged medium. The non-disturbed wavefield is sketched schematically in Fig. 11.16. The matrix element $\tau_{\Delta}^{(m)}$ contains only one term,

$$-K^2|\Delta\chi_0|^2 \int d^2\boldsymbol{r}_{\parallel} e^{-i(\boldsymbol{k}_{2\parallel}-\boldsymbol{k}_{1\parallel}).\boldsymbol{r}_{\parallel}} T_1^{(m)} T_2^{(m)} S_m(q_1;\boldsymbol{r}_{\parallel}), \tag{11.59}$$

describing the primary scattering process between the transmitted waves belonging to $|E_1^{(A)}\rangle$ and $E_2^{(-A)}\rangle$. Different degrees of approximation are discussed in more detail in [34, 39, 159]. As a first simple example, let us deal

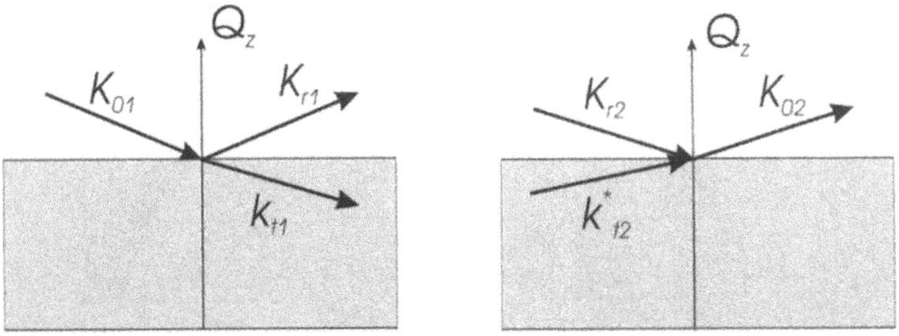

Fig. 11.16. Illustration of the non-perturbed states for the simplified DWBA. The model takes the reflection by the surface and the mean refraction in the medium into account. Specular reflection by inner interfaces is neglected.

with a single rough surface. The matrix element $\mathcal{T}_{12}^{(B)}$ consists of only one term, $\tau_\Delta^{(1)}$. The covariance of this term, $\mathrm{Cov}(\tau_\Delta^{(1)}, \tau_\Delta^{(1)})$ contains the covariance $\widetilde{Q}^{(11)}$ in the integral

$$\int \mathrm{d}^2\boldsymbol{r}_\parallel \int \mathrm{d}^2\boldsymbol{r}'_\parallel \mathrm{e}^{-\mathrm{i}\boldsymbol{Q}_\parallel\cdot(\boldsymbol{r}_\parallel-\boldsymbol{r}'_\parallel)}\widetilde{Q}^{(11)} = AK^4\frac{|\chi_0|^2}{|Q_{Tz}|^2}\mathrm{e}^{-\sigma^2(Q_{Tz}^2+(Q_{Tz}^*)^2)/2}$$
$$\times \int \mathrm{d}^2(\boldsymbol{r}_\parallel - \boldsymbol{r}'_\parallel)\left[\mathrm{e}^{|Q_{Tz}|^2 C(\boldsymbol{r}_\parallel-\boldsymbol{r}'_\parallel)} - 1\right], \quad (11.60)$$

and the transmission function $|T_1T_2|^2 = |t_1t_2|^2$, where $t_{1,2}$ are the Fresnel transmission coefficients of the surface corresponding to the undisturbed states $|E_1^{(A)}\rangle$ and $|E_2^{(-A)}\rangle$; Q_{Tz} is the vertical component of the scattering vector in the sample, A is the illuminated sample surface.

If $|Q_{Tz}\sigma|^2 \ll 1$, we can expand the exponential function with the correlation function C into a Taylor series, taking only two terms into account. The differential cross section for diffuse scattering is then proportional to the Fourier transform C^{FT} of the correlation function $C(\boldsymbol{r}_\parallel - \boldsymbol{r}'_\parallel)$:

$$\left(\frac{\mathrm{d}\sigma}{\mathrm{d}\Omega}\right)_{\mathrm{incoh}} = \frac{K^4}{16\pi^2}S|t_1t_2|^2\mathrm{e}^{-\frac{\sigma^2}{2}(Q_{Tz}^2+(Q_{Tz}^*)^2)}C^{\mathrm{FT}}(\boldsymbol{Q}_\parallel). \quad (11.61)$$

Consequently, by fitting the measured reciprocal-space distribution of the diffusely scattered intensity, the correlation function C of the surface can be obtained either by use of (11.60) or by an inverse Fourier transformation according to (11.61).

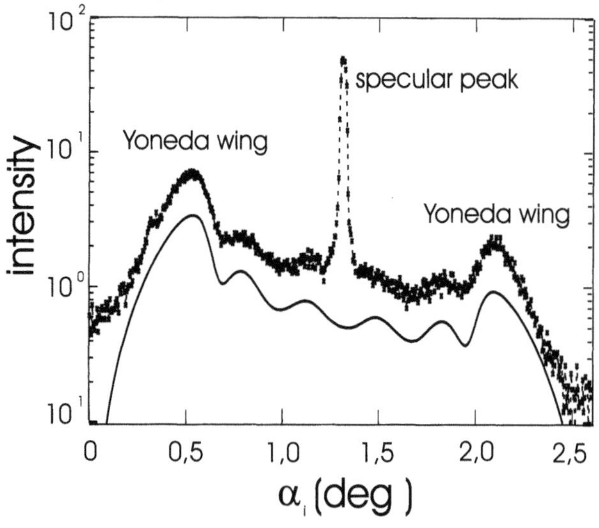

Fig. 11.17. (a) Measured and calculated ω scan of a thin tungsten layer on sapphire [39].

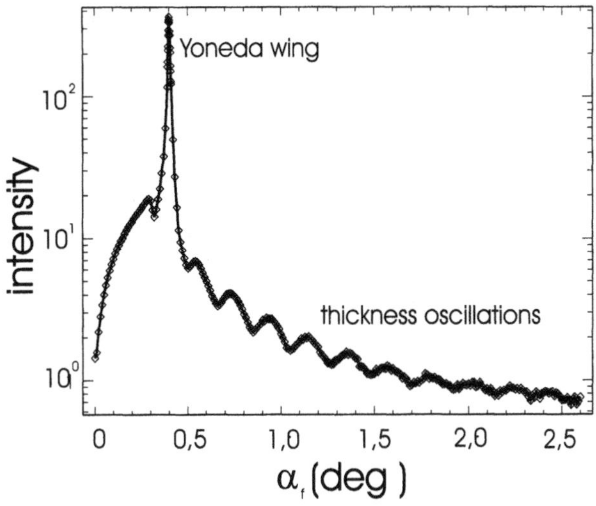

Fig. 11.18. Measured and calculated 2θ scans of a thin Ge layer on Si [310].

In many cases one is exclusively interested in the values of the parameters characterizing the surface roughness (σ, Λ and h), and it is not necessary to study to whole form of the correlation function $C(\boldsymbol{r}_{\|} - \boldsymbol{r}'_{\|})$. As shown in [303], the values of the parameters can easily be determined from the slopes of the asymptotic intensity distributions in reciprocal space. In a Q_x scan (i.e., at constant Q_z), for $Q_x\Lambda \ll 1$ the intensity drops as $Q_x^{-\gamma}$, where $\gamma = 2(1+h) - (Q_z\sigma)^2$; for $Q_x\Lambda < 1$ the intensity remains nearly constant. Therefore, from such a scan, we can determine both Λ and h if we know the r.m.s. roughness σ and if the range of measured values of Q_x is sufficiently large.

The diffuse intensity distribution along the Q_z-axis is nearly constant for $(Q_z\sigma)^2 < 1$ and decreases as $Q_z^{-\gamma}$ for larger Q_z. The value of the exponent γ depends on the resolution of the reflectometer in the Q_y direction (i.e., perpendicular to the scattering plane). If the reflectometer has an ideal resolution, then $\gamma = 2 + 2/h$. In the case of a reflectometer with a long detector slit (no resolution in the Q_y direction), then $\gamma = 2 + 1/h$.

In the case of a layered structure and given that the roughnesses of the interfaces and/or the scattering vectors are sufficiently small, we can expand the exponential functions in Eq. (11.58) and obtain the Fourier transform of the correlation functions C_{mn}, as in to the case of a single surface.

Let us consider two examples of a single layer with rough interface on a substrate. A measured and calculated ω-scan of a tungsten layer on sapphire and a 2θ-scan of a germanium layer on a silicon substrate are plotted in Figures 11.17 and 11.18. We find intense specular peaks at $\alpha_i = 2\Theta/2$. Maxima of diffuse scattering occur when α_i or α_f equals the critical angle α_c, i.e., at the Yoneda wings [395]. These peaks are caused by the maxima of the transmission functions $|t_1 t_2|^2$ in Eq. (11.61).

Beside the specular peak and the Yoneda wings, we observe additional dynamical fringes in the ω-scans. These fringes are created by the specular reflection of the diffusely scattered wave and/or by the diffuse scattering of the specularly reflected waves at the upper and the lower interfaces. The fringe period depends on the layer thickness. The Yoneda wings and these fringes are of purely dynamical nature. In contrast, the fringes in the 2θ-scan in Fig. 11.18 are a primary scattering effect; therefore, they can also be obtained within the kinematical scattering theory. They are caused by the replication of the roughness profile from the substrate surface to the layer surface.

In the following, we give an overview about the main features in the non-specular reflected intensities and discuss their physical origin. The diffuse x-ray scattering pattern is characterized by the *transmitted/reflected wave amplitudes* of the incident and final wavefields in the layers and by the 16 co-variances of the scattering processes, $\widetilde{Q}_{jk}^{(mn)}$, (see Fig. 11.15), for each pair of interfaces (m, n). We want to study the features of the intensity pattern from the aspect of whether they are particularities of scattering by the roughness profiles, caused by the correlation properties, or of the excited non-perturbed wave amplitudes. In other words, we want to distinguish between effects of the random disturbance potential and of the non-perturbed potential. The existence of the latter effects does not depend on the particular statistical roughness properties; we call them *dynamical scattering effects* . The former contain the structure information we are interested in the experiment. Knowing about the influence of the latter dynamical effects, it is often sufficient to apply the simpler description according to Eq. (11.59) without the risk of misinterpretations. Further it has been shown that use of comfortable non-coplanar scattering geometries may avoid the confusion caused by strong dynamical effects.

11.3.2 Resonant Diffuse Scattering

First we investigate the influence of the interface roughness correlation. One essential characteristic caused by the interplane correlation is the so-called *resonant diffuse scattering* (RDS). We simplify the discussion of these phenomena by using the simple model of vertical roughness correlation according to Eq. (11.28) In Fig. 11.19 we show some calculated reciprocal-space maps of the diffusely scattered intensity for a GaAs/AlAs superlattice, assuming this vertical replication model. The figure shows results for the cases of no vertical replication, partial replication and full replication.

In the first case, all interfaces scatter independently and the diffuse intensities of all individual interfaces are superimposed. In the other two cases, the roughness profiles of different interfaces are (at least partially) correlated, so

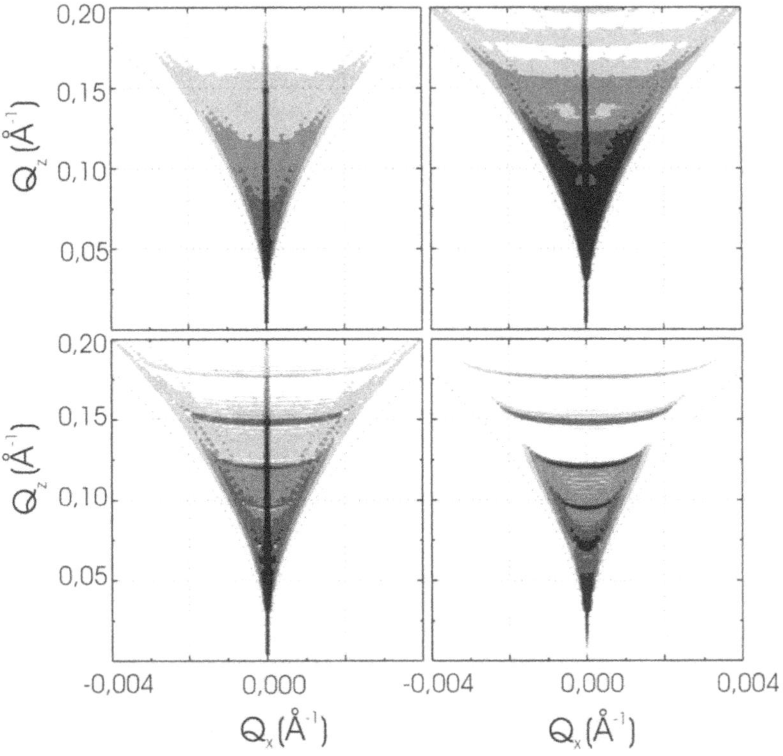

Fig. 11.19. Reciprocal-space maps of the diffusely scattered intensity calculated for a [GaAs(7 nm)/AlAs(15 nm)]×10 multilayer using the DWBA method and the simpler replication model according to Eq. (11.28). All the interfaces have the same r.m.s. roughness, 1 nm; the same correlation length, 50 nm; but different vertical correlation lengths, $\Lambda_\perp$. Upper left panel: no replication, $\Lambda_\perp = 0$ nm. Upper right panel: full replication $\Lambda_\perp = \infty$. Bottom left panel: $\Lambda_\perp = 100$ nm. Bottom right panel: full replication $\Lambda_\perp = \infty$, calculated by the simpler DWBA. The full lines represent the arcs of the Ewald spheres for the limiting cases of $\alpha_i = 0$ and $\alpha_f = 0$. Resonant diffuse scattering (RDS) disappears, if the roughness profiles are not replicated (upper left panel). Bragg-like resonance lines are visible in all maps calculated by the full DWBA. They are not reproduced by the simpler DWBA (bottom right panel).

that the phases of the waves scattered by these profiles are partially coherent. This leads to a concentration of the scattered intensity in the reciprocal plane into narrow sheets (*resonant diffuse scattering* – RDS). These sheets intersect the truncation rod (i.e., the Q_z-axis) in the coherent satellite maxima. If refraction could be neglected, these sheets would be straight and parallel to the Q_x and Q_y axes and would lie at

$$Q_z = K(\sin \alpha_i + \sin \alpha_f) = m\frac{2\pi}{D}, \tag{11.62}$$

where m is an integer, D is the multilayer period, α_i and α_f are the incident and exit angles. This situation, which is the case in a kinematical treatment, is schematized in Fig. 11.20.

Owing to refraction, the sheets are curved upward, forming typically a banana-like shape following the modified Bragg law

$$Q_z^{\mathrm{ML}} = K\left(\sqrt{\sin^2 \alpha_i + \langle \chi_0 \rangle_{\mathrm{ML}}} + \sqrt{\sin^2 \alpha_f + \langle \chi_0 \rangle_{\mathrm{ML}}}\right) = m\frac{2\pi}{D}, \tag{11.63}$$

where $\langle \chi_0 \rangle_{\mathrm{ML}} = \sum_{m=1}^{M} \chi_0^{(m)} T^{(m)}/D$ is the mean polarizability of the multilayer period (that contains the layers with the thicknesses $T^{(1)}, \ldots, T^{(M)}$) and Q_z^{ML} is the mean scattering vector in the multilayer period.

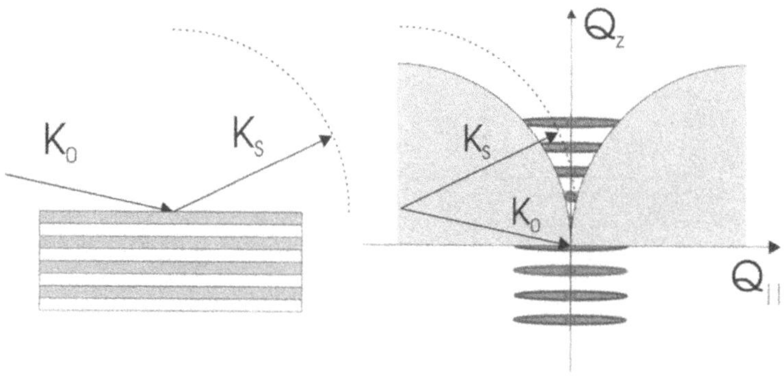

Fig. 11.20. Schematic representation of diffuse scattering by a periodic multilayer with interface roughness replication. The essential features are (1) the multilayer truncation rod and (2) horizontal sheets of resonant diffuse scattering crossing the rod in the satellite positions.

The length of the *RDS bananas* in Q_x-direction is inversely proportional to some effective correlation length related to the lateral correlation lengths Λ_m of the interfaces. The width of the sheets in the Q_z direction represents the degree of replication. In the simple model it is inversely proportional to $\Lambda_\perp$ and, if $\Lambda_\perp > ND$, to the total thickness ND of the multilayer. If no vertical replication is present, the sheets disappear, turning into a broad single maximum similarly to the case of a single surface. The RDS sheets can also be obtained within the kinematical scattering theory; their existence is not related to any kind of multiple scattering.

Resonant diffuse scattering has been experimentally observed in amorphous and polycrystalline as well as epitaxial multilayers (see the examples of Figs. 11.21 and 11.22). In Fig. 11.22 we show an experimental example of a GaAs/AlAs superlattice having the same structure assumed in Fig. 11.19. The RDS bananas are clearly visible and are bent because of refraction. Their existence and narrow vertical width give evidence for full roughness replication in this sample.

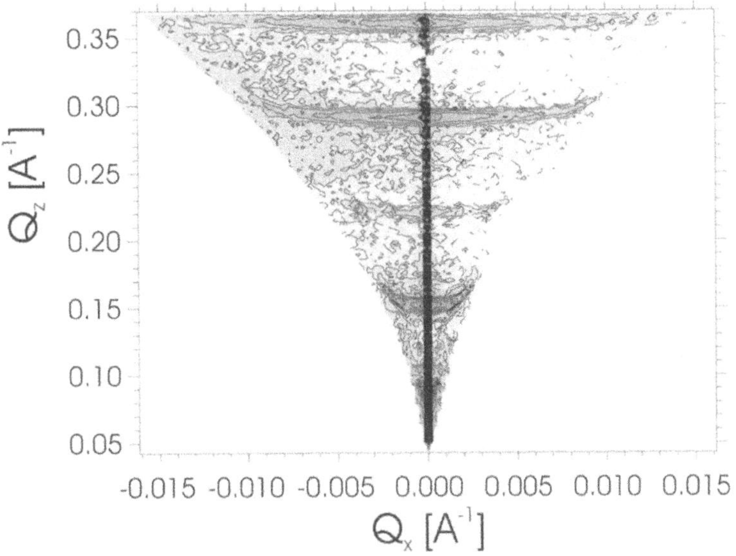

Fig. 11.21. X-ray reflectivity by a periodic multilayer with rough interfaces. Measured reciprocal-space map. Periodic multilayer [Si (3.0 nm),Nb (5.8 nm)]×10 starting from a rough Si substrate of $\sigma = 0.46$ nm and with interface roughness decreasing towards the free surface [39].

11.3.3 Dynamical Scattering Effects

In addition to the RDS sheets, other typical features can be observed in the reciprocal-space maps of periodic multilayers. In contrast to the sheets, these effects are of a purely dynamical nature. *Bragg-like resonant lines* occur if the incident or exit wave fulfills the diffraction condition corrected for refraction,

$$K\sqrt{\sin^2 \alpha_{i,f} + \chi_0} = m_{i,f}\frac{2\pi}{D}, \tag{11.64}$$

where $m_{i,f}$ are integers. In this case, dynamical scattering processes occur (diffuse scattering of the coherently reflected wave or coherent reflection of the diffusely scattered wave). It is easy to prove that the zero-order Bragg-like lines are identical to the Yoneda wings. The Bragg-like lines have a maximum, the so-called *Bragg-like peak (BL)*, if the incident and scattered waves fulfill

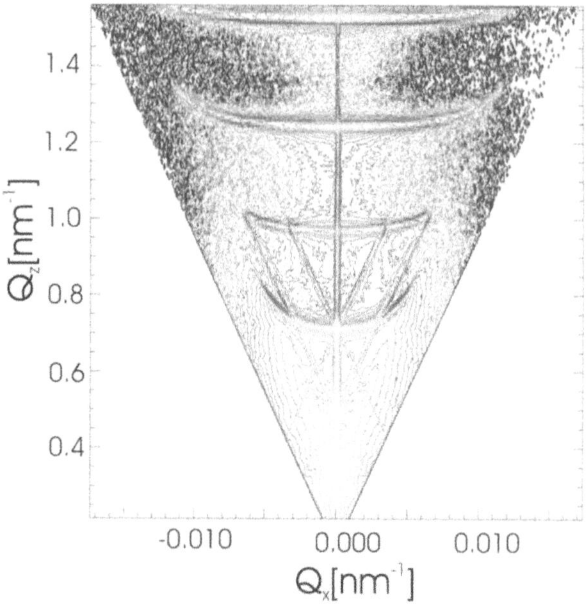

Fig. 11.22. X-ray reflectivity by periodic multilayers with rough interfaces. Reciprocal space maps of the diffusely scattered intensity measured for an AlAs/GaAs multilayer having the same structure as shown in Fig. 11.19.

the diffraction condition simultaneously and the Bragg-like lines intersect, i.e., at the positions

$$Q_{z,m_i m_f} = \sqrt{(m_i 2\pi/D)^2 - K^2 \langle\chi_0\rangle_{\mathrm{ML}}} + \sqrt{(m_f 2\pi/D)^2 - K^2 \langle\chi_0\rangle_{\mathrm{ML}}}$$

$$Q_{\|,m_i m_f} = \sqrt{K^2 - (m_i 2\pi/D)^2 + K^2 \langle\chi_0\rangle_{\mathrm{ML}}} + \sqrt{K^2 - (m_f 2\pi/D)^2 + K^2 \langle\chi_0\rangle_{\mathrm{ML}}}. \quad (11.65)$$

The Yoneda wings, Bragg-like lines and Bragg-like peaks are of dynamical origin and occur independently of the actual interface correlation function. However, their form and intensity depend on the interface correlation [184].

In the case of vertically replicated roughness, we see with (11.63–11.65) that all Bragg-like peaks of an even number $n_f + n_i$ are situated on RDS-sheets (see Fig. 11.23). These Bragg-like peaks are very pronounced with respect to the others. That can be interpreted by the concept of *Umweganregung* (excitation of a reflection by another reflection), well known from x-ray diffraction and outlined in Fig. 11.23. In our experimental map in Fig. 11.22, the Yoneda wings and the Bragg-like lines are well resolved. Along the RDS sheets we observe intense Bragg-like peaks. All the features are well reproduced in the intensity distribution calculated by the full DWBA treatment.

A more detailed discussion of the profile of Bragg-like peaks in dependence on the interlayer roughness correlations can be found in [181].

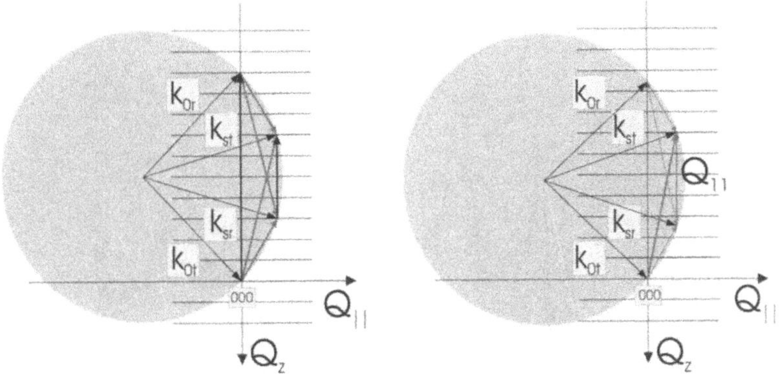

Fig. 11.23. Generation of Bragg-like peaks on the RDS-sheets and interpretation by the concept of *Umweganregung.* On the left side, both the incidence and final non-perturbed state fulfill the Bragg condition (11.62). Simultaneously all four diffuse scattering processes are in the situation of resonant diffuse scattering according to Eq. (11.63). On the right side, the condition (11.63) is fulfilled for the primary scattering process. The incident wave is out of the Bragg condition; consequently, the final state is also out of the Bragg condition. Additionally all three secondary diffuse scattering processes are out of resonance.

It is not always possible or necessary to measure a full, well-resolved intensity map; ω scans at different values of 2Θ and offset scans or 2Θ scans cross the map. Even one offset scan or 2Θ scan is sufficient to give evidence for the existence of RDS sheets and thus for vertical replication. However, one should be careful to avoid the misinterpretation of dynamic Bragg-like maxima as RDS sheets caused by replication.

11.3.4 Non-Coplanar X-Ray Reflection

Reciprocal-space maps of diffusely scattered intensity are usually measured in a coplanar geometry since this is simple to realize with conventional diffractometers and reflectometers. The accessible scattering vectors within this geometry are limited by the Ewald spheres for $\alpha_i = 0$ and $\alpha_f = 0$, which represent the horizon of the sample surface. Since the range of accessible values of Q_x is limited, the coplanar scattering geometry is restricted to large, mesoscopic correlation lengths Λ. This limitation has been overcome by employing a non-coplanar geometry according to Chapters 3 and 4 [301]. It requires a monochromatic beam collimated in two directions, such as those available from modern synchrotron radiation sources. Typically, the non-coplanar set-up used allows the detection of diffuse x-ray scattering over a large range of parallel momentum transfer up to 1 Å^{-1}, which allows the determination of correlation lengths down to a few angstroms.

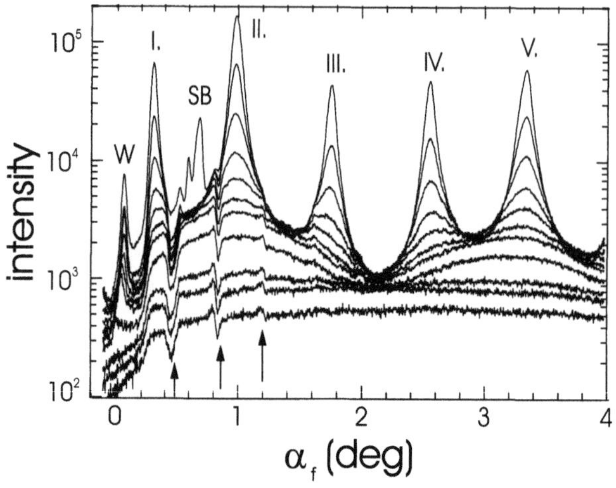

Fig. 11.24. Measured Q_z-scans for different values of Q_x of an amorphous W/Si multilayer. The decay of the RDS sheets for increasing Q_x is visible. The RDS sheets are indicated by roman numberals, and the specularly reflected beam by *SB*. The arrows show the positions of the Bragg-like peaks; and the letter *W*, the Yoneda wing [302].

Figure 11.24 shows, for illustration, a series of Q_z-scans (performed as α_f scans for fixed α_i) measured in a non-coplanar geometry from a W/Si multilayer at different values of $|\boldsymbol{Q}_{\parallel}|$. From the figure it can be seen that if $Q_{\parallel}$ increases, the width of the RDS sheets increases and their height decreases. Since the amplitudes of the undisturbed eigenstates depend only on Q_z and not on $Q_{\parallel}$, these changes can only be caused by a reduction of the vertical replication length $\Lambda_{\perp}$ for roughness fluctuations with smaller wavelengths (rapid roughness). In [303], the formula for the inter-plane correlation function (11.29) was used, which followed from the stochastic Edwards–Wilkinson equation describing the growth kinetics [301].

The decay of the intensity of the RDS sheet with increasing $Q_{\parallel}$ can be studied by using the values of the intensity of the sheet integrated along Q_z as a function of $Q_{\parallel}$ (Fig. 11.25). When this dependence is plotted on a double logarithmic scale, the validity of various growth models can be demonstrated from power-law fits.

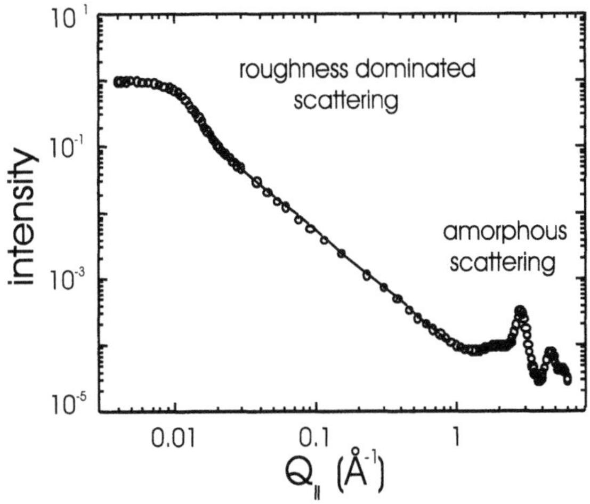

Fig. 11.25. Decrease of the intensity in the first RDS sheets as a function of Q_x for an amorphous W/Si multilayer. The measured data indicate a logarithmic scaling behavior of the correlation function, as predicted by the Edwards Wilkinson equation [301].

11.4 Interface Roughness in Surface-Sensitive Diffraction Methods

Most commonly, diffuse scattering in the small-angle scattering geometry is measured in the coplanar regime since it is easily performed with commercial reflectometers or diffractometers (see Chap. 2). In this geometry the accessible lateral momentum transfer is limited by the sample horizon (for CuK_α radiation $Q_x < 0.002$ Å^{-1} for a perpendicular momentum transfer $Q_z < 0.2$ Å^{-1}). Consequently, coplanar x-ray reflection is restricted to mesoscopic correlation lengths (cut-off lengths $\Lambda_{\parallel} > 300$ nm in our example). The more sophisticated non-coplanar reflection geometry can overcome these limitations [301, 302, 303]. However, the experimental requirements are much more demanding, including a two-dimensionally collimated beam and the ability to detect the scattered intensity in a non-coplanar geometry, usually not available on commercial diffractometers. Further difficulties arise for interface studies of compositional multilayers systems such as GaInAsP superlattices with a low refraction contrast at the interfaces. In consequence the weak diffuse scattering signal of the buried interfaces may be easily dominated by the comparatively strong scattering at the rough sample surface.

Therefore, in some cases of epitaxial multilayers it may be advantageous to study surface and interface roughness by surface sensitive x-ray *diffraction* methods such as *grazing-incidence diffraction* (GID) and *strongly asymmetric diffraction* (SAXRD) [39, 41, 351]. The methods are based on diffraction by the layer lattices under conditions of simultaneous specular reflection.

The present section discusses diffuse scattering in the SAXRD-mode. Here the limitations discussed for coplanar x-ray reflection concerning the measurable cut-off length are overcome by taking the advantage of a simple coplanar scattering geometry.

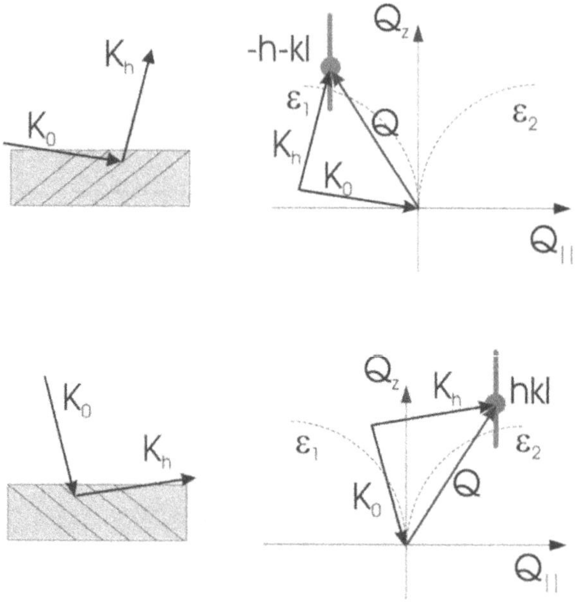

Fig. 11.26. Schematic representation of asymmetric x-ray diffraction in real space and in reciprocal space for the two cases of grazing incidence (left) and grazing exit (right). The spheres ε_1 and ε_2 separate the regions of Bragg (reflection) and Laue (transmission) geometry.

Despite this, the general principles of diffraction by rough multilayers are similar to those of x-ray reflection. All used theoretical treatments can be extended within the uniform formalism derived in Chapter 6. We make use of a distorted wave Born approximation in combination with the dynamical diffraction theory for perfect multilayers summarized in Chapter 6. The concept of *Umweganregung* illustrates graphically the different scattering processes and their influence on the diffuse scattering pattern.

Assuming pseudomorphically grown multilayers, crystal truncation rods through each reciprocal lattice point characterize the structure amplitude of a crystalline layer. In the case of a superlattice all truncation rods contain the fine structure of equidistant superlattice satellites. For the description of Bragg-diffraction in the two-beam approximation, only the truncation rods corresponding to the polarizability Fourier components with the indices $\boldsymbol{h}$, $-\boldsymbol{h}$ and 0 are of importance,

$$\chi^{(m)}(\boldsymbol{r}) = \Omega^{(m)}(\boldsymbol{r}) \sum_{\boldsymbol{g}=0,-\boldsymbol{h},\boldsymbol{h}} \chi_{\boldsymbol{g}}^{(m)}(\boldsymbol{r}) \mathrm{e}^{\mathrm{i}\boldsymbol{g}.\boldsymbol{r}}. \tag{11.66}$$

For multilayers with rough interfaces, the layer disturbances $\Delta\chi_{\text{layer}}^{(m)}(\boldsymbol{r})$ are

$$\Delta\chi_{\text{layer}}^{(m)}(\boldsymbol{r}) = \Omega_{\text{id}}^{(m)}(z) \sum_{\boldsymbol{g}=0,-\boldsymbol{h},\boldsymbol{h}} \mathrm{e}^{\mathrm{i}\boldsymbol{g}.\boldsymbol{r}} \mathrm{e}^{-\mathrm{i}\boldsymbol{g}.\langle \boldsymbol{u}(z)\rangle} \times$$

$$\times \left[\sum_n \left(\chi_{\boldsymbol{g}}^{(n)} - \chi_{\boldsymbol{g}}^{(m)} \right) \Omega^{(n)}(\boldsymbol{r}) \mathrm{e}^{-\mathrm{i}\boldsymbol{g}.\delta\boldsymbol{u}(\boldsymbol{r})} + \chi_{\boldsymbol{g}}^{(m)} \left(\mathrm{e}^{-\mathrm{i}\boldsymbol{g}.\delta\boldsymbol{u}(\boldsymbol{r})} - 1 \right) \right]. \tag{11.67}$$

Equation (11.67) considers the *morphological fluctuation* and the related elastic lattice distortion in the layer, due to the rough interface morphology. The morphological fluctuation causes a random variation of the Fourier components of the polarizability with the amplitude $(\chi_{\boldsymbol{g}}^{(m\pm1)} - \chi_{\boldsymbol{g}}^{(m)})$; the roughness-induced lattice distortion gives rise to the random contribution $\delta\boldsymbol{u}(\boldsymbol{r}) = \boldsymbol{u}(\boldsymbol{r}) - \langle \boldsymbol{u}(z)\rangle$ of the lattice displacement function. The first term in Eq. (11.67) is different from zero only in the upper and lower interface regions $z_m < z < z_m + U_{\text{max}}^{(m)}$ and $z_{m+1} + U_{\text{min}}^{(m+1)} < z < z_{m+1}$ and disappears between the interface regions in the remaining homogeneous part of the layer. In the general case both the morphological fluctuation and the random strain give rise to diffuse scattering.

The contribution of each layer disturbance to the scattered wavefield is within the first-order distorted-wave Born approximation described by the transition elements $\Delta\tau^{(m)}$ (see Sect. 11.3). It depends on the covariance of the layer structure amplitudes $\tilde{Q}_{jk}^{(mn)}$ (see Eq. (11.56)) and on the amplitudes of the non-disturbed wavefield of diffraction by the planar (non-perturbed) epitaxial multilayer (see Section 6.5). For each polarization, the non-disturbed wavefield in a given layer consists of eight plane waves (corresponding to four tie-points of the dispersion surface, with one transmitted and one diffracted wave for each tie-point):

$$E^{(m)}(\boldsymbol{r}) = \begin{pmatrix} \sum_{j=1}^4 T_j^{(m)} \mathrm{e}^{\mathrm{i}\boldsymbol{K}_{0\parallel}\boldsymbol{r}_\parallel} \mathrm{e}^{\mathrm{i}k_{0j,z}^{(m)}(z-z_m)} \\ +\sum_{j=1}^4 R_j^{(m)} \mathrm{e}^{\mathrm{i}\boldsymbol{K}_{\boldsymbol{h}\parallel}.\boldsymbol{r}_\parallel} \mathrm{e}^{\mathrm{i}k_{\boldsymbol{h}j,z}^{(m)}(z-z_m)} \end{pmatrix} \Omega_{pl}^{(m)}(z). \tag{11.68}$$

The number of possible diffuse scattering processes between two non-perturbed states at one interface increases up to 64 (see Eq. (11.50),

$$\tau_{\Delta}^{(m)} = -K^2 \left\{ \sum_{i,j=1}^{4} T_{1i}^{(m)} \widetilde{S}_{\Delta,\boldsymbol{h}}^{(m)}(\boldsymbol{k}_{20j} - \boldsymbol{k}_{10i} - \boldsymbol{h}) T_{2j}^{(m)} \right.$$

$$+ \sum_{i,j=1}^{4} R_{1i}^{(m)} \widetilde{S}_{\Delta,-\boldsymbol{h}}^{(m)}(\boldsymbol{k}_{2\boldsymbol{h}j} - \boldsymbol{k}_{1\boldsymbol{h}i} + \boldsymbol{h}) R_{2j}^{(m)}$$

$$+ \sum_{i,j=1}^{4} R_{1i}^{(m)} \widetilde{S}_{\Delta,0}^{(m)}(\boldsymbol{k}_{20j} - \boldsymbol{k}_{1\boldsymbol{h}i}) T_{2j}^{(m)}$$

$$+ \sum_{i,j=1}^{4} T_{1i}^{(m)} \widetilde{S}_{\Delta,0}^{(m)} (\boldsymbol{k}_{2\boldsymbol{h}j} - \boldsymbol{k}_{10i}) R_{2j}^{(m)} \Bigg\} , \qquad (11.69)$$

where the structure amplitudes of the layer disturbances $\widetilde{S}_{\Delta,\boldsymbol{g}}^{(m)}(\boldsymbol{Q}_{ij}^{(m)})$ correspond to the diffractions $\boldsymbol{h}, -\boldsymbol{h}, 0$. Fortunately a certain number of terms in Eq. (11.69) is almost negligible.

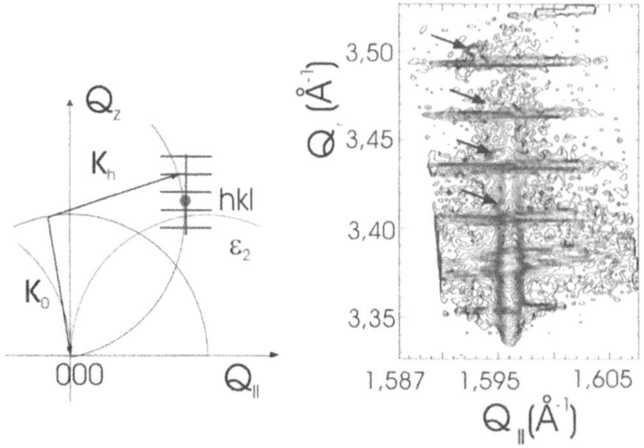

Fig. 11.27. Diffuse x-ray scattering by rough interfaces in the strongly asymmetric diffraction mode. Left: schematic situation in reciprocal-space. Right: reciprocal space map of the 113-reflection in a grazing exit geometry measured for $\Lambda = 1.47$ Å. The coherent crystal truncation rod is crossed by horizontal sheets of resonant diffuse scattering, indicating correlated roughness. The sheets are laterally not limited by the experimental geometry. The arrows mark multiple scattering features explained in the text.

If the roughness profile is replicated, the diffusely scattered intensity is concentrated in horizontal sheets of resonant diffuse scattering crossing the crystal truncation rods in the position of the diffraction satellites. The sheets arise now from partially coherent *diffraction and reflection* by the interface disturbances, as illustrated in Fig. 11.27.

Neglecting the strain fluctuations, the expression (11.68) contains the covariance functions that are formally similar to those derived for diffuse x-ray reflection in Eq. (11.58),

$$\widetilde{Q}_{ijkl}^{(mn)} = \frac{K^4 \Delta\chi_{\boldsymbol{g}}^{(m)} \Delta\chi_{\boldsymbol{g}'}^{(n)*}}{q_{ij,z}^{(m)} q_{kl,z}^{(n)*}}$$

$$\times \left[\chi_{U_m,U_n}(q_{ij,z}^{(m)}, q_{kl,z}^{(n)*}) - \chi_{U_m}(q_{ij,z}^{(m)}) \chi_{U_n}(q_{kl,z}^{(n)*}) \right] . \qquad (11.70)$$

However, now with the *reduced* scattering vectors $q_z^{(m)}$ of the corresponding scattering process in the layers, which depend on the *local* reciprocal lattice vectors $g_z^{(m)}$ in the layers by $q_z^{(m)} = Q_z^{(m)} - g_z^{(m)}$.

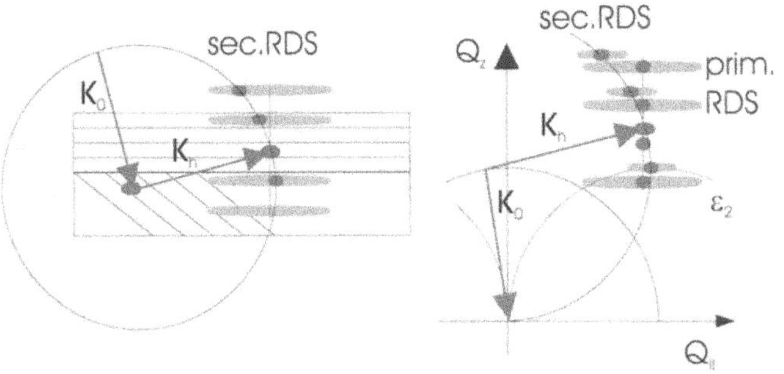

Fig. 11.28. Schematic illustration of Umweganregung, creating secondary resonant diffuse scattering. It is excited by substrate Bragg-diffraction (here in a grazing exit geometry). (a) Illustration in a mixed real-space/reciprocal- space representation: the incident wave is diffracted in the substrate. The intense diffracted wave penetrates the multilayer under grazing incidence from the bottom. In the multilayer it generates resonant diffuse scattering in forward direction. (b) The resulting situation in reciprocal-space including (1) primary RDS due to *diffraction* of the incident wave by the interface disturbances and (2) secondary RDS due to *small-angle scattering* (non-specular reflection) of the Bragg-diffracted substrate wave by interface disturbances in the multilayer.

As an example, Fig. 11.27 shows the scattering geometry and the corresponding measured reciprocal-space map of strongly asymmetric diffraction by a GaAs/AlAs superlattice (113-reflection, grazing exit geometry, wavelength $\lambda = 1.475$ Å). Along the Q_z-direction we see the superlattice crystal truncation rod. In our example the wavelength used defines the cutoff of the accessible range of the truncation rod by the sphere ε_2 at $Q_z = 3.35$ Å^{-1}. Consequently the -1st order satellite occurs near the Yoneda-like wing. At larger Q_z we observe the substrate Bragg peak and further satellite orders. Beside the crystal truncation rod elongated in Q_z-direction we observe horizontal sheets of resonant diffuse scattering in Q_x-direction, which cross the truncation rod in the superlattice diffraction satellites. Note that around the substrate peak and around the superlattice main maximum the RDS-sheets must disappear (not to be confused with the measured strong monochromator streaks in the map, which are inclined with respect to the Q_x-direction). The RDS-sheets of all positive satellites have nearly the same lateral extension. The sheets are not affected by banana-like bending characterizing the non-specular x–ray reflection patterns. The sheets are laterally not limited by the experimental geometry, allowing us to determine the lateral mean island length with better precision than by coplanar reflection.

The diffuse scattering signal contains contributions both of $\Delta\chi_0^{(m)}$ and $\Delta\chi_{\boldsymbol{h}}^{(n)}$. The primary diffuse scattering i.e., the scattering of the initial transmitted waves into the final transmitted waves described by $T_{1i}^{(m)}\widetilde{S}_{\Delta,\boldsymbol{h}}^{(m)}T_{2j}^{(m)}$, is caused by Bragg-diffraction from the interface disturbance at the reciprocal lattice point $\boldsymbol{h}$. Umweganregung *between the substrate and the multilayer* creates additional features in the maps. The intense Bragg-diffracted wavefield of the substrate is diffusely scattered by the rough interfaces of the multilayer. In the case of *grazing exit* geometry the strong Bragg-diffracted waves of the substrate cause strong resonant diffuse scattering of the type $R_{1i}^{(m)}\widetilde{S}_{\Delta,0}^{(m)}T_{2j}^{(m)}$. This effect is illustrated in Fig. 11.28. The wave diffracted by the substrate penetrates the multilayer from its bottom at grazing angles. In the superlattice it will be subsequently diffusely scattered in the forward direction (small-angle scattering). Enhanced small angle scattering takes place when the Ewald sphere crosses the sheets of resonant diffuse scattering. In the diffraction map these crossing positions have a constant Q_z-offset with respect to the *primary* RDS-sheets, which are generated by diffuse scattering within the superlattice. As a result we observe additional spots of *secondary resonant diffuse scattering* under participation of the substrate, as marked by arrows in the experimental map of Fig. 11.27.

Considering the complementary *grazing incidence* $(\overline{11}3)$ diffraction, the terms $T_{1i}^{(m)}\widetilde{S}_{\Delta,0}^{(m)}R_{2j}^{(m)}$ are now responsible for secondary RDS. Consequently we would find a mirrored pattern of dynamical features in the maps.

The main structure information is obtained by evaluating the primary RDS-sheets. Therefore, in most cases it is sufficient to reduce the numerical expenses on the calculation of the related scattering processes. The primary RDS depends on $\widetilde{S}_{\Delta,\boldsymbol{h}}^{(m)}$ and consequently on $\Delta\chi_{\boldsymbol{h}}^{(m)}$ rather than on $\Delta\chi_0^{(m)}$. That is why surface-sensitive diffraction can be applied to investigate interface roughness even in semiconductor compound superlattices without or with only a very low density modulation.

Part IV

X-Ray Scattering by Laterally Structured Semiconductor Nano-Structures

Artificial lateral patterning of solids in one and two dimensions on submicrometer scale holds great potential for the fabrication of micro- and optoelectronic devices (e.g. [45]) and x-ray optical elements. Mostly, the structures create one-dimensional or two-dimensional periodic arrays at the sample surface. To emphasize the periodicity, we call these structures *surface grating.*

One application of such structures in semiconductor technology is creation of quantum-wires and -dots (e.g. [54]). In the field of telecommunications, semiconductor nanostructures are increasingly used in devices such as distributed feedback (DFB) lasers, in which the selection of the laser modes is performed by a grating that replaces the usual Fabry-Perot cavity (see e.g., [358]). Often the patterned layers are simultaneously part of the active region of the laser. A lateral patterning of planar heterostructures can produce one-dimensional or quasi zero-dimensional quantum wells (quantum dots). The confinement of charge carriers in these low-dimensional systems substantially influences their energy spectrum and density of states and subsequently the emission characteristics.

Lateral structures have been produced by various approaches: Artificial periodic patterning can be realized by planar growth of a multilayer followed by an etching step (e.g. [201, 252]. A second route starts from a patterned substrate followed by epitaxial growth on the patterned surface (e.g., [49, 50])). A third route focuses on periodic band-gap modulation in planar layers induced by the strain field of patterned stressor layers [138] or by periodic ion implantation [282].

A fourth, alternative approach is based on lateral self-organization phenomena during growth of strained layers and superlattices. Such strain induced self-assembled interface patterning has been observed to show a remarkable lateral periodicity and vertical replication; this seems to be a fourth way for future device technology to develop.

Combining the second and fourth approaches highly regular strain driven compositional ordering can be arranged by growing ternary and quaternary alloys on artificially patterned substrates. Such a stimulated compositional modulation is successfully employed to produce periodic vertical quantum wells [51]. A promising approach seems to be the growing on a regular strain pattern generated by a highly periodic interface dislocation network at the interface of ultra-thin twist-bonded silicon wafers [65, 122].

Another type of laterally periodic crystalline structures are dynamically formed gratings induced by surface acoustic waves (see e.g., [287]). Other groups of mesoscopic and nanoscopic gratings are non-epitaxial patterns such as amorphous multilayer gratings, polymer gratings, and organic light-induced surface reliefs. The series of grating forms and applications could be continued.

The grating structure determines the properties of the patterned object. In particular, the vertical compositional profile, the quality of the grating shape, and the strain status are essential parameters to know. Structure characterization is often performed by microscopy methods (e.g. , scanning electron microscopy, atomic force microscopy, transmission electron microscopy). The microscopy methods give quite precise images of the sample structure, albeit within a very limited region. The need for non-destructive methods has also stimulated methodical developments within the field of x-ray scattering techniques (high resolution x-ray diffraction, grazing-incidence diffraction (see, e.g., [12, 13, 32, 35, 36, 37, 75, 82, 84, 85, 86, 87, 96, 97, 124, 133, 158, 164, 184, 217, 226, 241, 242, 273, 328, 329, 330, 331, 361, 362, 363, 373, 374, 381]) and x-ray reflection (see e.g. [34, 39, 130, 131, 178, 240, 241, 242, 243, 373, 374, 375, 376]). While electron microscopy has a high spatial resolution, x-ray diffraction methods have high resolution in the reciprocal space, detecting the angular distribution of the scattered intensity with a precision in the range of arc-seconds. Thus x-ray methods non-destructively provide information about the *statistical correlation properties* (from the micrometer range down to atomic scale) representative of a macroscopic part of the sample.

The structure information of x-ray methods concerns:

1. The form and perfection of the mesoscopic super-structure (grating periodicity, grating shape, layer thickness, compositional profile, miscut, misalignment of the grating stripes),
2. The quality of the surface and interfaces (interface roughness, its lateral correlation, and vertical replication properties), and
3. The crystalline properties (the lateral and vertical lattice misfit, elastic lattice distortion and grating-induced strain relaxation).

Unique advantages of x-ray techniques are their high strain sensitivity and the ability to study buried structures, both express the complementary character of the two groups of microscopy and x-ray scattering methods.

Chapter 12 starts with the scattering potential of gratings and develops a theory for all considered x-ray scattering methods within one uniform formalism applying results of the kinematical and semikinematical approaches described in Chapters 5 and 7. Important scattering processes will be illustrated by schematic representations in reciprocal space.

The second part of chapter 12 is focused on the determination of geometrical dimensions of the lateral superstructure and on the characterization of composition morphology, chapter 13 on the characterization of strain and strain relaxation. Measurement of complete reciprocal-space maps provides a detailed quantitative strain and shape analysis of layered gratings. In particular, the combination of complementary scattering techniques permits the comparison of morphological patterning and strain, e.g., characterizing phenomena of self-organized compositional ordering during the overgrowth of gratings.

Chapter 14 finally deals with structure investigation of self-organized arrays of quasi-zero-dimensional islands. After a brief description of the particular growth condition allowing self-organization, grazing-incidence small-angle scattering (GISAXS) and x-ray diffraction schemes are discussed as preferential scattering methods for the analysis of dot arrays.

The sample structure often cannot be deduced directly from the measured diffraction pattern; the fit of the experimental data must verify an expected structure model. Therefore, some general knowledge is of interest regarding how the particular structure of the laterally patterned object influences the diffraction patterns. The present chapters provide such knowledge by reporting on the theoretical background. Characteristic diffraction features are demonstrated by experimental examples.

The vertical and lateral mesoscopic superstructure of a pseudomorphic layered sample (with characteristic dimensions of a few nanometers up to a few microns) creates a two-dimensional fine structure of the diffracted intensity in the vicinity of all reciprocal lattice points. In detail this fine structure consists of the crystal truncation rod of the mean planar multilayer and of the so-called *grating truncation rods* GTRs generated by the lateral grating pattern. The lateral distance between the GTRs in reciprocal space represents the inverse grating period. The intensity profile along a GTR contains information about the grating shape, the inner compositional profile of the layers and the state of lattice strain in the sample.

Early x-ray studies of semiconductor gratings used double-crystal diffractometry [75, 96, 226, 361, 362], a method that provided integrated information about the lateral and vertical structure. The first high-resolution reciprocal space map of a simple surface grating imaged the fine structure around the reciprocal lattice point consisting of the grating rod pattern [124]. First applications of (coplanar) triple-crystal diffrcatometry were oncentrated on a pure shape analysis of simple surface gratings (e.g. [124, 330, 373, 374]) and superlattice gratings and dots [158, 363]. Subsequent experimental studies also investigated the strain field in the samples, detecting elastic strain relaxation in simple layered surface gratings and dots [37, 82, 85] and in buried wires [217, 328, 329, 331].

First quantitative strain studies were either based on elastic models that neglect the influence of the interface between the grating and the substrate [82], or, in the case of buried gratings, by empirically assuming a lateral strain gradient in terms of a power law [328, 329, 331]. X-ray grazing incidence diffraction studies have investigated the depth variation of lateral strain in surface gratings and buried gratings [36, 37, 86, 87, 217, 381] More recently, the *mean strain relaxation* in superlattice SiGe/Si quantum dots was measured by comparison of the isotropic strain model and numerical calculations with x-ray diffraction data, now taking into account the elastic interaction between the grating and the non-patterned region [84, 312].

The experimental determination of the spatial strain distribution in superlattice gratings by coupling x-ray diffraction theory and elasticity theory in a fitting procedure was reported in [37, 273, 403]. Strain models with analytical solutions have been developed for free-standing rectangular superlattice gratings and for rectangular buried gratings [97, 184].

X-ray reflection theory by gratings has been worked out by different groups based on dynamical and semidynamical treatments (see, e.g., [39, 96, 241, 242, 304, 305, 373]). The method has been used for the shape analysis of epitaxial gratings, the analysis of shape and morphology in amorphous layered gratings and organic gratings (see, e.g., [34, 39, 96, 130, 131, 178, 240, 241, 242, 243, 373, 374, 375]).

Theoretical description of x-ray scattering from self-organized arrays of one-dimensional wires of quasi-zero-dimensional islands is complicated by the statistical nature of these arrays. Therefore, the coherently scattered radiation is accompanied by diffuse scattering. In Chapter 14 we present some phenomenological statistical models that make it possible to simulate the diffuse scattering, and we discuss the methods of how to determine the parameters of these models from experimental data.

A review of especially surface-sensitive x-ray diffraction techniques was presented by [284]. The characterization of self-assembled nanostructures by diffuse x-ray scattering has been reviewed in [312]. Chason and Mayer [73] reviewed specular x-ray reflectivity from rough surfaces and interfaces. The recent results of x-ray scattering on self-organized nanostructures are also summarized in [346] and the references therein.

12 X-Ray Scattering by Artificially Lateral Semiconductor Nanostructures

A simple crystalline surface grating can be assumed to be a three-dimensional crystalline lattice modulated by a one- or two-dimensional mesoscopic superstructure with characteristic dimensions of a few nanometers up to a few microns. The scattering process includes the diffraction by the crystal lattice and by the superstructure. In reciprocal space the crystal lattice is represented by reciprocal lattice points. The superstructure of the grating creates a fine structure of the reciprocal lattice points Supposing, for example, a surface grating patterning a pseudomorphic, vertically layered sample (e.g., a planar multilayer), the fine structure consists of a central crystal truncation rod, corresponding to the mean vertically layered stack, surrounded by a so-called *grating rod* generated by the grating. Figure 12.1 shows this situation for several reciprocal lattice points in the case of a one-dimensional surface grating (for example an array of quantum wires).

The reciprocal lattice points and their grating rod patterns can be probed by x-ray reflectivity at the reciprocal lattice point (000), coplanar symmetric high-resolution diffraction at (00l), asymmetric coplanar diffraction at (hkl) or x-ray grazing-incidence diffraction at (hk0 and -hk0) assuming the mean surface normal is parallel to the crystallographic direction [001]. Naturally, an amorphous grating contains these grating rods also, but they are accessible at the origin of the reciprocal space only.

The lateral distance between the grating rods in reciprocal space represents the inverse grating period . The intensity profile along a grating rod contains information about the grating shape, the vertical compositional profile of the layers and the state of lattice strain in the sample.

The coplanar x-ray diffraction of grating structures was usually treated within the limits of kinematical and semikinematical theory of x-ray diffraction. Higher-order distorted wave Born approximations were applied in order to describe multiple diffraction effects [32, 124]. X-ray grazing incidence diffraction by gratings was evaluated by use of the first order distorted wave Born approximation (DWBA) [36]. Fully dynamical treatments were used for the case of x-ray reflection by gratings (e.g. [96, 241, 242, 373]). The scattering theory for grating structures of amorphous materials was developed in [269, 376]. In this chapter the different theoretical approaches are outlined, and their particularities discussed with the help of illustrations in reciprocal space.

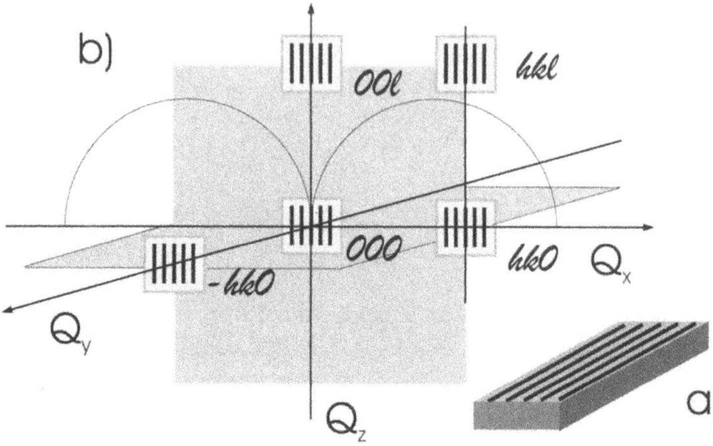

Fig. 12.1. Real-space representation (a) and schematic illustration of the respective reciprocal-space structure (b) of a one-dimensional surface grating. Each reciprocal lattice point shows a lateral fine structure consisting of grating truncation rods.

12.1 The Scattering Potential and the Structure Amplitude

Here we deal with laterally patterned structures in one or two dimensions. The *first* main structural property that these objects have in common is the *lateral grating periodicity* of the polarizability with the period $\boldsymbol{D}_G$,

$$\chi(\boldsymbol{r}) = \chi(\boldsymbol{r} + \boldsymbol{D}_G). \tag{12.1}$$

Thus we can develop the polarizability in a Fourier series with the reciprocal grating vectors $\boldsymbol{G} = \boldsymbol{e}_x G_x + \boldsymbol{e}_y G_y = \boldsymbol{e}_x(2\pi n/D_x) + \boldsymbol{e}_y(2\pi m/D_y)$ (n and m being integers), choosing the z-axis parallel to the mean surface normal:

$$\chi(\boldsymbol{r}) = \sum_{\boldsymbol{G}} \chi_G(z) \mathrm{e}^{\mathrm{i}\boldsymbol{G}.\boldsymbol{r}}. \tag{12.2}$$

The *second* common property is the mean *crystalline periodicity*. As a consequence, the polarizability can be developed in a Fourier series over the averaged reciprocal lattice vectors $\boldsymbol{g}$:

$$\chi(\boldsymbol{r}) = \sum_{\boldsymbol{g}} \tilde{\chi_{\boldsymbol{g}}}(\boldsymbol{r}) \mathrm{e}^{\mathrm{i}\boldsymbol{g}.\boldsymbol{r}}. \tag{12.3}$$

In contrast to the Fourier components for an infinite crystal the Fourier components $\tilde{\chi}_{\boldsymbol{g}}(\boldsymbol{r})$ depend on the position $\boldsymbol{r}$ as well, since they contain the super-periodicity of the grating. Consequently, one can express them by the Fourier series

$$\tilde{\chi}_{\boldsymbol{g}}(\boldsymbol{r}) = \sum_{\boldsymbol{H}} \tilde{\chi}_{\boldsymbol{gH}}(z) \mathrm{e}^{\mathrm{i}\boldsymbol{H}.\boldsymbol{r}}. \tag{12.4}$$

In the representation of Eq. (12.4) each reciprocal grating vector $\boldsymbol{H}$ corresponds to a lateral satellite belonging to the in-plane component of the reciprocal lattice vector $\boldsymbol{g}$.

In this book we consider gratings of mesoscopic period length, which is usually 2 - 3 orders of magnitude larger than the atomic distances. Then scattering by grating rods shows measurable intensity only in the vicinity of the reciprocal lattice points of the crystalline lattice. Subsequently, we can neglect the overlap of the contributions of different reciprocal lattice points to a certain grating rod.

Within the two-beam-case approximation (which treats the wavefields of the direct beam and of one selected diffracted beam only) we divide the scattering potential into two contributions:

$$\hat{\mathbf{V}}(\boldsymbol{r}) = \hat{\mathbf{V}}_0(\boldsymbol{r}) + \hat{\mathbf{V}}_{\boldsymbol{h}}(\boldsymbol{r}), \tag{12.5}$$

or expressed by the polarizability

$$\hat{\mathbf{V}}(\boldsymbol{r}) = -K^2\chi(\boldsymbol{r}) \approx -K^2 \left(\tilde{\chi}_0(\boldsymbol{r}) + \tilde{\chi}_{\boldsymbol{h}}(\boldsymbol{r})\mathrm{e}^{\mathrm{i}\boldsymbol{h}.\boldsymbol{r}}\right) \tag{12.6}$$

$$= -K^2 \sum_{\boldsymbol{G}} \left[\tilde{\chi}_{0\boldsymbol{G}}(z)\mathrm{e}^{\mathrm{i}\boldsymbol{G}.\boldsymbol{r}} + \tilde{\chi}_{\boldsymbol{hG}}(z)\mathrm{e}^{\mathrm{i}(\boldsymbol{G}+\boldsymbol{h}).\boldsymbol{r}}\right]. \tag{12.7}$$

Here the first terms on the right side of equations (12.5) and (12.6) each describe the modulation of the mean electron density (modulation of the mean refractive index) related to the origin of reciprocal space. The second term is the scattering potential for kinematical diffraction related to the reciprocal lattice vector $\boldsymbol{h}$.

Since we have included the shape function of the diffracting grating in the polarizability coefficients $\tilde{\chi}_{\boldsymbol{g}}(\boldsymbol{r})$, the amplitude of the diffracted wave is proportional to the Fourier transformation of the polarizability (the so-called *grating structure amplitude*),

$$S(\boldsymbol{Q}) = \int \mathrm{d}^3\boldsymbol{r}\, \chi(\boldsymbol{r})\mathrm{e}^{-\mathrm{i}\boldsymbol{Q}.\boldsymbol{r}} = S_0(\boldsymbol{Q}) + S_{\boldsymbol{h}}(\boldsymbol{Q}). \tag{12.8}$$

Substituting Eq. (12.6) in Eq. (12.8), we find (as in the case of planar layers in Chap. 5)

$$\begin{aligned} S_{\boldsymbol{h}}(\boldsymbol{Q}) &= 4\pi^2 \textstyle\sum_{\boldsymbol{H}} \delta^{(2)}(\boldsymbol{Q}_\parallel - \boldsymbol{h}_\parallel - \boldsymbol{H}) \int \mathrm{d}z\, \tilde{\chi}_{\boldsymbol{hH}}(z)\, \mathrm{e}^{-\mathrm{i}(Q_z - h_z)z} \\ S_0(\boldsymbol{Q}) &= 4\pi^2 \textstyle\sum_{\boldsymbol{H}} \delta^{(2)}(\boldsymbol{Q}_\parallel - \boldsymbol{H}) \int \mathrm{d}z\, \tilde{\chi}_{0\boldsymbol{H}}(z)\, \mathrm{e}^{-\mathrm{i}Q_z z}. \end{aligned} \tag{12.9}$$

From Eq. (12.9) it follows that the structure amplitude is characterized by the generation of a fine structure around each reciprocal lattice point. This fine structure consists of *grating rods* (GTR), which are perpendicular to the mean sample surface, that is, parallel to Q_z. Their positions in the (Q_x, Q_y) plane are defined by the reciprocal-lattice vector and the reciprocal grating vectors

$$\boldsymbol{Q}_{\|} = \boldsymbol{h}_{\|} + \boldsymbol{H}. \tag{12.10}$$

For an array of wires (a one-dimensional grating), the grating rods are equidistantly arranged in a plane perpendicular to the sample surface and perpendicular to the direction of wire elongation. The intersection points of the grating rods with the (Q_x, Q_y) plane form a one-dimensional dot lattice. For a two-dimensional surface grating (array of dots), the reciprocal lattice points have a three-dimensional fine structure, which is formed by parallely arranged grating rods. The intersection points of the grating rods with the (Q_x, Q_y) plane form a two-dimensional dot lattice. The structure amplitude of each grating rod is characterized by the *grating rod form factor*

$$F_{\boldsymbol{hH}}(Q_z) = \int \mathrm{d}z \, \chi_{\boldsymbol{hH}}(z) \, \mathrm{e}^{-\mathrm{i}(Q_z - h_z)z}. \tag{12.11}$$

In the following we shall study diffraction by surface gratings based on planar multilayers (see Fig. 12.2) in more detail. Thus the *third* common property of the gratings is the vertically *layered set-up* of patterned and non-patterned layers. Representing a perfectly periodic patterned array, a grating is characterized by its lateral periodicity $\boldsymbol{D}_G$ and by the vertical arrangement of layers which also are laterally patterned. The vertical layer arrangement in a multilayer grating can be treated in a way similar to that done for a planar multilayer by writing the polarizability of the layered stack as a sum over the individual layers as in Sect. 5.4 or 9.4. Each horizontal (patterned and non-patterned) layer is considered in terms of its (planar) layer size function $\Omega_{\mathrm{pl}}^{(m)}$, where m is the layer index. The grating shape is described by the corresponding shape function $\Omega_{\mathrm{SG}}(\boldsymbol{r})$, this function equals unity in the grating volume and zero in the trenches. For patterned layers we obtain the shape function of the wires in each layer

$$\Omega_{\mathrm{wire}}^{(m)}(\boldsymbol{r}) = \Omega_{\mathrm{SG}}(\boldsymbol{r}) \Omega_{\mathrm{pl}}^{(m)}(z). \tag{12.12}$$

In a multilayer grating, the kinematical scattering potential in Eqs. (12.5)–(12.8) sums up the contributions of all individual layers $\hat{\mathbf{V}}_{\boldsymbol{h}}(\boldsymbol{r}) = \sum_m \hat{\mathbf{v}}_{\boldsymbol{h}}^{(m)}$ with

$$\hat{\mathbf{v}}_{\boldsymbol{h}}^{(m)} = -K^2 \tilde{\chi}_{\boldsymbol{h}}(\boldsymbol{r}) \mathrm{e}^{\mathrm{i}\boldsymbol{h}.\boldsymbol{r}} \Omega_{\mathrm{pl}}^{(m)} = -K^2 \chi_{\boldsymbol{h}}^{(m)} \mathrm{e}^{\mathrm{i}\boldsymbol{h}^{(m)}(\boldsymbol{r} - \boldsymbol{u}^{(m)}(\boldsymbol{r}))} \Omega_{\mathrm{wire}}^{(m)}(\boldsymbol{r}), \tag{12.13}$$

which are characterized by

1. the wire shape function $\Omega_{\mathrm{wire}}^{(m)}$ of layer (m) constituting the grating,
2. the local Fourier component $\chi_{\boldsymbol{h}}^{(m)}$ of the polarizability corresponding to the individual layer lattice, and

3. the lattice displacement function $\boldsymbol{u}^{(m)}(\boldsymbol{r})$, which we define with respect to the local reciprocal lattice vector $\boldsymbol{h}^{(m)}$ of the planar layer before being patterned.

Please do not confuse the space-dependent function $\tilde{\chi_{\boldsymbol{h}}}(\boldsymbol{r})$ and the constant Fourier components of the layer polarizability $\chi_{\boldsymbol{h}}^{(m)}$.

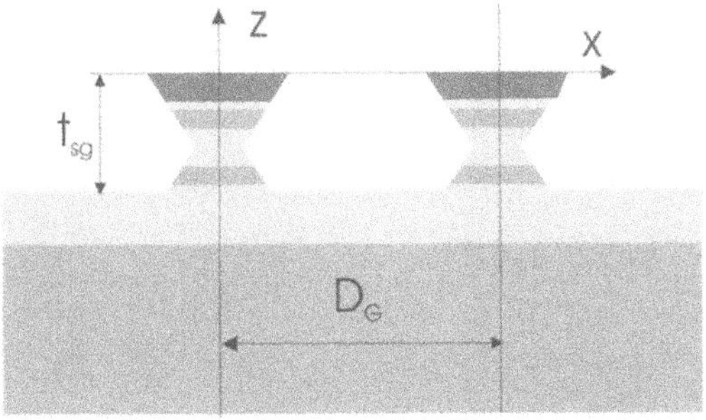

Fig. 12.2. Schematic arrangement of a multilayer surface grating.

The lattice displacement functions of all layers can be written as a sum of the lattice displacement due to elastic distortion in the planar layer before etching (see Sect. 9.1 for the tetragonal distortion in planar multilayers) and its variation due to grating-induced strain relaxation:

$$\boldsymbol{u}(\boldsymbol{r}) = \boldsymbol{u}_0(z) + \Delta\boldsymbol{u}(\boldsymbol{r}). \tag{12.14}$$

Separating grating independent from grating-dependent functions in the kinematical diffraction potential and expressing the latter by a two-dimensional Fourier series, we find

$$\hat{\mathbf{v}}_{\boldsymbol{h}}^{(m)} = -K^2 \mathrm{e}^{\mathrm{i}\boldsymbol{h}^{(m)}[\boldsymbol{r}-\boldsymbol{u}_0^{(m)}(\boldsymbol{r})]} \sum_{\boldsymbol{H}} \chi_{\boldsymbol{hH}}^{(m)}(z) \mathrm{e}^{\mathrm{i}\boldsymbol{H}.\boldsymbol{r}_\parallel} \tag{12.15}$$

with the grating Fourier components

$$\chi_{\boldsymbol{hH}}^{(m)}(z) = \frac{\chi_{\boldsymbol{h}}^{(m)}}{D_x D_y} \int_{D_x/2}^{D_x/2} \int_{D_y/2}^{D_y/2} \mathrm{d}x\, \mathrm{d}y\, \mathrm{e}^{-\mathrm{i}\boldsymbol{H}.\boldsymbol{r}_\parallel} \mathrm{e}^{\mathrm{i}\boldsymbol{h}^{(m)}\Delta\boldsymbol{u}(\boldsymbol{r})} \Omega_{\mathrm{wire}}^{(m)}(\boldsymbol{r}). \tag{12.16}$$

The grating form factors of the whole multilayer belonging to a certain GTR vector $\boldsymbol{H}$ add up to the grating form factors of all individual layers. This provides a phase shift which depends on the vertical position of the respective layer within the multilayer stack and the lattice displacement between the sublayers, similar to the procedure for a planar multilayer presented in Sect. 5.4:

$$F_{\boldsymbol{hH}}(Q_z) = \sum_{m=1}^{N} \mathrm{e}^{-\mathrm{i}Q_z z_m} \mathrm{e}^{\mathrm{i}h_z^{(m)} u_{0z}^{(m)}(z_m)} \widetilde{F}_{\boldsymbol{hH}}^{(m)}(Q_z) \tag{12.17}$$

with

$$\widetilde{F}_{hH}^{(m)}(Q_z) = \int_{z_{m+1}}^{z_m} dz \chi_{hH}^{(m)}(z) e^{-iq_z^{(m)}(z-z_m)}. \qquad (12.18)$$

Here we used the reduced scattering vector $q_z^{(m)} = Q_z - h_z^{(m)}$.

We may distinguish the contribution of the periodic surface shape and of periodic strain on the grating form factors of the layers by replacing the shape functions $\Omega_{\mathrm{a}}^{(m)}(\boldsymbol{r})$ and the laterally periodic functions of the grating-induced displacement $U^{(m)}(\boldsymbol{r}) \equiv \exp(-i\boldsymbol{h}^{(m)} \Delta \boldsymbol{u}^{(m)}(\boldsymbol{r}))$ by their Fourier series

$$U^{(m)}(\boldsymbol{r}) = \sum_{\boldsymbol{H}} U_{\boldsymbol{H}}^{(m)}(z) e^{i\boldsymbol{H}.\boldsymbol{r}}$$

$$\Omega_{\mathrm{wire}}^{(m)}(\boldsymbol{r}) = \sum_{\boldsymbol{H}} \Omega_{\boldsymbol{H}}^{(m)}(z) e^{i\boldsymbol{H}.\boldsymbol{r}}, \qquad (12.19)$$

and substituting these in Eq. (12.16), we obtain for (12.18),

$$\widetilde{F}_{hH}^{(m)}(Q_z) = \chi_{h}^{(m)} \sum_{\boldsymbol{M}} \int_{z_{m+1}}^{z_m} dz \; e^{-iq_z^{(m)}(z-z_m)} U_{\boldsymbol{M}}^{(m)}(z) \Omega_{\boldsymbol{H}-\boldsymbol{M}}^{(m)}(z). \qquad (12.20)$$

Since the surface patterning does not induce additional residual strain, all non-zero-order Fourier components of the lattice displacement function $U_M^{(m)}$ vanish. In the case of a pure strain grating without a surface relief, the components of the shape function vanish in Eq. (12.20).

The symmetry of the scattered wavefield corresponds to the symmetry of the scattering potential. The lateral periodicity of the grating scattering potential, developed in Fourier series in Eq. (12.2), predicts that the coherently scattered intensity from a plane incident wave is exclusively concentrated along grating rods

$$E(\boldsymbol{r}) = \sum_{\boldsymbol{G}} \int d^2\boldsymbol{K}_{\parallel} E_{\boldsymbol{K}_{\parallel}}(\boldsymbol{r}) \delta(\boldsymbol{K}_{\parallel} - \boldsymbol{K}_{0\parallel} - \boldsymbol{G})$$

$$= \sum_{\boldsymbol{G}} e^{i\boldsymbol{K}_{\boldsymbol{H}\parallel}\boldsymbol{r}_{\parallel}} \left[T_{\boldsymbol{G}}(z) + R_{\boldsymbol{G}}(z)\right]. \qquad (12.21)$$

Here we introduced the wave amplitudes of the transmitted and reflected waves $T_{\boldsymbol{G}}(z)$ and $R_{\boldsymbol{G}}(z)$.

The formal solution of the Bloch waves in Eq. (12.21) is independent of the level of approximation (e.g., kinematical theory, different distorted-wave-Born approximations, semikinematical treatments, or the dynamical theory). The differences of these theories only concern the z-components of the diffracted wave vectors and consequently the diffracted wave amplitudes *along* the grating rods.

Up to now, Eq. (12.21) has described explicitly the grating periodicity only, whereas the crystalline lattice structure is considered implicitly. On

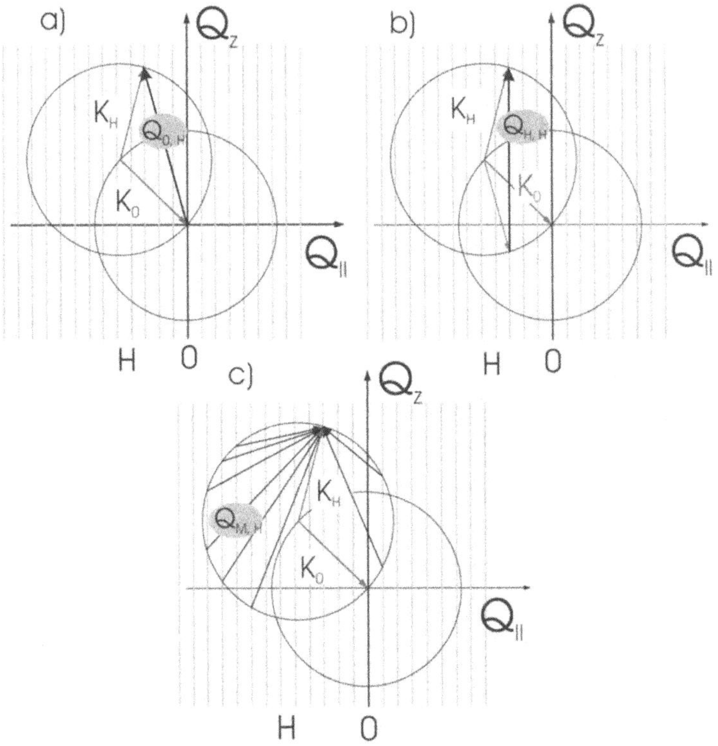

Fig. 12.3. Schematic illustration of different scattering processes: (a) primary scattering (between the incident and the considered diffracted wave), (b) intra-rod scattering (between waves corresponding to the same rod), (c) inter-rod scattering (between waves corresponding to different rods).

one hand, the lateral lattice periodicity of a planar pseudomorphic epitaxial multilayer is violated by the superstructure of the surface grating. On the other hand, a perfectly periodic crystalline grating requires commensurate properties of the lateral grating and lattice periodicities. Given such commensurate conditions the reciprocal grating vectors of Eq. (12.21) can be interpreted as satellites to any reciprocal lattice vector according to the redefinition $\boldsymbol{G} = \boldsymbol{G}' + \boldsymbol{g}_{\|}$.

Using the kinematical and semikinematical theories, in most cases we will relate the reciprocal grating vector as satellite to the next nearest reciprocal lattice vector of the host lattice. In those cases we will add the corresponding reciprocal lattice vector as an index to the amplitudes of the reflected and transmitted waves, e.g., $T_{0,\boldsymbol{H}}$, instead of $T_{\boldsymbol{G}}$, where $\boldsymbol{G} = \boldsymbol{H}$ or $R_{\boldsymbol{h}\boldsymbol{H}}$, instead of $R_{\boldsymbol{G}}$, where $\boldsymbol{G} = \boldsymbol{h}_{\|} + \boldsymbol{H}$.

Usually the grating will not be commensurate with the host lattice. However, for our examples, these effects will be negligible in most cases. For very small grating periods non-commensurate gratings will lead to a splitting of grating rods between two neighboring reciprocal lattice points, similar to cases known from non-commensurate superlattices.

12.2 Kinematical Theory

For simplicity we focus here on one-dimensionally patterned devices (wire gratings), defining the direction of patterning to be along the x-axis.

Assuming an incident plane wave, the expression for the kinematically scattered wavefield can be derived by use of the results of Section 4.2 (Eqs. (4.15) to (4.18)), replacing there the *scattering operator* $\hat{\mathbf{T}}(\boldsymbol{r})$ by the *kinematic* approximation $\hat{\mathbf{T}} \approx \hat{\mathbf{V}}$. Then, substituting Eqs. (12.5)–(12.8), we find for the kinematically diffracted wave amplitude,

$$E^{(\mathrm{kin})}(\boldsymbol{r}) = -\frac{K^2}{4\pi^2}\int \mathrm{d}^2\boldsymbol{K}_{\|}\frac{\mathrm{e}^{\mathrm{i}\boldsymbol{K}.\boldsymbol{r}}}{2\mathrm{i}K_z}\left(S_0(\boldsymbol{Q}) + \S_{\boldsymbol{h}}(\boldsymbol{Q})\right). \tag{12.22}$$

Expressing the structure factors $\widetilde{S}_{0,\boldsymbol{h}}$ by the use of formulas of the previous section and after some rearrangement for the kinematical Bragg-case (the Bragg-diffracted waves leave the entrance surface), we obtain

$$E^{(\mathrm{kin})}(\boldsymbol{r}) = \sum_{\boldsymbol{H}}\left[T^{(\mathrm{kin})}_{0\boldsymbol{H}}(z)\mathrm{e}^{\mathrm{i}(\boldsymbol{K}_{0\|}+\boldsymbol{H})\boldsymbol{r}_{\|}} + R^{(\mathrm{kin})}_{\boldsymbol{hH}}(z)\mathrm{e}^{\mathrm{i}(\boldsymbol{K}_{0\|}+\boldsymbol{h}_{\|}+\boldsymbol{H})\boldsymbol{r}_{\|}}\right], \tag{12.23}$$

a sum of one transmitted and one Bragg-diffracted wavefield. Both wavefields consist of a fan of coherent grating-diffracted waves (see Fig. 12.4). One, with the kinematical amplitudes of transmission $T^{(\mathrm{kin})}_{0\boldsymbol{H}}(z)$ corresponds to the groups of grating rods near the origin of reciprocal space, the other one with the reflection amplitudes $R^{(\mathrm{kin})}_{\boldsymbol{hH}}(z)$ to the grating rods near the the reciprocal lattice point $\boldsymbol{h}$. The Bragg-diffracted wave of a certain rod at $\boldsymbol{h}_{\|} + \boldsymbol{H}$ above the grating (at $z > 0$) has, the kinematical amplitude

$$R^{(\mathrm{kin})}_{\boldsymbol{hH}}(z{>}0) = K^2\frac{\mathrm{e}^{\mathrm{i}K_{\boldsymbol{hH}z}z}}{2\mathrm{i}K_{\boldsymbol{hH}z}}F_{\boldsymbol{hH}}(Q_z). \tag{12.24}$$

The transmitted wave amplitude of a single rod $\boldsymbol{H}$ below the grating (at $z{<}z_{\mathrm{SG}}$) is

$$T^{(\mathrm{kin})}_{0\boldsymbol{H}}(z{<}z_{\mathrm{SG}}) = K^2\frac{\mathrm{e}^{\mathrm{i}K_{\boldsymbol{hH}z}z}}{2\mathrm{i}K_{\boldsymbol{hH}z}}F_{0\boldsymbol{H}}(Q_z). \tag{12.25}$$

The kinematical amplitudes of transmission and reflection of the grating rods resemble formally those of a crystal truncation rod derived in Sect. 5.1. Usually, due to absorption in the substrate, we can measure the intensity of the Bragg-diffracted waves only. In an idealized triple-crystal experiment (plane

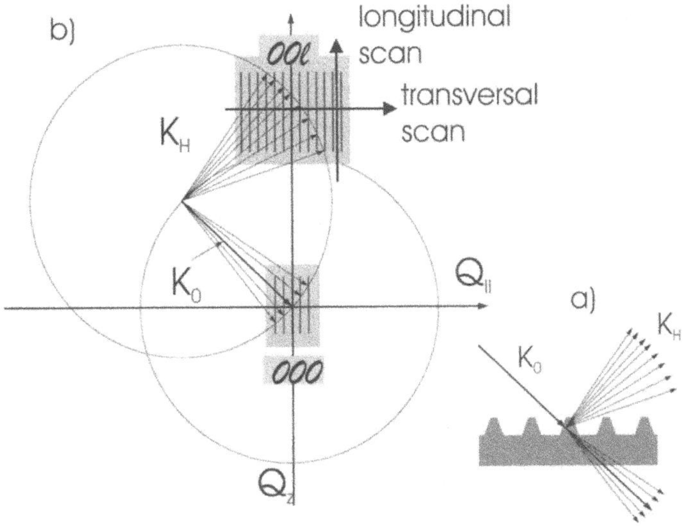

Fig. 12.4. Schematic representation of coplanar diffraction by surface gratings, real space (a), reciprocal space (b). One incoming plane wave is diffracted into a fan of coherently transmitted and reflected waves. Additionally, two different scan directions (longitudinal and transverse scan, as introduced in Sect. 3.2) are shown to probe the grating.

incident wave, infinite sample size, and open slits) the *grating reflectivity* of a grating rod is given by

$$\mathcal{R}_{\boldsymbol{hH}} = \frac{|R_{\boldsymbol{hH}}(z>0)|^2}{|E_0|^2} \left|\frac{K_{\boldsymbol{H}z}}{K_{0z}}\right| . \tag{12.26}$$

12.3 Dynamical Theory

The dynamical theory has been worked out and applied to the important case of x-ray reflection. However, the treatment developed there is sufficiently general to be adopted also to diffraction by the grating. The complete treatment is rather lengthy and expensive. Here only the basic formulas are outlined for a rectangularly shaped grating; the interested reader is refereed to [242].

Actually the theoretical procedure is quite similar to the multiple-beam case of conventional x-ray diffraction by planar crystals (see Chap. 6). We can apply Ewald's concept, after which the mathematical form of the wavefield propagating in a medium keeps the translational symmetry of the scattering medium. Therefore, by use of Ewalds's concept, the solution for a periodic medium can be expressed as a superposition of a discrete number of plane waves. By use of of Eqs. (4.12) to (4.18) and (12.21), we can write

$$E(\boldsymbol{r}) = \sum_{\boldsymbol{H}} \mathrm{e}^{\mathrm{i}\boldsymbol{K}_{\boldsymbol{H}\parallel}.\boldsymbol{r}_{\parallel}} E_{\boldsymbol{H}}(z) = \frac{\mathrm{i}E_0}{8\pi^2} \int \mathrm{d}^2\boldsymbol{K}_{\parallel} \frac{\mathrm{e}^{\mathrm{i}\boldsymbol{K}.\boldsymbol{r}}}{K_z} \langle \boldsymbol{K}|\hat{\mathbf{V}}|\boldsymbol{E}\rangle . \tag{12.27}$$

The wave amplitudes corresponding to each grating rod may be decomposed into a sum of a transmitted and a reflected wave amplitude $E_{\boldsymbol{H}}(z) = [T_{\boldsymbol{H}}(z) + R_{\boldsymbol{H}}(z)]$. By splitting the potential into two parts $\hat{\mathbf{V}} = \hat{\mathbf{V}}_A + \hat{\mathbf{V}}_B$, we obtain for Eq. (12.27) ctually, (see Eq. (7.3))

$$\sum_{H} \mathrm{e}^{\mathrm{i}\boldsymbol{K}_{H\parallel}\boldsymbol{r}_{\parallel}} \left[T_H(z) + R_H(z)\right] =$$

$$-\frac{\mathrm{i}E_0}{8\pi^2} \int \mathrm{d}^2\boldsymbol{K}_{\parallel} \frac{\mathrm{e}^{\mathrm{i}\boldsymbol{K}.\boldsymbol{r}}}{K_z} \left[\left\langle \boldsymbol{E}_K^{(-A)} \middle| \hat{\mathbf{V}}_A \middle| \boldsymbol{K}_0 \right\rangle + \left\langle \boldsymbol{E}_K^{(-A)} \middle| \hat{\mathbf{V}}_B \middle| \boldsymbol{E} \right\rangle\right] \tag{12.28}$$

It is convenient to divide the scattering potential in such a way that $\hat{\mathbf{V}}_A$ describes the susceptibility of the lateral average over the whole grating structure $-K^2 \langle\chi_0\rangle_{\parallel}(z)$. The respective non-perturbed solutions $\left|\boldsymbol{E}_{K_0}^{(-A)}\right\rangle$ describe the wavefield generated by refraction and specular reflection in a virtual *planar* multilayer (the minus sign denotes the time-inverted state). The potential $\hat{\mathbf{V}}_B$ is simply the laterally periodic perturbation of the planar multilayer set-up. In other words, it may include properties of the grating and of the crystal lattice. Equation (12.28) still contains on the right side the total wavefield $|\boldsymbol{E}\rangle$ which is unknown. As a starting point for approximations, Eq. (12.28) will be useful to explain the physical meaning of dynamical scattering in comparison with other approximations (kinematical theory, DWBA in the sections 12.2 - 12.4). However, for the further theoretical treatment it is suitable to continue by use of the wave equation Eq. (4.11), which is for the case of gratings (substituting the grating scattering potential of Eq. 12.2)

$$(\triangle + K^2(1+\chi_0))E(\boldsymbol{r}) + K^2 \sum_{\boldsymbol{H}\neq 0} \chi_{\boldsymbol{H}}(z)\mathrm{e}^{\mathrm{i}\boldsymbol{H}.\boldsymbol{r}} E(\boldsymbol{r}). \tag{12.29}$$

Further it is convenient to introduce the wave vector components *inside* the medium with laterally averaged susceptibilities according to the spherical dispersion relation

$$k_{\boldsymbol{H}z} = \sqrt{K^2(1+\chi_0) - \boldsymbol{k}_{\boldsymbol{H}\parallel}^2}\,. \tag{12.30}$$

The wave equation (12.29) decomposes into a set of differential equations corresponding to the Fourier components $\boldsymbol{H}$ [113]:

$$k_{\boldsymbol{H}z}^2 E_{\boldsymbol{H}}(z) + \frac{\mathrm{d}^2 E_{\boldsymbol{H}}(z)}{\mathrm{d}z^2} + K^2 \sum_{\boldsymbol{G},\boldsymbol{G}\neq\boldsymbol{H}} E_{\boldsymbol{G}}(z)\chi_{\boldsymbol{H}-\boldsymbol{G}} = 0\,. \tag{12.31}$$

Next we search for the solutions of Eq. (12.31). If we choose a plane wave representation for the particular solutions of each layer (m), $E_{\boldsymbol{H},n}^{(m)}(z) = E_{\boldsymbol{H},n}^{(m)} \mathrm{e}^{\mathrm{i}\kappa_{nz}^{(m)} z}$, (the index n enumerates the particular solutions), we obtain the system of equations [304]

$$E_{\boldsymbol{H},n}^{(m)} = \frac{K^2}{\kappa_{nz}^{(m),2} - k_{\boldsymbol{H}z}^{(m),2}} \sum_{\boldsymbol{G},\boldsymbol{G}\neq\boldsymbol{H}} E_{\boldsymbol{G},n}^{(m)} \chi_{\boldsymbol{H}-\boldsymbol{G}} \tag{12.32}$$

for each component $\boldsymbol{H}$.

Now we can write the wave amplitudes $E_{\boldsymbol{H}}(z)$ in each layer (m) as a linear combination of all *particular* solutions, $E_{\boldsymbol{H},n}^{(m)}(z)$,

$$E_{\boldsymbol{H}}^{(m)}(z) = \sum_n \left[\mathrm{e}^{\mathrm{i}\kappa_{nz}^{(m)}(z-z_{m+1})} \tilde{T}_n^{(m)} + \mathrm{e}^{-\mathrm{i}\kappa_{nz}^{(m)}(z-z_{m+1})} \tilde{R}_n^{(m)} \right] E_{\boldsymbol{H},n}^{(m)}. \tag{12.33}$$

The coefficients $\tilde{T}_n^{(m)}$ and $\tilde{R}_n^{(m)}$ are here the wave amplitudes related to the particular solutions n.

Similar to the dynamical diffraction by crystals, the amplitudes of the particular solution n corresponding to a grating rod $\boldsymbol{H}$ in Eq. (12.32) are proportional to the resonance factor $K^2/(\kappa_{nz}^2 - k_{\boldsymbol{H}z}^2)$. It is worth noting that the dispersion surface corresponding to a $\kappa_{\boldsymbol{H}z}$ differ from those of the components $k_{\boldsymbol{H}z}$, which are spherical [11]. Therefore, their differences $\kappa_{nz}^2 - k_{\boldsymbol{H}z}^2$ do never vanish and the amplitudes $E_{\boldsymbol{H},n}^{(m)}$ are always finite.

The particular solutions associated to a certain truncation rod are influenced by dynamical contributions of all other grating rods (so called *inter-rod scattering)*, (see the illustration in Fig. 12.3). The treatment also includes all grating rods which are not directly excited by the Ewald sphere of the incident wave.

The dynamical theory can be formulated as a matrix eigenvalue approach [11] based on an extension of the Abèles matrix formalism (originally developed for diffraction by planar multilayers) to the case of multilayer gratings. Here the influence of the inter-rod scattering is expressed elegantly in a transparent way after introducing generalized Fresnel reflection and transmission coefficients [242], similar to the conventional Fresnel coefficients for reflection and transmission in stratified media,

$$r_{\boldsymbol{HG}}^{(m)} = \frac{\kappa_{\boldsymbol{H},z}^{(m-1)} - \kappa_{\boldsymbol{G},z}^{(m)}}{\kappa_{\boldsymbol{H},z}^{(m-1)} + \kappa_{\boldsymbol{G},z}^{(m)}} \tag{12.34}$$

and

$$t_{\boldsymbol{HG}}^{(m)} = \frac{2\kappa_{\boldsymbol{H},z}^{(m-1)}}{\kappa_{\boldsymbol{H},z}^{(m-1)} + \kappa_{\boldsymbol{G},z}^{(m)}} . \tag{12.35}$$

Here the indices (m) of the generalized Fresnel coefficients correspond to the interface (m) formed between the layers (m) and $(m-1)$. The generalized Fresnel coefficients determine the propagation by transmission and diffraction e.g., of a wave in layer $(m-1)$ with the wave vector $k_{\boldsymbol{H}z}^{(m-1)}$, into a wave with the wave vector $k_{\boldsymbol{G}z}^{(m)}$ in the layer (m).

The dynamical treatment considers the conservation of energy. The dynamical theory takes all kinds of multiple scattering into account. Also it considers ordinary refraction in the layers and total external reflection by the interfaces. It includes multiple scattering within a certain grating rod (so called *intra-rod scattering*) as well as *inter-rod scattering* between different grating rods of the same reciprocal lattice point and between the grating rods of different reciprocal lattice points (000 and hkl). For example, strong diffraction into a certain grating rod causes extinction and can significantly influence the intensity profile of other grating rods.

The calculation in terms of the dynamical theory is cumbersome and requires a time-consuming numerical procedure. Fortunately, in many cases only a small number of strongly scattering grating rods exists which have to be considered. Often one searches for simpler, approximate treatments, which shall be numerically less demanding, more transparent, and sufficiently precise. Unfortunately, for all these approximations, the law of energy conservation is violated. The validity of various theoretical approximations is limited to certain regions in reciprocal space and depends on the actual sample set-up and on the special conditions of scattering. Therefore, the choice of approximation should be done carefully.

The simplest approach is the kinematical theory, developed in the previous section and applicable for many cases. Since it neglects refraction and any kind of multiple reflection, it fails for scattering geometries of grazing incidence or exit. In the wide-angle range, dynamical diffraction by the substrate and or strongly diffracting planar multilayers may essentially influence the diffraction intensities and give rise to strong inter-rod scattering. Also the latter effects are all neglected in the kinematical treatment.

Perturbation methods such as the distorted wave Born approximation (DWBA) can explain the experimental results in most relevant cases. The aim of using this approach is to explain the main dynamical features by a simple and transparent procedure. Within this book we restrict ourselves to principles of first-order DWBA. This approach allows us, for example, to take intra-rod scattering into account among all grating rods and inter-rod scattering from the central specular rod 000 toward all grating rods and also around other reciprocal lattice points hkl.

In the following sections we outline two different treatments. Section 12.4 focuses on x-ray diffraction methods under conditions of grazing incidence or exit, which are mainly influenced by specular interface reflection. The procedure is quite similar to that employed for the description of x-ray scattering by rough planar multilayers (see Sect. 11.4).

For x-ray diffraction at angles of several degrees, multiple scattering effects due to the specular reflection are negligible. However, the diffraction along the crystal truncation rod of the substrate and/or strongly diffracting planar multilayers may influence the grating diffraction intensities essentially. Therefore one can use another splitting of the scattering potential within the framework of DWBA, which will be outlined in Sect. 11.3. Inter-rod scattering among different grating diffraction rods is neglected within the DWBA of first order. That would play a role, first, in weak nearly forbidden grating rod orders [32, 242] and, second, in the case of strongly refracting, but weakly diffracting surface gratings or relatively strong diffracting (thick) planar multilayers. An extension to second-order DBWA reproduces qualitatively nearly all dynamical scattering features of inter-truncation rod scattering used for XRD in [32].

12.4 Distorted Wave-Born Approximation for Grazing-Incidence Diffraction

As shown in Chap. 7, the DWBA consists in the replacement of the unknown solution for the total wavefield $|\mathbf{E}\rangle$ of Eq. (12.28) by the scattering of the non-disturbed wavefield (obtained by solving the wave equation with the non-disturbed scattering potential $\hat{\mathbf{V}}_A$) by the disturbance $\hat{\mathbf{V}}_B$ of the scattering potential

$$E(\boldsymbol{r}) = \sum_{\boldsymbol{H}} \mathrm{e}^{\mathrm{i}\boldsymbol{K}_{\boldsymbol{H}\parallel}\boldsymbol{r}_{\parallel}} \left[T_{\boldsymbol{H}}(z) + R_{\boldsymbol{H}}(z)\right]$$

$$= -\frac{\mathrm{i}E_0}{8\pi^2}\int \mathrm{d}^2\boldsymbol{K}_{\parallel}\frac{\mathrm{e}^{\mathrm{i}\boldsymbol{K}\boldsymbol{r}}}{K_z}\left[\left\langle \boldsymbol{E}_K^{(-A)}\middle|\hat{\mathbf{V}}_A\middle|\boldsymbol{K}_0\right\rangle + \left\langle \boldsymbol{E}_K^{(-A)}\middle|\hat{\mathbf{V}}_B\middle|\boldsymbol{E}_K^{(A)}\right\rangle\right] . \quad (12.36)$$

Keeping the non-disturbed part of the scattering potential $\hat{\mathbf{V}}_A$ as in the previous section, $\hat{\mathbf{V}}_A = -K^2 \langle\chi_0\rangle_{\parallel}(z)$, the disturbance within the two-beam case is

$$\hat{\mathbf{V}}_B = \sum_m \left[\Delta\hat{\mathbf{v}}_0^{(m)}(\boldsymbol{r}) + \hat{\mathbf{v}}_{\boldsymbol{h}}^{(m)}(\boldsymbol{r})\right] , \quad (12.37)$$

with

$$\Delta\hat{\mathbf{v}}_0^{(m)}(\boldsymbol{r}) = -K^2\left(\chi_0^{(m)}(\boldsymbol{r}) - \langle\chi_0\rangle_{\parallel}^{(m)}(z)\right) , \quad (12.38)$$

and (with 12.13)

$$\hat{\mathbf{v}}_{\mathrm{h}}^{(m)}(\boldsymbol{r}) = -K^2\chi_{\boldsymbol{h}}^{(m)}\mathrm{e}^{\mathrm{i}\boldsymbol{h}^{(m)}(\boldsymbol{r}-\boldsymbol{u}^{(m)}(\boldsymbol{r}))}\Omega_{\mathrm{wire}}^{(m)}(\boldsymbol{r}) . \quad (12.39)$$

Then the first term under the integral of Eq. (12.36) describes the reflection by the mean depth profile of electron density in the multilayer stack dynamically. The second term can be interpreted as a kinematical scattering of this dynamical wavefield by the perturbation potential. The latter includes the reflection by the grating and x-ray grazing incidence diffraction by the crystalline lattice modulated by the grating. The total wavefield can be written in the form

$$E(\boldsymbol{r}) = E_{\mathrm{XR}}(\boldsymbol{r}) + E_{\mathrm{GID}}(\boldsymbol{r}) \quad (12.40)$$

$$= \sum_{\boldsymbol{G}}\left\{\begin{array}{l}\left[T_{0,\boldsymbol{G}}(z) + R_{0,\boldsymbol{G}}(z)\right]\mathrm{e}^{\mathrm{i}(\boldsymbol{K}_{0\parallel}+\boldsymbol{G})\boldsymbol{r}_{\parallel}} + \\ \left[T_{\boldsymbol{h},\boldsymbol{G}}(z) + R_{\boldsymbol{h},\boldsymbol{G}}(z)\right]\mathrm{e}^{\mathrm{i}(\boldsymbol{K}_{0\parallel}+\boldsymbol{h}+\boldsymbol{G})\boldsymbol{r}_{\parallel}}\end{array}\right\} , \quad (12.41)$$

where $E_{\mathrm{XR}}(\boldsymbol{r})$ is the part of the wavefield to be studied by x-ray reflection and $E_{\mathrm{GID}}(\boldsymbol{r})$ the corresponding part studied by the x-ray grazing incidence diffraction. Respectively, the first parenthesis of Eq.(12.40) corresponds to the transmitted and reflected wave amplitudes of the grating rods in the vicinity of the specular rod, the second parenthesis describes those near the crystal truncation rod of GID. We only investigate by grazing-incidence diffraction the part $E_{\mathrm{GID}}(\boldsymbol{r})$, and there only the reflected-diffracted wavefield *above* the sample surface (since our detector position will be above the sample). Comparing equations 12.36 and 12.41 for this case, we are consequently interested in the calculation of

$$E_{\mathrm{GID}}(\boldsymbol{r}_{\|}, z > 0) = \sum_{\boldsymbol{G}} R_{\boldsymbol{h},\boldsymbol{G}}(z > 0)\mathrm{e}^{\mathrm{i}(\boldsymbol{K}_{0\|}+\boldsymbol{h}+\boldsymbol{G})\boldsymbol{r}_{\|}}$$

$$= \int \mathrm{d}^2\boldsymbol{K}_{\|} \frac{\mathrm{e}^{\mathrm{i}\boldsymbol{K}\boldsymbol{r}}}{8\pi^2\mathrm{i}K_z} \left\langle \boldsymbol{E}_K^{(-A)} \middle| \hat{\mathbf{V}}_{\boldsymbol{h}} \middle| \boldsymbol{E}_{K_0}^{(A)} \right\rangle . \quad (12.42)$$

The non-perturbed states $\left|E_K^{(-A)}\right\rangle$ and $\left|E_K^{(A)}\right\rangle$ of the *distorted* wavefield (distorted due to specular reflection by horizontal interfaces) have been derived in Sect. 11.3, Eq. (11.50). They consist of plane homogeneous waves in each layer with the transmitted and reflected wave amplitudes $\tilde{T}_{K_0}^{(m)}$, $\tilde{R}_{K_0}^{(m)}$, $\tilde{T}_K^{(m)}$, $\tilde{R}_K^{(m)}$ of the initial and final *non-perturbed* states (after specular reflection from the horizontal multilayer). In contrast to Eq. (11.50), they are marked by a *tilde* in this section in order to avoid confusion with transmitted and reflected wave amplitudes of the total wavefield $T_{\boldsymbol{h}\boldsymbol{G}}(z)$ and $R_{\boldsymbol{h},\boldsymbol{G}}(z)$ (after grazing-incidence diffraction).

We may describe the contributions of each layer to the diffracted wavefield by introducing the transition elements $\tau_{\boldsymbol{h}}^{(m)} \equiv \left\langle \boldsymbol{E}_K^{(-A)} \middle| \hat{\mathbf{V}}_{\boldsymbol{h}} \middle| \boldsymbol{E}_{K_0}^{A} \right\rangle$, which express the transition probability from one non-disturbed state into another state due to scattering by the disturbance potential of grazing incidence diffraction $\hat{\mathbf{v}}_{\boldsymbol{h}}^{(m)}$.

The transition elements describe the diffraction by both the crystalline layer lattice and its modulation caused by the grating pattern. Consequently each transition element $\tau_{\boldsymbol{h}}^{(m)}$ can again be decomposed into a sum over the contributions of all grating rods (including the crystal truncation rod). Putting Eq. (12.39) by means of equations (12.15-12.18) into Eq. (12.42) we obtain

$$\tau_{\boldsymbol{h}}^{(m)} = \sum_{\boldsymbol{H}} \tilde{\tau}_{\boldsymbol{h},\boldsymbol{H}}^{(m)}(Q_z)\delta(\boldsymbol{Q}_{\|} - \boldsymbol{h}_{\|} - \boldsymbol{H}) \quad (12.43)$$

with

$$\tilde{\tau}_{\boldsymbol{h},\boldsymbol{H}}^{(m)}(Q_z) = -4\pi^2K^2 \Big[\tilde{T}_1^{(m)}\tilde{F}_{\boldsymbol{h},\boldsymbol{H}}^{(m)}(q_{11}^{(m)})\tilde{T}_2^{(m)} + \tilde{R}_1^{(m)}\tilde{F}_{h,H}^{(m)}(q_{12}^{(m)})\tilde{T}_2^{(m)} +$$

$$+ \tilde{T}_1^{(m)}\tilde{F}_{\boldsymbol{h},\boldsymbol{H}}^{(m)}(q_{21}^{(m)})\tilde{R}_2^{(m)} + \tilde{R}_1^{(m)}\tilde{F}_{\boldsymbol{h},\boldsymbol{H}}^{(m)}(q_{22}^{(m)})\tilde{R}_2^{(m)}\Big] . \quad (12.44)$$

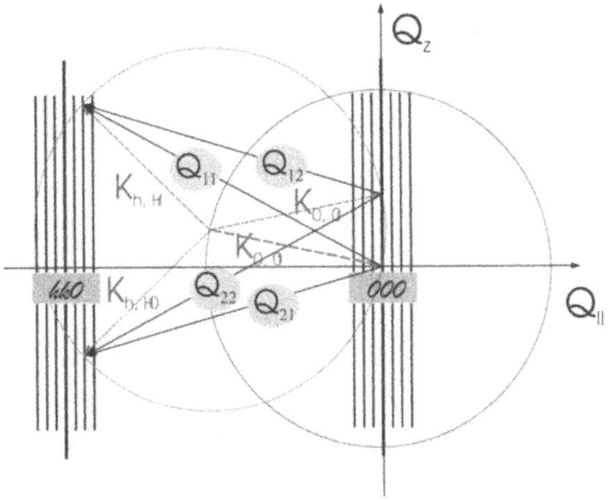

Fig. 12.5. Schematic illustration of the wave vectors and the scattering vectors of grazing-incidence diffraction by gratings treated within the DWBA.

Here the $q_{ij}^{(m)}$ are the z-components of the reduced scattering vectors $\boldsymbol{q}_{ij}^{(m)} = \boldsymbol{Q}_{ij}^{(m)} - \boldsymbol{h} - \boldsymbol{H}$, taken in the medium. The corresponding scattering vectors $\boldsymbol{Q}_{ij}^{(m)}$ are illustrated in Fig. 12.3 and discussed below. They are related to the transmitted and reflected wave vectors of the initial and final non-perturbed states $\left|E_1^{(-A)}\right\rangle$ and $\left|E_2^{(A)}\right\rangle$ (see Fig. 12.5). The diffraction by each grating rod considers four scattering processes, similar to the DWBA derived for diffuse scattering by planar multilayers discussed in Sect. 11.3. Their amplitudes are proportional to the grating form factors $\tilde{F}_{\boldsymbol{h},\boldsymbol{H}}^{(m)}$ and to the transmitted and specular reflected wave amplitudes of the initial and final distorted wavefields $\tilde{T}_1^{(m)}$, $\tilde{R}_1^{(m)}$ $\tilde{T}_2^{(m)}$, and $\tilde{R}_2^{(m)}$.

Within this approach the wave amplitudes of all grating rods are influenced by the specular reflection. All other possible inter-rod scattering processes (see Fig. 12.3) are neglected. In other words, the treatment of Eq. (12.44) includes *refraction* in the medium as well as extinction due to *multiple reflection* by the horizontal surface and interfaces. Multiple *diffraction* and any influence of the diffraction process on the incident distorted wavefield of specular reflection are neglected. The first term in Eq. (12.44) determines the primary scattering Q_{11}; the other three terms Q_{12}, Q_{21}, Q_{22} are of purely dynamical nature (secondary scattering processes).

The approximation works reasonably well except in a small region very close to the reciprocal lattice point of the host lattice, where the intensity is very large. In these regions, one would obtain more precise results by replacing the non-perturbed states Eq. (11.50) by those of the extended dynamical theory for planar multilayers, Eq. (11.68) [350], but keeping the formal treatment described here. Then each distorted wavefield contains eight plane waves, increasing the number of scattering processes in Eq. (12.44) from 4 to 64. Finally, by putting Eq. (12.43) into Eq. (12.44) we can derive the diffracted wave amplitude of a single grating rod at the position of the analyzer crystal, which is (with the exception of a phase term)

$$R_{\boldsymbol{h},\boldsymbol{H}}(z>0) = \frac{-\mathrm{i}E_0 \mathrm{e}^{\mathrm{i}K_z z}}{8\pi^2 K_z} \sum_m \tilde{\tau}_{\boldsymbol{H}}^{(m)}(Q_z)\,. \tag{12.45}$$

In buried gratings of material systems with small density modulation (e.g. the system InGaAsP),indexInGaAsP or generally for sufficiently large incident and exit angles α_i and α_f, the reflection amplitudes of the non-perturbed states, $\tilde{R}_{1/2}^{(m)}$, are below the sample surface very low and therefore negligible. Then only primary scattering (first term in Eq. (12.43)) plays a role in the diffraction pattern, and Eq. (12.45) can be simplified by

$$R_{\boldsymbol{h},\boldsymbol{H}}(z) = \frac{-\mathrm{i}E_0 K^2 \mathrm{e}^{\mathrm{i}K_z z}}{2K_z} t(\alpha_i)t(\alpha_f) \times \left[\sum_m F_{\boldsymbol{h},\boldsymbol{H}}^{(m)}\left(q_z^{(m)}\right) \mathrm{e}^{-\mathrm{i}\sum_{j=1}^{m-1} q_z^{(j)} \boldsymbol{r}_\parallel}\right] \tag{12.46}$$

with the Fresnel transmission coefficients of the mean surface $t(\alpha_i)$ and $t(\alpha_f)$ and the reduced scattering vectors in the medium

$$q_z^{(m)} = K^2 \left[\sqrt{\sin^2\alpha_i + \chi_0} + \sqrt{\sin^2\alpha_f + \chi_0}\right]\,. \tag{12.47}$$

This approximation corresponds to a similar one that was derived earlier for description of diffuse x-ray scattering by multilayers with rough interfaces in Sect. 11.3 (see there Fig. 11.16 and Eq. (11.59)). Notice again that in a real experiment the actual experimental conditions, the limited sample size, and the lateral and vertical size of the incident and exit slits have to be considered in the data analysis.

12.5 Distorted Wave-Born Approximation for X-Ray Diffraction

When we investigate the grating in conventional Bragg-diffraction geometries at large angles with respect to the sample surface, specular reflection drops off and related effects as described in Sect. 12.4 do not play a role in the measured maps. Then, in many relevant cases, the kinematical theory provides sufficient agreement with the experiment, assuming the grating thickness is much smaller than the extinction length. However, also in such cases, multiple diffraction, namely, so-called *Umweganregung* between the grating and the substrate or additional buffer layers, may create intense features deforming the kinematical diffraction pattern of the grating, also far from the dynamical substrate Bragg-peak. Such multiple diffraction effects may play an even an larger role if the Bragg-diffraction by the crystalline lattice of the grating is relatively weak compared to the grating diffraction in transmission. This takes place, for example, with the material system GaAs/GaAlAs at the nearly forbidden 002-Bragg-reflection. There the grating rods in the vicinity of 002 are very weak compared to the grating rods of transmission near 000. Subsequently, multiple diffraction between the dynamically diffracting substrate and the grating transmission rods may dominate the regular kinematical Bragg-diffraction by the grating in certain region of the reciprocal space (or of the rocking curve). For the case of a simple grating on a substrate,

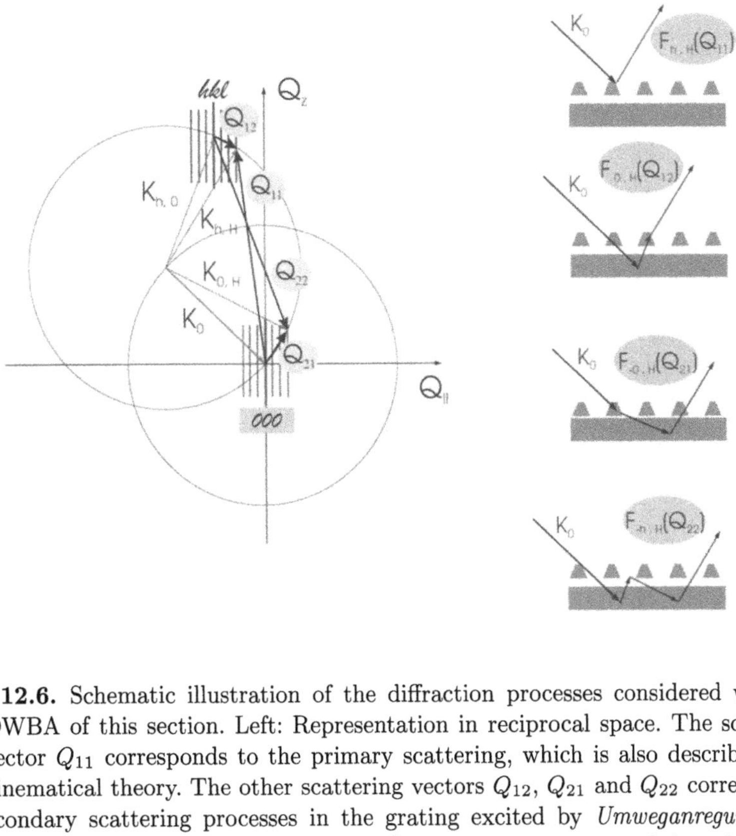

Fig. 12.6. Schematic illustration of the diffraction processes considered within the DWBA of this section. Left: Representation in reciprocal space. The scattering vector Q_{11} corresponds to the primary scattering, which is also described by the kinematical theory. The other scattering vectors Q_{12}, Q_{21} and Q_{22} correspond to secondary scattering processes in the grating excited by *Umweganregung* involving Bragg-diffraction in the substrate (the scattering vectors of the substrate are not drawn in the figur). Right: Representation in real space. The drawing on top illustrates primary scattering by the grating. The lower three drawings illustrate secondary scattering processes of *Umweganregung* between the grating and the substrate. For example, the third process describes *Umweganregung* between the grating diffracted wave in transmission and Bragg-diffraction in the substrate.

some possible multiple diffraction processes are illustrated in Fig. 12.6.

In this section we present one simple approximation based on the DWBA approach which allow us to describe some of the most frequent multiple scattering effects. It is an extension of the conventional semikinematical diffraction by planar multilayers described in Sect. 7.2.4 for the case of grating diffraction. The scattering potential will be divided such that the diffraction by the substrate and possible thick non-patterned layers are treated dynamically, but all patterned layers in the grating are described kinematically. Finally, the coupling between both wave fields is handled within DWBA. Thus the non-perturbed potential only includes the planar part of the multilayer $\hat{\mathbf{V}}_A = \sum_n \hat{\mathbf{v}}^{(n)}_{\text{planar}}(\boldsymbol{r})$ and the perturbation potential now sums over the patterned layers of the grating $\hat{\mathbf{V}}_B = \sum_m \hat{\mathbf{v}}^{(m)}_{\text{grating}}(\boldsymbol{r})$, with

$$\hat{\mathbf{v}}^{(m)}_{\text{grating}}(\boldsymbol{r}) = \left[\hat{\mathbf{v}}^{(m)}_0(\boldsymbol{r}) + \hat{\mathbf{v}}^{(m)}_{\boldsymbol{h}}(\boldsymbol{r})\right], \tag{12.48}$$

$$\hat{\mathbf{v}}^{(m)}_0(\boldsymbol{r}) = -K^2\chi^{(m)}_0(\boldsymbol{r}) \tag{12.49}$$

and $\hat{\mathbf{v}}^{(m)}_0(\boldsymbol{r})$ of Eq. 12.39.

Consequently, now the initial and final non-disturbed states $\left|E_1^{(-A)}\right\rangle$ and $\left|E_2^{(-A)}\right\rangle$ describe the disturbed wavefield after dynamical diffraction by the non-patterned part below the surface grating (e.g., a non-patterned multilayer including the substrate). Since the potential $\hat{\mathbf{V}}_A$ vanishes in the patterned region, the present non-disturbed states are in the grating (that is outside of the planar part of the sample) simply the sum of two homogeneous waves. Therefore, it is sufficient to know their wave amplitudes just at one position, which we take at the top of the grating, at $z = 0$:

$$\boldsymbol{E}_K^{(-A)}(\boldsymbol{r}_\parallel, z = 0) = \tilde{T}\mathrm{e}^{\mathrm{i}(\boldsymbol{K}_\parallel)\boldsymbol{r}_\parallel} + \tilde{R}\mathrm{e}^{\mathrm{i}(\boldsymbol{K}_\parallel + \boldsymbol{h})\boldsymbol{r}_\parallel}\,. \tag{12.50}$$

Equation (12.50) contains the incident wave amplitude $\tilde{T} = E_0$ and the Bragg-diffracted wave amplitude of the planar multilayer $\tilde{R}$. Treating now the diffraction by the grating, we put the equations (12.48) and (12.50) in the general DWBA-expression for gratings, Eq. (12.36). Since we may neglect all effects of specular reflection, the number of considered waves in the wavefield of the grating is reduced compared with the equations (12.23) and (12.40) to

$$\begin{aligned} E(\boldsymbol{r}) &= E_{\mathrm{XR}}(\boldsymbol{r}) + E_{\mathrm{XRD}}(\boldsymbol{r}) \\ &= \sum_{\boldsymbol{G}} (T_{0,\boldsymbol{G}}(z)\mathrm{e}^{\mathrm{i}(\boldsymbol{K}_{0\parallel}+\boldsymbol{G})\boldsymbol{r}_\parallel} + R_{\boldsymbol{h},\boldsymbol{G}}(z)\mathrm{e}^{\mathrm{i}(\boldsymbol{K}_{0\parallel}+\boldsymbol{h}+\boldsymbol{G})\boldsymbol{r}_\parallel}). \end{aligned} \tag{12.51}$$

where $E_{\mathrm{TR}}(\boldsymbol{r})$ is the part of the wavefield to be studied by small-angle scattering (x-ray transmission) and $E_{\mathrm{XRD}}(\boldsymbol{r})$ the part we are interested in by

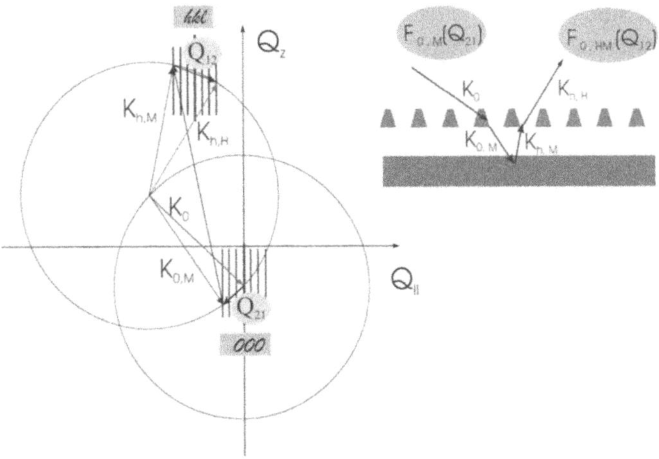

Fig. 12.7. Schematic illustration of one possible process of *Umweganregung*, examined by the second-order DWBA, described in the text.

x-ray Bragg-diffraction. Respectively, the first parenthesis of Eq. (12.40) corresponds to the transmitted wave amplitudes of the grating rods in the vicinity of the origin of reciprocal space; the second parenthesis describes the reflected waves near the crystal truncation rod of the reciprocal lattice point hkl.

Repeating the procedure of the previous section (Eqs. (12.42)–(12.46), we calculate the diffracted wavefield above the sample, $E_{\mathrm{XRD}}(\boldsymbol{r}_{\|}, z > 0)$, obtaining for diffracted wave amplitude of a single grating rod formally

$$R_{\boldsymbol{h},\boldsymbol{H}}(z{>}0) = \frac{-\mathrm{i}E_0 K^2 \mathrm{e}^{\mathrm{i}K_{\boldsymbol{H}z}z}}{2K_z} \Big[\tilde{T}_0 F_{\boldsymbol{h},\boldsymbol{H}}(q_{11}) \tilde{T}_{\boldsymbol{H}} + \tilde{R}_0 F_{0,\boldsymbol{H}}(q_{12}) \tilde{T}_{\boldsymbol{H}} + $$

$$+ \tilde{T}_0 F_{0,\boldsymbol{H}}(q_{21}) \tilde{R}_{\boldsymbol{H}} + \tilde{R}_0 F_{-\boldsymbol{h},\boldsymbol{H}}(q_{22}) \tilde{R}_{\boldsymbol{H}} \Big] \, . \tag{12.52}$$

Here the $q_{ij}^{(m)}$ are again the z-components of the reduced scattering vectors $\boldsymbol{q}_{ij}^{(m)} = \boldsymbol{Q}_{ij}^{(m)} - \boldsymbol{h} - \boldsymbol{H}$. The corresponding scattering vectors $\boldsymbol{Q}_{ij}^{(m)}$ are illustrated in Fig. 12.6. Please note that in Eq. (12.52) the scattering processes correspond to structure amplitudes of different reciprocal lattice points 0, $\boldsymbol{h}$ and $-\boldsymbol{h}$, in contrast to the DWBA for GID in the previous section. The first term with reduced scattering vector q_{11} describes the Bragg-diffraction by the grating rod fine structure of the reciprocal lattice point $\boldsymbol{h}$ as the primary scattering process. The other terms are caused by *Umweganregung* involving dynamical Bragg-diffraction in the planar part of the sample including the substrate. The terms with reduced scattering vectors of mixed indices $\boldsymbol{q}_{ij}$ in Eq. (12.52) or, respectively, $\boldsymbol{Q}_{ij}$ in Fig. 12.6 appear to be due to forward diffraction in transmission by the grating (see Fig. 12.6). The last remaining term with indices q_{22} involves Bragg-diffraction from the back of the grating, corresponding to the grating rods around $-\boldsymbol{h}$. In Fig. 12.6 this is represented by the downward direction of the scattering vector $\boldsymbol{Q}_{22}$.

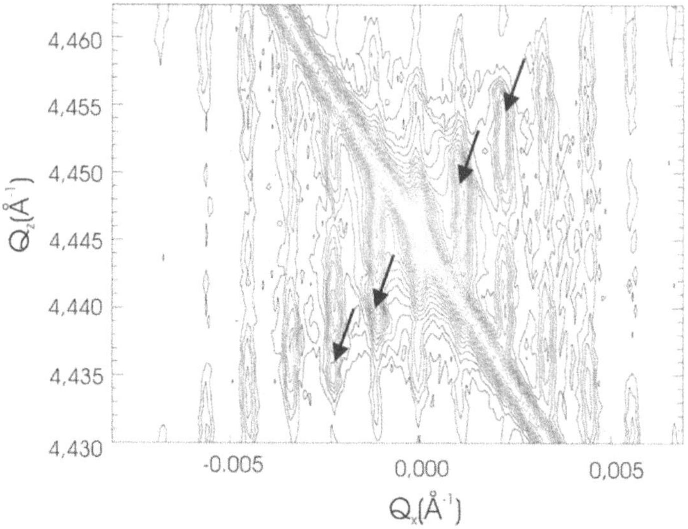

Fig. 12.8. Reciprocal-space map of a trapezoidal surface grating in GaAs, 004 diffraction. The map is measured with a conventional x-ray tube. The intensity distribution in the map shows the grating rod pattern. The trapezoidal shape concentrates the intensity of the grating rods in form of a cross pattern. Additionally, we observe the analyzer streak as an artifact of the instrumental resolution function and dynamical peaks due to *Umweganregung* between the grating and the substrate. The latter are marked by arrows.

By the second iteration of the DWBA, the scattering processeses of the kind drawn in Fig. 12.7 are also taken into account. Fortunately, these effects are important only in regions where the primary scattering is very weak. This could account, e.g., for nearly forbidden reflections, where the main contribution comes from forward diffraction in transmission in the grating. This would also be the case if we took an amorphous grating pattern on a diffracting multilayer or substrate.

An example showing strong dynamical features of *Umweganregung* is shown in Fig. 12.8. Some intense dynamical features of *Umweganregung* involving the grating and the substrate are marked by arrows. Measurable *Umweganregung* peaks are mostly situated along lines crossing the intense Bragg-peaks (e.g., the substrate Bragg peak) or multilayer peaks. These lines are perpendicular to the incident and the diffracted wave vector. In other words, they appear there, where in a triple-crystal diffractometer experiment analyzer or monochromator streaks would also be expected (see Fig. 3.7).

Actually, the scattering arrangement *grating - substrate - grating* behaves like an internal coherent triple-crystal diffractometer.

12.6 Determination of the Lateral Superstructure

12.6.1 Grating Period and the Etching Depth

Lateral structures on a mesoscopic scale with a lateral periodicity and a depth between tens of nanometers and a few microns create grating truncation rods equidistantly arranged around the main truncation rod, with spacings of about $2\pi \times 10^{-1}$ nm^{-1} to $2\pi \times 10^{-3}$ nm^{-1}, as illustrated by the schematic drawing in Fig. 12.4. An incoming plane monochromatic incident wave excites simultaneously a fan of coherently diffracted plane waves with measurable intensity, determined by the intersection of the Ewald sphere with the grating rods.

The first X-ray diffraction studies of gratings employed *double-crystal diffractometers* (see, e.g., 2.5). Without slits the detector integrates the intensities of different grating truncation rods.

A conventional *triple-axis diffractometer* allows two-dimensionally resolved measurements separating contributions of different grating truncation rods to the detected intensity (see, e.g., Fig. 12.8). A transverse Q_x scan crosses the grating rods at points with equal Q_z positions (see Fig. 12.4), giving rise to transverse satellites around the central peak of the crystal truncation rod. Such transverse satellites are visible in Fig. 12.9), where we plot a Q_z scan in grazing-incidence geometry. The grating period D_G can be easily determined from the spacing Δq_x of the neighboring transverse satellites.

In a geometry of symmetrical Bragg-diffraction, a transverse scan is approximately carried out by a simple ω-scan (see Chap. 3 and Fig. 12.4). For this geometry the grating period can be determined from the angular spacing of the neighboring transverse satellites by the simple relation

$$D_G = \frac{2\pi}{\Delta q_x} \overset{\text{symm. reflection}}{=} \frac{2\pi}{Q_z \Delta\omega_{\text{rod}}} = \frac{\lambda}{2\Delta\omega_{\text{rod}} \sin\Theta} . \tag{12.53}$$

A Q_z-scan along the central truncation rod ($\omega/2\Theta$ scan for a symmetrical reflection) contains information about the compositional depth profile and the related strain profile averaged over the grating period. In the case of a simple surface grating, thickness fringes appear around the substrate Bragg maximum, similarly to the case of a planar mono-crystalline layer. The mean grating depth can be estimated from the fringe spacing, where for symmetrical reflection the simple relation

$$t_G = \frac{2\pi}{\Delta q_z} \overset{\text{symm. reflection}}{=} \frac{\lambda}{2\Delta\omega_{\text{fringes}} \cos\Theta} \tag{12.54}$$

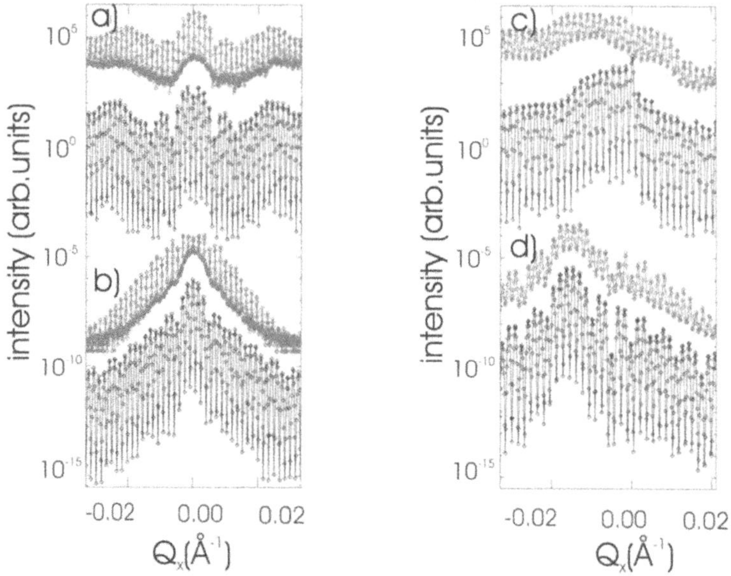

Fig. 12.9. GID by a strained layer GaInAs/InP surface grating. Measured and simulated Q_x-scans: left, the shape-sensitive $2\bar{2}0$-reflection proves the trapezoidal shape; right, the strain-sensitive 220-reflection, indicating depth-dependent elastic strain relaxation toward the sample surface. The incidence angle is $\alpha_i = 0.27°$ in (a) and (c) and 0.19° in (b) and (d).

holds for the angular space between neighboring thickness fringes.

A large number of sharp transversal grating satellites is a mark of quality for a good lateral grating periodicity. The required angular resolution of the experimental set-up depends on the grating periodicity, the grating depth, and the choice of the reflection in reciprocal space.

12.6.2 Reciprocal-Space Mapping

The coherently scattered intensity is concentrated along the grating rods. Thus, it would be sufficient to measure the intensity profile along each grating rod by a single Q_z-scan. However, this is usually difficult, since the exact orientation of the grating rods with respect to the crystallographic directions might be affected by slight misalignment of the wire elongation and by miscut of the substrate surface, leading to experimental errors if not taken into account correctly.

By coupling transverse and Q_z-scans, a complete two-dimensional map around a reciprocal lattice point can be measured (see Fig. 12.8). Reciprocal space maps are much more suitable as single scans, as long as the orientation

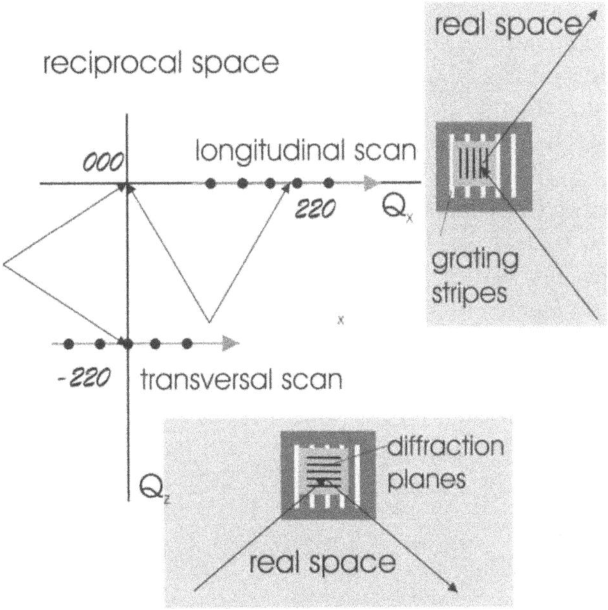

Fig. 12.10. X-ray grazing incidence diffraction by a surface grating. Schematic illustration of the projected in-plane scattering geometry in real and reciprocal space, showing the intersection points of the grating rods with the Q_x, Q_y plane.

of the superstructure in not exactly known (and that is mostly the case). The map ensures that the intensity pattern is not distorted by misalignment of the grating. The map images the grating rod structure and gives evidence for the quality of the lateral periodicity. It allows us to get detailed information about the internal grating structure from the coherent intensity distribution along the grating rods and the diffuse scattering between the grating rods. However, the measuring time increases with the number of measuring points and one has to find a compromise for the applied instrumental resolution. The measuring time can essentially be reduced by abandoning analyzer crystals and making use of one- and two-dimensional position-sensitive detectors.

X-ray grazing-incidence diffraction measures the diffraction patterns of Bragg reflections with very small vertical components of the reciprocal lattice vector. For [001]-oriented cubic materials these are the $hk0$-reflections.

The angles of the incident and exit wave vectors with respect to the surface α_i and α_f and the angles of the in-plane wave vector components with respect to the diffraction plane, θ_i and θ_f, are linked through [33]

$$\sin\left(\theta_{i/f}\right) = \frac{\pm\left(K_{0z}^2 - K_{Hz}^2\right) - \left|\boldsymbol{h}_\parallel + \boldsymbol{H}\right|^2}{-2\left|\boldsymbol{h}_\parallel + \boldsymbol{H}\right|\sqrt{K^2 - K_{i/fz}^2}}. \quad (12.55)$$

The grating rod patterns can be detected by applying *non-coplanar* triple-crystal diffractometry. The schematic set-up is shown in Fig. 2.6 in Chapter 2. An analyzer crystal is mounted in front of the detector slits in order to obtain high resolution, mainly in the Q_x/Q_y -plane, where a moderate resolution in Q_z is achieved by the slits of entrance and exit or by use of a position sensitive detector.

Considering that a non-coplanar scattering geometry is described by four angles $(\theta_i, \alpha_i, \theta_f, \alpha_f)$ (see Fig. 3.9) we are free (in contrast to coplanar scattering geometries) to tune the information depth by different sets of (α_i, α_f) keeping $Q_z = K(\sin\alpha_i + \sin\alpha_f)$ constant.

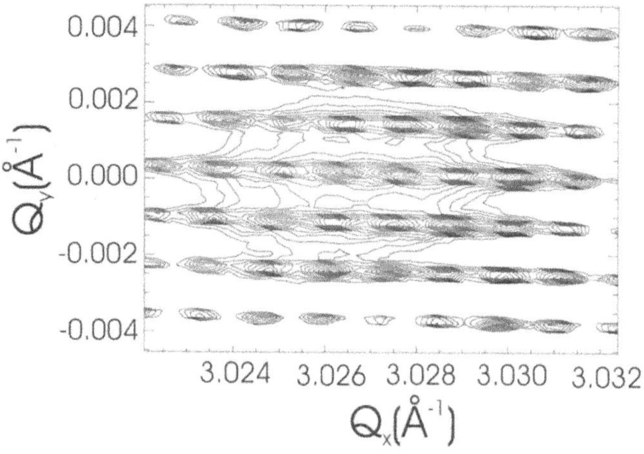

Fig. 12.11. Measured Q_xQ_y-map of a grating array of etched Si/Ge quantum dots (e.g. a two dimensionally patterned multilayer surface grating). The map shows the intersection points of the grating rods, which form a two-dimensional pattern in reciprocal space. The direction of two crystallographic orientations and the off-orientation of patterning are indicated by arrows.

For many applications (e.g., strain analysis (see Chap. 13) or self-organized morphological patterning in strained superlattices on vicinal surfaces (see Chap. 14) it is very useful to compare complementary in-plane reflections with reciprocal lattice vectors either *parallel* or *perpendicular* to the grating grooves. Consequently the intersection points of the grating rods are arranged on lines either parallel or perpendicular to the reciprocal lattice vector. In the schema of Fig. 12.10 the grating rods of the $2\bar{2}0$ reflection are crossed by a *transversal* scan (perpendicular to the reciprocal lattice vector). In the complementary 220-reflection the diffraction planes are parallel with respect to the grooves, and the GTRs are crossed by a *longitudinal* scan (in the direction of the reciprocal lattice vector).

By coupling transversal and longitudinal scans, a complete Q_x, Q_y-mapping can be measured for a given α_i (i.e. fixed incident penetration depth) and a given Q_z. As an example, Fig. 12.11 plots a Q_x, Q_y-map measured from a two-dimensional grating formed by an etched Si/Ge quantum-dot array . Accordingly we observe the well resolved two-dimensional pattern of grating rod intersection points. Changing Q_z, the set-up permits *three-dimensionally* resolved measurements for fixed α_i. However, three-dimensional maps are very time consuming. Mostly it is sufficient to record Q_x, Q_y–maps (so-called in-plane maps, which are perpendicular to the surface normal) and Q_x, Q_z–maps, which are perpendicular to the direction of surface patterning. See for illustration the examples drawn in Figs. 12.8 , 12.13, 12.11, and 12.15. Supposing a proper sample alignment, even well-defined single scans along the grating rods can give valuable structure information (see, e.g., Fig. 13.6 in Chap. 13).

12.6.3 Orientation of the Grating Pattern

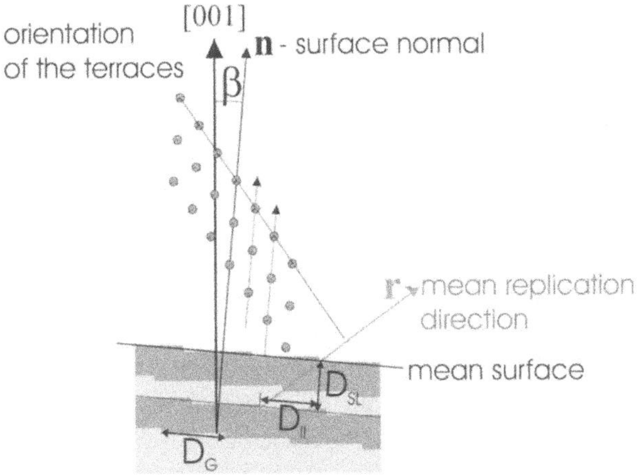

Fig. 12.12. Illustration of an off-orientated one-dimensional lateral grating in a common representation of both real and reciprocal space. The grating is formed by regular interface steps in a vertical superlattice. We assume perfect lateral periodicity of the interface steps. The lateral period of interface steps defines the distance of grating rods in reciprocal space. The grating rods are oriented in direction of the mean surface normal, which is tilted with respect to the crystallographic direction of the facets by the miscut angle β. The vertical periodicity of the superlattice creates a pattern of superlattice-satellites along all grating truncation rods. The satellites of these rods form branches, which are inclined toward the surface normal in dependence on the replication direction of the interface step pattern.

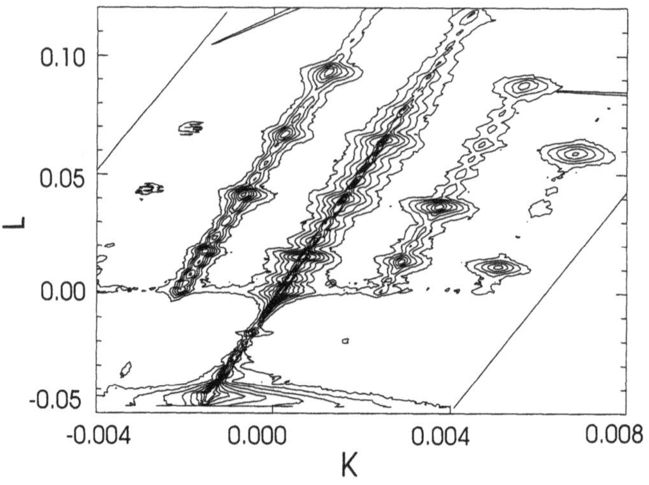

Fig. 12.13. Q_xQ_z–map of a grating formed by a superlattice with stair-like interfaces on an off-orientation substrate (220-reflection measured by grazing incidence diffraction). The grating truncation rods are inclined with respect to the crystallographic orientation of the surface facets (for details see section 14.3)

Let us consider a one-dimensional array of wires and recollect the three-dimensional character of the diffraction pattern. The fine structure of reciprocal lattice points is formed by the grating rod patterns. Their orientation in reciprocal space is determined by the orientation, first, of the wires and, second, of the averaged grating surface with respect to the crystallographic frame of the substrate. In our description the reference frame for our formulas is determined by the mesoscopic superstructure: The normal of the mean surface shows in the z-direction, the grating stripes go along the y-axis, the x-axis marks the direction of the grating periodicity. Consequently, all truncation rods are aligned along Q_z, which is perpendicular to the mean macroscopic sample surface. Samples discussed here are mostly based on intentionally [001]-oriented substrates. A slight miscut of the substrate and a related off-orientation of the growth direction of subsequent epitaxial layers lead to an inclination of the rods with respect to the crystallographic [001] direction toward the direction of interface steps caused by the off-orientation (see for illustration the schema of Fig. 12.12). As an experimental example, Fig. 12.13 shows a measured map of a strained layer superlattice grown on a miscut substrate. The grating is formed by remarkably periodic step-bunching at the multilayer interfaces, a process described in detail in Chapter 14. Here we concentrate only on the coherent part, assuming a perfect staircase-like interface patterning. All grating rods are inclined by the miscut angle with respect to the crystallographic frame. It is important to notice that when performing only single Q_z-scans, the inclination must be taken into account.

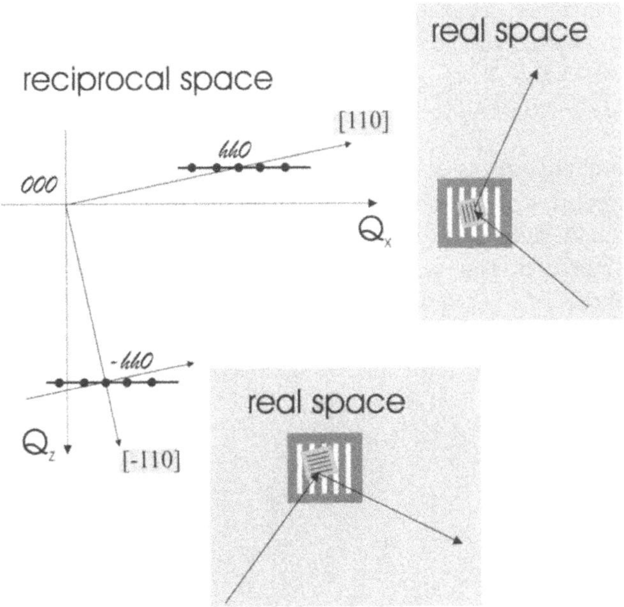

Fig. 12.14. Schematic illustration of wire misalignment in real and reciprocal space.

Figure 12.14 illustrates the influence of a slight *in-plane* wire misalignment with respect to the crystallographic orientation. The intersection points of the grating rods with the Q_xQ_y plane are aligned in the direction of patterning which is perpendicular to the wires. A misorientation of the wires with respect to the crystallographic direction creates a non-zero angle between the longitudinal or transversal in-plane line scans with respect to the lines connecting the intersection points of the grating rods. On the one hand, this effect deforms the recorded diffraction pattern, probably leading to misinterpretations if not taken into account. On the other hand, by measuring the angle between the line connecting the intersection points (often called grating satellite reflections) and the crystallographic directions, the inclination of the wires can be determined with high precision. In Fig. 12.15 we plot a Q_xQ_y-map of a multilayer surface grating, from which the wire misalignment with respect to the [110]-direction could be determined with high precision $\gamma = 1.62° \pm 0.02°$.

12.6.4 Grating Shape

Let us consider a pattern etched directly into a substrate. It generates a simple surface grating (the patterned part consists of the same material as the substrate). In such a case grating-induced lattice strain is expected to be negligible. Then the grating form factor depends only on the z-dependent Fourier components of the shape function

$$F_{\boldsymbol{h},\boldsymbol{H}}(Q_z) = \chi_{\boldsymbol{h}} \int \mathrm{d}z \; \Omega_{\boldsymbol{H}}(z) \mathrm{e}^{-\mathrm{i}q_z z} . \tag{12.56}$$

Most etching processes act strongly anisotropic for different crystallographic directions. As a result, quite precise geometrical forms (grooves and mesas of square shape, triangular or trapezoidal shape) can be achieved. The wires are limited by their side walls $L_-(z) + lD < x < L_+(z) + lD$. Here l is an integer and $L_-(z)$ and $L_+(z)$ are the positions of the left and right side walls at height z, which can be expressed by the top path b and by the slope functions $p_\pm(z) = \pm dx/dz$ of the left and right side walls (see Fig. 12.18):

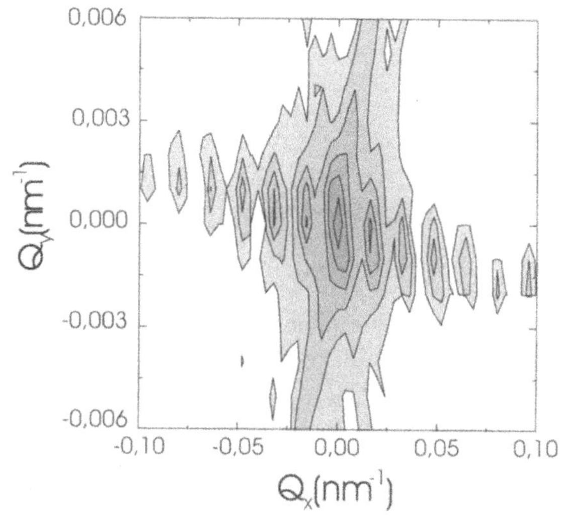

Fig. 12.15. Measured Q_x, Q_y-map of a GaAs/InGaAs/GaAs quantum well surface grating. The direction of one crystallographic orientation and the off-orientation of patterning are indicated by arrows.

$$L_+(z) = -\left(\frac{b}{2} + \int_0^z \mathrm{d}z' p_+(z')\right) , \quad L_-(z) = \frac{b}{2} + \int_0^z \mathrm{d}z' p_-(z') . \tag{12.57}$$

The Fourier components $\Omega_{\boldsymbol{H}}(z)$ can be written in the form

$$\Omega_{\boldsymbol{H}}(z) = \frac{1}{D_{\boldsymbol{G}}} \int_{L_+(z)}^{L_-(z)} dx \; \mathrm{e}^{-\mathrm{i}Hx} . \tag{12.58}$$

Introducing the reduced vector components

$$q^+ = q_z + Hp_+ \,, \; q^- = q_z - Hp_- \,, \tag{12.59}$$

we obtain

$$F_{\boldsymbol{h},\boldsymbol{H}}(Q_z) = \frac{\mathrm{i}\chi_{\boldsymbol{h}}}{HD_G}\left(\mathrm{e}^{-\mathrm{i}Hb/2}\int_0^{t_{sg}} \mathrm{e}^{-\mathrm{i}\int_0^z q^+(z')\mathrm{d}z'}\mathrm{d}z - \right.$$

$$\left. - \,\mathrm{e}^{\mathrm{i}Hb/2}\int_0^{t_{sg}} \mathrm{e}^{-\mathrm{i}\int_0^z q^-(z')\mathrm{d}z'}\mathrm{d}z\right) . \tag{12.60}$$

For a grating of the depth t_g with linear slopes ($p_\pm$ = const, trapezoidal shape) Eq.(12.60) simplifies to

$$F_{\boldsymbol{h},\boldsymbol{H}}(Q_z) = \frac{\mathrm{i}\chi_{\boldsymbol{h}}}{HD_G}\left(\mathrm{e}^{-\mathrm{i}Hb/2}\frac{\mathrm{e}^{-\mathrm{i}q^+t}-1}{-\mathrm{i}q^+} - \mathrm{e}^{\mathrm{i}Hb/2}\frac{\mathrm{e}^{\mathrm{i}q^-t}-1}{-\mathrm{i}q^-}\right) \tag{12.61}$$

$$\equiv \frac{\mathrm{i}}{HD_G}\left(\mathrm{e}^{-\mathrm{i}Hb/2}F^+_{\boldsymbol{h},\boldsymbol{H}} - \mathrm{e}^{\mathrm{i}Hb/2}F^-_{\boldsymbol{h},\boldsymbol{H}}\right) . \tag{12.62}$$

In equations (12.60) and (12.61) we express the grating rod structure amplitude as the difference of two amplitudes corresponding to the reduced scattering vectors q^+ and q^-, respectively. For triangular shape (the top path $b = 0$) those amplitudes are proportional to the difference of the grating form factors $F^+_{\boldsymbol{h},\boldsymbol{H}}$ and $F^-_{\boldsymbol{h},\boldsymbol{H}}$, corresponding to the scattering vectors q^+ and q^-, respectively. For all grating rods of rectangular shape the condition $q^+ = q^- = q_z$ holds and Eq. (12.61) reduces to the simple product [363, 374]

$$F_{\boldsymbol{h},\boldsymbol{H}}(Q_z) = \frac{\chi_h}{D_G}\,\mathrm{e}^{\mathrm{i}q_z t_{sg}/2}\,\frac{\sin(q_z t_{sg}/2)}{q_z/2}\,\frac{\sin(Hb/2)}{H/2} . \tag{12.63}$$

Taking the inclined shape of a parallelogram, the slope functions behave as $p_+ = -p_-$ and consequently we find $q^+ = q^- \neq q_z$ for all non-zero-order grating rods. Thus we can formally apply (12.63) by replacing q_z with q^+.

In the simple example of a trapeze, the difference between the reduced scattering vectors q^+ and q^- increases linearly with the order of the grating rod, proportionally to the side wall slopes p_+ and p_-. In conventional XRD this behavior gives rise to a progressive splitting of the main maxima on the grating rods, leading to the characteristic cross pattern of the scattered intensity of a trapezoidal surface grating (see Fig. 12.16 and [124]). The opening angle γ is a direct measure for the angle between the side walls:

$$\gamma = \beta_l + \beta_r, \; p_{l,r} = \tan\beta_{l,r} \,. \tag{12.64}$$

For asymmetrically shaped trapezoidal gratings the two crossing branches have different inclination angles $\beta_l \neq \beta_r$ (see Fig. 12.16 a,b).

The experimental diffraction map of a trapezoidal grating contains, in addition to the characteristic cross pattern, analyzer or monochromator streaks as artifacts of the instrumental resolution function and multiple scattering peaks. The latter are generated by multiple diffraction of the kind illustrated

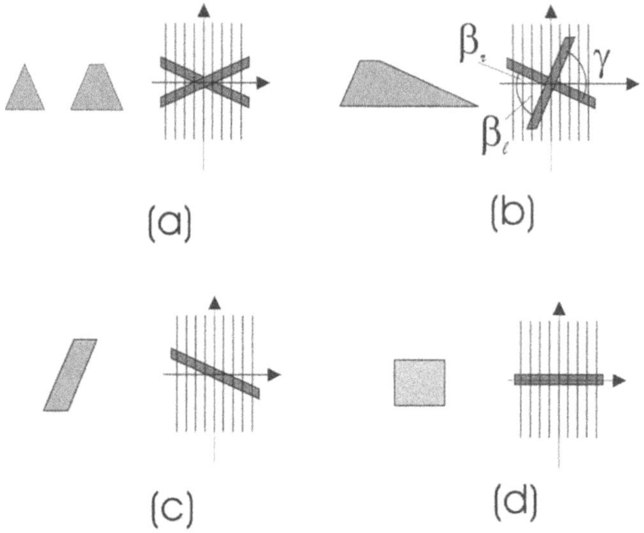

Fig. 12.16. Schematic illustration of the intensity peak positions for a simple surface grating as a function of the grating shape. The intensity patterns of simple surface gratings are exclusively characterized by the Fourier components of the shape function. The trapezoidal shape, for example, creates a typical cross pattern of enhanced intensity in the maps, which can be understood as a splitting of the main peaks we observe in the grating rods of a rectangular shape.

in Fig. 12.6. They are localized along a line with a direction perpendicular to the incident or diffracted wave vectors.

The shape of the non-strained grating influences the intensity patterns of all reciprocal lattice points in the same way. The measured intensity patterns may be modified by the scattering geometry. However, diffraction under conditions of grazing incidence gives rise to some particularities in the diffraction maps, as shown in Fig. 12.17 for the two cases of rectangular and trapezoidal grating shapes. Firstly, only the upper part of the cross patterns is accessible in this scattering geometry. Thus, taking our trapezoidal example, one cannot observe the complete cross pattern of trapezoidal gratings, but only the upper part. Secondly, the grating rod profiles show also all characteristic features of a GID-crystal truncation rod of planar thin layers, such as so-called Yoneda-like wings. The Yoneda-like wings of a grating do not occur at the critical angle $\theta_{\mathrm{crit}} = 2\sqrt{\delta}$ as usually for non-patterned surfaces but at a smaller angle θ_{SG} which corresponds to the mean density, averaged over the surface grating. Above the Yoneda-like wings occur thickness fringes related to the etching depth. The grating shape can be determined by the phase shift of the thickness fringes along the rod. For non-rectangular grating shapes a characteristic phase shift of the thickness fringes can be observed, which de-

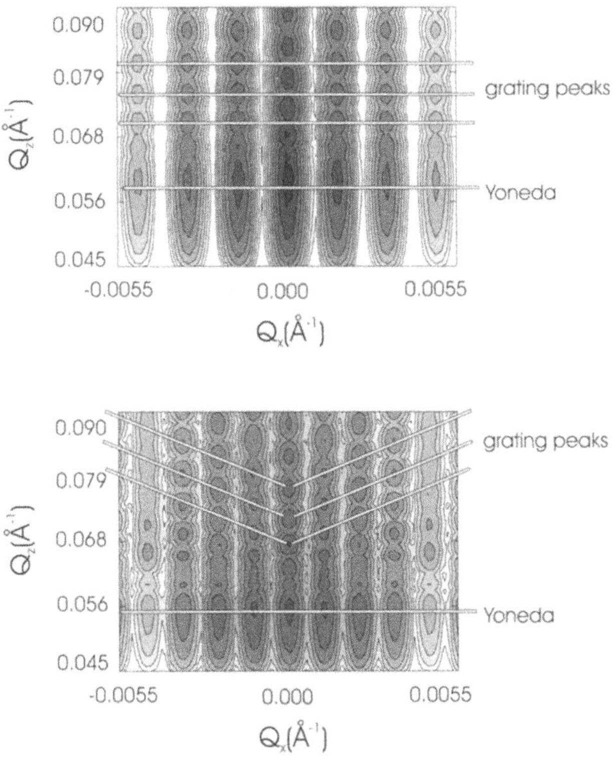

Fig. 12.17. Calculated GID reciprocal space maps of simple surface gratings (220-reflection, in-plane resolution $\Delta q_x = 0.002$ nm^{-1}): Top: rectangular shape, $D =$ 390 nm, top path 250 nm, and grating depth 100 nm. Bottom: trapezoidal shape, $D = 550$ nm, top path 135 nm, groove path 85 nm, and grating depth 100 nm. The intensity is plotted in arbitrary units on a logarithmic scale, with a dynamic range of 10^0 to 10^6 and ten contour levels. The thin lines mark the alignment of grating peaks and the Yoneda wings of the truncation rods.

pends on the grating diffraction order and which is related to the wire slopes. For better visibility the shift of thickness fringes Fig. 12.17 is marked by thin lines. For the rectangular shape the fringes of all grating rods are on lines parallel to the sample surface. For the trapezoidal shape the fringes of neighboring grating rods are shifted.

Depending on the slope of the wire side walls the mean refraction in non-rectangular gratings is strongly depth dependent, influencing the truncation rod profile near the Yoneda-like wings. Therefore it might be necessary to consider the depth dependence of the refractive index in the numerical procedure. Depth-dependent refraction influences mainly the position and form of the Yoneda-like wings between the critical angle of total reflection for the top of the surface grating and the one for the substrate.

Fitting the Q_z-shift of the thickness fringes and the relative intensity ratios between the different grating rods and the averaged slope of the trapeze, the top path and the groove can be estimated. This can be seen in Fig. 13.6 of Chap. 13.

12.7 Superlattice Surface Gratings

In electronic devices, the single strained layer described in the previous section is often replaced by a periodic multilayer, and then we obtain a superlattice surface grating.

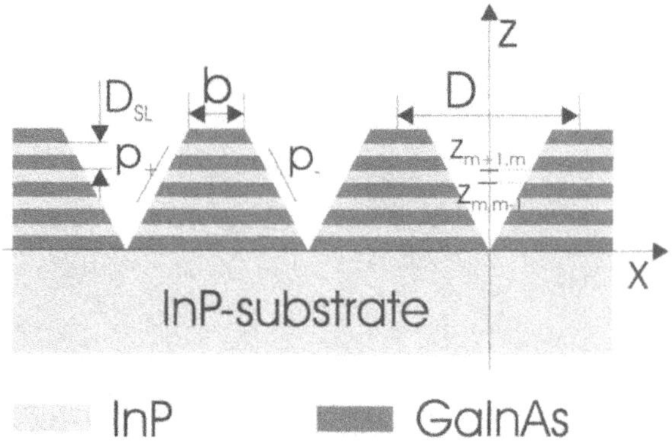

Fig. 12.18. Schematic representation of the sample set-up of a trapezoidal GaInAs multilayer grating on a InP-substrate.

For purpose of calculation the contributions of all layers have to be added together in order to determine the structure amplitude of the layer stack: As we do in the treatment of a single layer, we can write each layer structure amplitude as a sum of two terms, with different reduced scattering vectors $q^{+(m)}$ and $q^{-(m)}$. These amplitudes involve the slope of the layer, its lattice vectors and its displacement fields, if we suppose identical slopes of the side walls for all layers of the same layer type (see Fig. 12.18). Neglecting for the time being any strain relaxation (the strain analysis is described in detail in Chap. 13), the structure amplitude of a one-dimensional superlattice surface grating is [40]

$$F_{\boldsymbol{h}H}^{\mathrm{ML}}(Q_z) = \frac{\mathrm{i}}{HD_G}\left[\mathrm{e}^{-\mathrm{i}Hb_0/2}F_{\boldsymbol{h}H}^{+\mathrm{ML}}(q_z^+) - \mathrm{e}^{\mathrm{i}Hb_0/2}F_{\boldsymbol{h}H}^{-\mathrm{ML}}(q_z^-)\right] , \qquad (12.65)$$

where the expressions $F_{\boldsymbol{H}}^{\pm \mathrm{ML}}$ are formally identical to the structure amplitude of a planar superlattice.

$$F_{\boldsymbol{h}H}^{\pm \mathrm{ML}}(q_z^{\pm}) = f^{SL\pm}\left(q_z^{\pm}\right) \frac{\sin\left(q_z^{\pm} M_{\mathrm{ML}} D_{\mathrm{ML}}/2\right)}{\sin\left(q_z^{\pm} D_{\mathrm{ML}}/2\right)} \mathrm{e}^{-\mathrm{i}q_z^{\pm} M_{\mathrm{ML}} D_{\mathrm{ML}}/2} \tag{12.66}$$

with the form factor of one multilayer period $f^{SL\pm}(q_z^{\pm})$. This makes the calculation very convenient, since the conventional procedures and experience for the diffraction of planar multilayers can be applied. Note, however, that the reduced scattering vectors $q_z^{\pm}$ contain, additionally the information of the grating shape (via $p^{\pm}$) and the state of lattice strain relaxation (via Δh_x and Δh_z, discussed in Chap. 13.

12.8 Shape and the Morphological Set-Up of a Multilayer Grating

In a planar superlattice the crystal truncation rod is characterized by the substrate Bragg-peak and vertical superlattice-satellite reflections. After lateral patterning each satellite reflection becomes structured in a similar way as the Bragg-reflection of simple surface gratings. As a result each satellite is surrounded by a grating rod pattern whose profile is influenced by the grating shape. The effects are quite similar to those of a simple layer grating and are illustrated in Fig. 12.16.

In the case of a rectangular grating shape all grating rods contain vertically equidistant superlattice satellites forming horizontal branches in the map. For a parallelogram shape the satellite branches are inclined with respect to the Q_x-direction according to the replication direction of the grating. For a symmetrical trapezoidal shape each vertical satellite splits into two branches, forming a cross pattern similar to that in the diffraction patterns of simple trapezoidal gratings [37] (see Fig. 12.19). For trapezoids with large slopes, the cross patterns of different satellites may overlap. If the shape is an asymmetrical trapezoid, the diffraction pattern is asymmetric too In case of a stair-like pattern the off-orientation angle of the interfaces (miscut) β and the mean replication direction of the interface profile with the is the lateral shift of the stair-like pattern $\mathbf{r}$ $D_{\|}$ during the growth of one superlattice period are of interest, also the lateral terrace length between the macrosteps and their height defining the lateral grating period, (see fig. 12.12). For perfect lateral periodicity, the whole diffracted intensity would be concentrated along coherent grating truncation rods. All grating rods are *perpendicular* to the *averaged* surface. They are consequently *inclined* with respect to the *crystallographic* orientation by an angle equal to that of the off-orientation. As in the case of a planar superlattice, superlattice satellites modulate the grating rods. The satellites of the same order form branches, which are inclined with respect to the sample surface, similarly to a parallelogram-shaped grating.

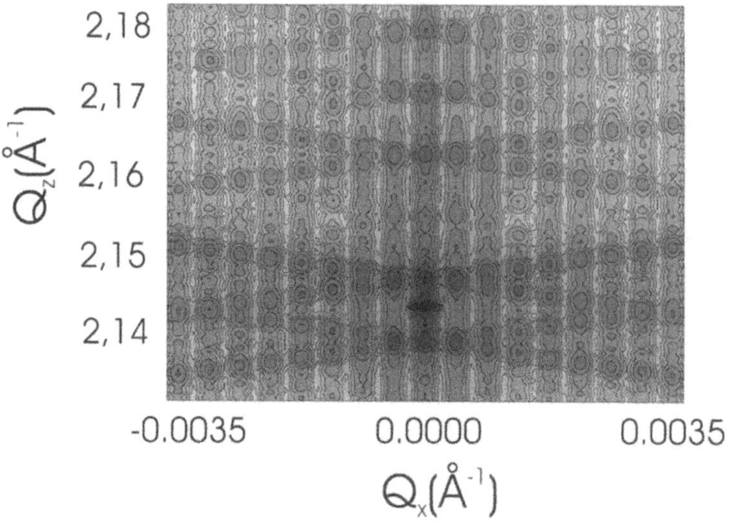

Fig. 12.19. Calculated 002-diffraction map of a GaInAs/InP multilayer for a symmetrically trapezoidal grating shape.

Neglecting any strain modulation, the inclination angle of the branches represents the angle between the direction of maximal interface replication and the surface normal. However, the center of mass of the diffracted intensity along the branches follows the direction of the crystallographic orientation of the terraces. For more detailed discussions see [41, 124, 223].

12.9 Non-Epitaxial Gratings

The examples shown above were based on diffraction methods (e.g., symmetrical and asymmetrical diffraction, grazing-incidence diffraction). Fields of their application are restricted to mono-crystalline gratings, i.e., epitaxially grown nanostructures. In contrast, the methods of specular and non-specular x-ray reflection and grazing-incidence small-angle scattering (GISAXS) are additionally applicable for non-epitaxial gratings. The scattering patterns of these methods are sensitive to the diffraction by the nanoscopic surface pattern and to statistical correlation properties of the interface morphology.

Figure 12.20 shows a measured map of coplanar reflection for an amorphous W/Si multilayer grating. The typical features are (a) the central rod of specular reflection, (b) the surrounding grating rod pattern corresponding to the periodic surface grating, and (c) diffuse scattering between the rods.

The central rod of specular reflection contains all the features of the laterally averaged layered stack: i.e., total external reflection at very small angles and multilayer-Bragg-reflections corresponding to the vertically periodic set-up of the multilayer and thickness fringes.

The intensity along the grating rods represents both the shape and the layered set-up in a way similar to that described in the previous section. The grating of the present example has a rectangular shape. Therefore the grating rod intensity has a structure similar to the central rod of specular reflection, with the multilayer Bragg-reflections and thickness fringes.

The diffuse scattering between the grating rods come from surface and interface roughness. The diffusely scattered intensity is manly concentrated in slightly bent horizontal sheets crossing the multilayer Bragg-peaks. These sheets of resonant diffuse scattering are known from correlated interface roughness in planar multilayers, discussed in Sect. 11.4. The appearance of such sheets gives evidence of partially replicated interface roughness in growth direction.

In the near neighborhood of the grating rods the experienced reader may discover additional less intense parasitic grating rods. Their existence indicates a laterally cyclic deviation of the wire widths, which arises from periodic variations during patterning by electron beam evaporation. In the corresponding transversal scan in Fig. 12.21 the parasitic rods are responsible for the appearance of additional small satellite peaks surrounding the peaks of the grating rods.

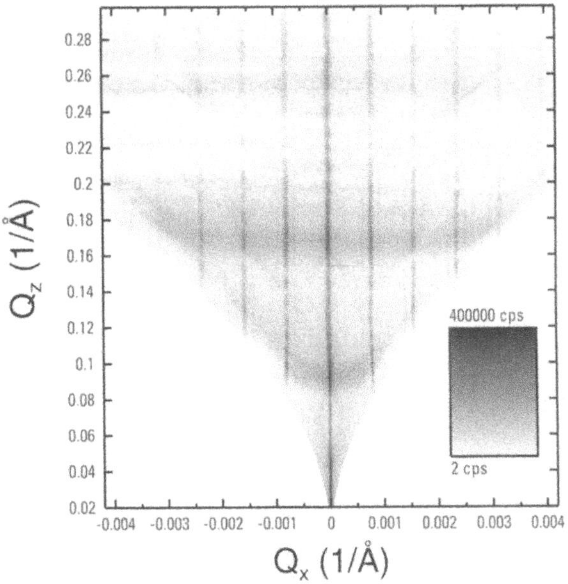

Fig. 12.20. X-ray reflection by a multilayered 10x[W 1 nm/ Si 7 nm]-grating. 000-reciprocal space map measured by coplanar non-specular x-ray reflection.

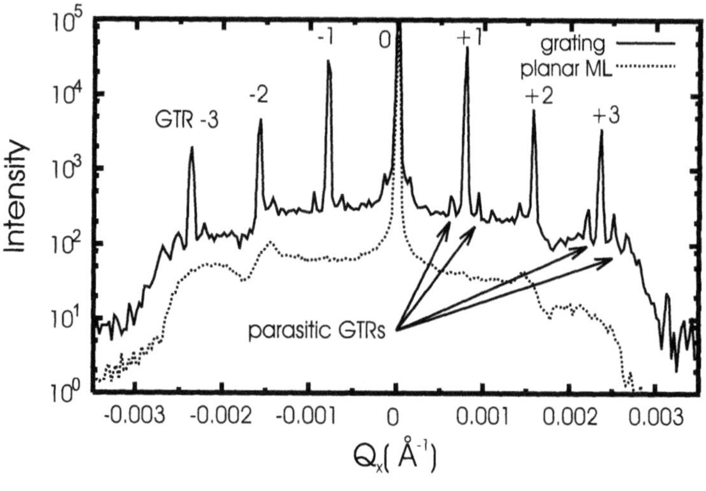

Fig. 12.21. X-ray reflection by a multilayered 10x[W 1 nm/ Si 7 nm]. Non-specular x-ray reflection. Transversal scans before patterning (lower curve) and after patterning (upper curve).

In x-ray reflection methods, multiple scattering effects are usually more dominant compared to wide-angle diffraction methods, which renders the theoretical description more difficult. The dynamical theory of x-ray reflection, distorted-wave Born approximations of first and second orders, and various semikinematical approaches have been employed [34, 240, 241, 242, 373, 374, 375] whereas the required theoretical treatment depends strongly on the particular application. Generally and similarly to reflection by planar multilayers, all influences of intra-rod scattering (see Fig. 12.3) - such as total external reflection; refraction correction; and, near the multilayer Bragg peaks, multiple specular reflection - may influence the reflection pattern and should consequently be considered in the evaluation. Inter-rod scattering can be essential, e.g., in (kinematically) forbidden truncation rods (that is the case, e.g., for all even-order grating rods of rectangular gratings with a top to groove ratio of 1:1). Applying a non-coplanar GISAXS geometry as introduced in Sect. 11.4, the effect of total external reflection by the grating side walls leads to strong inter-rod scattering (for a detailed description see, e.g., [242, 243].

Due to the high sensitivity of x-ray reflection for surface and interface roughness, this influence on grating reflectivity has been studied within nearly all theoretical treatments. Within the matrix approach of the dynamical theory [240, 241, 242] the generalized Fresnel coefficients 12.34 and 12.35 corresponding to the non-specular grating rods are corrected by static Debye-Waller-like factors quite similar to the well-known expressions for specular reflection from rough planar multilayers,

$$r_{HG}^{(m)} = r_{HG}^{(m,\text{flat})} e^{-2k_{Hz}k_{Gz}\sigma_m^2} \quad \text{and} \quad t_{HG}^{(m)} = t_{HG}^{(m,\text{flat})} e^{(k_{Gz}-k_{Gz})^2\sigma_m^2/2}, \tag{12.67}$$

with the meaning of the vector components k_{Gz} given in Sect. 12.3.

However, in contrast to planar multilayers, roughness in gratings reduces the scattered intensity also for incident angles below the critical angle of total reflection. Besides, weak grating rods and strong grating rods are differently sensitive to surface and interface roughnesses. Figure 12.21 shows examples of measured grating rod profiles, fitted including interface roughness and rough wires .

In our examples a number of applications are focused on the characterization of mesoscopic lateral patterning of amorphous or polycrystalline multilayers. Other applications concern the study of soft organic layers deposited on patterned substrates and on surface relief patterns inscribed on polymer films (e.g., [130, 131, 178, 243, 376]).

The example shown in Fig. 12.22 displays a reciprocal-space mapping of a polymer surface relief grating prepared via holographic exposure of the amorphous polymer film by the light of an Ar^+ laser [129]. The blue light acts at the azobenzene moieties attached to the polymer main chains and induces a periodic change of the molecular conformation from a streched *trans* to a tilted *cis* form. This process creates *free volume* and produces pressure toward the neighbored molecular unit of the polymer. The amount of induced conformation changes is maximum in areas of maximum gradient of the electric field vector of the actinic light. This results in a periodic lateral force which is, considering further cooperative processes, the origin of surface

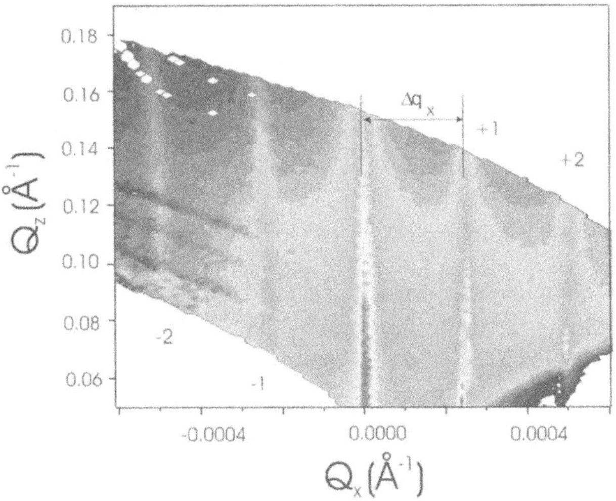

Fig. 12.22. X-ray reciprocal space map of a sinusoidal polymer surface relief grating.

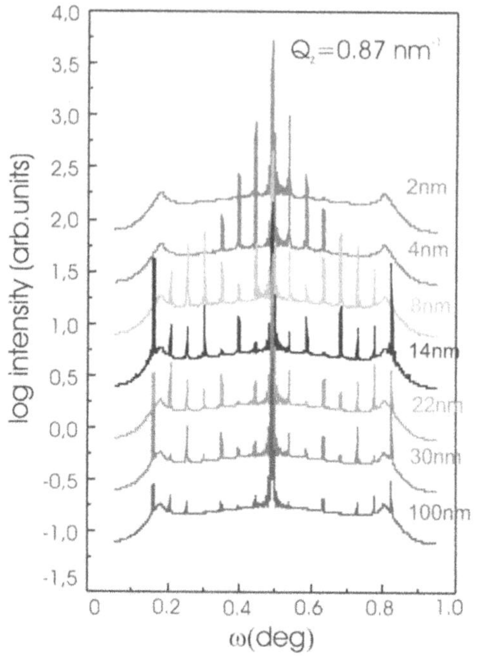

Fig. 12.23. Non-specular reflection curves (Q_x-scans) simulated for different heights of a light-induced sinusoidal surface relief grating of polymers.

relief grating formation. The advantage of this technology is that one can study the grating formation in situ. Here the x-ray reflectivity can be used as a probe for the formation velocity. Due to the perfectly sinusoidal shape of the surface relief the x-ray scattering has to be described by a series of Bessel functions [269]. Therefore x-ray reflectivity reveals great sensitivity for shallow grating heights which appear even at the onset of the grating formation.

Figure 12.23 show a series of ω scans through a reciprocal space map for different heights of the surface relief [132]. The incidence angle is $\alpha_i = 0.5°$. Grating peaks clearly appear for grating amplitudes of less than 2 nm. That is far below the sensitivity for visible light scattering. Due to the character of the Bessel function additional grating peaks appear. They increase and decrease later with the grating height. Subsequently each grating rod passes through a maximum which corresponds to a certain grating height. This can be used for studying the dynamics of grating formation or its erasure depending, e.g., on the temperature [130, 131, 132, 152].

13 Strain Analysis in Periodic Nanostructures

The advantage of x-ray diffraction lies in its unique capability to detect the lattice deformation. This is the reason that nowadays strain analysis is one of the main applications of x-ray diffraction to epitaxial structures.

In pseudomorphically grown epitaxial structures the lattice mismatch between the layers and the substrate is completely accommodated by elastic strain. The incorporated elastic strain in the layers and in the substrate varies during the technological steps of planar growth, patterning, and subsequent overgrowing.

Let us consider the example of layered gratings of the zinc-blend structure type. Pseudomorphic growth requires complete in-plane lattice accommodation, resulting in a vanishing lateral lattice misfit at the horizontal layer interfaces. Planar growth on a well-oriented [001]-substrate, for example, is accompanied by tetragonal distortion of the layer lattices. The strain state in each homogeneous layer is characterized by the vertical lattice misfit with respect to the substrate (see Sect. 9.1). The subsequent formation of a periodic surface pattern, e.g., by selective etching, is accompanied by strain relaxation. This is due to the increase of the free surface of the patterned layer with respect to the substrate interface. The free side walls of the wires enable the strain to relax partially in an elastic way. Often subsequent regrowth buries the grating structure into an embedding top layer. The strain evolution during the burying step is characterized by partially declining the elastic relaxation effect caused by the counterbalance of opposite strain in the embedding layer. Finally, in a buried superlattice grating the strain state is laterally and vertically strongly modulated. The strain has strong influence on the growth and may even stimulate compositional ordering. Finally it essentially affects the band structure. Therefore many efforts will be made to optimize strain-induced band gap engineering, using different design routes. Examples are stress-optimized gratings using symmetrically strained layer superlattices. Other routes intentionally generate stress gratings by the use of stressor layers, periodic ion implantation, or periodic dislocation networks. For x-ray-optical applications dynamically induced strain gratings are of particular interest, employing piezoelectric effects.

According to Eqs. (12.13)–(12.20) the x-ray scattering potential is directly related to the displacement field. A mean strain relaxation of a single-layer grating on a substrate can be detected through the angular shift of the layer diffraction maximum with respect to that of the substrate, by simply comparing the crystal truncation rods of a symmetrical Bragg-diffraction geometry measured before and after etching. The grating rod pattern contains valuable information about the lateral and vertical distribution of lattice strain. Since the phase retrieval is not solved in scattering experiments, the complete actual strain field is usually estimated by a trial-and-error data evaluation. X-ray diffraction methods do rather prove the strain models. Often it is suitable to combine x-ray methods and electron-microscopical methods, whereby the latter gives good starting parameters for the sample geometry. The fast and successful data fitting of a series of x-ray diffraction maps starts with a good intuition on the expected structure of sample. For this reason the influences of strain on the diffraction patterns will be discussed in the following for different examples. The chapter starts from the introduction in strain analysis of surface gratings, continues with the investigation of strain evolution due to embedding and investigates pure strain gratings induced in planar structures.

13.1 Strain Analysis in Surface Gratings

This section first deals with surface gratings that have subsequently been etched into the top of the layered stack. In the beginning we present some simplified elastic models but finish with a completely quantitative analysis of measured diffraction patterns based on a complete elastic description. We discuss particularities of different diffraction geometries applying symmetrical and asymmetrical diffraction and grazing incidence diffraction.

We start with the simple case of a thin planar layer on top of the (planar) substrate. For comparison, the first model shall remain the layer fully tetragonally strained. That means the grating pattern is assumed to change the strain state in the layer not at all. In a second simple model, we allow elastic relaxation to occur, however, with the simplification that the substrate shall remain non-strained. In the direction along the wires the grating keeps the in-plane lattice parameters of the substrate. In the direction of patterning we assume the grating to behave as if it were free-standing. As a result the whole grating relaxes uniformly in an orthorhombic state. In a third model, we consider the effect of non-uniform relaxation on the diffraction maps, the presently most complete but most complex description.

Currently the discussion is restricted to one-dimensional gratings (wire structures).

13.1.1 Simple Strain Models

Section 12.1 has formally described the influence of strain on the scattering potential . In the full-strain model all non-zero Fourier components U_M in Eq. (12.19) vanish and Eq. (12.20) simplifies to

$$\widetilde{F}_{\boldsymbol{h},\boldsymbol{H}}^{(m)}(Q_z) = \chi_{\boldsymbol{h}\mathrm{a}}^{(m)} \int \mathrm{d}z e^{-\mathrm{i}q_z^{(m)}(z-z_m)} U_{\boldsymbol{h},0}^{(m)}(z)\, \Omega_{\boldsymbol{H}}^{(m)}(z)\,. \tag{13.1}$$

The local reciprocal lattice vector in the layer is characterized by the non-zero vertical misfit with respect to the substrate

$$\frac{\Delta h_z^{(m)}}{h_z^{\mathrm{sub}}} = -\frac{\Delta d_\perp^{(m)}}{d_\perp^{\mathrm{sub}}}\,. \tag{13.2}$$

If we modify the reduced scattering vectors in Eq. (12.59) in Chap. 12 by

$$q^{(m)+} = q_z^{(m)} + \Delta h_z^{(m)} + Hp_+^{(m)},\; q^{(m)-} = q_z^{(m)} + \Delta h_z^{(m)} - Hp_-^{(m)}\,., \tag{13.3}$$

we can proceed as in the case of a simple surface grating from Sect. 12.2, reproducing formally Eq. (12.61).

Concluding the tetragonal strain leads to a vertical shift of the diffraction pattern with respect to the substrate Bragg-peak, Δh_z, where the diffraction pattern itself is exclusively determined by the grating shape (see Fig. 13.2). The most simple model of strain relaxation assumes orthorhombic lattice distortion in the grating above a transition region, which is supposed to be thin enough to be neglected. The model also neglects any deformation in the substrate, $\boldsymbol{u}^{\mathrm{sub}}(\boldsymbol{r}) = 0$.

In the direction parallel to the wires y, the grating material is supposed to keep the same in-plane lattice parameter as the substrate, leading to the layer strain component in wire direction $\epsilon_{yy}^{\mathrm{layer}} = \frac{a_{\mathrm{sub}} - a_{\mathrm{layer}}}{a_{\mathrm{layer}}}$. The stress can partially relax in the direction of patterning, i.e., along the x-direction. The surface normal is parallel z. The model supposes vanishing stress components at the free surfaces and consequently $\sigma_{zz} = 0$ and $\sigma_{xx} = 0$ in the wires. For the other strain-relevant components one finds the relations (see also Sect. 9.1)

$$\epsilon_{zz}^{\mathrm{layer}} = -\frac{c_{12}}{c_{11}}\left(\epsilon_{xx} + \epsilon_{yy}\right) \quad \text{and} \quad \epsilon_{xx}^{\mathrm{layer}} = -\frac{c_{12}}{c_{11} + c_{12}}\,\epsilon_{yy}\,. \tag{13.4}$$

The above-mentioned relations are based on elastic constants c_{11}, c_{12}, c_{44}, related to the same coordinate system as the strain and stress tensor.

Usually the coordinate system is defined as $e_x \parallel [100]$, $e_y \parallel [010]$, $e_z \parallel [001]$. In many of our experimental examples the wires are oriented along the crystallographic [110] direction. Then, applying Eq. (13.4), the elastic constants must be expressed in the corresponding coordinate system $e_x \parallel [110]$, $e_y \parallel [1-10]$, $e_z \parallel [001]$. Based on the transformation of a fourth-rank tensor [10] the relations for the elastic constants c_{ij} in the old and c'_{ij} in the new system can be found [274].

Since the wires relax *elastically*, the lattice is still *pseudomorph* to the substrate. We want to describe the lattice deformation by the displacement field with respect to the substrate reference lattice. The displacement component in the direction of patterning is a periodic function $u_x(\boldsymbol{r}) = u_x(\boldsymbol{r}+\boldsymbol{D_G})$. We can therefore replace $u_x(\boldsymbol{r})$ by $u_x(x-x_j, z)$, where x_j is that position at the interface of the wire (j) to the substrate, where the lateral displacement vanishes, $\Delta u_x(x_j, 0) = 0$. For pseudomorphic gratings the difference of the displacement-free points of two arbitrary wires $x_j - x_i = nD_G$ holds, where n is an integer. If we introduce wire shape functions $\Omega_{\text{wire(j)}}(x-x_j)$ for all individual wires, which form together the grating with the shape functions $\Omega_{\text{layer}}(\boldsymbol{r}) = \sum_j \Omega_{\text{wire(j)}}(x-x_j)$, the model of uniform relaxation predicts

$$u_x(\boldsymbol{r}) = \sum_j (x-x_j)\tilde{\epsilon}_{xx}\Omega_{\text{wire(j)}}(x-x_j)$$

expressed by the in-plane lattice misfit $\tilde{\epsilon}_{xx} = \epsilon_{xx} - \epsilon_{yy}$.

The influence of uniform strain relaxation on the grating form factors of Eq.(12.16) can also be described by use of the difference between the reciprocal lattice vectors in the layer (after strain relaxation) and in the substrate, $\Delta\boldsymbol{h} = \boldsymbol{h}^{\text{layer}} - \boldsymbol{h}^{\text{sub}}$. Then we may conveniently calculate the grating form factors of Eq.(12.16) in the form

$$\tilde{\chi}_{\boldsymbol{h},\boldsymbol{H}}(z) = \frac{\chi_{\boldsymbol{h}}^{\text{layer}}}{D_G}\int_{-D_G/2}^{D_G/2} \mathrm{d}x \mathrm{e}^{-\mathrm{i}(\boldsymbol{H}+\Delta\boldsymbol{h}_{\|})\boldsymbol{r}_{\|}}\mathrm{e}^{-\mathrm{i}\Delta h_z(z-z_0)}\Omega^{layer}(\boldsymbol{r}) . \qquad (13.5)$$

Further we can proceed as in Sect. 12.6 and obtain similar expressions to Eq. (12.61) if we replace q^+ and q^- by

$$q^+ = q_z + \Delta h_z + (H_x + \Delta h_x)p_+, \; q^- = q_z - (H_x + \Delta h_x)p_- . \qquad (13.6)$$

Please note that the positions of the central truncation rod and of all grating rods are *not* affected by *elastic* strain relaxation. Although the relaxation leads to different mean lateral lattice parameters in the wires and the substrate, the atomic positions in the wires x_j at the interface are pseudomorphically connected to the substrate lattice. Consequently, the grating period is still commensurate to the substrate lattice. Let us consider the lattice displacement with respect to the substrate lattice. In a plastically relaxed planar layer the lateral components of the lattice displacement diverge when going to infinity. In contrast, the elastic displacement field in the layer lattice of the grating converges. It oscillates with the grating period. The laterally averaged displacement is zero in spite of a non-zero, averaged lateral strain in the wires. That is the reason that the position of the grating rods is determined by the substrate lattice, despite the existing lattice misfit. Although the elastically relaxed wires may have largely different lateral lattice parameters compared to the substrate, their contributions to the diffracted wavefield interfere constructively along the rods which are defined by the *substrate lattice* and the grating period only.

However, keeping the position of the grating rods in reciprocal space fixed, the uniform strain relaxation of the wires leads to a shift of the wire Bragg peaks with respect to the full-strained lattice and consequently of the whole enveloping intensity pattern in the maps (e.g. the typical cross pattern in our example of trapezoidal grating shape) with respect to the pattern of the full-strain model.

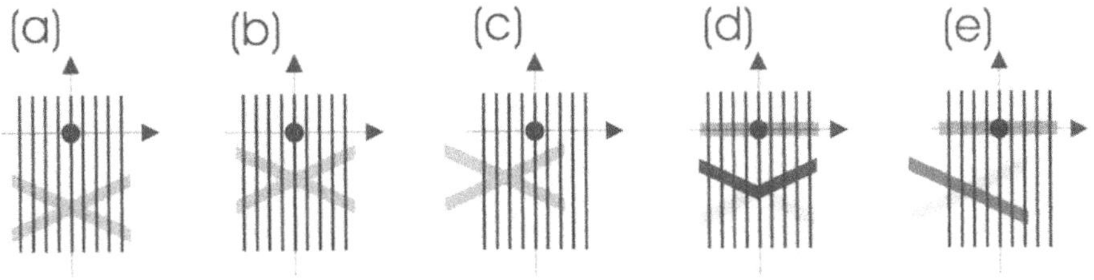

Fig. 13.1. Illustration of the influence of elastic strain relaxation of the reciprocal space map of a trapezoidal grating. Fully strained layer (a), uniform relaxation in symmetrical diffraction (b) and in asymmetrical diffraction (c), non-uniform strain relaxation in the grating and periodic strain in the substrate in symmetrical diffraction (d), and in asymmetrical diffraction (e). The oblique lines present the positions of the intensity maxima on the grating truncation rods; the positions of the rods are not changed.

Since the whole grating relaxes into a uniform strain state, the form of this enveloping intensity pattern remains exclusively determined from the shape of the grating. Fig. 13.1(a–c) illustrates this shift of the envelope marked schematically by the cross pattern of a trapezoidal grating. For symmetric reflections this shift goes in Q_z-direction only. For asymmetric reflections the relaxation shift contains components in Q_x-direction too, leading to an asymmetry in the intensity of the grating rods of positive and negative diffraction order (counted with respect to both sides of the crystal truncation rod) (see Fig. 13.1(c)). Summarizing, the grating rod position is determined by the in-plane lattice parameters of the substrate; their distance is determined by the grating period. Uniform relaxation shifts but does not deform the enveloping intensity pattern.

The model of uniform orthorhombic relaxation describes the actual situation reasonably well, if any non-uniform transition layer between the substrate and the grating is thin with respect to the uniformly relaxed part of the wires. This is fulfilled, for example, for rectangular-shaped gratings with large grating depths if compared to the top path (see, e.g., Fig. 13.4). In many other examples, e.g. triangular or trapezoidal-shaped wires, the model mostly fails (see the actual strain fields of Fig. 13.3). In conclusion it can be stressed that the model of orthorhombic relaxation allows a simple calculation of the diffraction maps. However, from the viewpoint of elasticity theory, it is a strong simplification.

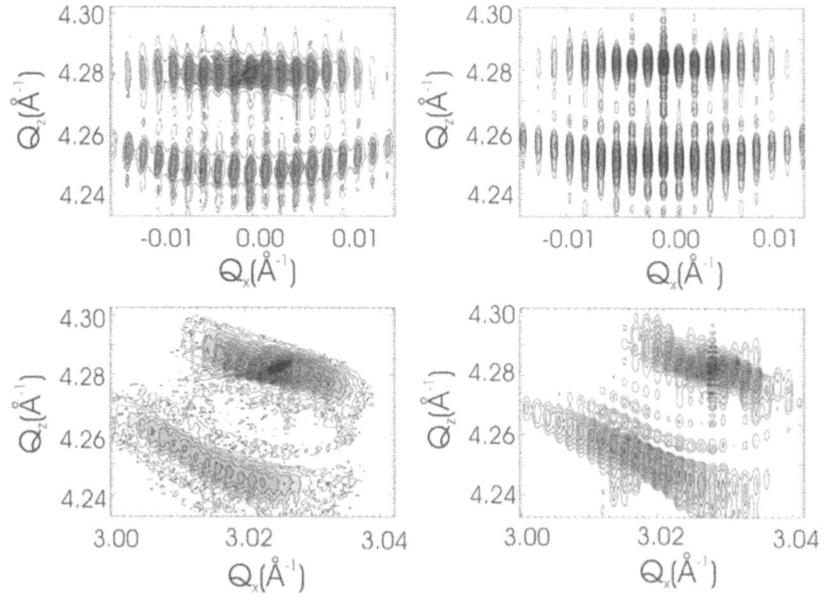

Fig. 13.2. Measured and calculated reciprocal-space maps of a trapezoidal InGaAs grating on InP substrate. Measured (top left) and calculated (top right) map of the symmetrical 004-reflection. Measured (bottom left) and calculated (bottom right) map of the asymmetrical 224-reflection (for the scattering geometry of small incident and large exit angles). The patterns of the lower part of both maps arise mainly from the surface grating; those in the upper part from the strain grating in the substrate. The simulation corresponds to the lattice distortion fields shown in Fig. 13.3.

Therefore the experimental diffraction maps obtain several features which cannot be explained by this model. Some of the differences will be summarized by the examples of Figs. 13.2 and 13.4, which plots measured 004- and 224 maps of strained-layer surface gratings with trapezoidal and rectangular shape, respectively:

1. *Shape-dependence of strain relaxation:* The measured distances between Bragg peaks of the wire and the substrate vary with the grating shape. In the trapezoidal case the corresponding misfit does not confirm the assumption of one dimensional *free standing* wires.
2. *Strain-grating in the non-patterned region*: In the experimental maps of Fig. 13.2 and 13.4 we observe grating rod patterns not only around the reciprocal-lattice point of the wires (around the wire Bragg peaks) but

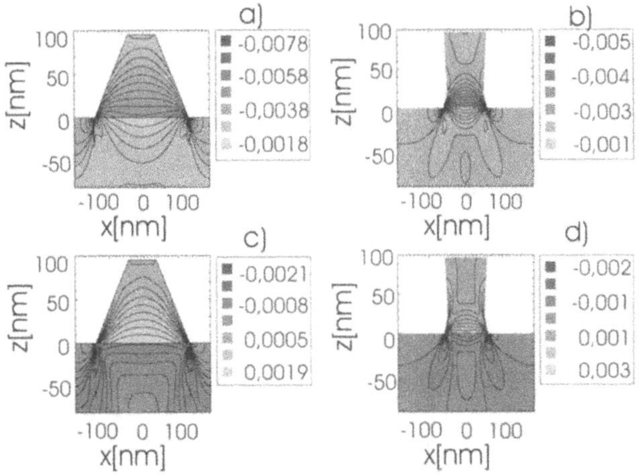

Fig. 13.3. Simulated components of the strain tensor in $Ga_{0.4}In_{0.6}As$-surface gratings on InP substrate corresponding to the best fit with the experimental results. Left panel: trapezoidal grating. Shape and strain profile correspond to the simulated maps in 13.2. Right panel: nearly rectangular grating (shape and strain profile correspond to the simulated maps in Fig. 13.4). Top: the ε_{zz}-component. Bottom: the ε_{xx}-component. The values in the grating have their maximum near the interface and decrease near the top. In the substrate we can observe a lateral modulation of strain with average $\langle \varepsilon_{zz} \rangle = 0$.

also around the substrate reciprocal-lattice point. The appearance of the latter rods gives evidence for the existence of a mesoscopic *strain grating* in the substrate, within an estimated depth quite similar to the etching depth. In the maps of asymmetric reflection the patterns around the substrate Bragg peak remain well centered around the *crystal* truncation rod, in contrast to the grating rods of the layer. The findings correspond to a periodic strain field with a vanishing averaged misfit in the substrate (and therefore vanishing average strain).

3. *Non-uniform strain fields in the wires:* Strong deviations in the form of the measured enveloping intensity patterns are observed in comparing the experimental findings near the layer Bragg peak with maps calculated by use of the model of uniform orthorhombic relaxation (expecting, that the intensity patterns would be determined by the grating shape, only). In the measured maps of a trapezoidal grating (Fig. 13.2), for example, we do not detect the characteristic expected cross pattern. Such strong deformations indicate strong deviations of the actual strain field from the assumption of uniform strain relaxation, which cannot be explained by the above-mentioned model.

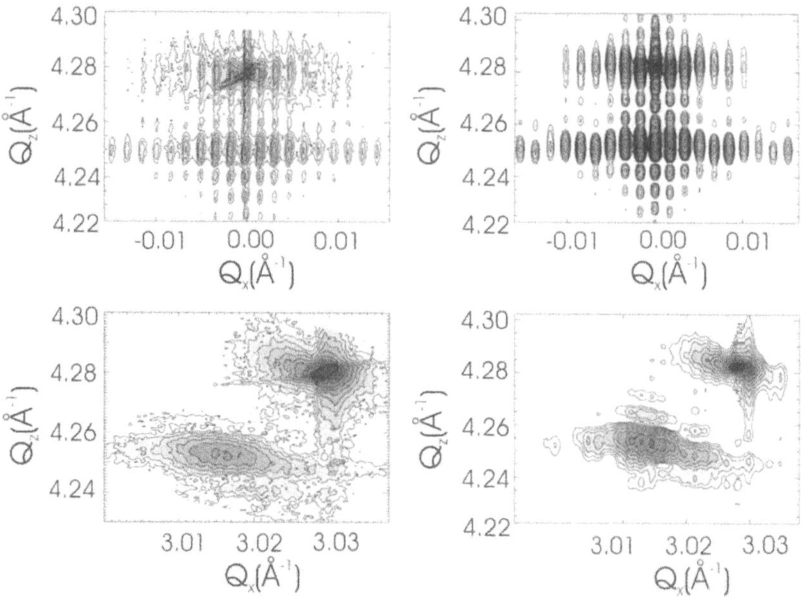

Fig. 13.4. Measured and calculated reciprocal-space maps of a nearly rectangular InGaAs grating on InP substrate. (a) Measured and (b) calculated map of the symmetrical 004-reflection. (c) Measured and (d) calculated map of the asymmetrical 224-reflection (for the scattering geometry of small incident and large exit angles). The patterns of the lower part of both maps arise mainly from the surface grating, those in the upper part from the strain grating in the substrate. The calculations correspond to the shape and lattice distortion fields shown in Fig. 13.3.

13.1.2 Full Quantitative Strain Analysis by Coupling Elasticity Theory and X-Ray Diffraction

In order to explain the above-mentioned experimental findings, we connect the elasticity theory with the diffraction theory [273], studying the mechanism of non-uniform strain relaxation and the relation between non-uniform strain fields and the diffraction patterns.

In a general elastic description, the free energy of the multilayer grating is composed from contributions of bulk and interface terms. The bulk elastic free energy density takes into account the state of deformation in each layer; interface terms include the interaction energy density at the interfaces between the neighboring layers (see [273] for ore details). The accompanying boundary conditions must also take into consideration the condition of zero normal stress at the free grating surface, the balance of the forces (stresses) at all interfaces, and the grating periodicity. We can pass from the discrete model to a continuum approach if the displacement of neighboring lattice points is always smoothly changing. That allows us to apply continuum elasticity theory.

In studies of strained surface gratings the applicability of the continuum elasticity theory to simple gratings with sub-micrometer dimensions has been tested by quantitative comparison between x-ray diffraction experiments and simulations based on elasticity theory (see, e.g., [37, 38]). The lattice deformation varies in length scales much larger than the range of intermolecular forces, i.e., the displacement vector changes smoothly over nanoscopic lengths. The elastic coherence length of the lattice strain field is of macroscopic dimensions. Assuming dislocation-free samples, the lattice displacement functions are continuous at the interfaces. In other words, the normal stresses at each interface are continuous, and the lattices match perfectly together.

The complete set of differential equations and boundary equations can be solved numerically, e.g., by use of the method of finite-elements (FEM). Using this method, the whole sample is decomposed into small cells, each of uniform strain. The cell segments become deformed while the total strain energy is minimized. In this way we obtain the components of the strain tensor at each of the microscopic cells. It should be mentioned also that an analytical solution has been found for the case of a rectangular cross section of wires buried into an embedding material, but it is restricted to elastic isotropy, which can be used instead of numerical calculations for those cases [184].

In all the following examples we combine the diffraction theory with the numerical solution of the elasticity theory described above. First, the wire has to be subdivided into a number of cell segments. Then the diffracted wave amplitudes from all cells must be calculated and finally summed up coherently. Within the kinematical theory this can be done by calculating the structure factor of each segment (marked by the index n) $\tilde{F}_{\boldsymbol{hH}}^{(n,\,m)}(Q_z)$ depending on its actual cell shape $\Omega^{(n,m)}$, composition, and its local strain tensor. The strain tensor also includes shear strain in and rotation of the segments. In order to obtain the grating structure factors of the patterned and non-patterned layers, defined in Eq. (12.18) the structure factors of a patterned layer are added up, taking into account the segment position $\boldsymbol{r}_n$ and its lattice displacement $\boldsymbol{u}(\boldsymbol{r}_n)$:

$$\tilde{F}_{\boldsymbol{hH}}^{(m)}(Q_z) = \sum \mathrm{e}^{-\mathrm{i}\boldsymbol{H}\boldsymbol{r}_n}\, \mathrm{e}^{\mathrm{i}\boldsymbol{h}[\boldsymbol{r}_n - \boldsymbol{u}(\boldsymbol{r}_n)]}\, \tilde{F}_{\boldsymbol{hH}}^{(n,m)}(Q_z)\,. \tag{13.7}$$

Remember that diffraction by a periodic lattice strain modulation in the substrate or other non-patterned layers must also be considered. The contribution of the substrate can be split into the dynamical diffraction by the non-disturbed part and the kinematical diffraction by the distorted upper part of the substrate (within a certain deformation depth Δt^{sub}).

Based on semikinematical diffraction theory and the above-described elasticity model, the whole x-ray diffraction map can be simulated. The lattice deformation can be determined by comparing measurements and calculated reciprocal lattice maps as a function of the actual grating set-up (shape and composition). From the best fit we determine the actual strain distribution in the samples. The results for the vertical and lateral components of the strain tensor for the samples in Figs. 13.2 and 13.4 are shown in Fig. 13.5. For the examples discussed the calculated diffraction maps show a very good agreement with the experiments, reproducing all essential details of the maps. Moreover, x-ray results confirmed the validity of continuum elasticity theory for such nanoscopically patterned semiconductor surface structures.

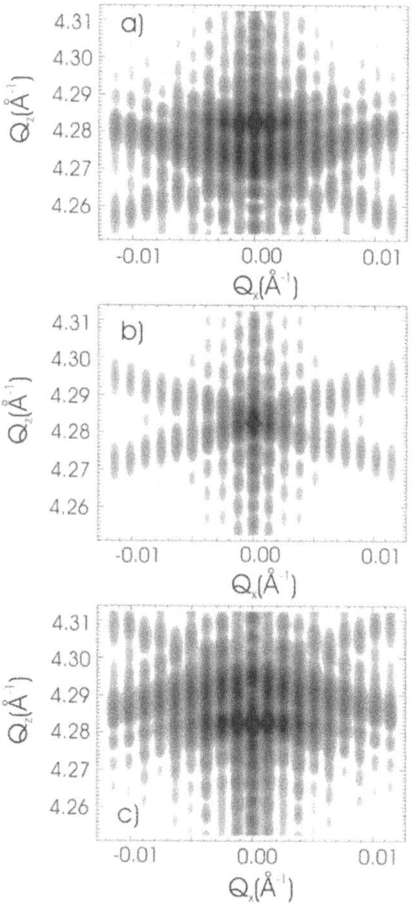

Fig. 13.5. Calculated 004-diffraction maps of trapezoidal $Ga_xIn_{1-x}As$ surface gratings on InP for three different strain states corresponding to different Ga content in the $Ga_xIn_{1-x}As$-grating: (a) $x = 0.44$, (b) $x = 0.47$, (c) $x = 0.50$.

In our example of a grating with trapezoidal shape, the appearance of a distortion of the (001) diffraction planes in the wires can be detected by the deformation of the typical cross pattern for such a shape. For illustration of this deformation effect we present in Fig. 13.5 the calculated maps for three different layer strain states: a lattice-matched InGaAs grating on InP (with gallium content of 0.47) in comparison with the maps for a gallium content of 0.44 and 0.5. For the grating (see Fig. 13.5b) no strain is expected because the layer composition corresponds to a lattice parameter which equals with the substrate. The characteristic cross pattern is centered at the substrate peak. In the sample shown in Fig. 13.5a, the compressive lateral strain at the substrate/layer interface and its relaxation toward the free grating surface leads to a mainly concave bending of the (001) planes in the layer and gives rise to the asymmetry of the upper and lower branch intensities in the map. The inverse asymmetry effect occurs in the map for the sample with the tensile interface strain accompanied by a mainly convex bending of the (001) planes (see Fig. 13.5c) .

One problem concerning the interpretation of the diffraction patterns of gratings consists in the separation of the effects of strain, grating shape, and compositional profile due to their complex influence on the scattering potential (see Eq. (12.13)). The combination of various diffraction methods such as high-resolution coplanar x-ray diffraction and high-resolution x-ray grazing-incidence diffraction helps to distinguish the different effects. *Symmetrical* coplanar diffraction provides the reciprocal-space map of the $00L$ Bragg reflections (we suppose (001)-oriented growth), where only the *vertical* component of the lattice strain plays a role. *Asymmetrical* diffraction investigates HKL reflections, which contain both the *vertical and lateral* strain components (see Sect. 9.1).

GID measures the diffraction patterns of the reciprocal-lattice vectors $HK0$ parallel to the surface, which are sensitive exclusively to the *lateral* strain components, since $h_z \approx 0$. Let us consider the case where the gratings are patterned in the [110] direction and the wires are oriented along the $[\bar{1}10]$ direction. The projection of reciprocal space onto the (Q_x, Q_y) plane is shown schematically in Fig. 12.10.

One advantage of this non-coplanar diffraction geometry lies in its ability to measure a reflection, where the reciprocal-lattice vector is perpendicular to the direction of grating-induced displacement. In this case

$$\boldsymbol{h}\Delta\boldsymbol{u}(\boldsymbol{r}) = 0 \tag{13.8}$$

holds for $\bar{h}h0$ reflections, $U^{(m)}(\boldsymbol{r}) = 1$, and all $U^{(m)}_{M\neq 0}$ vanish. For such reflections the grating truncation rods are totally insensitive to both the vertical and the lateral grating-induced strain. The complementary $hh0$ reflections are sensitive to the influence of the lateral lattice strain, since in this case, $\boldsymbol{h}$ is perpendicular to the grating stripes and parallel to the in-plane lattice displacement:

$$h\Delta u(r) = |h_{\|}| \, |\Delta u_{\|}(r)| \, . \tag{13.9}$$

This allows us to separate the effect of the pure *compositional surface grating* from that originated by the a periodic strain modulation. Additionally, GID allows the investigation of the depth dependence of grating structure and the strain by varying the angle of incidence α_i near the critical angle α_c (see Sect. 2.4).

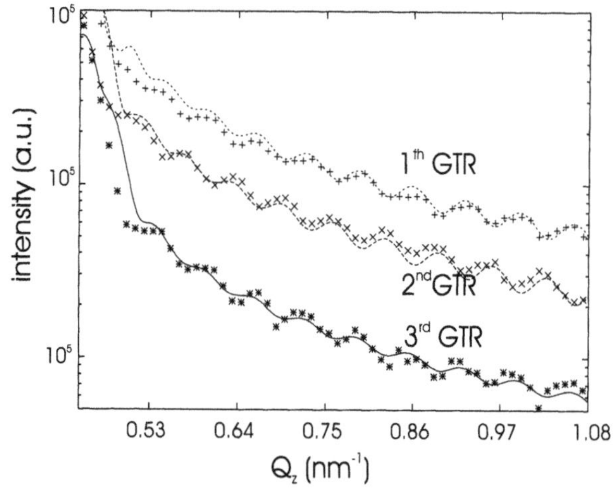

Fig. 13.6. Grazing-incidence diffraction from a strained-layer GaInAs/InP surface grating: three measured grating truncation rod profiles in $2\bar{2}0$-diffraction for constant incident angle α_i. This diffraction geometry, with the scattering vector nearly parallel to the wires, allows the direct determination of the grating without any distortion of the diffraction data due to the residual strain.

Figure 12.9 in Chapter 12 shows a transverse $\bar{2}20$ scan of the trapezoidal strained-layer surface grating studied before by coplanar diffraction and shown in Fig. 13.2. The intensities of the positive and negative grating satellites are fully symmetrical with respect to the central truncation rod, giving evidence of a symmetrical shape of the grating. The intensity profile along the grating truncation rods (Q_z scans in Fig. 13.6) shows a grating fringe pattern similar to that of a simple surface grating without any influence of strain; the fringe period is exclusively related to the etching depth and shape (which is here equal to the layer thickness). With higher grating satellite order, the thickness fringes are progressively shifted to larger Q_z, as is characteristic for a trapezoidal shape. In contrast to the transverse scans of the $\bar{2}20$ reflection, the longitudinal scans of the 220 reflection show a strongly asymmetric satellite pattern (Fig. 12.10, curves b and c).

The differences with respect to the $\bar{2}20$ reflection curves are caused exclusively by the *lateral* (or so-called in-plane) strain, indicating elastic strain relaxation. For decreasing angles of incidence (decreasing penetration depth) an increasing asymmetry of the satellite intensities occurs. The mean center of the envelope shifts toward lower Q_x, since the elastic relaxation increases towards the grating surface.

13.2 Strain in Superlattice Surface Gratings

In order to demonstrate the influence of grating-induced strain on the reciprocal-space maps of superlattice gratings, the discussion is modelled on the example of a $Ga_xIn_{1-x}As/InP$ multilayer with a nominal set-up similar to that of the experimentally studied sample in Fig. 12.19.

By changing the gallium concentration of distinct ternary layers, the lattice mismatch between the layers can be varied. For a gallium content of $x = 0.47$, $Ga_xIn_{1-x}As$ is lattice matched to InP. Together with the thickness ratio for both layers of the multilayer period, the Ga-content in the ternary layer also determines the net (average) misfit of the whole multilayer grating with respect to the substrate surface. With the growing net multilayer misfit the importance of the substrate-grating interface increases, influencing the strain field in all layers and creating a strain field in the planar region too (in the present case in the substrate). When starting from Ga concentrations $x < 0.47$ and gradually increasing the Ga contents to $x > 0.47$, the stress in the ternary layer switches from compressive to tensile (see Fig. 13.7). Simultaneously the displacement field switches from a concave bending of the diffraction planes in the major part of the deformed region toward a convex bending. Let us demonstrate the influence of the bending on the diffraction patterns in a more detail. Therefore, the reader may focus on the changes within the typical envelope functions of grating rod peaks, which in the non-strained case form a characteristic cross pattern, as discussed in Sect. 12.5. The cross patterns consist of an upper and a lower branch of wings. Introducing compressive stress by decreasing the gallium content in the ternary layer *below* $x < 0.47$, we first observe an increase of intensity in the upper branches of the symmetrical reflections, as 004, whereas the lower branches intensity is reduced (Fig. 13.7b), until they disappear completely. Finally, when we further reduce the gallium content, the upper branch also becomes deformed (Fig. 13.7a). In the opposite direction, i.e., increasing the gallium content in the ternary layers $x > 0.47$, we observe the reverse effects, now generating an asymmetry in favor of the lower branches (Fig. 13.7d).

As mentioned above, the symmetric diffraction maps are exclusively sensitive to the vertical strain component and the bending of the diffraction planes, which are parallel to the surface in the non-strained state. The asymmetric diffraction maps also probe the lateral strain components. In Fig. 13.8 we show the asymmetric diffraction maps at the 224-reflection for the same series of material compositions, shown in Fig. 13.7. As with the symmetrical reflection, we find a strain-induced asymmetry in the upper and lower

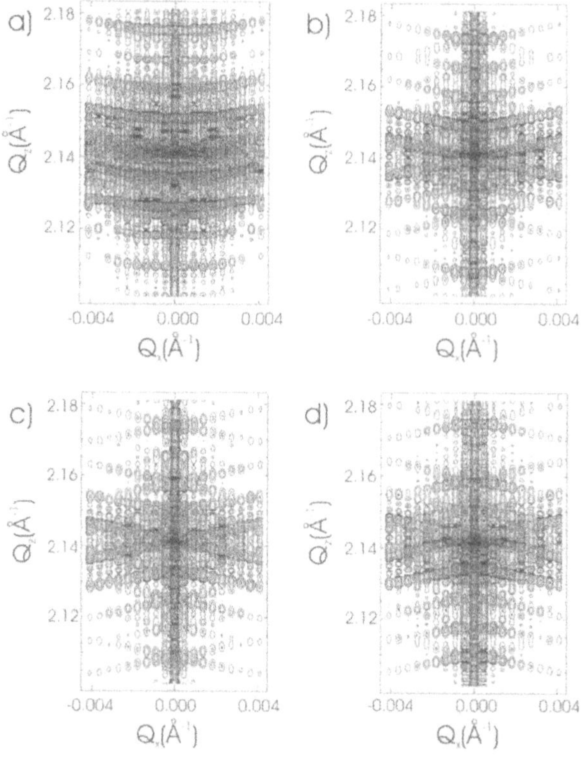

Fig. 13.7. Influence of different amounts of strain for trapezoidal multilayered $Ga_xIn_{1-x}As/InP$ gratings with different values of strain originated by different a Ga-concentration in the ternary layers for x=0.40 (a), 0.46 (b), 0.47 (c), and 0.48 (d) to the 004 reflection.

branches of the envelope function of grating peaks. In addition we observe a shift of the center of mass of the branches from smaller Q_x to larger Q_x, indicating the influence of increasing lateral strain components.

Similar features can be found in other the reciprocal-space maps of other material systems. Here we show an experimental example of a GaInAsP/InP superlattice grating, grown by chemical beam epitaxy (see Fig. 13.9a). It consists of five multilayer periods. After holographic lithography was performed, a periodic surface pattern was etched. The exact shape of the resulting pattern depends on the crystallographic orientation of the etching mask. The best agreement with the measured symmetrical 002-map and asymmetrical 224-map is plotted in Fig. 13.9. It corresponds to certain parameters describing the grating set-up and the spatial variation of the vertical and lateral components of the strain tensor in Fig. 13.10. Note that the same compositional profile and the same set of grating parameters allows us to reproduce

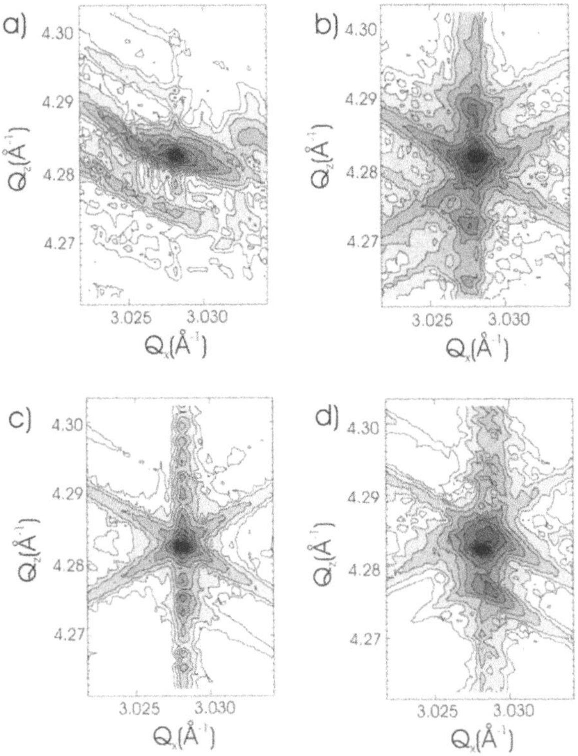

Fig. 13.8. Influence of different strain values on the asymmetrical reciprocal space maps measured in vicinity of the 224-diffraction of the sample shown in fig. of the same structure and x-values as shown in Fig. 13.7.

both the experimental symmetrical 002- and the asymmetrical 224-maps. In Fig. 13.10 the best fit for the most important components of the strain field, as well as the resulting displacement field, are given. The quantitative agreement between experimental and calculated reciprocal-space maps is based on the semikinematical diffraction theory and elasticity theory and confirms the applicability of the presented procedure. The general outcome of our strain studies can be summarized as follows:

1. The essential influence of the grating-substrate interface on the strain field and the relaxation of the net misfit in the multilayer toward the free surface.
2. The influence of the vertical compositional profile, which manifests in the stress relaxation of the GaInAs-layers accompanied by the stress of opposite sign induced in the InP-layers. Both are increasing toward the free surface under the effect of the elastic relaxation process.

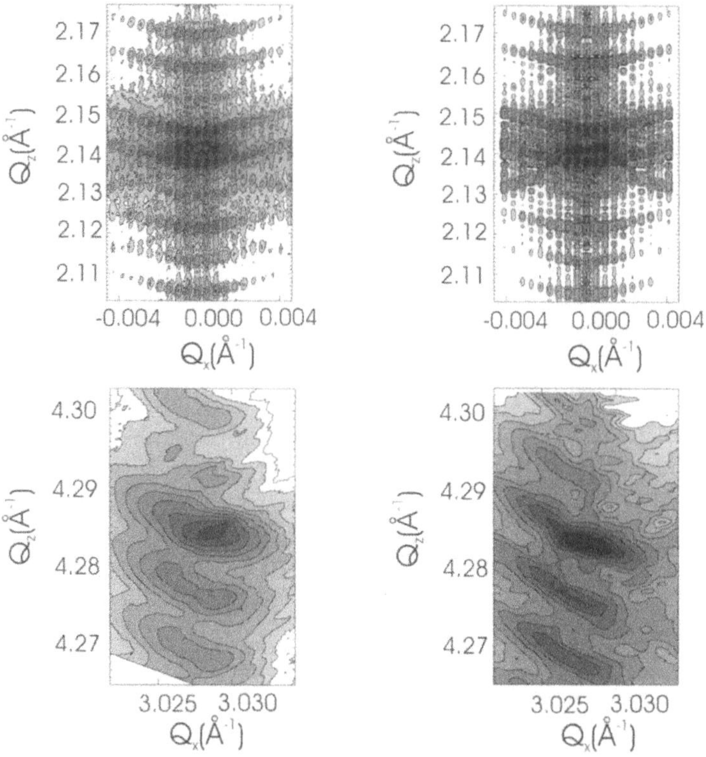

Fig. 13.9. Measured (left) and fitted (right) reciprocal space maps of the GaInAs/InP multilayer grating: above 002-reflection, below 224-reflection.

13.3 Quantum Dots

In the previous section, we studied diffraction by quantum *wires*, which are characterized by a one-dimensional grating structure. Two-dimensionally patterned surface arrays are fabricated, for example, to obtain periodically arranged quantum *dot* structures. Since the extension of x-ray diffraction methods to such two-dimensional grating structures is straightforward, the application of the simulation methods mentioned above to two-dimensional structures is straightforward as well. This will be demonstrated by two examples. Here we restrict ourselves to artificially patterned nanostructures. The x-ray diffraction of self-organized lateral nanostructures is described in Chapter 14.

Considering [001]-oriented structures, the grating rod structure around each reciprocal-lattice point now proceeds in two directions of the plane in the (Q_x, Q_y) of reciprocal space. Thus, we need three-dimensionally resolved measurements in order to resolve the intensity pattern along a single grating

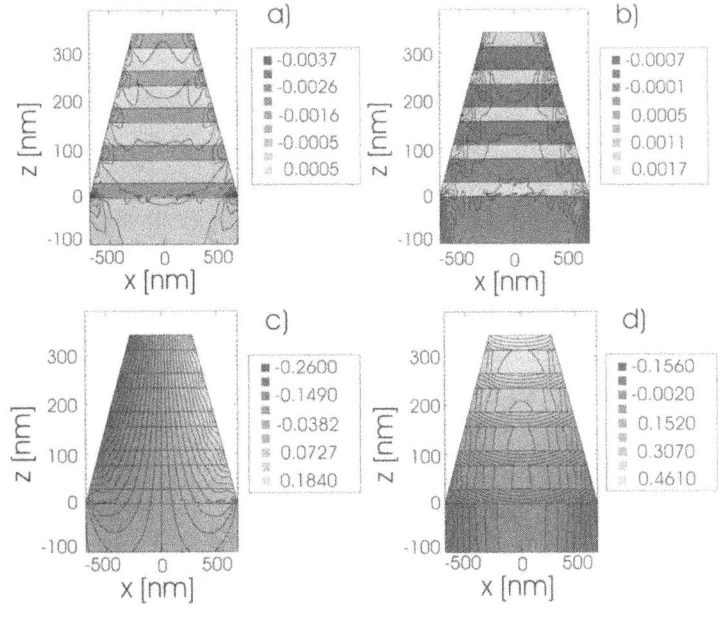

Fig. 13.10. Strain (above) and displacement field (below) determined by simulation to (see Fig. 13.9) the measured reciprocal space maps: lateral components (left) and vertical components (right).

truncation rod. Figure 12.11 shows a (Q_x, Q_y) map of the 220 reflection from a GeSi quantum dot structure measured in GID-geometry. The intersections of the grating rods with the plane Q_z = const. are clearly resolved. The grating period is equal in both directions (Here the asymmetric shape of the truncation rod maxima is a consequence of the asymmetric resolution function of the experimental scattering geometry and not related with the sample structure [87]. From the map, the directions of the grating periodicity can be determined, which in our example slightly deviate from the crystallographic directions.

The projections of the grating truncation rods on the (Q_x, Q_z) plane can be measured by conventional symmetrical and asymmetrical diffraction. This is demonstrated for the example of the symmetrical 004 and asymmetrical 113 maps of a GaAs/AlAs quantum dot structure (Fig. 13.11) [162]. Here we find a large number of well-resolved grating truncation rods, indicating the high perfection of the grating periodicity. From the intensity ratio of the grating truncation rods, the dot radius within the (x, y) plane can be estimated. As in case of wire gratings, the diffraction maps are strain sensitive. In our example an elastic strain relaxation is evident from the slight shift of the intensity envelope in the 113 map. Additionally, there is a large cloud of diffuse scattering of approximately elliptical shape in both maps. The diffuse

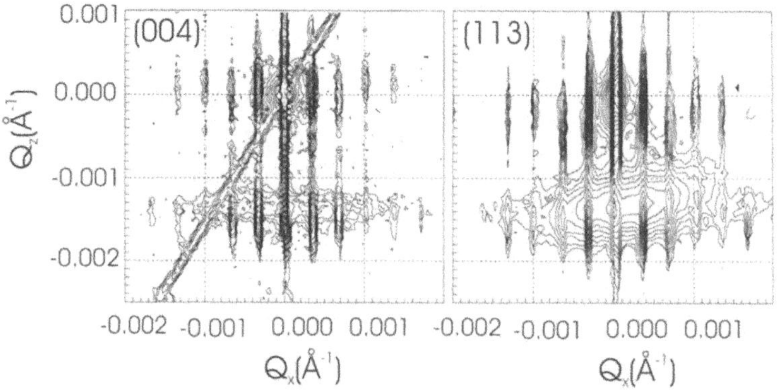

Fig. 13.11. Measured symmetrical 004 and asymmetrical 113 maps of a GaAs/AlAs quantum dot structure. The ellipses of diffuse scattering are several times larger than the grating truncation rods. These arise from laterally non-replicated random-like strain fluctuations. In the asymmetrical map the ellipse of diffuse scattering is inclined with respect to the sample normal, indicating the presence of shear stress in the dots [162].

scattering is generated from random strains. The displacement field in each layer consists of non-randomly and randomly distributed components is

$$\boldsymbol{u}(\boldsymbol{r}) = \langle \boldsymbol{u}(z) \rangle + \delta \boldsymbol{u}(\boldsymbol{r}) . \tag{13.10}$$

The random-like strain component is obviously not correlated between different dots, since the transverse diameter of the ellipse is several times larger than the spacing between the grating truncation rods.

13.4 Strain Evolution Due to Embedding

Frequently, the etching process of a surface pattern is followed by a burying growth step. By post-patterning growth, the quantum well structure or the surface grating will be completely embedded in the substrate material. The laboratories involved in the conception of optoelectronic components which include buried surface grating in their engineering are in need of a non-destructive method for strain characterization.

Here we focus on the strain evolution in surface gratings and buried gratings, including the expansion of the strain field into the embedding media below and above the patterned region.

Our special interest in the non-destructive x-ray investigation of buried gratings has two main causes:

1. The conventional surface sensitive methods such as SEM, AFM, are not available here as non-destructive methods.

2. The technological step of embedding a laterally patterned surface layer or multilayer in the substrate material involves a strain evolution caused by regrowth. At the moment x-ray diffraction is the only method, available to study this strain evolution before and after burying.

Additional technological questions of interest consider the characterization of the state of planarization of the surface grating pattern to form a plane, mono-crystalline surface and the investigation of the strain evolution before and after the technological step of burying.

Post-patterning growth buries the surface grating or the quantum wires in the embedding material, which itself forms in its turn an inverse grating. In other words, the structured layers consist of two wires of different materials, $\Omega^{(a,m)}(\boldsymbol{r})$ and $\Omega^{(b,m)}(\boldsymbol{r})$, with

$$\Omega^{(b,m)}(\boldsymbol{r}) = (1 - \Omega_{\mathrm{SG}}(\boldsymbol{r}_{\|}, z))\Omega_{\mathrm{pl}}^{(m)}(z) .$$

The grating Fourier components of the polarizability (12.16) modify to

$$\chi_{\boldsymbol{hH}}^{(m)}(z) = \frac{1}{D}\int_{-D/2}^{D/2} \mathrm{d}^2\boldsymbol{r}_{\|}\mathrm{e}^{-\mathrm{i}Hx}\mathrm{e}^{-\mathrm{i}\boldsymbol{h}\Delta\boldsymbol{u}(\boldsymbol{r})}$$
$$\times\, [\chi_{\boldsymbol{h}}^{(a,m)}\Omega^{(a,m)}(\boldsymbol{r}) + \chi_{\boldsymbol{h}}^{(b,m)}\Omega^{(b,m)}(\boldsymbol{r})] . \qquad (13.11)$$

X-ray grazing incidence diffraction especially allows detailed and depth selective investigation and the comparison of strain-insensitive (and therefore often called "morphological") reflections with highly strain sensitive reflections.

For the grating rod form factors of the GID-reflection, we obtain

$$\widetilde{F}_{\boldsymbol{hH}}^{(m)}(Q_z) = (\chi_{\boldsymbol{h}}^{(a,m)} - \chi_{\boldsymbol{h}}^{(b,m)}) \sum_{\boldsymbol{M}\neq\boldsymbol{H}} \int \mathrm{d}z\mathrm{e}^{-\mathrm{i}q_z(z-z_m)}U_M^{(m)}(z)\Omega_{\boldsymbol{H}-\boldsymbol{M}}^{(m)}(z)$$
$$+ \int \mathrm{d}z\mathrm{e}^{-\mathrm{i}q_z(z-z_m)}U_H^{(m)}(z)\langle\chi_{\boldsymbol{h}}^{(m)}\rangle(z) . \qquad (13.12)$$

The x-rays are diffracted by the buried grating and the now inverse, embedding grating. The crystal truncation rod depends on the mean *scattering power* $\langle\chi_{\boldsymbol{h}}^{(m)}\rangle$ averaged over the grating period. The non-zero grating rods depend on the compositional contrast $(\chi_{\boldsymbol{h}}^{(a,m)} - \chi_{\boldsymbol{h}}^{(b,m)})$ (diffraction power contrast) between both wire materials.

According to our experimental examples we expect the grating to be patterned in the [1-10] direction. For the strain-insensitive *morphological* $(h\bar{h}0)$ reflections, Eq. (13.12) simplifies to

$$\begin{aligned} \widetilde{F}_{\boldsymbol{hH}}^{(m)}(Q_z) &= (\chi_{\boldsymbol{h}}^{(a,m)} - \chi_{\boldsymbol{h}}^{(b,m)}) \int \mathrm{d}z\, \mathrm{e}^{-\mathrm{i}q_z(z-z_m)}\, \Omega_{\boldsymbol{H}}^{(a,m)}(z) \\ \widetilde{F}_{\boldsymbol{h}0}^{(m)}(Q_z) &= \int \mathrm{d}z\, \langle\chi_{\boldsymbol{h}}^{(m)}\rangle\, \mathrm{e}^{-\mathrm{i}q_z(z-z_m)} . \end{aligned} \qquad (13.13)$$

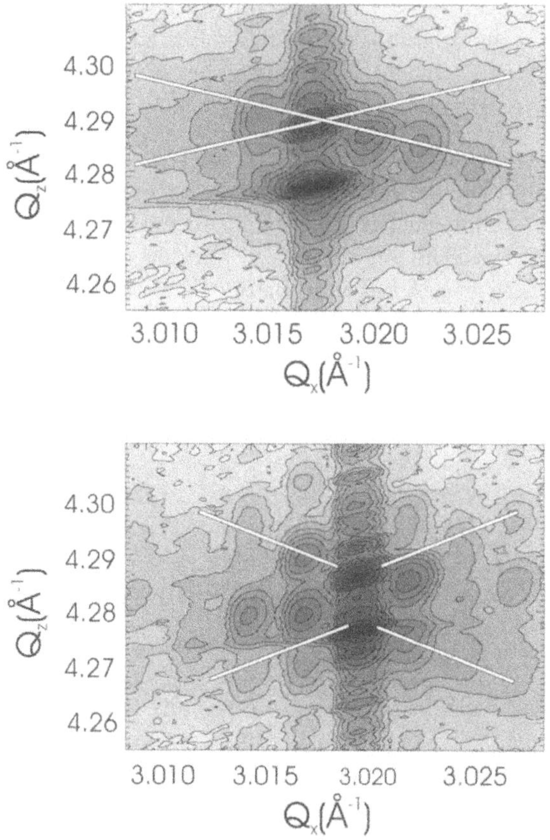

Fig. 13.12. Measured reciprocal-space maps of a free-standing strained layer InGaAsP surface grating (a) at the asymmetrical 224-diffraction, and after the overgrowth with (b) with InP. The white bars mark the directions of trancations rods created from the trapezoidal side walls.

If the surface is completely levelled out, the morphologically planar substrate and all planar layers, including the embedding surface layer, do not contribute to the grating rods apart from the central crystal truncation rod.

If the grating is deeply buried in the embedding material, the contribution of the compositional grating to the Bragg-diffracted amplitudes depends on the information depth, which is determined by the incident and exit angles α_i and α_f. Often the maps are recorded for a fixed incident angle α_i. Then, within a single map, the information depth varies with α_f (see Sect. 2.4) and therefore with Q_z. With decreasing Q_z there follows a strongly reduced sensitivity for the morphology in the buried compositional grating, which opens the possibility of studying the small Q_z the region near the surface separately. More generally speaking, by choosing small incident or exit angles, we can investigate the state of planarization of the embedding material by discriminating the contribution of the morphological buried grating through depth selectivity.

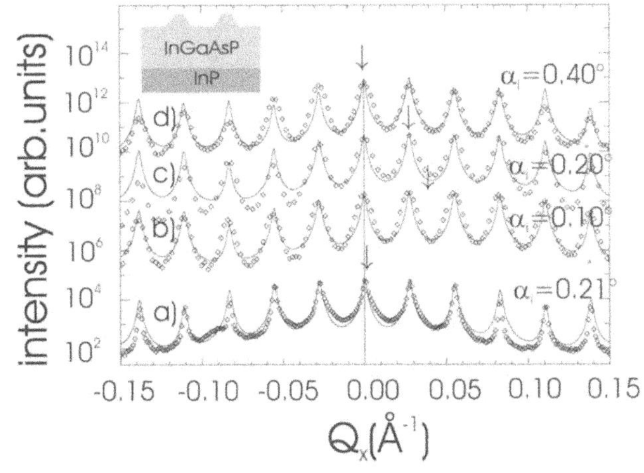

Fig. 13.13. Measured (points) and calculated (lines) Q_x-scans of the strained layer surface grating GaInAsP on InP-substrate. Transversal scan of the $2\bar{2}0$-diffraction (curve a) and longitudinal scans of the 220-diffraction for different incident angles α_i (curves b–d). The shift of the maximum of the envelope in the longitudinal scans (indicated by arrows) gives evidence of the depth-dependent lateral strain relaxation.

Let us discuss the strain evolution in a more detail. The effort of the grating material to relax elastically will be balanced by a counteraction of the embedding material. As a result, the state of lattice relaxation in a grating changes while burying . The effect of restraining has been investigated experimentally by high resolution x-ray diffraction and grazing incidence diffraction (see, e.g., [217, 224, 329, 381]).

In order to be highly sensitive to strain and less sensitive to compositional morphology it is interesting to employ such Bragg diffractions, where variations of the composition of the studied system create only a weak contrast of the *scattering power* $(\chi_{\boldsymbol{h}}^{(a,m)} - \chi_{\boldsymbol{h}}^{(b,m)})$ (i.e., strong 220-Bragg diffractions of the material system GaInAsP/InP): The *contrast problem* has already been addressed by Figs. 9.10 for HXRD and 8.20 for GID.

Then within the measurable dynamical range the grating rods are only weakly affected by the morphology of the buried grating. For x-rays the material appears nearly homogeneous in composition. They *see* mainly a strain

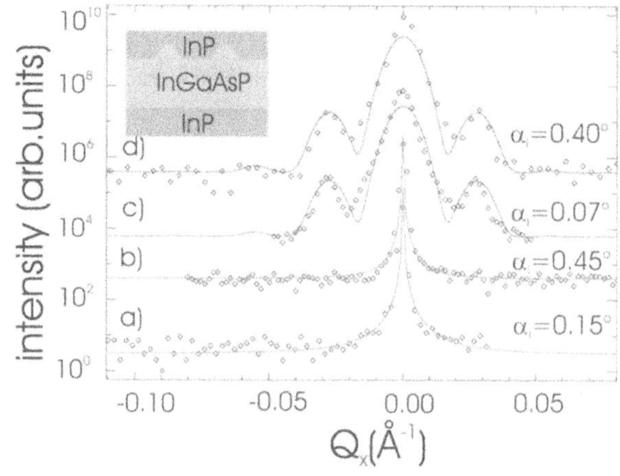

Fig. 13.14. GID by a GaInAsP grating embedded in InP: Transversal scans of the "morphological" $2\bar{2}0$-diffraction (curves a and b) and longitudinal scans of the strain-sensitive 220-diffraction (curves b–d) for two different incident angles α_i.

grating. For negligible compositional contrast the grating form factor of Eq. (12.20) for such cases simplifies to

$$\widetilde{F}_{hH}^{(m)}(Q_z) = \int \mathrm{d}z_{\|}\, \mathrm{e}^{-\mathrm{i}q_z(z-z_m)}\, \langle \chi_{h}^{(m)} \rangle\, U_{H}^{(m)}(z)\,. \tag{13.14}$$

A pure strain grating is created, for example, by a laterally periodic ion-implantation or wafer-bonding, described in Sects. 13.4.1 and 13.5.1.

Up to now we have discussed the methodical aspects of the x-ray analysis of buried structures. In the following we illustrate the method with some examples relevant for research and application in optoelectronics.

Technological patterning of periodic gratings at the surface of III-V semiconducting strained layers is frequently required for the engineering of optoelectronic devices for telecommunication applications.

The high sensitivity of the method even for weak strain fields in low-strained buried gratings is demonstrated by x-ray results for studying the strain evolution in GaInAsP/InP low-strain grating structures for distributed feedback lasers in telecommunication applications. After etching into the planar layer we observe a non-uniform strain distribution, which characterizes the relaxation phenomena due to the corrugation shape at the layer surface, and, after burying, their subsequent partial reversion caused by the counteraction of the embedding material (Fig. 13.12). The difference between the mean vertical lattice parameters of both materials allows some separation of their main contributions in the maps. The upper part of the map (large Q_z) is mainly influenced by the GaInAsP grating, and the lower part by the InP-substrate in Fig. 13.12a and additionally the inverse (embedding) InP

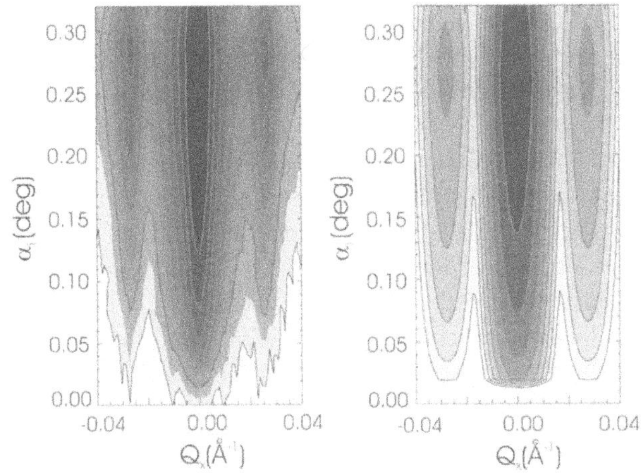

Fig. 13.15. GID by buried grating: longitudinal scans of the strain-sensitive 220-diffraction for different α_i, represented as a contour plot: measurement (left) and simulation (right). Apart from the main maximum at $\Delta Q = 0$ intense satellite maxima at $\Delta Q \approx \pm 0.028\ \text{nm}^{-1}$ are visible, even for very small α_i.

grating in Fig. 13.12b. There the GaInAsP grating is restrained because of the elastic counteraction of the embedding material. However, the grating does not return completely to the fully strained state of the planar structure before etching. The still slightly increased intensity of the right-hand grating rods with respect to the left-hand grating rods close to the reciprocal-lattice point of InGaAsP indicates the remaining slight elastic relaxation in the grating. This is related to compressive stress in the inverse InP grating, which is

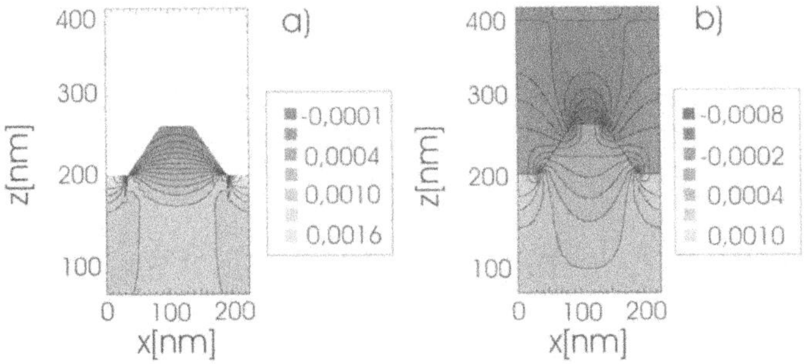

Fig. 13.16. Numerical simulation of the elastic strain field for the surface grating (a) and the buried grating (b): contour plot of the lateral strain component. Only the InGaAsP layer and (for the buried grating) the embedding InP-layer are represented. The difference in strain between two neighboring contour lines is $\Delta\varepsilon_{xx} = 1 \cdot 10^{-4}$.

detected by the reverse behavior of the asymmetry in the grating rod pattern near the InP reciprocal-lattice point (here the right-hand rods are more pronounced). The tiny lines in Fig.13.12 indicate the essential changes in the maps.

Finally the GID-mode allows depth-selective measurements. Figure 13.13 compares one transversal scan of the *morphological* (strain-insensitive) reflection and three longitudinal scans recorded under different angles of incidence α_i of the free standing *surface* grating. In the longitudinal scans we notice an increasing shift of the center of mass toward larger Q_x with decreasing angle of incidence. The results confirm the partial relaxation of the compressive in-plane stress in the surface grating, which increases toward the free surface.

Considering the *buried* grating, depth-selective measurements allow us to determine whether the embedding process is finished successfully by producing a planar monocrystalline surface without any remaining periodic surface corrugation. In this case, the intense grating rods found in the free-standing grating (curve a) are to disappear in the $\bar{h}h0$ reflection for small α_i or α_f (curves a and b in Fig. 13.14). Only the central crystal truncation rod remains visible. The burying step is completed by producing a nearly plane surface layer of the substrate material. Finally, depth-selective measurements in the *strain-sensitive* mode permit us to give evidence for, first, a remaining strain grating and, second, the possible expansion of the strain grating beyond the borders of the morphological grating into the morphologically homogeneous and planar regions. The strain field in the grating gives rise to grating satellites in the strain-sensitive mode. In Fig. 13.15 we draw depth selective 220-longitudinal scans of the buried grating for different α_i in the form of a Q_x/α_i-map. For all angles of incidence, grating satellites are discovered. Even at angles far below the critical angle of total external reflection the GTRs still appear. These experimental results confirm that the periodic strain modulation is not limited to the morphological grating wires (InGaAsP/InP), but expands in the planar top layer and is still verifiable near the sample surface. All findings are confirmed, fitting the experimental results based on the elasticity theory described in Sect. 13.1.2. Figure 13.16) [224] illustrates the above-discussed results, showing the calculated strain fields before and after the embedding process for the above discussed example.

13.4.1 Strain Optimization and Strain-Induced Band Gap Engineering

Strain optimization and the careful investigation of non-homogeneous strain relaxation phenomena in relation to the different technological treatments is a crucial task since the strain status influences electronic and optical properties of devices.

One example is technological patterning for the engineering of optoelectronic devices for telecommunication applications (e.g., gain coupled distributed feedback lasers). It is based on the planar growth of superlattices, holographic lithography with subsequent etching followed by a burying growth step [255, 359]. In some cases the patterned multilayer may correspond to the active part of the laser. This latter point enhances the sensitivity of the device's performances to strain relaxation and distribution. One route to minimize the grating-induced strain is based on the growth of symmetrically strained superlattices. The aim is to achieve a nearly zero effective mismatch of the superlattice compared to the substrate, despite the large lattice mismatch of the individual layers. This has been realized, e.g., by the growth of quaternary InGaAsP/InGaAsP superlattice periods with alternating alloy composition, causing alternating tensile strain and compressive strain (1%). From a successful growth the smallest possible lateral misfit in the wires with respect to the substrate is expected. Consequently, there is an interest in x-ray methods which are able to determine the remaining lateral rest strain, which can be caused even by slight deviations of compositional and thickness in the strained layers.

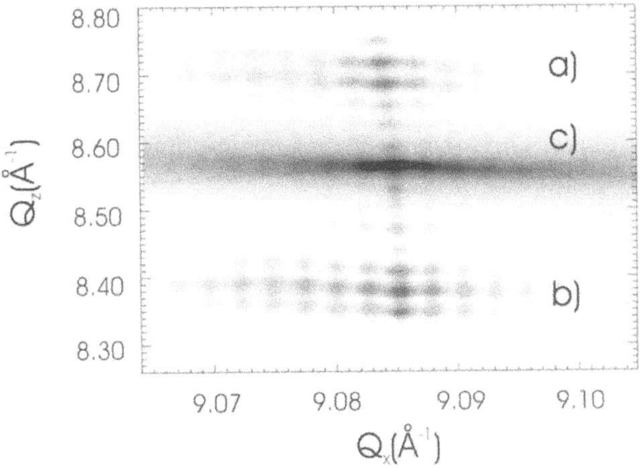

Fig. 13.17. Measured reciprocal-space maps of a multilayer grating based on symmetrically strained GaInAsP layers and buried in InP. Asymmetrical 668 reflection.

In order to arrange high strain sensitivity of the x-ray measurements, high index *hkl* reflections with large components of the reciprocal lattice vectors are suitable for the investigation. This will be demonstrated at the example shown in Fig. 13.17. It shows a measured 668-map of a GaInAsP-superlattice with layers of alternating misfit ± 0.1, giving rise to alternating compressive and tensile stresses in the multilayer, but keeping the averaged misfit with respect to the substrate lattice small. Especially *large* h_z-values of the employed reflections enable us to distinguish the different layer types in reciprocal space. Namely, the tensile layers produce satellites along the rods *above* the substrate Bragg-peak (see Fig. 13.17a); the compressively stressed layers *below* the substrate Bragg-peak (Fig. 13.17b); and the contributions of the inverse embedding InP-grating are visible *around* the substrate Bragg-peak (Fig. 13.17c). The measured shift to the left side of the grating rod intensity corresponding to both the tensile and the compressive layers gives evidence of a small non-zero lateral misfit. It indicates a remaining mismatch of the superlattice with respect to the substrate due to incomplete stress-compensation between the tensile and compressively stressed layers in the multilayer. In consequence, slight strain relaxation occurs after patterning, which is only partially reduced by the embedding layer. In the result one has to expect a remaining *laterally* and *vertically* periodic strain modulation in the active part of the lasers with all influences on the electronic band structure.

Many studies have been published about optimization strain in the active layers. X-ray strain results may be correlated with photoluminescence by measuring strain-driven photoluminescence shifts in such samples (see, e.g., [381]).

Up to now stress investigations in buried structures have been stimulated by the attempt to design devices with reduced non-uniform strain contribution near the active region of the devices. The opposite concept of laterally patterned stressor layers on top of single quantum wells is based on the idea of intentionally introduced periodic strain fields. The patterned stressor layer causes a periodic strain which extends through the underlying non-patterned quantum wells and initiates deliberated periodic band edge variations leading to periodic carrier confinement. In contrast to the approach of patterned active layers described above, this method avoids the creation of non-radiative defects, which usually occur at the interfaces of etched wire structures after regrowth [138, 400]).

Another route is the laterally periodic modulation of the electronic band gap in multi-quantum well structures by focused ion beam implantation [137]. Heavy ion implantation is associated with a local increase of the lattice parameter caused by the creation of structural defects. Owing to the lateral constraint of nanostructured samples the lattice expansion in the implanted stripes is accompanied by lattice compression in between. The x-ray grazing-incidence diffraction method provides a separate inspection of the induced strain and the damage profile as a function of depth below the sample surface. The diffraction intensity of such a pure strain grating can be calculated using Eq. (13.14).

Figure 13.18 shows results obtained by such an example. The strain insensitive GID-scans (see Fig.13.18b) do not show any indication of lateral compositional patterning. The features at larger Q_x, respective h values are caused by non-periodic defects close to the sample surface. On the other hand, the corresponding strain-sensitive scans clearly give evidence of the existence of the strain grating. Because the number and intensity of grating

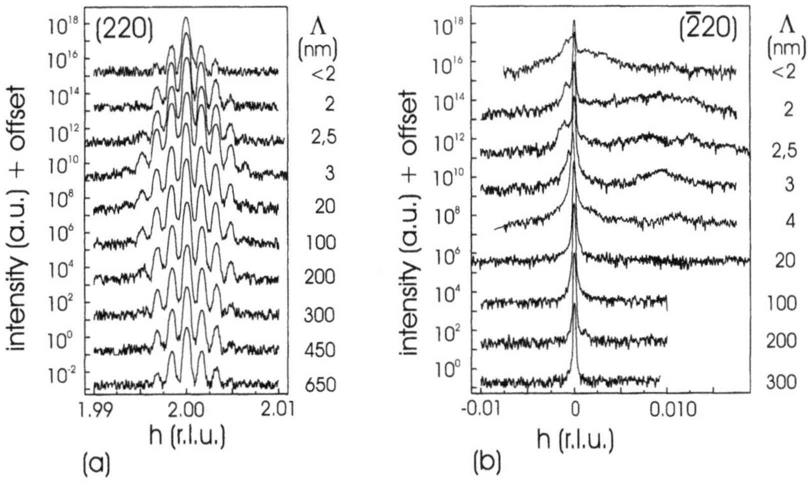

Fig. 13.18. Grazing incidence diffraction by lateral strain gratings induced by focused ion beam implantation after rapid thermal annealing. While the strain-insensitive (transversal) scans (b) show no indication for a compositional grating, the strain-sensitive (longitudinal) scans (a) indicate the existence of a pure strain grating (e.g., a laterally periodic deformation field induced by a periodic implantation process).

peaks is nearly independent of the probed angle of incidence, the induced strain modulation might be extended up to the maximum penetration depth. This changes after rapid annealing of the sample at 650°C, where the depth of the strain modulation becomes reduced to about 250 nm [137].

13.4.2 Strain-Induced Morphological Ordering in Buried Gratings

So far in all our examples the laterally etched nanostructures were embedded by post-patterning growth of *binary* material. In contrast, the embedding process by a solid solution may be additionally affected by local strains [88], giving rise to compositional ordering. Such grating-stimulated self-organization can be observed when a periodic surface pattern is overgrown with a ternary layer. Depending on growth conditions, vertical quantum wells (VQW) of different thickness are obtained [51]. This behavior can be explained by the growth-rate anisotropy of different lattice planes, where the gradients of the surface chemical potential due to curvature-, entropy-, and strain-related contributions act as the driving force for self-organization. As found for a GaAs wire structure overgrown by about 150 nm $Al_{0.3}Ga_{0.7}As$, the transversal scans of the strain-insensitive reflection show low intense first-order side peaks at small incident angles α_i (see [88]). They can be explained by the presence of a compositional modulation , which appears in addition to the modulation of the in-plane strain visible in the radial, strain-sensitive scans.

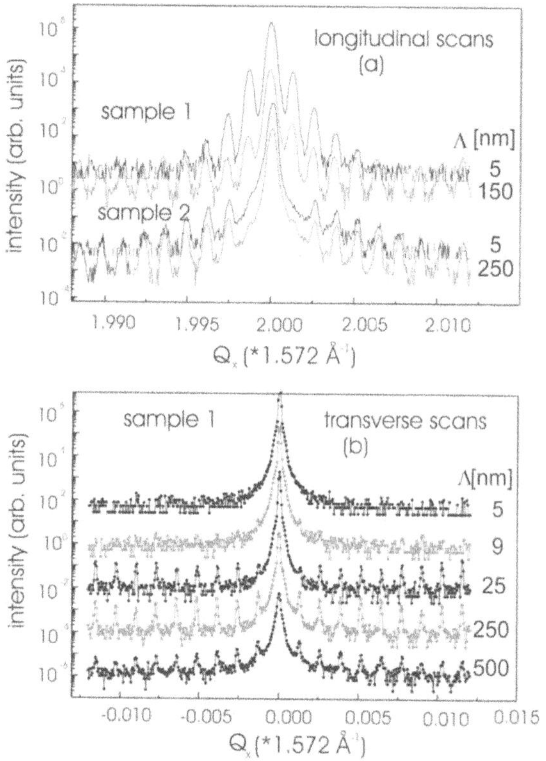

Fig. 13.19. X-ray grazing incidence diffraction of a strain-induced compositional modulation generated during overgrowth of a GaAs/InGaAs surface grating by a $Al_xGa_{1-x}As$ with $x = 0.2$ (sample 1) and $x = 0.7$ (sample 2) overlayer. The strain-sensitive longitudinal scans (top graph) indicate the existence of a periodic deformation field up to the planar sample surface. The strain-insensitive transversal scans (bottom graph) give evidence for a strain-induced formation of a compositional modulation, a vertical quantum well pattern, up to the near surface region. Λ marks the information depth.

Expressed with respect to the non-strained GaAs lattice, the average in-plane strain is slightly compressive in the depth close to the former GaAs wires but slightly tensile close to the surface. Obviously the incorporation coefficient of aluminum during the overgrowth process depends on local lattice strain. Owing to the lattice-parameter difference between GaAs and AlGaAs, the lattice of the solid solution becomes compressed in regions between the GaAs wires. This effect induces strain in vertical direction with the periodicity of the former GaAs surface grating. Scanning electron micrographs taken from the cleavage plane verified the existence of a periodic Al-distribution which has its maximum between the former wires. This is seen in Fig. 13.19 [267]. The amount of induced compositional modulation is obviously driven by the initial in-plane strain of the surface grating. Overgrowing an initially strained GaAs/$Ga_{0.9}In_{0.1}As$/GaAs surface grating by $Al_{0.1}Ga_{0.1}As$ an increased number of grating peaks in the strain-insensitive scan together with even much higher grating rod peak intensities compared with the sample mentioned above are found [267]. The existence of the VQW could be verified by SEM using the composition-sensitive backscattered electron mode, which is sensitive to the Al concentration. The VQWs are visible as bright stripes of low Al concentration, reaching from the bottom of the valleys up to the planar surface (see Fig. 13.20). At a growth temperature of 770°C and a sample with $x = 0.7$, the VQWs appear as uniform small stripes about 14 nm wide.

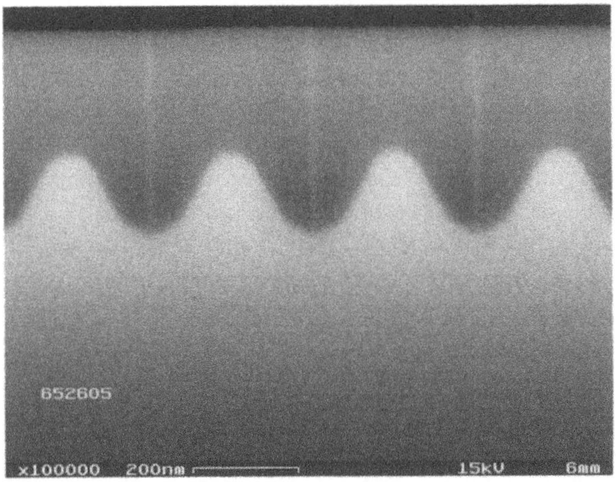

Fig. 13.20. SEM viewgraph of the cleavage plane of a GaAs/InGaAs surface grating after overgrowth with $Al_{0.7}Ga_{0.3}As$ (sample 2 in Fig. 13.19). The VQW is visible as bright stripe of low Al concentration.

13.5 Induced Strain Gratings in Planar Structures

13.5.1 Periodic Dislocation Network in Wafer-Bonded Samples

One challenge in the technology of quantum wires and quantum boxes lies in the establishment of technologies which, on the one hand are simpler and less costly than lithographic methods, and on the other hand show a very improved size and positional uniformity comparing with the self-organized patterning technologies described in Chapter 14.

In the case of self-organized nanostructures the growth starts on an unprepared substrate. During growth, a strain-driven morphological pattern develops, statistically regular with properties which sensitively depend on the miscut, mechanical stress, and growth conditions. Up to now those patterns show statistical distribution of sizes and positions of the dots or wires, which is unfavorable for electronic applications. From the growth of strained layers on substrate surfaces with a slight miscut, it is known that strain-induced patterning can be additionally stimulated by an intentional stepped surface profile (see Chap. 14). A expensive route to achieve highly regular self-organized compositional ordering in combination with growth on lithographically patterned substrates has been discussed in the previous section.

A very promising way to stimulate self-organized growth with highly lateral perfection is based on the idea of generating a highly periodic strain field on top of a planar substrate. Such a strain field can be induced by an interface dislocation network. A grid of regularly spaced dislocations with high perfection of periodicity can be created successfully by the use of wafer bonding, i.e., so-called twist-bonding techniques (see, e.g., [65, 122]).

Assuming an interface between two parallel (001) surfaces bonded with a slight non-zero twist angle ξ to each other, but supposing zero miscut, a square grid of pure or dissociated screw-dislocations, so-called twist-interfacial dislocations, is generated in the (001)-plane. The Burgers vectors of the dislocations are of the $a/2 < 110 >$ type. The dislocations are localized at the bonded interface with no emergence to the surface (see e.g.[122, 297]). The expected stress effect of the buried dislocations is rather weak. However, a periodic strain field may extend to the surface, if the film thickness is thin enough, i.e., approximately half the period of the dislocation array. Then the remaining strain may induce controlled surface patterning and regular self-organized growth of quantum dots with narrow size distribution [66].

The respective diffraction pattern can be described by a pure strain grating. In this case the structure amplitude is

$$S_{\boldsymbol{h}}(\boldsymbol{Q}) = \chi_{\boldsymbol{h}} \int \mathrm{d}^3\boldsymbol{r}\, \mathrm{e}^{-\mathrm{i}\boldsymbol{Q}\boldsymbol{r}}\, \mathrm{e}^{\mathrm{i}\boldsymbol{h}[\boldsymbol{r}-\boldsymbol{u}(\boldsymbol{r})]} . \tag{13.15}$$

The dislocation network creates a displacement field $\Delta\boldsymbol{u}^{\mathrm{disloc}}(\boldsymbol{r})$ in the silicon substrate and in the bonded silicon layer. We define the displacement function with respect to the non-deformed substrate reference lattice before bonding. The displacement at a certain position $\Delta\boldsymbol{u}^{\mathrm{disloc}}(\boldsymbol{r})$ is the sum of the contributions generated by all dislocations.

Assuming a regular grid of dislocations, the displacement field is a two-dimensionally periodic function, $\Delta\boldsymbol{u}^{\mathrm{disloc}}(\boldsymbol{r}) = \Delta\boldsymbol{u}^{\mathrm{disloc}}(\boldsymbol{r}+\boldsymbol{D})$. Hence, the strain function $U(\boldsymbol{r}) = \exp[-\mathrm{i}\boldsymbol{h}\boldsymbol{u}(\boldsymbol{r})]$ can be developed in a Fourier series with the Fourier components $U^{\mathrm{disloc}}_{\boldsymbol{h},\boldsymbol{G}}(z)$ (see 12.19). The translation vector $\boldsymbol{D}$ is identical to the translation vector of the Moire fringe pattern, which can be observed from the superposition of two twisted lattices and is directly related to the twist angle ξ.

The displacement field in the layer takes the distortion field generated by the dislocations and the mean rotation of the layer lattice with respect to the

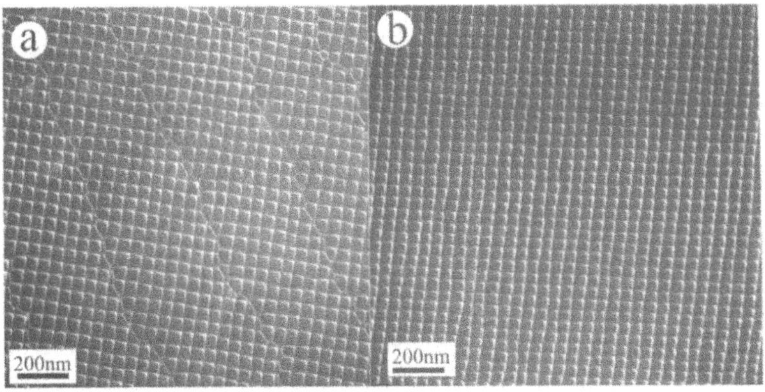

Fig. 13.21. Transmission electron micrograph of wafer-bonded silicon: Twisted silicon crystals with flat parallel 001-surfaces and a square grid of twist induced dislocations, superposed by tilt-induced dislocations (mixed bonding) (a) and of twist-induced dislocations without tilt-induced dislocations (pure twist bonding) (b). [123, 298]

substrate into account, $\boldsymbol{u}^{\text{layer}}(\boldsymbol{r}) = \Delta\boldsymbol{u}^{\text{disloc}}(\boldsymbol{r}) + \Delta\boldsymbol{u}^{\text{rot}}(\boldsymbol{r})$. The contribution of the rotation is for small twist angles $\Delta\boldsymbol{u}^{\text{rot}}(\boldsymbol{r}) = \boldsymbol{\xi}\times(\boldsymbol{r} - \boldsymbol{r}_a)$, assuming $\boldsymbol{r}_a$ an arbitrary point on the axis of rotation. The dislocation network is bounded to the bonding interface. The scattering amplitude adds up the contributions of the strain field extending in the substrate and in the layer,

$$S_{\boldsymbol{h}}(\boldsymbol{Q}) = S_{\boldsymbol{h}}^{\text{sub}}(\boldsymbol{Q}) + S_{\boldsymbol{h}}^{\text{layer}}(\boldsymbol{Q}). \tag{13.16}$$

Using the substitution $\boldsymbol{h}' = \boldsymbol{h} + \xi \times \boldsymbol{h}$ for the mean local reciprocal lattice vector in the layer, we find

$$S_{\boldsymbol{h}}(\boldsymbol{Q}) = 4\pi^2\chi_{\boldsymbol{h}}\left[\sum_{\boldsymbol{G}}\delta^{(2)}(\boldsymbol{Q}_\parallel - \boldsymbol{h}_\parallel - \boldsymbol{G})\int_{-\infty}^{z_{\text{sub}}} \mathrm{d}z\ \mathrm{e}^{-\mathrm{i}(Q_z-h_z)z}\ U_{\boldsymbol{h},\boldsymbol{G}}^{\text{disloc}}(z)\right]$$
$$+4\pi^2\chi_{\boldsymbol{h}}\left[\sum_{\boldsymbol{G}}\delta^{(2)}(\boldsymbol{Q}_\parallel - \boldsymbol{h}'_\parallel - \boldsymbol{G})\int_{z_{\text{sub}}}^{0} \mathrm{d}z\ \mathrm{e}^{-\mathrm{i}(Q_z-h_z)z}\ U_{\boldsymbol{h},\boldsymbol{G}}^{\text{disloc}}(z)\right]. \tag{13.17}$$

The strain field of the dislocation grid gives rise to two sets of grating rods, the first group is assigned to the substrate, the second one to the layer. Both have the same period and the same reciprocal grating vectors. The layer rods are shifted with respect to the substrate rods by $\xi \times \boldsymbol{h}$. Since the periodicity of twist-induced dislocations is directly determined by the twist angle, there always exists a certain grating rod vector fulfilling the condition $\xi \times \boldsymbol{h} = \boldsymbol{H}$. Consequently, the positions of both groups of satellites of the substrate lattice and the layer lattice are commensurate.

$$S_{\boldsymbol{h}}(\boldsymbol{Q}) = 4\pi^2\chi_{\boldsymbol{h}}\sum_{\boldsymbol{G}}\delta^{(2)}(\boldsymbol{Q}_\parallel - \boldsymbol{h}_\parallel - \boldsymbol{G})\int_{-\infty}^{z_{\text{sub}}} \mathrm{d}z\ \mathrm{e}^{-\mathrm{i}(Q_z-h_z)z}\ U_{\boldsymbol{h},\boldsymbol{G}}^{\text{disloc}}(z)$$

$$+4\pi^2\chi_{\boldsymbol{h}}\sum_{\boldsymbol{G}}\delta^{(2)}(\boldsymbol{Q}_{\parallel}-\boldsymbol{h}_{\parallel}-\boldsymbol{G})\int_{z_{\rm sub}}^{0}\mathrm{d}z\ \mathrm{e}^{-\mathrm{i}(Q_z-h_z)z}\ U^{\rm disloc}_{\boldsymbol{h},\boldsymbol{G}-\boldsymbol{H}}(z). \quad (13.18)$$

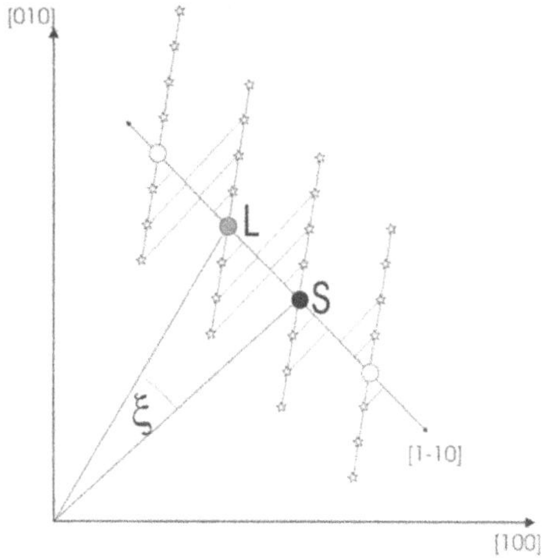

Fig. 13.22. Schematic illustration of the intersection points of grating rods with the Q_xQ_y-plane induced by a twist dislocation grid for mixed (twist and tilt) bonding. Open circles correspond to the strain field induced by the grid of twist-interfacial dislocations. Only the measured peaks of the full square network of intersection points are drawn. Full circles noted L (S) correspond to the layer and the substrate Bragg peaks. Open stars mark the position of satellite peaks caused by the tilt-induced dislocation network. The curves in Fig. 13.23 and Fig. 13.22 are measured along a line connecting the open and filled circles. Measured with an open detector, tilt-induced satellite peaks appear at positions, which are marked by thin projection lines from the satellite peaks on the line of the measured scan.

In the case of a remaining miscut, the influence of a non-zero tilt angle between the two twisted silicon crystals has to be considered in experimental examples. In such cases, in addition to the twist-induced dislocation network, 60° dislocations or mixed dislocations (so-called tilt-interfacial dislocations) are induced. Examples of twist-induced dislocation grids together with and without tilt-interfacial dislocations are seen in the transmission electron micrographs in Fig. 13.21.

For the general case of including twist-induced and tilt-induced dislocations the resulting deformation field will represent the contributions of the superposing dislocation networks. A first contribution to a periodic strain field arises from the orthogonal array of twist-induced dislocations that are dissociated at low ψ in two 30 grad partial dislocations separated by an intrinsic stacking fault. The strain field becomes superposed by a second contribution generated by an array of tilt-induced dislocation lines consisting of mixed dislocation segments. Additionally, interaction between both types of dislocations causes a shift of the screw array of twist-induced dislocations by half of its (

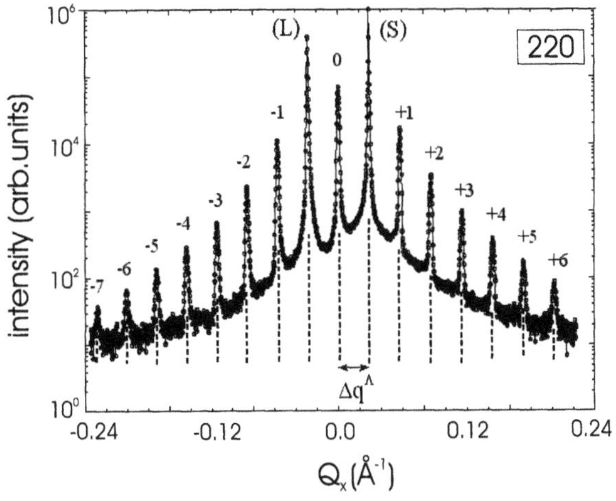

Fig. 13.23. X-ray grazing-incidence diffraction by an ultrathin bonded Si sample with a periodic dislocation grid of pure twist-induced dislocations. Transversal scan of the 220 in-plane reflection. S and L correspond to the peaks of the substrate and the thin layer [117].

X-ray grazing-incidence diffraction is the method of choice for the investigation of the related strain field of the thin twist-bonded surface layer. Figures 13.23 and 13.24 show examples of transversal GID-scans of the 220 reflection, measured at grazing incidence slightly above $\alpha_i \approx \alpha_c$. A pattern of very sharp satellite peaks appears in the plots arising from the intersection of the Ewald sphere with the grating rods. The sharpness of the peaks proves the existence of a strain field with remarkably perfect periodicity over a long range. In Figure 13.23 the peaks correspond to the lateral satellite reflections caused by the deformation field due to pure twist bonding.

In Figure 13.24 the deformation field of two superposing dislocation networks of twist- and tilt-induced dislocations gives rise to two groups of in-plane satellite reflections. The origin and the positions of the various peaks can be explained by drawing the situation in reciprocal space (Fig. 13.22). The deformation field of two superposing dislocation networks gives rise to two groups of in-plane satellite reflections. The first group related to the twist-induced dislocation network was explained above. The contribution of the tilt-induced dislocations to the strain field creates additional satellite reflections, which surround all diffraction peaks of the twist-induced dislocation network. The tilt-induced satellites are equidistantly spaced perpendicularly

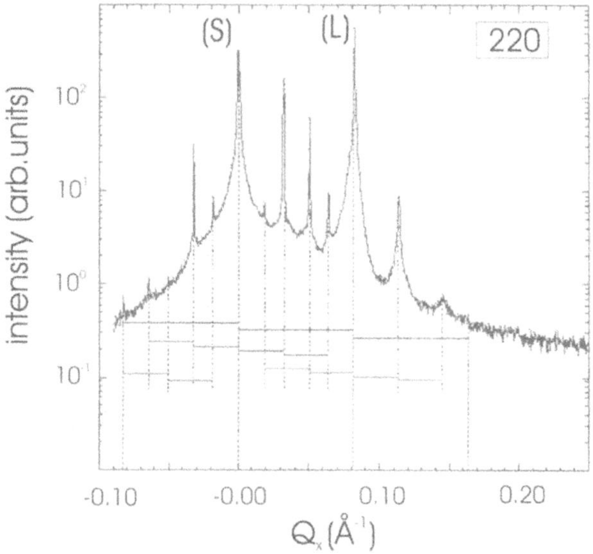

Fig. 13.24. X-ray grazing-incidence diffraction by an ultrathin bonded Si sample with a periodic dislocation grid of twist-induced and tilt-induced dislocations. Transversal scan of the 220 in-plane reflection. S and L correspond to the peaks of the substrate and the thin layer. The lower part of each curve stresses the regularity of the networks and allows the indexation given in Fig. 13.22 [116].

to the direction of the miscut. An open detector allows us to measure some of those tilt-induced satellite reflections projected in the direction of the measured transversal GID-scans. In our case we may observe up to three groups of satellites: one centered around the reflection of the crystal truncation rod of the substrate, another around the corresponding layer peak with one further satellite groups centered at higher-order grating rods of the twist-induced rod pattern.

13.5.2 Dynamical Strain Gratings

Another type of laterally periodic structure with a highly regular lateral ordering is created by dynamically formed gratings, which are generated at the surface of crystalline substrates under the influence of, e.g., piezoelectrically induced surface acoustic waves (SAW). One possible application is to use them in order to store light (delay its propagation) in novel optoelectronic devices [287]; others are in x-ray optical elements (monochromators and beam deviators). Surface acoustic waves in crystals have been studied by a variety of methods including x-ray diffractometry and x-ray topography in a stroboscopic mode (see e.g. [111, 149, 294, 295, 296, 306, 388, 404]).

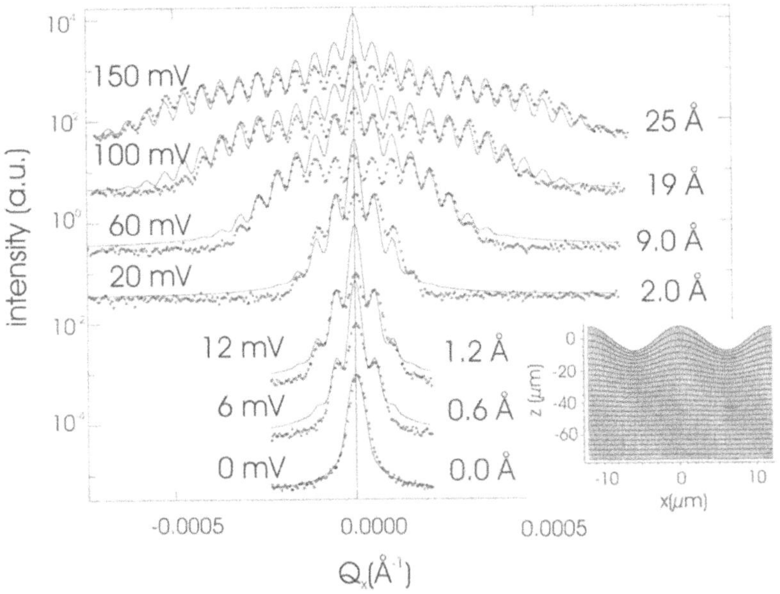

Fig. 13.25. The amplitude of rocking curves measured with open detector on an acoustically excited $LiNbO_3$ crystal [223]. The voltage applied to the piezoelectric transducer (the best-fit acoustic wave amplitude) for each curve is given on the left; the corresponding grating height is shown at the right of the curves. The distortion field of a surface acoustic waves is shown as an inset. The vertical displacement in the image is exaggerated by a factor 104 for better visibility.

X-ray diffraction from these acoustic gratings can be described with an experimental approach similar to the one used for etched gratings [223]. A surface acoustic wave corresponds to a laterally periodic deformation field in a crystal, which gives rise to grating rods. Surface acoustic waves can occur in two different forms: traveling waves or standing waves. In the former case the shape of the wavefield remains constant in time and propagates along the crystal surface. In the latter case a spatially immobile wavefield oscillates in time, with an amplitude that varies periodically between 0 and some maximum amplitude. Therefore the x-ray diffraction pattern from standing SAW varies periodically with the same frequency as the acoustic wave. This effect has been observed experimentally [294, 296].

Figure 13.25 shows a series of rocking curves measured on standing acoustic waves of an acoustically excited $LiNbO_3$ crystal and the best fit by use of 13.14 within a semikinematical model [223]. The number of diffraction satellites increases with increasing voltage, indicating an increasing acoustic wave amplitude. The fitting procedure yields additionally a number of parameters: lateral period, penetration depth of the acoustic wave, acoustic wave amplitude at the surface, width of the resolution function. The simulated curves fit well for higher rod orders. For lower rod orders the simulated curves give systematically higher intensities than their experimental counterparts. That is because the semikinematical approach overestimates the diffracted intensity. The period and the penetration depth of the acoustic waves are in the range of 10 μm, that is, above the extinction length for strong reflections. Thus for the low-order rods extinction and inter-rod scattering (see Sec. 12.3), which are not considered in the evaluation become essential.

14 X-Ray Scattering from Self-Organized Structures

A promising method for fabricating low-dimensional systems is based on processes of self-organization taking place during epitaxial growth under suitable growth conditions. In contrast to the lithographically based techniques, the self-organization method can pattern large areas of the substrate (even whole substrate wafers) and it is also less time-consuming. The disadvantage of this method lies in its statistical nature. Resulting arrays of nano-objects are not exactly periodic and homogeneous. In this context, x-ray methods as tools for the investigation of the structure quality of the arrays are of extraordinary importance.

For a self-organizing growth, two growth scenarios are utilized, namely, the step-flow growth and the Stranski-Krastanow growth mode. These growth modes are discussed thoroughly in several reviews [53, 125, 253, 327, 367].

In contrast to the lithographic technique, the self-organization process can pattern the whole surface of the wafer, and it is suitable for a device production. Its substantial drawback is that the resulting array of quantum wires or quantum dots is not ideally periodic, and it has a certain dispersion of the sizes of the objects. During the overgrowth of the objects, their shape and chemical composition may change due to interdiffusion between the objects and the capping layer above them. All these problems increase the importance of a structural investigation of nanostructures. X-ray scattering is the only method capable for a non-destructive investigation of shapes and chemical composition of free-standing and buried nanostructures. Small-angle x-ray scattering studies the positions of the objects, and their shapes and sizes. X-ray diffraction is also sensitive to the strains in the objects and around them; and using a suitable structure model, we can determine the chemical composition of the objects.

14.1 Self-Organizing Growth Modes

The step-flow growth mode can be described as a lateral movement of monolayer steps on a vicinal growing surface (see Fig. 14.1). The atoms are (1) deposited onto the growing surface, (2) move along the surface until they reach the neighboring monolayer step, where (3) they can attach. It is also possible that the moving adatoms leap "upstairs" to the next flat terrace

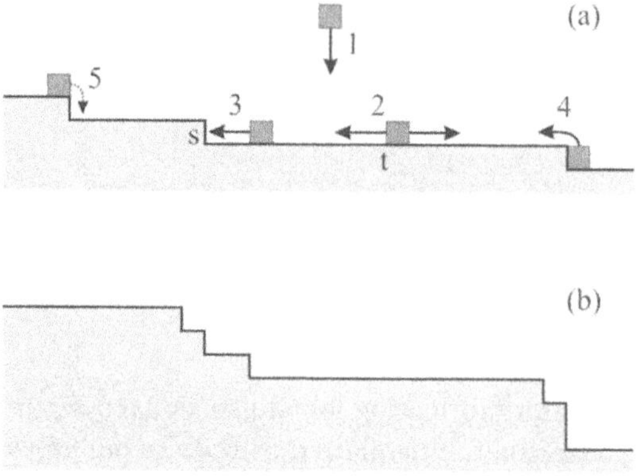

Fig. 14.1. Processes occurring at a surface growing in the step-flow mode (a) and the sketch of the surface with step bunches (b).

(4). The jump "downstairs" (5) is usually assumed forbidden due to the so-called Ehrlich-Schwoebel barrier [324] . Therefore, the monolayer step "s" collects all the atoms deposited on the neighboring terrace "t". The larger the terrace, the faster the movement of the step is; this process is therefore self-controlling. The movement of the steps can be described using the concept of effective repulsive forces acting between the steps [231, 232]. The resulting surface consists of a nearly periodic sequence of flat terraces and monolayer steps; this surface has a minimum surface energy.

However, under suitable growth conditions, a tendency for bunching of the steps is observed. This step-bunching process takes place if the growing layer is elastically compressed, i.e., during heteroepitaxy [7, 220, 263, 368]. On the surface, a sequence of mesoscopic atomically flat terraces divided by bunches of several tens of monolayer steps occurs (Fig. 14.1b). The energy of such a surface is larger than the energy of a surface with a periodic distribution of individual monolayer steps; this increase is overcompensated by a decrease of the volume elastic energy of the lattice due to elastic relaxation of the lattice at the step bunches. The tendency of step bunching occurs also during homoepitaxy; this process is probably caused by a reconstruction of the vicinal surface [309].

As an example, we present in Fig. 14.2 the surface of a SiGe layer deposited onto a (001)Si substrate visualized by atomic force microscopy (AFM). Such a surface can act as a template of a self-organized growth of quantum dots [367]; this growth mode can also be used for a growth of one-dimensional quantum wires in a semiconductor multilayer structure.

The driving force of the bunching of monolayer steps is the relief of the elastic energy during the growth of strained heteroepitaxial layer. This driving force acts also in the Stranski-Krastanov growth mode. In the first stage of this growth, a flat thin heteroepitaxial layer is grown pseudomorphically on

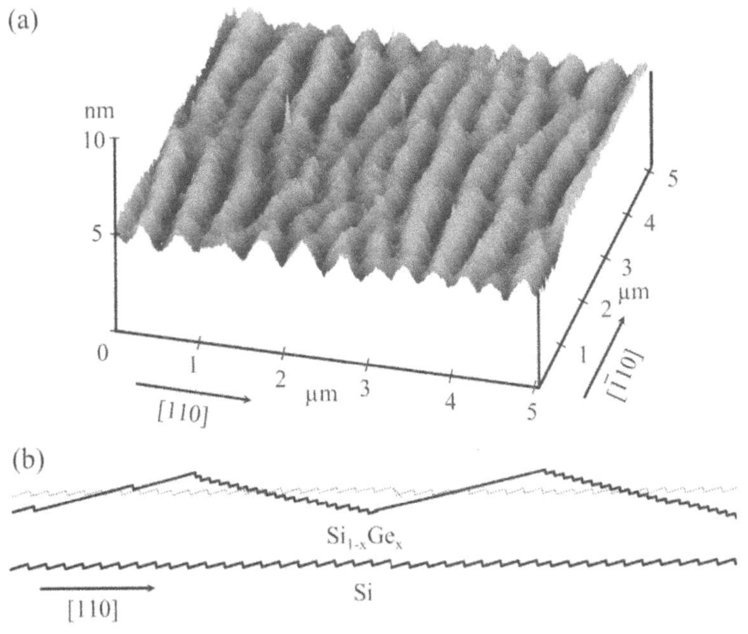

Fig. 14.2. AFM of the surface of a SiGe layer deposited on a Si substrate (a) and the sketch of the surface morphology (b) [263].

a substrate by the step-flow growth (so-called wetting layer). The growing layer is elastically deformed and the amount of the elastic energy stored in the layer increases during the layer growth. At a certain critical thickness (few monolayers) a wavy surface morphology is energetically more favorable, resulting in a sequence of small islands sitting at the surface of the wetting layer. The lattice in the islands elastically relaxes reducing the deformation energy [16, 143, 179, 336, 341]. The two steps in the Stranski-Krastanow growth are sketched schematically in Fig. 14.3.

Since the surface morphology is a result of an interplay between the relief of the elastic energy and an increase of the surface energy, the mean size L_{crit} of the islands depends on the elastic stress σ in the growing layer, on the surface energy density γ, on the Young elasticity modulus E and the Poisson ratio ν by [341]

$$k_{\text{crit}} = \frac{2\pi}{L_{\text{crit}}} = \frac{3\sigma^2(1-\nu^2)}{2E\gamma}. \tag{14.1}$$

The shape of the islands is formed mainly by the direction dependence of the surface energy γ. This dependence is substantially affected by surface reconstruction of the crystal lattice. For instance, the islands on the surface of a Ge layer on (001) Si substrate have two possible geometries, namely, a pyramid with a square or an orthogonal base and {105} side walls, and a

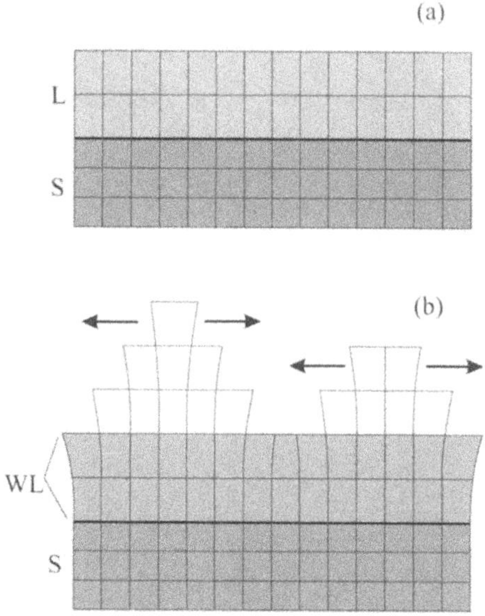

Fig. 14.3. The sketch of the Stranski-Krastanow growth mode. In first stage (a) a flat strained layer is grown, and a wavy morphology (b) occurs if the layer thickness exceeds a critical value. The lattice in the islands is laterally relaxed. WL denotes the wetting layer, L is the layer and S denotes the substrate.

"dome" with a nearly circular base and side walls {105}, {113} and {15 3 23} [91, 186, 183, 236, 366]. Figure 14.4 presents an in situ scanning tunnelling microscope (STM) picture of a Ge surface, where both pyramids and domes can be seen.

In a superlattice, the positions of the islands at a given interface are influenced by the island positions underneath due to the propagation of local strains from the buried islands towards the free surface [166, 326, 369, 394, 401]. The islands at different interfaces are therefore correlated. In a SiGe/Si superlattice, for instance, the positions of the SiGe islands are vertically aligned; the lateral correlation of the positions of the islands at the same interface is rather weak [83, 165, 192, 317]. Similar behavior was also found for InAs dots in an InAs/GaAs superlattice [46]. In a PbSe/PbEuTe superlattice, an oblique correlation of the island positions is observed [166, 339, 340] (see Fig. 14.5).

Increasing the number of layers in the superlattice, the island arrays become more regular and the dispersion of the island sizes decreases, until the critical thickness for a plastic relaxation in the superlattice is reached. Above this thickness, misfit dislocations occur that affect the nucleation of the islands [367].

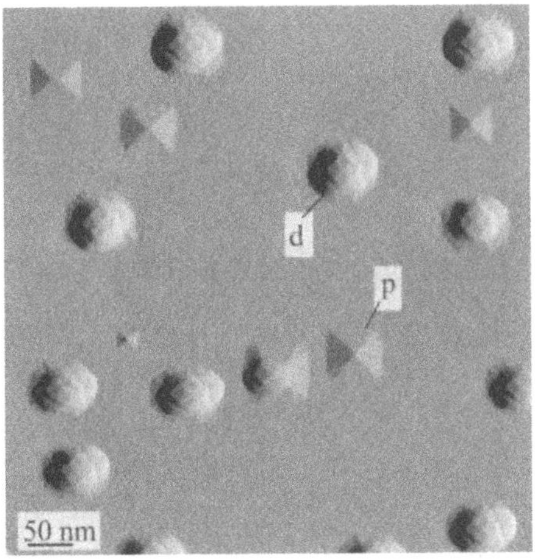

Fig. 14.4. The STM picture of a Ge surface, d and p denote the domes and the pyramids, respectively [154].

The main problem in x-ray scattering from self-organized nanostructures is the very weak signal, since the objects are extremely small. In most cases, it is necessary to suppress the signal from the substrate using surface sensitive scattering geometries such as grazing-incidence small angle scattering (GISAXS) and grazing-incidence diffraction (GID). Surface-sensitive x-ray methods are described also in an extended review [284], the recent results in x-ray scattering from free-standing self-organized quantum dots can be found in [312]. In the next sections of this chapter we formulate the principles of the theoretical descriptions of the scattering and present also several experimental examples.

14.2 Small-Angle X-Ray Scattering from Self-Organized Nanostructures

For the theoretical description of small-angle scattering from self-organized nanostructures, we can use the kinematical theory (Chap. 5) or the more exact semikinematical theory (Chap. 7), we set $\boldsymbol{h} = 0$ in the corresponding formulas. The expression (5.45) for the correlation function C reads

$$C(\boldsymbol{r}, \boldsymbol{r}') = \langle \chi_0(\boldsymbol{r})(\chi_0(\boldsymbol{r}'))^* \rangle . \tag{14.2}$$

The zero-th Fourier coefficient of the polarizability χ_0 is proportional to the local electron density, therefore the small-angle scattering method is sensitive to the *shape* but not to the strains in and around the nanostructures. Let us denote as $\Omega(\boldsymbol{r})$ the shape function of an individual object (a quantum dot or

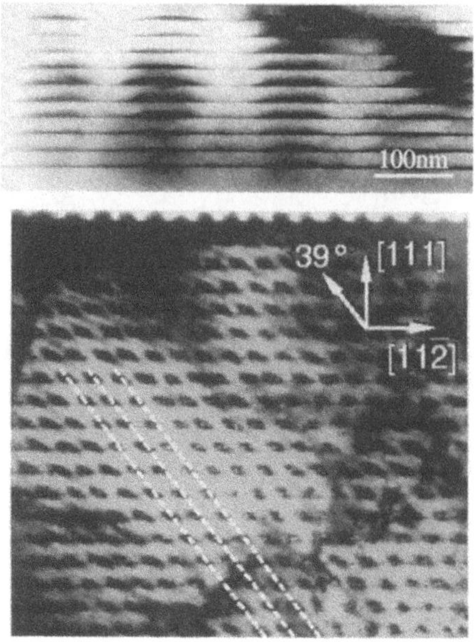

Fig. 14.5. The transmission electron micrographs (TEM) of cross sections of semiconductor superlattices with self-organized islands at the interfaces. In the left panel, a cross section of an SiGe/Si superlattice is shown, the SiGe islands (black) are vertically correlated [165]. The right panel shows the cross-section of a PbSe/PbEuTe superlattice, the islands depicted as black spots are obliquely aligned along the white dashed lines [339].

a quantum wire) and $\boldsymbol{R}_n$ is the random position vector of the n-th object. The covariance $M(\boldsymbol{r}-\boldsymbol{r}')$ is

$$M(\boldsymbol{r},\boldsymbol{r}') = |\Delta\chi_0|^2 \left[\sum_{m,n} \langle \Omega(\boldsymbol{r}-\boldsymbol{R}_m)\Omega(\boldsymbol{r}'-\boldsymbol{R}_n)\rangle - \right.$$

$$\left. - \sum_{m,n} \langle \Omega(\boldsymbol{r}-\boldsymbol{R}_m)\rangle \langle \Omega(\boldsymbol{r}'-\boldsymbol{R}_n)\rangle \right], \tag{14.3}$$

where $\Delta\chi_0 = \chi_0^{\mathrm{object}} - \chi_0^{\mathrm{matrix}}$ is the contrast in χ_0 between the object and its neighborhood. Let us assume, for the sake of simplicity, that all the objects have the same form, they lie at the same depth $Z = Z_m$, and their lateral positions $\boldsymbol{R}_{\parallel m}$ are random. Using Eq. (5.49), we obtain the reciprocal space distribution of the incoherently scattered intensity as

$$J_{\mathrm{incoh}}(\boldsymbol{Q}) = I_i \frac{K^6}{16\pi^2 A} \left|\Omega^{\mathrm{FT}}(\boldsymbol{Q})\right|^2 G_{\parallel}(\boldsymbol{Q}_{\parallel}). \tag{14.4}$$

Here $\Omega^{\mathrm{FT}}(\boldsymbol{Q})$ is the Fourier transformation of the shape function, A is the irradiated sample surface, and

$$G_{\|}(\boldsymbol{Q}_{\|}) = \left\langle \sum_{m,n} \mathrm{e}^{-\mathrm{i}\boldsymbol{Q}_{\|}.(\boldsymbol{R}_{\|m}-\boldsymbol{R}_{\|n})} \right\rangle \tag{14.5}$$

is the *lateral geometrical factor* of the objects. Since we have used the kinematical approximation, i.e., we have neglected absorption, the scattered intensity does not depend on the common depth Z of the objects below the surface. If we use the semikinematical approximation, this formula will be modified; we multiply the expression by the Yoneda term $|t_1 t_2|^2$ (see Chap. 7) and we replace the scattering vector $\boldsymbol{Q}$ by its value $\boldsymbol{Q}_T$ corrected by refraction. Q_T is complex, its imaginary part represents absorption, and we add the absorption term

$$\mathrm{e}^{-\mathrm{Im}(Q_{Tz})Z}$$

in Eq. (14.5). In deriving Eq. (14.4) we have assumed that

$$\langle \sum_m \exp(-\mathrm{i}\boldsymbol{Q}_{\|}.\boldsymbol{R}_m) \rangle = 0$$

for any $\boldsymbol{Q}_{\|} \neq 0$.

If the objects are not ordered, the lateral geometrical factor is constant. In this case, the intensity distribution in reciprocal space is entirely determined by the shape of a single object. For quantum dots with pyramidal shapes, the Fourier transformation of the shape function exhibits maxima along the normals to the side walls of the pyramid. These maxima are analogous to the coherent CTR maximum of the ample surface; from their angle the angle of the side walls can be determined [402]. In the following we present several models of the object arrangement and we derive the corresponding formulas for the lateral geometrical factor.

14.2.1 Short-Range-Order Model

In this model the position of a particular object depends only on the positions of the neighboring objects. Here we derive only the one-dimensional version of the model, a two-dimensional model can be obtained by an analogous way. The lateral position of the m-th object is X_m. We denote $L_m = X_m - X_{m-1}$ and we assume that the distances L_m, L_n are statistically independent for $m \neq n$ (see Fig. 14.6). If the total number N of the objects is very large, the lateral geometrical factor is [77, 276]

$$G_{\|}(Q_{\|}) = N\left[1 + 2\mathrm{Re}\left(\frac{\xi}{1-\xi}\right)\right](1-\delta_{Q_{\|},0}) + N^2\delta_{Q_{\|},0}, \tag{14.6}$$

where

$$\xi = \left\langle \mathrm{e}^{-\mathrm{i}Q_{\|}L} \right\rangle$$

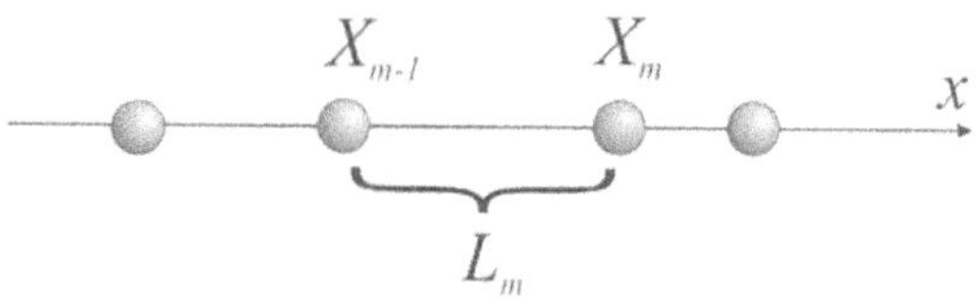

Fig. 14.6. The sketch of the short-range-order model; the objects are denoted by small spheres.

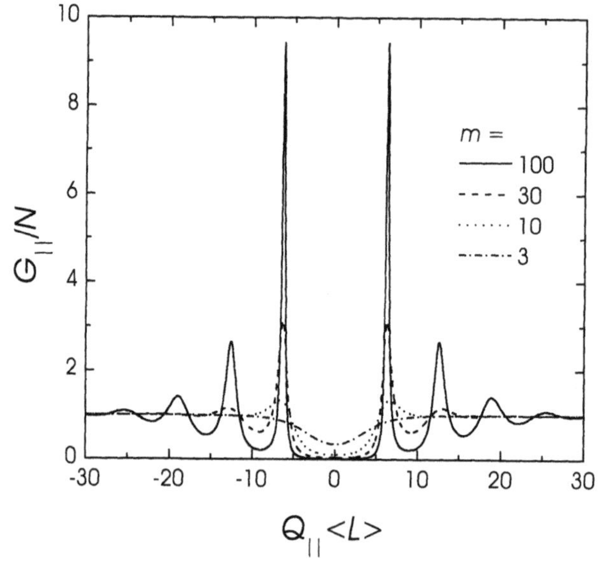

Fig. 14.7. The lateral geometrical factor of the objects, one-dimensional short-range-order model calculated for various orders m of the gamma distributions of the object distances L. The central δ-like maximum is not displayed.

is the characteristic function of the probability distribution of L.

In Fig. 14.7 we show the geometrical factors calculated for various orders m of the gamma distribution of the distance L. The dispersion of L depends on m by $\sigma_L = \langle L \rangle / \sqrt{m}$. The geometrical factor exhibits maxima in points

$$Q_{\|p} \approx p\frac{2\pi}{\langle L \rangle}, \ p = 0, \pm 1, \pm 2, \ldots$$

The better ordered the objects (i.e., the larger m), the narrower the maxima are. The central maximum with $p = 0$ is infinitely narrow and in the resulting intensity distribution it coincides with the coherent CTR peak. For a given m the width of the maxima *increases* with increasing $|p|$, the full width at half-maximum is approximately proportional to p^2. The height of the maxima is roughly inversely proportional to p^2; therefore, the integrate intensity of the maximum is approximately independent on p. According to Eq. (14.4), the distribution of the scattered intensity is proportional to the product of $G_{\|}$ with the square of the Fourier transformation of the shape function of the objects. Therefore, from the integrated intensities of the maxima, the shape of the objects can be deduced.

As an example, we present here GISAXS measurements with $\lambda = 1.54$ Å on a SiGe/Si superlattice [343]. The superlattice consists of 20 periods; each contains a 2.5-nm-thick $Si_{0.55}Ge_{0.45}$ layer and a 10-nm-thick Si spacer layer. The superlattice has been grown by molecular beam epitaxy on a (001) Si substrate with a 2° miscut along [100]. At the surfaces of the SiGe layers, SiGe self-organized quantum dots appeared. In Fig. 14.8 (left panel) we present a two-dimensional intensity distribution in the reciprocal plane $Q_z = \text{const} = 0.1$ Å^{-1}. Besides the central coherent maximum, several lateral maxima can be seen caused by a lateral arrangement of the dots.

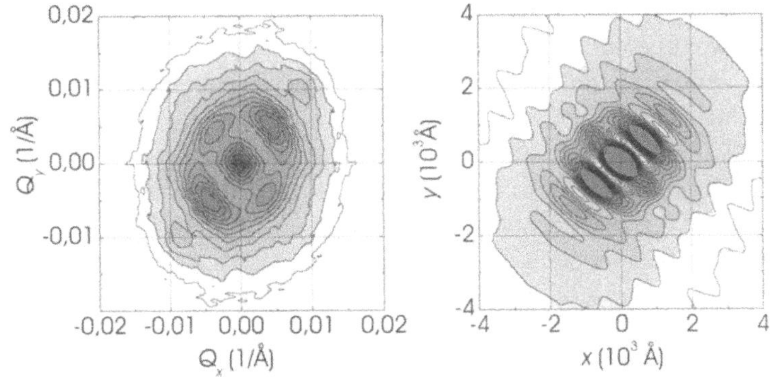

Fig. 14.8. The two-dimensional intensity distribution (left) measured in the GISAXS geometry for constant Q_z and its two-dimensional Fourier transformation (right) [343].

We can extract information about its shape and the lateral geometrical factor $G_{\|}$ using the Fourier filtration of the measured data. In Fig. 14.8 (right panel), the two-dimensional Fourier transformation $J^{\mathrm{FT}}(\boldsymbol{r}_{\|}; Q_z)$ of the measured data is plotted. We define $w(\boldsymbol{r}_{\|})$, the probability of finding a dot with its center in point $\boldsymbol{r}_{\|}$ under the condition that the origin $(0,0)$ is occupied by another dot center. For completely non-correlated dots $w = N/A = n$ holds, n is the area density of the dots. We denote as

$$p(\boldsymbol{r}_{\|}) = w(\boldsymbol{r}_{\|}) - n \tag{14.7}$$

the pair correlation function of the dots. In the short-range-order model, its Fourier transformation is (see Eq. (14.6))

$$p^{\mathrm{FT}}(\boldsymbol{Q}_{\|}) = 2\mathrm{Re}\left(\frac{\xi}{1-\xi}\right).$$

The Fourier transformation of the intensity distribution is

$$J^{\mathrm{FT}}_{\mathrm{incoh}}(\boldsymbol{r}_\parallel;Q_z) = N\Phi(\boldsymbol{r}_\parallel;Q_z)\otimes[\delta^{(2)}(\boldsymbol{r}_\parallel)+p(\boldsymbol{r}_\parallel)], \tag{14.8}$$

where

$$\Phi(\boldsymbol{r}_\parallel;Q_z) = \int \mathrm{d}^3\boldsymbol{r}' \int \mathrm{d}^3\boldsymbol{r}''\Omega(\boldsymbol{r}')\Omega(\boldsymbol{r}'')\mathrm{e}^{-\mathrm{i}Q_z(z'-z'')}\delta^{(2)}(\boldsymbol{r}'_\parallel - \boldsymbol{r}''_\parallel - \boldsymbol{r}_\parallel)$$

and $\otimes$ denotes convolution. Since the dots do not intersect, the pair correlation function for small $\boldsymbol{r}_\parallel$ is $-n$ and the second term in Eq. (14.8) represents a constant contribution to J^{FT}. Therefore, the central maximum of J^{FT} is determined mainly by the first term in Eq. (14.8), i.e., by the dot shape. For large $\boldsymbol{r}_\parallel$, this first term is zero and J^{FT} is given by the second term in Eq. (14.8). If the dots are much smaller than their distance, the convolution in this term can be replaced by the product $\Phi(\boldsymbol{r}_\parallel = 0; Q_z).p(\boldsymbol{r}_\parallel)$, i.e., from the values of J^{FT} for large $\boldsymbol{r}_\parallel$, the pair correlation function follows.

In Fig. 14.9 we have shown line scans extracted from the distribution of J^{FT} for two different azimuths ϕ with respect to the Q_x-axis, along with their fits. The central maximum has been fitted assuming that the dot has a form of a half-ellipsoid; for the more distant parts of the scans, the short-range-order model has been used. From the fits we have determined the half-axes of the dots to (25±2) nm, (20±2) nm (horizontal axes), and (2.3±0.3) nm (vertical axis), as well as their mean distance and its dispersion. In the azimuth $\phi = 45°$, the mean distance of the dots is (100 ± 5) nm, its dispersion is about 25 nm; in the other azimuth ($\phi = 135°$) the distance is (12 ± 5) nm with the dispersion about 40 nm. From this finding it follows that the dots are better ordered in the azimuth $\phi = 45°$; this azimuthal is parallel to the miscut. Therefore, the step bunches created at the interfaces during the growth induce better ordering of the self-organized dots.

14.2.2 Long-Range-Order Model

In a one-dimensional long-range-order model, we define a periodic ideal lattice with period L and we assume that the objects are displaced from the lattice points by random displacements U_m. We assume $\langle U_m\rangle = 0$ and $\langle U_m U_n\rangle = \sigma^2\delta_{mn}$, where σ is the root mean square displacement of the objects form their ideal positions (see Fig. 14.10). The lateral geometrical factor for N objects is [392]

$$G_\parallel(Q_\parallel) = N(1-D^2) + D^2\left|\sum_{m=1}^{N}\mathrm{e}^{-\mathrm{i}Q_\parallel Lm}\right|^2, \tag{14.9}$$

where

$$D = \left\langle E^{-\mathrm{i}Q_\parallel U}\right\rangle \equiv \mathrm{e}^{-\sigma^2 Q_\parallel^2/2}.$$

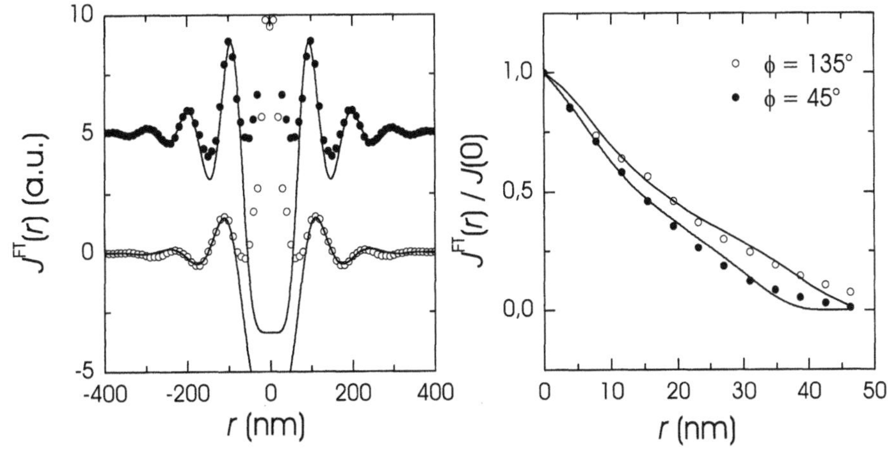

Fig. 14.9. The linear scans extracted from the distribution of J^{FT} in Fig. 14.8 in the azimuths 45° and 135° with respect to the Q_x-axis (points) with their fits (lines). In the left panel, the distant parts of the scans are plotted, the 45°-scan is shifted upward, the right panel shows the central maximum [343].

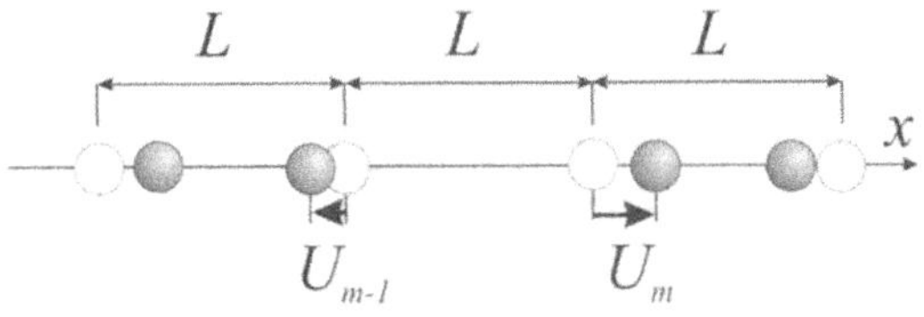

Fig. 14.10. The sketch of the long-range-order model, the gray spheres denote the actual positions of the objects displaced from their ideal positions (white spheres).

The factor D is analogous to the static Debye-Waller factor defined in Chapter 5 and its formula has been obtained assuming a normal distribution of the random displacements U_m.

The lateral geometrical factor is depicted in Fig. 14.11 for various values of σ. It exhibits a sequence of lateral maxima for

$$Q_{\|p} = p\frac{2\pi}{L}, \; p = 0, \pm 1, \pm 2, \ldots$$

In contrast to the short-range-order model, the width of the maxima is constant and independent of σ; it is inversely proportional to the number N of the objects. For an actual structure, N denotes the number of objects occurring in a *coherent domain*, the objects in different domains are not correlated. Then, the width of the lateral maxima is inversely proportional to the mean size of the domains. The height of the maxima decreases with increasing σ. Between the maxima, the value of the geometrical grows with increasing σ.

If the long-range-order model is valid, the integrated intensity is not directly determined by the shape of the objects.

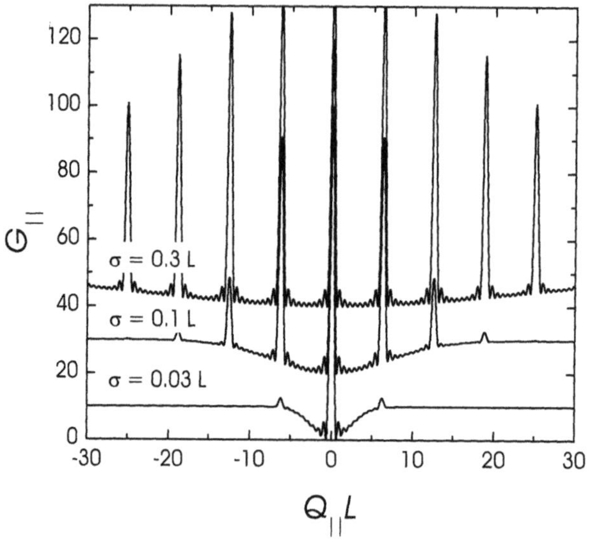

Fig. 14.11. The geometrical factor calculated for a one-dimensional long-range-order model for $N = 10$ objects and various values of σ. The curves are shifted vertically, for clarity.

14.2.3 Two-Dimensional Gas of Objects

In the last example of the lateral arrangement of self-organized objects we assume that the objects are randomly distributed at a plane. The pair correlation function $p(\boldsymbol{r}_{||})$ defined in Eq. (14.7) in Sect. 14.2.1 depends only on the length of $\boldsymbol{r}_{||}$ and not on its direction. Then, the lateral geometrical factor is [142]

$$G_{||}(\boldsymbol{Q}_{||}) = N\left[1 + 4\pi^2 n\delta^{(2)}(\boldsymbol{Q}_{||}) + 2\pi\int_0^\infty \mathrm{d}x\, x\, p(x) J_0(Q_{||}x)\right], \qquad (14.10)$$

where J_0 is the Bessel function of the zero-th order, and n is the area density of the objects. The second term in the square brackets coincides with the coherent crystal truncation rod.

The form of the pair correlation function depends on the interaction between the objects. As a simple example, we consider an *ideal gas* of cylindrical objects with radius R. We assume that the objects cannot penetrate, i.e., we put

$$p(x) = \begin{cases} -n & |x| < 2R \\ 0 & |x| \geq 2R \end{cases}. \qquad (14.11)$$

Then

$$G_{\|}(Q_{\|}) = N\left[1 + 4\pi^2 n\delta^{(2)}(\boldsymbol{Q}_{\|}) - 4\Theta\frac{J_1(2Q_{\|}R)}{Q_{\|}R}\right].$$

The term with the Bessel function J_1 contains the coverage Θ of the plane by the objects and this term is responsible for lateral maxima of the geometrical factor that are *not caused* by a correlation in the position of the objects (see Fig. 14.12). Therefore, the existence of lateral maxima in the reciprocal space distribution of the scattered intensity does not necessarily mean that the scattering objects have correlated positions.

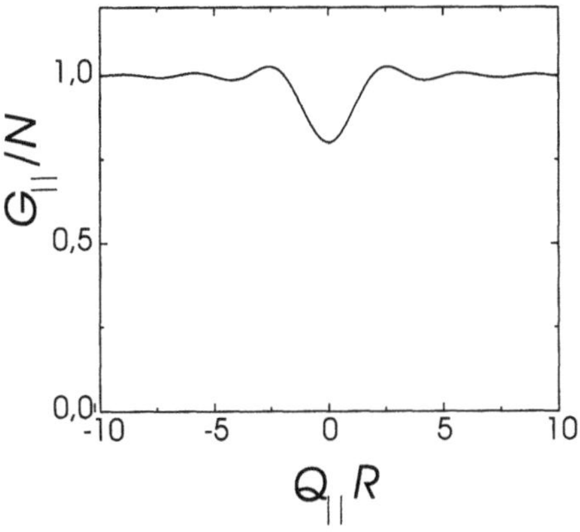

Fig. 14.12. The geometrical factor calculated for a two-dimensional ideal gas of the cylindrical objects with radius R. The central δ-like maximum is omitted.

We have used the two-dimensional-gas model for the description of diffuse scattering from Ge self-organized quantum dots grown in a Si/C/Ge superlattice. The superlattice has been grown by molecular beam epitaxy and it consists of 50 periods; each period contains 9.6 nm Si and 2.4 monolayers Ge. In each period, a very thin carbon layer (0.2 monolayers) has been added on the top of Si [342]. Due to a very small solubility of C in Si, the carbon atoms create small clusters in Si, inducing an inhomogeneous strain distribution on the Si surface. Due to this strain distribution, the Ge dots formed during the Stranski-Krastanow growth of the Ge layers are extremely small. Since their positions are determined mainly by the random position of the carbon clusters, the Ge dots are completely uncorrelated and a two-dimensional gas model can be used [106, 214, 313, 314, 315, 316].

Small-angle x-ray scattering from the dots has been measured in the GISAXS geometry using the wavelength 0.154 nm. In Fig. 14.13 we show the Q_y-scans measured for various azimuthal directions ϕ of the plane of in-

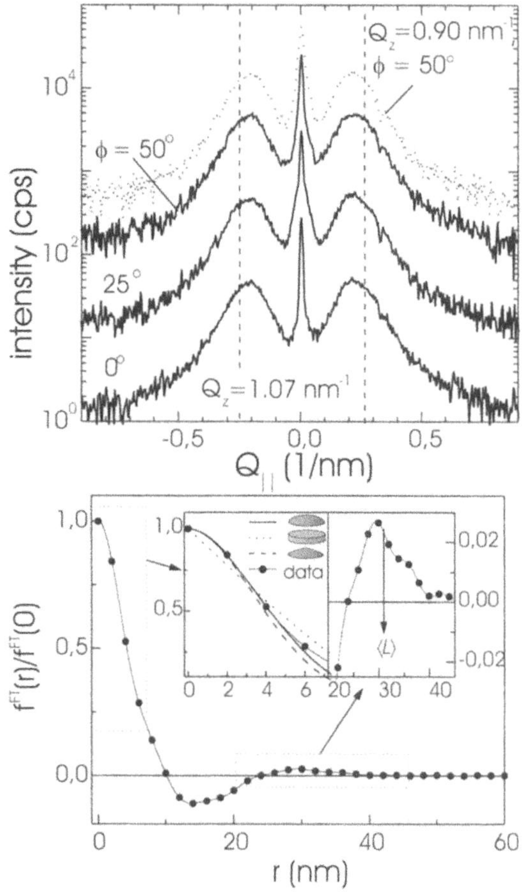

Fig. 14.13. The GISAXS scans (upper panel) measured on the Si/C/Ge multilayer for different azimuths (full lines) and two different values of Q_z (full and dashed lines), the Fourier transformation of the measured intensity (lower panel) and its fits by various shapes of the dots. The vertical lines denote the constant position of the lateral maxima, the lines are shifted vertically for clarity [342].

cidence and various values of Q_z. The intensity distribution does not depend on ϕ, and in the measured range, it is also independent of Q_z. Thus, the quantum dots are cylindrically isotropic and their positions are correlated neither laterally nor vertically.

If we assume that the dots are much smaller than the mean distance between them, we can extract information on the shape and on the correlation function $p(x)$ using the Fourier filtration of the measured intensity distribu-

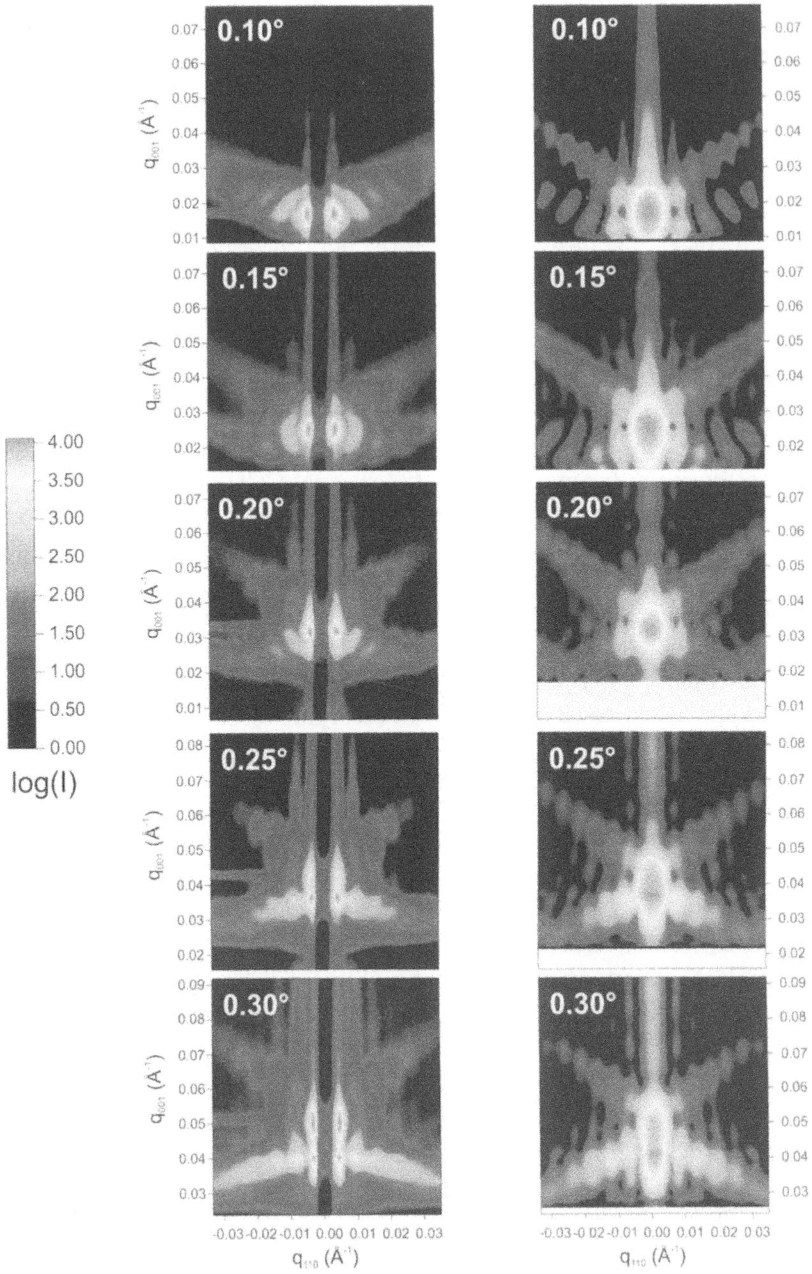

Fig. 14.14. The GISAXS intensity maps (left) measured on a two-dimensional array of SiGe free-standing quantum dots for different incidence angles an their simulations using the semikinematical approach [312].

tion $J(\boldsymbol{Q}_{\parallel})$ in the similar way as in the experimental example shown in Sect. 14.2.1. We have fitted the central peak by several shape models of the dots (see Fig. 14.13, the lower panel), the best agreement was obtained for a lens-shaped dot with the height of (1.7 ± 0.3) nm and radius (5.7 ± 0.3) nm. From the data it also follows that there is a zone around each dot in the distances from 10 nm to 22 nm, where the probability of finding another dot is smaller than in the non-correlated case; this zone corresponds to the negative part of J^{FT}. The mean distance $\langle L \rangle$ between the dots corresponds to the maximum of $p(x)$ (see the insert in the right panel of Fig. 14.13), $\langle L \rangle \approx 30$ nm.

A two-dimensional array of SiGe islands grown on a Si substrate by liquid phase epitaxy exhibits an excellent homogeneity in sizes and shapes [311, 312]. A set of GISAXS intensity patterns recorded form such an island array at different incidence angles is shown in Fig. 14.14 along with their simulations. The measurements have been carried out using an intense very well collimated synchrotron x-ray beam and a two-dimensional CCD detector placed about 4 m from the sample. The oblique intensity stripes correspond to the maxima of the Fourier transform Ω^{FT} of the shape function. Due to the very narrow distribution of the sizes, the measured intensity shows rapid oscillations – the side maxima of $|\Omega^{\mathrm{FT}}(\boldsymbol{Q})|^2$. The simulations were performed by means of the semikinematical approach according to Chapter 7, taking a semi-infinite substrate with a flat surface as the undisturbed system. In the calculation, four scattering processes were assumed, according to the discussion in Section 7.2.2. The positions of the islands were assumed to be completely uncorrelated so that the lateral correlation function $G_{\parallel}(\boldsymbol{Q}_{\parallel})$ is assumed to be constant.

Recently, GISAXS studies have been performed in situ during the growth of nanostructures. These studies require a combination of a synchrotron x-ray source with an appropriate primary optics and a goniometer stage combined with a growth chamber. In [177] a combination is described of a molecular-beam epitaxy with synchrotron radiation. This device was used in several studies of the growth kinetics of semiconductor epitaxy [63, 64]. A similar instrument was used in [279] for an in situ growth studies of metallic islands.

14.3 X-Ray Diffraction from Self-Organized Nanostructures

In contrast to the small-angle x-ray scattering, x-ray diffraction from self-organized nanostructures is mainly affected by strain fields in and around the nanostructures. The main problem in analyzing diffraction data from nanostructures is to distinguish the influence of the strain from the chemical contrast depending on the shape of the objects. The best way to solve this problem is to combine small-angle-scattering and diffraction data taken on the same sample. Analyzing the diffraction data, we use the shapes and positions of the nano-objects determined from small-angle scattering as an input; this facilitates the unambiguous determination of the strains. If the small-angle scattering data are not available, it is rather complicated to distinguish both effects. In this case, a detailed numerical fit of the data with a suitable structure model is necessary. If the objects are sufficiently large and/or the the strains are large, one can distinguish the size and the strain using the iso-strain approximation [189, 190, 191].

A general expression for the covariance $M(\boldsymbol{r}-\boldsymbol{r}')$ defined in Eq. (5.45) is complicated, if the positions of the objects are correlated. For a non-correlated case, we use the expression for M derived in the case of randomly placed weak volume defects in Chapter 10 (Eqs. (10.10,10.11)):

$$M(\boldsymbol{r},\boldsymbol{r}') = \int \mathrm{d}^3\boldsymbol{r}'' n(\boldsymbol{r}'')\Psi(\boldsymbol{r}-\boldsymbol{r}'')\Psi^*(\boldsymbol{r}'-\boldsymbol{r}''), \tag{14.12}$$

and

$$\Psi(\boldsymbol{r}) = \chi_{\boldsymbol{h}}\left(1-\mathrm{e}^{-\mathrm{i}\boldsymbol{h}.\boldsymbol{v}(\boldsymbol{r})}\right) + \Delta\chi_{\boldsymbol{h}}\Omega(\boldsymbol{r})\mathrm{e}^{-\mathrm{i}\boldsymbol{h}.\boldsymbol{v}(\boldsymbol{r})}. \tag{14.13}$$

Here $\boldsymbol{v}(\boldsymbol{r})$ is the displacement field around a single object, $\Omega(\boldsymbol{r})$ is its shape function, $n(\boldsymbol{r})$ is the volume density of the objects, and $\Delta\chi_{\boldsymbol{h}}$ is the difference between the $\boldsymbol{h}$-th Fourier coefficient of the crystal polarizability of the object and the surrounding crystal matrix.

The covariance M can be expressed in a simple way also for the case of correlated objects, if one can assume that the displacement field of different objects do not overlap. Then

$$M(\boldsymbol{r},\boldsymbol{r}') \approx \sum_{m,n} \langle \Psi(\boldsymbol{r}-\boldsymbol{R}_m)\Psi^*(\boldsymbol{r}'-\boldsymbol{R}_n)\rangle - \\ - \sum_{m,n} \langle \Psi(\boldsymbol{r}-\boldsymbol{R}_m)\rangle \langle \Psi^*(\boldsymbol{r}'-\boldsymbol{R}_n)\rangle \tag{14.14}$$

and the expression for the incoherently scattered intensity distribution is analogous to Eq. (14.4):

$$J_{\mathrm{incoh}}(\boldsymbol{q}) \approx I_i \frac{K^6}{16\pi^2 A} \left|\Psi^{\mathrm{FT}}(\boldsymbol{q})\right|^2 G(\boldsymbol{q}), \tag{14.15}$$

where

$$G(\boldsymbol{q}) = \left\langle \sum_{m,n} \mathrm{e}^{-\mathrm{i}\boldsymbol{q}.(\boldsymbol{R}_m-\boldsymbol{R}_n)} \right\rangle \tag{14.16}$$

is the geometrical factor of the self-organized objects and $\boldsymbol{q} = \boldsymbol{Q} - \boldsymbol{h}$ is the reduced scattering vector with respect to the reciprocal lattice point $\boldsymbol{h}$ of the non-deformed lattice. In contrast to Eq. (14.5), here we take into account also various depths of the objects Z_m below the surface. In deriving Eq. (14.15) we have again assumed $\langle \sum_m \mathrm{e}^{-\mathrm{i}\boldsymbol{Q}\boldsymbol{R}_m} \rangle = 0$ for any $\boldsymbol{Q} \neq 0$.

For the calculation of the geometrical factor, short-range and long-range order models can be used in the same way as for small-angle scattering. For the calculation of the displacement field $\boldsymbol{v}(\boldsymbol{r})$ around a single buried object the approach described in Sect. 10.3 can be used if the sample surface is flat. For the objects at the free surface, or if the surface is corrugated, the analytic method of Sect. 10.3 is not applicable. In this case, a numerical calculation based on the finite-element solution of the elastic equilibrium equations can be applied [46, 71, 237, 272, 275]. The displacement field of a self-organized object can also be obtained using an atomistic approach, minimizing the potential energy of the lattice and using semi-empirical interatomic potentials [90, 185, 218, 230, 251, 275]. Several studies have been devoted to the comparison of the strain fields around self-assembled quantum dots calculated by various methods with experimental data obtained by a image analysis of the contrast of transmission electron micrographs [71, 180, 221, 245, 293]. These works confirmed the validity of the conventional elasticity method even for the smallest self-assembled objects.

If the deformation fields of different objects *do* overlap, Eqs. (14.14) and (14.15) are not valid. In this case, the incoherently scattered intensity can be calculated relatively easily if the objects create an ideally periodic lattice. We calculate the total displacement field $\boldsymbol{u}(\boldsymbol{r})$ caused by the lattice of the objects using the continuum elasticity approach according to Sect. 10.3; the function $\Omega(\boldsymbol{r})$ occurring in Eq. (10.19) is replaced by the shape function $\Omega_{\text{latt}}(\boldsymbol{r})$ of the *lattice* of the objects. Let us assume that the objects create an ideal two-dimensional lattice with the basis vectors $\boldsymbol{a}_{1,2}$ parallel to the sample surface in the depth Z below the surface, and we construct its reciprocal lattice with the basis vectors $\boldsymbol{b}_{1,2}$, $\boldsymbol{a}_j.\boldsymbol{b}_k = 2\pi\delta_{jk}$. We denote the vectors of this reciprocal lattice by

$$\boldsymbol{G} = G_1\boldsymbol{b}_1 + G_2\boldsymbol{b}_2,\ G_{1,2} = 0, \pm 1, \pm 2, \ldots$$

The total displacement $\boldsymbol{u}(\boldsymbol{r})$ is periodic along the horizontal axes and we can express it as a two-dimensional Fourier series

$$\boldsymbol{u}(\boldsymbol{r}) = \sum_{\boldsymbol{G}} \boldsymbol{u}_{\boldsymbol{G}}(z; Z)\mathrm{e}^{\mathrm{i}\boldsymbol{G}.\boldsymbol{r}_{\parallel}}.$$

The coefficients $\boldsymbol{u}_{\boldsymbol{G}}(z; Z)$ can be obtained using Eqs. (10.18), (10.19), and (10.22) choosing $\boldsymbol{\kappa} = \boldsymbol{G}_{\parallel}$ in Eq. (10.22). The resulting expression for the reciprocal-space distribution of the diffusely scattered intensity is

$$J_{\text{incoh}}(\boldsymbol{q}) = \text{const.} \sum_{\boldsymbol{G}} \delta^{(2)}(\boldsymbol{q}_{\parallel} - \boldsymbol{G}) \left|F_{\boldsymbol{G}}(q_z)\right|^2, \tag{14.17}$$

where

$$F_{\boldsymbol{G}}(q_z) = \int_{-\infty}^{0} \mathrm{d}z \int_S \mathrm{d}^2\boldsymbol{r}_{\parallel} \mathrm{e}^{-\mathrm{i}q_z z} \mathrm{e}^{-\mathrm{i}\boldsymbol{G}.\boldsymbol{r}_{\parallel}} \times$$
$$\times \left[\left(\chi_{\boldsymbol{h}} + \Delta\chi_{\boldsymbol{h}} \sum_j \Omega(\boldsymbol{r}_{\parallel}, z - Z_j)\right) \phi(\boldsymbol{r}) - \chi_{\boldsymbol{h}}\right] \tag{14.18}$$

and

$$\phi(\boldsymbol{r}) = \exp\left[-\frac{\mathrm{i}}{S}\sum_{j,\boldsymbol{G}'} \boldsymbol{h}.\boldsymbol{u}_{\boldsymbol{G}'}(z;Z_j)\mathrm{e}^{\mathrm{i}\boldsymbol{G}'.\boldsymbol{r}_\parallel}\right]. \tag{14.19}$$

In these expressions, the sum $\sum_j$ runs over the depths Z_j of the two-dimensional arrays of the objects and $S = |\boldsymbol{a}_1 \times \boldsymbol{a}_2|$ is the area of the primitive unit cell of the two-dimensional array.

From Eq. (14.17) it follows that the intensity distribution consists of a periodic sequence of infinitely narrow lateral maxima analogous to crystal truncation rods. In experimental data, these δ-like peaks are broadened due to finite resolution of the diffractometer. The position of the lateral maxima is determined by the reciprocal lattice points $\boldsymbol{G}$; the heights of the maxima depend both on the shape and the strain field in and around the objects. Formula (14.17) can be modified for the case of slightly disordered positions of the objects. Then, the scattered intensity distribution can be approximately expressed as

$$J_{\text{incoh}}(\boldsymbol{q}) \approx \text{const.} \sum_{\boldsymbol{G}} \mathcal{R}_{\boldsymbol{G}}(\boldsymbol{q}_\parallel - \boldsymbol{G})\,|F_{\boldsymbol{G}}(q_z)|^2, \tag{14.20}$$

where the functions $\mathcal{R}_{\boldsymbol{G}}$ are chosen so that they account both for a limited resolution of the diffractometer and for the broadening of the lateral maxima due to the disorder of the object positions. If this disorder can be modelled by the short-range-order model, suitable functions $\mathcal{R}_G$ are Lorentzians, whose widths depend on $\boldsymbol{G}$ as

$$w_{\boldsymbol{G}} = w_0 + w_1|\boldsymbol{G}|^2 \tag{14.21}$$

and the heights of these functions are inversely proportional to $|w_{\boldsymbol{G}}|^2$. The constants $w_{0,1}$ correspond to the device resolution and the position disorder, respectively. It is necessary to point out that Eq. (14.20) is approximative, it is valid only if the neighboring lateral maxima do not overlap, i.e., if the objects are well-ordered.

We present several examples demonstrating the applicability of the theory above. In the first example, we show intensity maps measured in a coplanar symmetric geometry on superlattices PbSe/PbEuTe [166, 168, 339, 340]. The samples have been grown by molecular beam epitaxy on a cleaved BaF_2 substrate with thick (111)PbTe buffer layers. All superlattice periods consisted of five PbSe monolayers and $Pb_{1-x}Eu_xTe$ spacer layers with various thicknesses – namely, 36 nm (sample A), 46 nm (sample B), and 66 nm (sample C). The period was repeated 100 times. The Eu content has been adjusted so that the structure was strain symmetrized, i.e., the average lattice constant of the superlattice equaled that of the PbTe buffer. Because of the -5.4% lattice mismatch between PbSe and PbTe, island growth in the Stranski-Krastanow growth mode occurs if the critical thickness of about 1.5 monolayers of PbSe is exceeded. During the overgrowth of the PbSe islands by PbEuTe, a complete replanarization of the growth front occurs after the growth of about

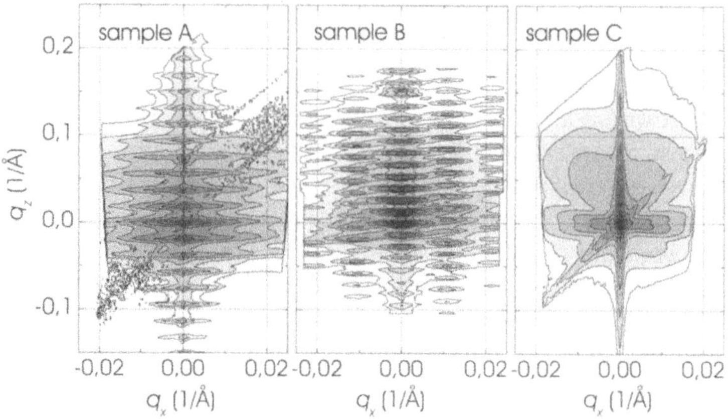

Fig. 14.15. Intensity maps of PbSe/PbEuTe superlattices taken in the symmetric 111 diffraction, coplanar geometry. The scales of the q_x and q_z axes are different [168, 339].

20 nm PbEuTe. It has been demonstrated both theoretically and experimentally [166, 339] that the arrangement of the PbSe islands is determined by local strains at the growing surface due to the buried islands. For superlattice periods smaller than about 40 nm, the islands are arranged vertically and their lateral ordering is very weak (sample A). If the period is between 40 and 60 nm (sample B), a nearly perfect three-dimensional ordering occurs; the islands create a trigonal lattice very similar to a fcc lattice. A transmission electron micrograph of the cross section of sample B in Fig. 14.5 (right panel) demonstrates this kind of ordering. If the superlatice period exceeds about

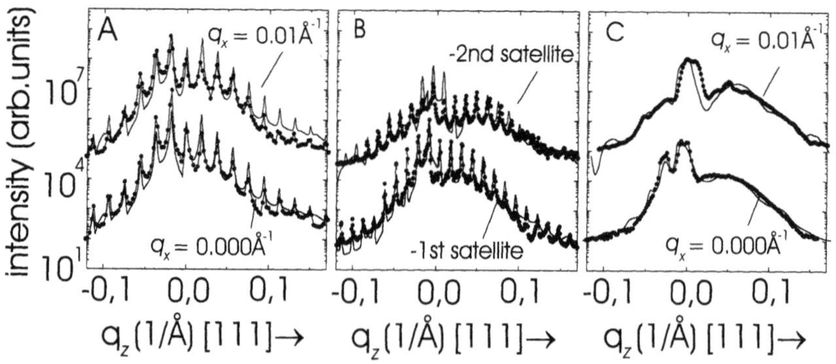

Fig. 14.16. The vertical scans along the lines q_x = const. extracted from the intensity maps in Fig. 14.15 (points) and their theoretical simulations (lines).

60 nm, the local strain modulation at the growing surface is so shallow that no ordering of the island positions takes place (sample C).

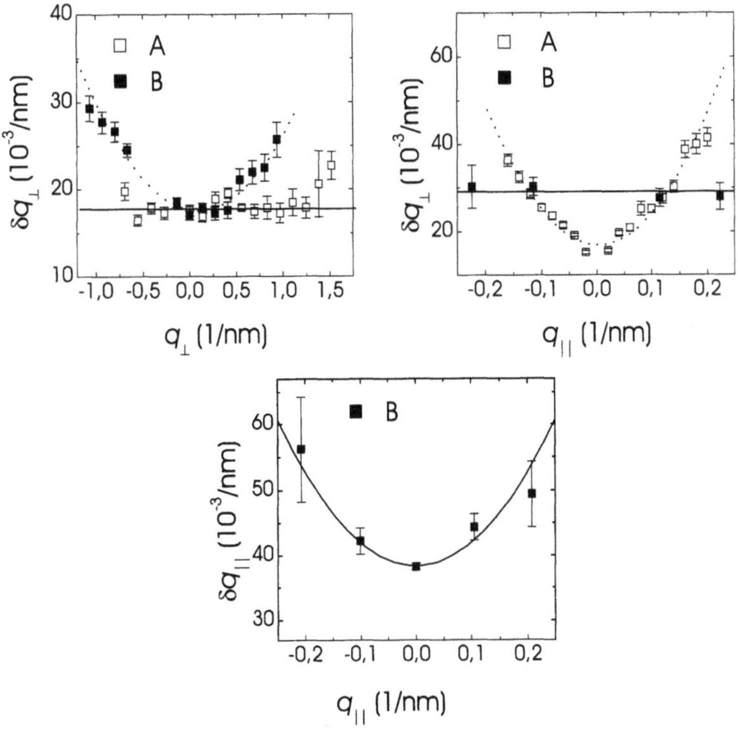

Fig. 14.17. The widths of the intensity maxima in vertical ($\delta q_\perp$) and horizontal directions ($\delta q_{\|}$) plotted as functions of $q_\perp$ and $q_{\|}$ for samples A and B and their fits by parabolic functions.

The intensity maps measured in symmetric 111 diffraction (wavelength 0.154 nm) are shown in Fig. 14.15. In sample A, the island positions are vertically correlated, and their lateral correlation is weak. Therefore, the geometrical factor $G(\boldsymbol{q})$ occurring in Eqs. (14.15,14.16) exhibits narrow maxima along q_z and no lateral maxima along $\boldsymbol{q}_{\|}$. This is the reason that the scattered intensity is concentrated along horizontal sheets, the period of the sheets is $2\pi/D$, where D is the superlattice period. This effect is fully analogous to intensity sheets observed in diffuse x-ray reflection in the case of a multilayer with vertically correlated roughness of interfaces (see Chapter 11). The intensity map of sample B shows a two-dimensional periodic pattern of maxima; these maxima correspond to the cross sections of the scattering plane with the reciprocal lattice points

$$\boldsymbol{G} = G_1\boldsymbol{b}_1 + G_2\boldsymbol{b}_2 + G_3\boldsymbol{b}_3, \ G_{1,2,3} = 0, \pm 1, \pm 2, \ldots$$

of the *three-dimensional* lattice reciprocal to the trigonal lattice of the islands. The islands in sample C are not correlated completely; therefore the scattered intensity is a incoherent superposition of the contributions of individual islands and no intensity maxima appear.

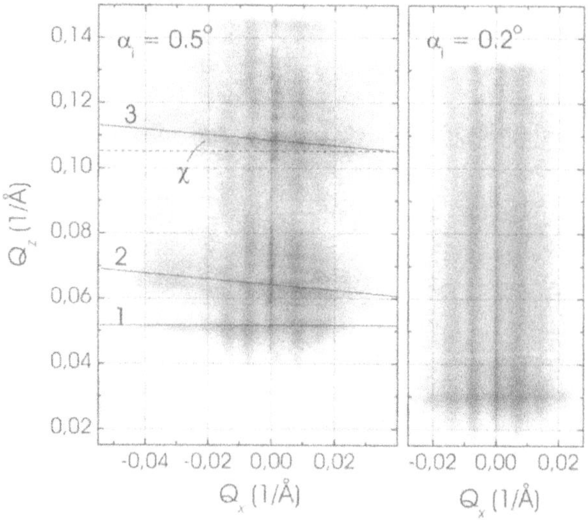

Fig. 14.18. The intensity distribution measured for two values of the angle of incidence α_i in the plane perpendicular to the self-organized quantum wires in a SiGe/Si superlattice, 400 diffraction, GID geometry [286].

Using the theory above (Eqs. (14.15), (14.16), and the short-range-order model (14.6)), we have simulated the intensity distribution along the lines parallel to the q_z-axis for various q_x. In the simulations, we have used the shapes and correlations of the islands determined from small-angle scattering and the only parameter adjusted during the simulation was the chemical composition of the islands – we have assumed that the islands are composed of $PbSe_{1-x}Te_x$. The results of the simulations are plotted in Fig. 14.16. The best agreement of the measured data with the theory (shown in this figure) has been achieved for $x = 0.6 \pm 0.1$. Therefore, the chemical composition of the islands changes considerably during their overgrowth.

In order to check the validity of the short-range-order model, we have plotted the widths $\delta q_{\|}$ and $\delta q_{\perp}$ of the intensity maxima as functions of their coordinates $q_{\|} \equiv q_x$ and $q_{\perp} \equiv q_z$ (Fig. 14.17) [168]. For sample A, the vertical width $q_{\perp}$ of the maxima depends quadratically on $q_{\|}$, indicating that the vertical correlation of the lateral positions of the islands obeys this model (see also [188]). The width $q_{\perp}$ does not depend on q_z and it corresponds to the experimental resolution; therefore the vertical correlation of the vertical dot positions is perfect and it follows from the superlattice periodicity. The lateral width $\delta q_{\|}$ is very large for this sample. For sample B, $\delta q_{\|}$ and $\delta q_{\perp}$ depend quadratically on $q_{\|}$ and $q_{\perp}$, respectively; therefore the short-range-order model applies for the lateral correlation of lateral positions and vertical correlation of vertical positions of the islands. On the other hand, the $\delta q_{\perp}$ is nearly independent on $q_{\|}$, and the model is not valid.

The second example shows the intensity maps measured in the GID geometry on a SiGe/Si superlattice [286]. The sample consists of 20 periods of nominally 2.5-nm-thick $Si_{0.55}Ge_{0.45}$ layer and a 10-nm Si spacer grown by molecular beam epitaxy on a (001) Si substrate with a large 3.5° miscut towards [100]. During the growth, the step-bunching mechanism occurred, giving rise to a quasi-periodic sequence of terraces and step bunches, sketched schematically in Fig. 14.1b. This structure induced the growth of a periodic one-dimensional array of self-organized SiGe quantum wires elongated parallel to [010], i.e., perpendicular to the miscut direction. This one-dimensional structure is visible in the intensity maps measured in the GID geometry in diffraction 400, where the diffraction vector $\boldsymbol{h}$ was perpendicular to the wires (Fig. 14.18). The intensity has been measured in the $(q_x q_z)$ plane, where the q_x-axis is parallel to $\boldsymbol{h}$ and q_z is perpendicular to the surface (the radial plane, see Chap. 3).

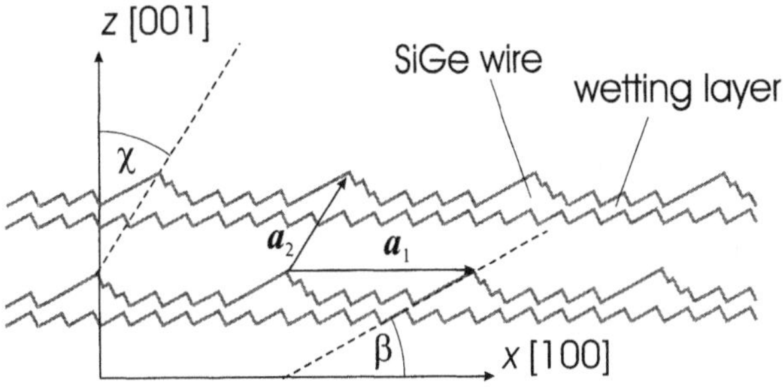

Fig. 14.19. The sketch of the arrangement of the quantum wires.

During the measurement, the angle of incidence α_i of the primary beam was kept constant. In the map in the left panel, this angle was larger than the critical angle α_c of total external reflection, so that the penetration depth of the primary radiation was larger than the superlattice thickness. In the right panel, the incidence angle was smaller than α_c and the penetration depth was smaller than the superlattice period D. Therefore, the correlation of the positions of the wires at different interfaces affects the intensity distribution in the left panel only, where the incoming x-ray beam can "feel" the vertical superlattice periodicity.

The positions of the wires at different interfaces are correlated, and they create a two-dimensional array with the basis vectors $\boldsymbol{a}_{1,2}$ sketched schematically in Fig. 14.19. This correlation gives rise to intensity maxima corresponding to the points of the lattice reciprocal to the array of the wires, lying in the given reciprocal plane. The horizontal distance of these maxima is $2\pi/\langle a_1 \rangle$, where $\langle a_1 \rangle$ is the mean lateral distance of the wires. The positions of the wires at different interfaces are not correlated vertically, but the correlation direction makes the angle χ with the vertical axis. Therefore, the intensity maxima in reciprocal plane are arranged along lines inclined by χ from the horizontal direction. These lines are denoted in Fig. 14.18 by numbers 2 and 3. The line 1 in this figure connects the intensity maxima, where the angle of exit α_f equals the critical angle α_c. These maxima have nothing to do with the positions of the wires and they are analogous to the Yoneda peaks introduced in Sect. 7.2.1. Figure 14.18 shows the intensity distribution in the plane $(q_x q_z)$ perpendicular to the wires and parallel to the diffraction vector. In the intensity distribution in the plane $(q_y q_z)$ parallel to the wires, no intensity maxima of this kind are present; the scattering centers are indeed arranged in the (xz) plane.

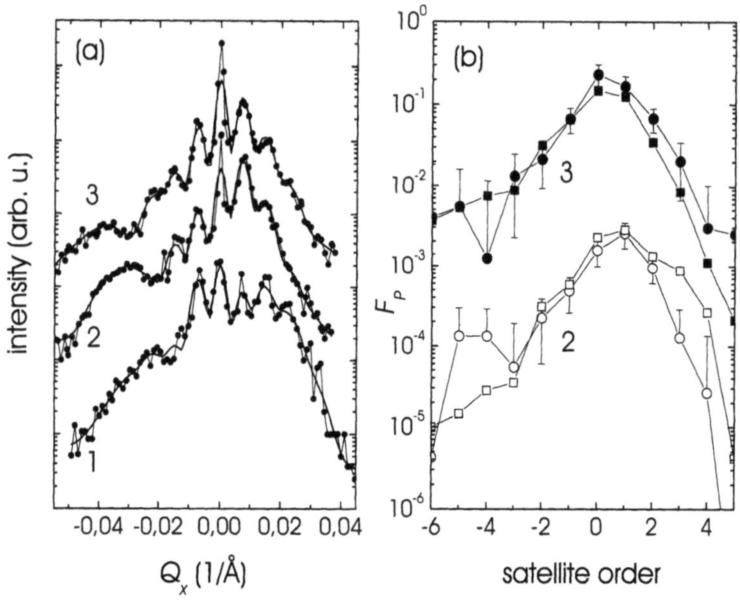

Fig. 14.20. The linear scans extracted from the intensity map in Fig. 14.18 (points) and their fit by the short-range-order model (lines) (a); the fit of the integrated intensities of the lateral maxima (b).

In order to determine the chemical composition of the wires, we have extracted linear scans from the intensity map measured at $\alpha_i = 0.5°$ along lines 2 and 3 in Fig. 14.18. The scans, shown in Fig. 14.20, have been fitted by a sequence of Lorentzian maxima, the widths of which obey the quadratic law (14.21) following from the short-range-order model. The integral intensities of the maxima are proportional to $|F_G(q_z)|^2$ defined in Eq. (14.18), and we have fitted them using the theoretical approach described above. From the fit we have estimated the mean Ge content in the wires to $x_{\mathrm{Ge}} = (20 \pm 10\%)$. Therefore, the Ge concentration in the wires is substantially smaller than the nominal value 45%, probably due to interdiffusion during the overgrowth.

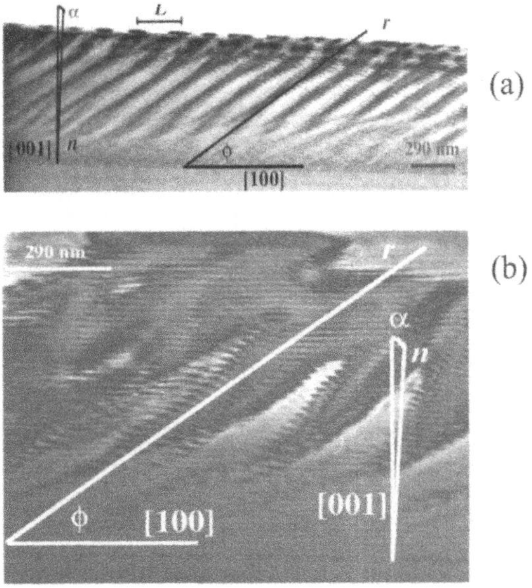

Fig. 14.21. Cross-sectional transmission electron micrographs of GaInAs/GaAs/GaPAs/GaAs superlattices grown on a miscut (001) GaAs substrates. The strain amplitudes of the multilayers are $\pm 7 \times 10^{-3}$ (a) and $\pm 1.2 \times 10^{-2}$ (b). The lateral modulation of the thicknesses of individual layers is visible by oblique black-and-white stripes [134]. The angle ϕ denotes the direction of the oblique replication of the step bunches at different interfaces, α is the substrate miscut angle.

In a series of papers [35, 134] a spontaneous lateral modulation of the thicknesses of individual layers in GaInAs/GaAs/GaPAs/GaAs superlattices has been investigated both by coplanar x-ray diffraction and GID. The chemical compositions of the layers were chosen so that the lattice mismatches of both GaInAs and GaPAs layers with respect to the GaAs substrate have the same magnitudes and opposite signs (symmetrically strained layers). The thin GaAs layers in the multilayer period prevent the intermixing during the MOVPE growth. Owing to local strains at a growing surface, an intense step-bunching process takes place during the epitaxial growth of these structures, which leads to a lateral modulation of the layer thicknesses in the direction [100] that coincides to the azimuthal direction of the miscut. Two samples have been investigated having the same miscut of about 2° of the (001) planes with respect to the surface, the samples differ in the vertical strain amplitude, $\pm 7 \times 10^{-3}$ and $\pm 1.2 \times 10^{-2}$.

In Fig. 14.21 cross-sectional transmission electron micrographs are presented for both samples showing that the lateral modulation of different layers in the superlattice stack is correlated in an oblique direction. In contrast to the previous example of SiGe/Si step bunches, the step-bunches in the III-V ternary layers are nearly perfectly periodic and they are much better developed. The periodicity of the bunches is much better developed in the sample with smaller strain (panel(a)). In the other sample, the direction of the oblique replication changes during the growth (panel (b)). Grazing-incidence diffraction intensity maps of these samples have been taken with the wavelength 1.215Å and they are show in Fig. 14.22.

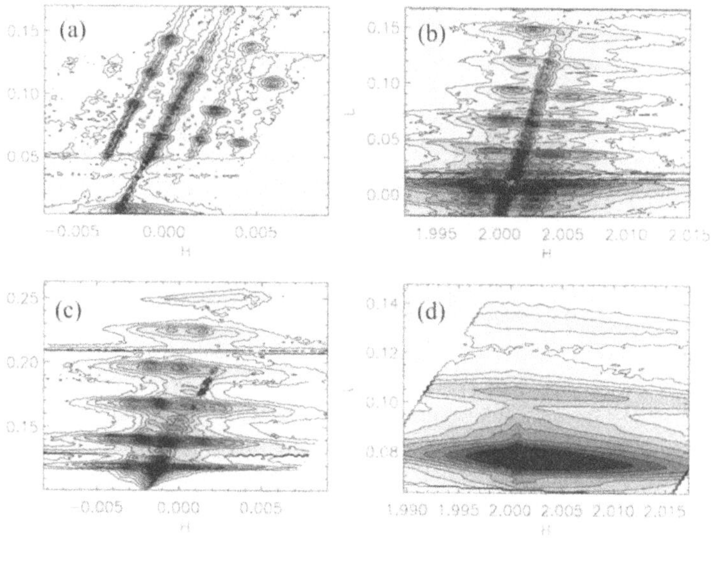

Fig. 14.22. Reciprocal-space map of the intensity diffracted in the GID geometry in the strain-insensitive diffraction 020 (a and c) and in the strain-sensitive diffraction 200 (b and d). GaInAs/GaAs/GaPAs/GaAs superlattices with the strain amplitude of $\pm 7 \times 10^{-3}$ (a and b) and $\pm 1.2 \times 10^{-2}$ (c and d) [134].

The reciprocal planes, where intensity was measured, were always parallel to the modulation direction, so that the lateral intensity satellites are visible. The distance of the satellites is inversely proportional to the mean distance of the step bunches. In diffraction 020 (panels a and c) the diffraction vector was perpendicular to the modulation direction (i.e., parallel to the step bunches). Since the local displacement field $\boldsymbol{u}(\boldsymbol{r})$ caused by the step bunches is perpendicular to the bunches, $\boldsymbol{h}.\boldsymbol{u}(\boldsymbol{r}) = 0$ in this case; and the measured intensity distribution is caused *only* by the chemical contrast, i.e., by the inhomogeneities in $\chi_{\boldsymbol{h}}(\boldsymbol{r})$. In diffraction 200 (panels b and d), the diffraction vector $\boldsymbol{h}$ is perpendicular to the step bunches, i.e., $\boldsymbol{h}.\boldsymbol{u}(\boldsymbol{r}) \neq 0$. In this case, the measured contrast is determined mainly by the displacement field. From the figure it is obvious that the chemical contrast yields much better developed intensity satellites than the strain contrast. The satellites are sharper in the sample with smaller vertical strain amplitude; this finding corresponds with the transmission electron micrographs in Fig. 14.21.

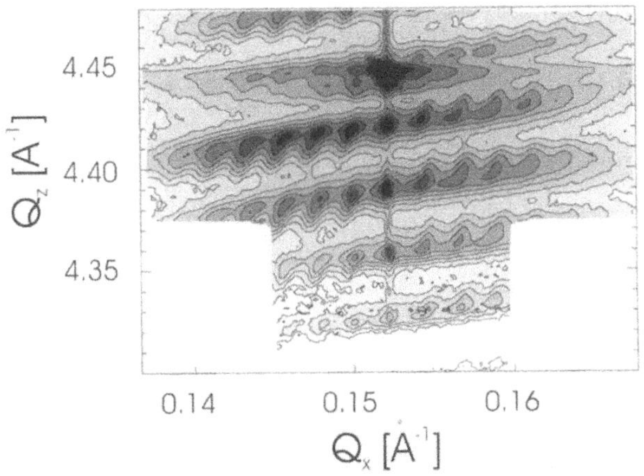

Fig. 14.23. Reciprocal space map of the intensity diffracted in coplanar 004 diffraction, GaInAs/GaAs/GaPAs/GaAs superlattice with the strain amplitude of $\pm 1.2 \times 10^{-2}$ [134].

The changes of the direction of oblique replication in the sample with larger strain amplitude visible in Fig. 14.21(b) affects the shape of the satellite maxima in coplanar diffraction 004 (see Fig. 14.23). The rows of the satellite maxima are inclined by the angle ϕ of the mean oblique replication from the vertical direction and the maxima are "tear-shaped" due to the change of the angle ϕ during the growth.

For quasi zero-dimensional quantum dots where the displacement field is nearly cylindrically symmetric, the angular scans in the GID geometry are nearly not affected by the strains in the sample. Then the intensity distribution can be calculated in the same way as for small-angle scattering, using Eqs. (14.4) and (14.5). This is shown in Fig. 14.24, where we present the GID measurement on a 19-period SiGe/Si superlattice [187]. Each period was constituted nominally by 5.5 Ge monolayers on a 30-nm-thick Si spacer layer.

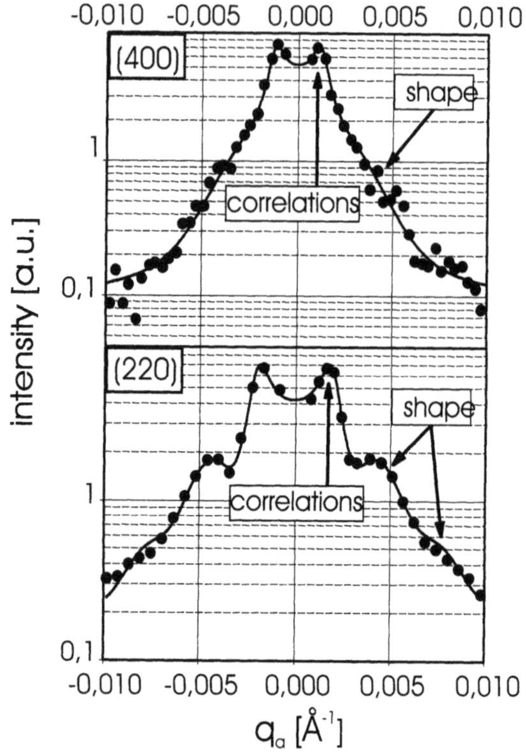

Fig. 14.24. The angular scans measured for two diffractions 400 and 220 in a GID geometry, SiGe/Si superlattice. q_a is the component of the scattering vector $\boldsymbol{Q}$ perpendicular to the diffraction vector [187].

At the SiGe/Si interfaces, the self-organized quantum dots created a disordered quadratic array, and the measurements have been carried out for two diffractions 400 and 220 with different azimuthal directions of the diffraction vector. The resulting angular scans shown in the figure are not sensitive to the strains in the dots, and they reflect the correlation of their positions and their shape. In the scans, several intensity maxima can be resolved; a numerical fit to the theory revealed that the maxima with smallest $|\boldsymbol{Q}_{\|}|$ stem from the correlation function, the other are caused by the dot shape. From the fit, the azimuthal dependence of the geometrical factor follows, indicating that the positions of the islands are better correlated in certain crystallographic directions.

X-ray diffraction can be used also for the investigation of self-organized objects situated at the free surface of the sample. In this case, the function $\Psi(\boldsymbol{r})$ defined in Eq. (14.13) comprises the shape of the object and the strain field in it, as well as the strain field induced by the object in the substrate underneath. In Fig. 14.25 we present the intensity map measured in coplanar

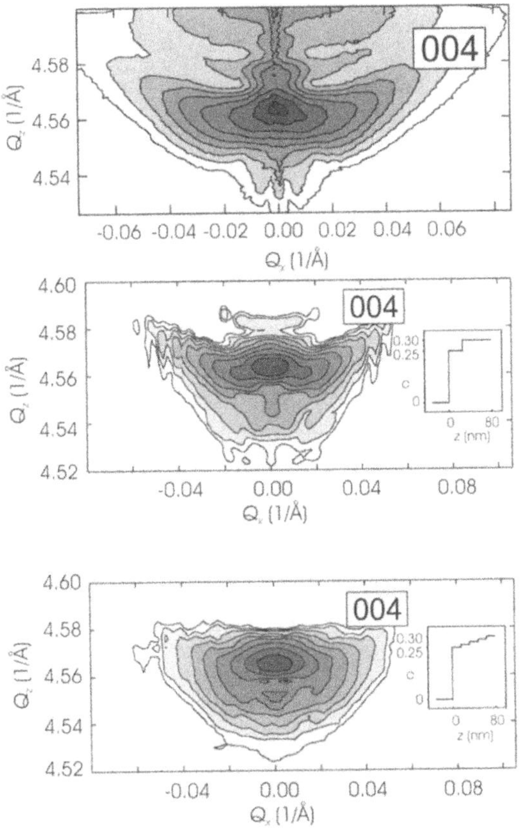

Fig. 14.25. Coplanar intensity map measured in the symmetric 004 diffraction on a Si sample with SiGe self-organized islands on the surface (left) and its simulation using two choices of the vertical Ge profile in the island (right). The assumed Ge profiles are shown in the insets [389].

symmetric 004 diffraction on a Si sample with a single SiGe layer on the top with self-organized SiGe islands. The layer has been grown by liquid-phase epitaxy method [389], the resulting islands have {111} side facets and a (001) top facet. The measured intensity map has been simulated numerically using Eq. (14.15) and assuming completely uncorrelated positions of the islands ($G(\boldsymbol{q}) = n$). From the figure it follows that the best agreement with the experimental data has been achieved for a relatively abrupt change in the vertical profile of the Ge concentration. Similar studies have been performed, among others, in [153, 205, 345, 348].

Similar studies have been published in [153] for free-standing Ge self-organized islands at a Si(001) surface. From a series of reciprocal space maps in the coplanar asymmetric 224 diffraction measured before and after over-

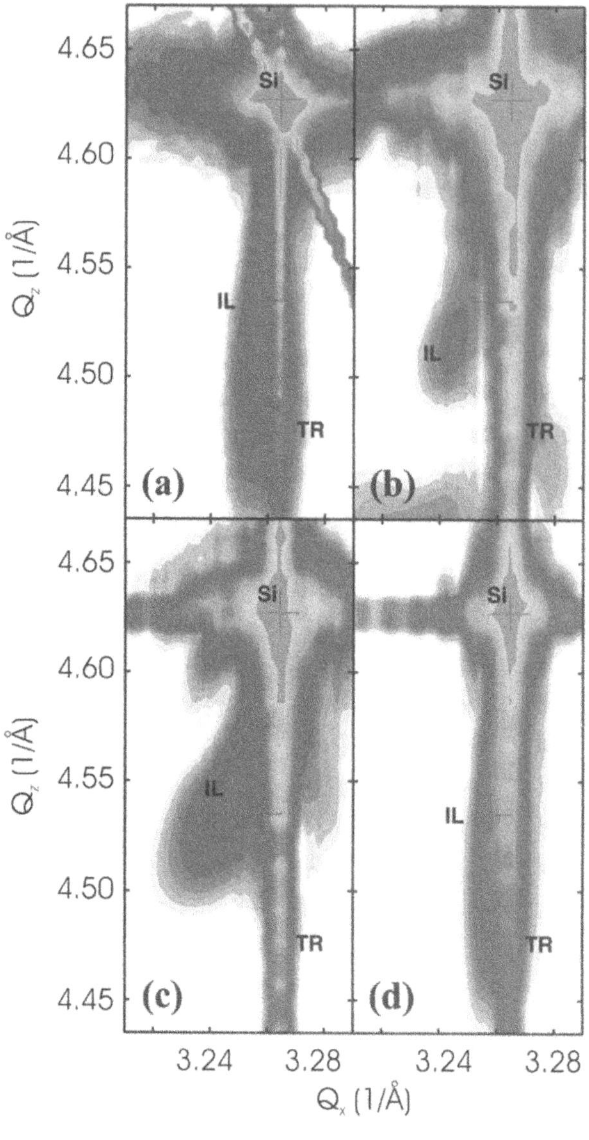

Fig. 14.26. The experimental reciprocal-space map of Ge quantum dots capped by a thick Si layer (a) (coplanar diffraction 224). The intensity distribution in (b) was simulated, assuming the same shape and chemical composition as for the free-standing islands. In (c) the Ge concentration profile was optimized, and finally in (d) both the concentration profile and shape were optimized [153]. "TR" and "IL" denote the coherent truncation rod and the island maximum, respectively.

growth of relatively thick Si capping layer (160 nm), the changes in the chemical composition and shape of the dots could be determined . Figure 14.26 shows an experimental reciprocal space map after the overgrowth (a) and series of simulations (b-d). In (b) the parameters of the simulation were obtained from the fit of the reciprocal space map measured before the Si capping, (c) shows the simulation results where the Ge concentration profile was optimized. In (d), both the shape and the Ge concentration profile were optimized. The resulting Ge vertical concentration profile and the profiles of the strain components $\epsilon_{xx,zz}$ are plotted in Fig. 14.27.

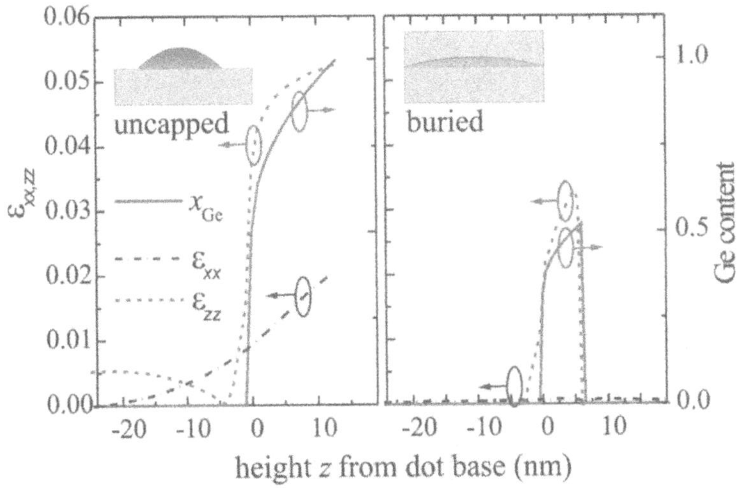

Fig. 14.27. The vertical profiles of the Ge concentration and the components of the strain tensor for a free-standing island (left) and for an island capped by a thick Si layer (right) [153].

Due to the capping, the average Ge content drastically decreases and the islands considerably flatten. The strain status of the capped islands approaches that of a homogeneous two-dimensional layer, since the lateral elastic relaxation ϵ_{xx} of the lattice in the island is substantially decreased.

In [189, 190, 191] a method has been found that enables us to determine both the shape and the strains in free-standing self-organized islands *directly* from a three-dimensional intensity distribution in reciprocal space measured in the GID geometry. This so called x-ray iso-strain method can be used for relatively large islands with a large lateral elastic relaxation, if the condition

$$h\epsilon(z) \gg \frac{2\pi}{L(z)} \tag{14.22}$$

is valid. Here h denotes the length of the diffraction vector, $\epsilon(z)$ is the lateral elastic strain in the island in in height z above the free surface with respect

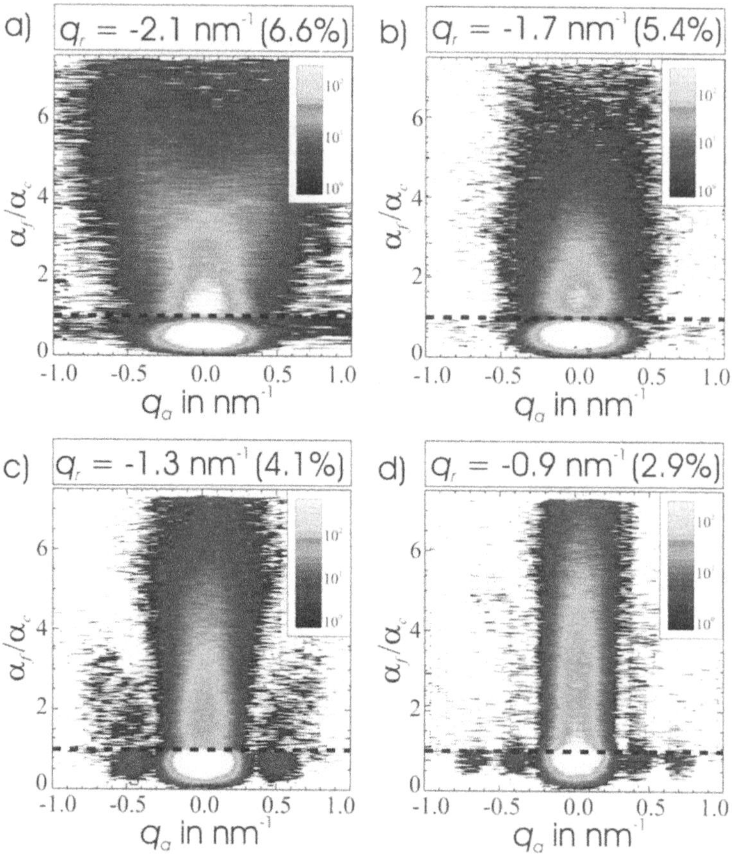

Fig. 14.28. The series of two-dimensional intensity maps measured in 220 GID diffraction on a sample with self-organized free-standing InAs quantum dots [191]; see the text for the details.

to the substrate, and $L(z)$ is the width of the island in the height z. If the condition (14.22) is fulfilled, the intensity in a given point (q_x, q_y, q_z) in reciprocal space stems from the volume in the island, where the local diffraction vector $\boldsymbol{h}_{\text{def}} = \boldsymbol{h} - \nabla(\boldsymbol{h}.\boldsymbol{u})$ obeys the condition $\boldsymbol{q} = \boldsymbol{h}^{\text{def}} - \boldsymbol{h}$ (the iso-strain volume). Assuming, for instance, a laterally homogeneous lateral elastic strain in the island structure (i.e., the strain depends on z only), the angular width of the intensity maximum for given radial coordinate q_x (the q_x axis is parallel to $\boldsymbol{h}$) is inversely proportional to the size $L^{(\epsilon)}(z)$ of the volume, where the lateral strain is $\epsilon = -q_x/(h\delta)$.

In the works cited above, the iso-strain method was used for non-capped islands deposited on a free surface. In this case, four scattering processes shown in Fig. 7.5 occur in the DWBA approach that determine the resulting intensity. The coherent superposition of the primary and specularly reflected waves gives rise to a standing wave that modulates the intensity of the non-disturbed wavefield. Using this modulation, the vertical position of the iso-strain volume can be determined.

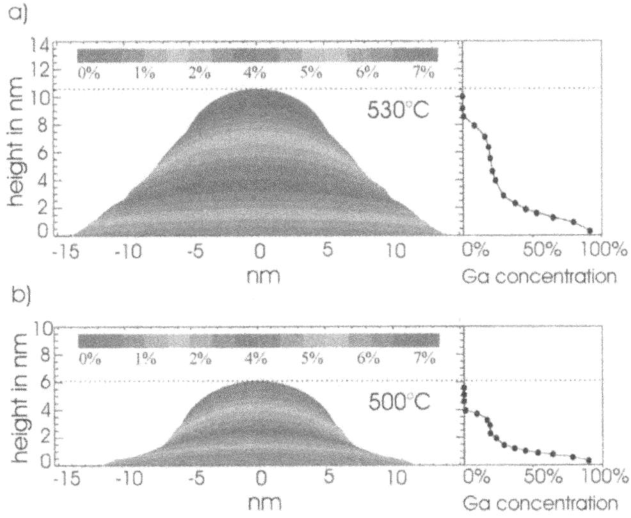

Fig. 14.29. The distribution of the local lateral strain in the island determined from the GID measurement (left), the vertical In profile following from the GID (middle), and the local lateral strain calculated using an atomistic approach (right) using the In profile in the middle panel [191].

The method is demonstrated in Figs. 14.28 and 14.29, showing the GID measurements on InAs self-organized islands grown on the top of an GaAs/AlAs short period superlattice [191]. In Fig. 14.28 the intensity measured in 220 diffraction was depicted in a series of two-dimensional (q_y, α_f) maps measured for various q_x and constant angle of incidence α_i. In Fig. 14.28 the coordinate q_y perpendicular to the diffraction vector $\boldsymbol{h}$ is denoted as q_a; the coordinate q_x (denoted q_r in the figure) is parallel to $\boldsymbol{h}$. In the vertical axes of the maps, the values α_f/α_c are plotted; α_c is the critical angle of total reflection, α_f is the angle of exit. Each map depicts the intensity distribution stemming from the region of the island with a given value of the lateral strain ϵ. From this three-dimensional intensity distribution both the shape and the vertical distribution of In atoms was obtained directly using the approach described above. The resulting distribution of the local value of lateral strain and the In profile are plotted in Fig. 14.29 along with the strain distribution calculated numerically using an atomistic approach. A remarkably good coincidence was achieved.

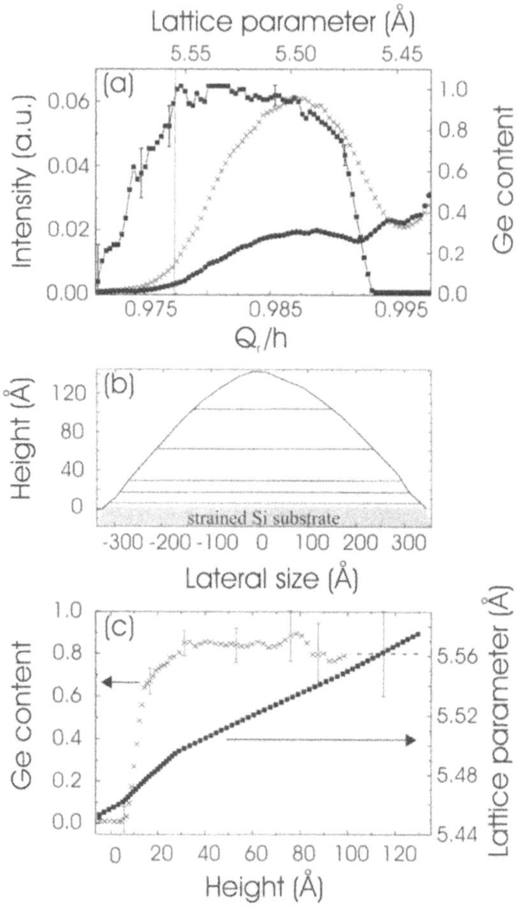

Fig. 14.30. (a) Radial scans in GID geometry around the (800) Bragg reflection, recorded at 11043 eV (crosses) and 11103 eV (dots). The intensity ratio is plotted as squares; (b) lateral dimension and (c) Ge content of SiGe islands and the lateral lattice parameter as functions of the z-coordinate obtained from anomalous x-ray scattering (from [320]).

Recently, a new method has been developed enabling a direct determination of chemical composition of quantum dots using the GID method [228, 319, 320]. The method is based on a strong energy dependence of the polarizability coefficient χ_h close to the absorption edge of an element present in the crystal lattice of the dot (so called "anomalous-scattering method").

For SiGe dots, for instance, the measurements close to the Kα edge of Ge at 11.1036 keV and far away are performed. Close to the Ge absorption edge, the real part of the atomic scattering factor f_{Ge} exhibits a deep minimum and the scattered intensity stems mainly from the Si atoms. Far from the absorption edge, the scattering from Ge atoms prevails, due to the difference in the atomic numbers of Ge and Si. The method is demonstrated in Fig. 14.30, where the radial (q_x) scans of SiGe quantum dots are plotted measured in GID geometry for two different wavelengths (panel (a)). From the intensity scans along q_y (i.e., in the direction in reciprocal space perpendicular to the diffraction vector) for various values of q_x, the size $L^{(\epsilon)}$ of the iso-strain volume could be determined in the same way as in the iso-strain method described above (panel b in Fig. 14.30). From the ratio of the intensities measured at the absorption edge and far from it in a given point q_x, one determines the chemical composition $x_{\mathrm{Ge}}(\epsilon)$ of the iso-strain volume corresponding to the given value of $\epsilon = -q_x/h$. Combining the functions $L^{(\epsilon)}(z)$ and $x_{\mathrm{Ge}}(\epsilon)$ both the shape and the vertical profile of the chemical composition could be determined (panels b,c in Fig. 14.30).

References

1. D. L. Abernathy, G. Gr"ubel: J. Synchrotron Rad. **5**, 37 (1998). T. Thurn-Albrecht, W. Steffen, A. Patkowski, G. Maier, E.W. Fischer, G. Grübel, D.L. Abernathy: Phys. Rev. Lett. **77**, 5437 (1996).
2. S. Adachi: J. Appl. Phys. **53** 8773 (1982); **58** R1 (1985).
3. A.M. Afanasev, M.K. Melkoyan: Acta Cryst. **39**, 207 (1983).
4. A.M. Afanasev, O.G. Melikyan: phys. stat. sol. (a) **122**, 459 (1990).
5. P.A. Aleksandrov, A.M. Afanasev, S.A. Stepanov: phys. stat. sol. (a) **86**, 143 (1984).
6. P.A. Aleksandrov, A.M. Afanasev, S.A. Stepanov: Sov. Phys. Cryst. **29**, 119 (1984).
7. O.L. Alerhand, D. Vanderbilt, R.D. Meade, and J.D. Joannopoulos, Phys. Rev. Lett. **61**, 1973 (1988).
8. J. Als-Nielsen, D. McMorrow: *Elements of Modern X-Ray Physics* (John Wiley & Sons, Ltd., New York 2001).
9. Amptek Firm Description (Amptek Inc. Bedford, MA 01730-2204 U.S.A. 1996).
10. E. Anastassakis, E. Liarokapis: phys. stat. sol. **149**, K1 (1988).
11. V.V. Aristov, A.I. Erko, V.V. Martynov: X-Ray Sci. Technol. **3**, 211 (1992).
12. V.V. Aristov, A.Yu. Nikulin, A.A. Snigirev, P. Zaumseil: phys. stat. sol. (a) **95**, 81 (1986).
13. V.V. Aristov, U. Winter, A.Yu. Nikulin, S.V. Redkin, A.A. Snigirev, P. Zaumseil, V.A. Yunkin: phys. stat. sol. (a) **108**, 651 (1988).
14. U.W. Arndt: J. Appl. Cryst. **19**, 145 (1986).
15. B.M. Arora, K.S. Chandrasekaran, M.R. Gokhale, G. Nair, G. Venugopal Rao, G. Amarenda, B. Viswanathan: J.Appl.Phys. **87**, 8444 (2000).
16. R.J. Asaro, W.A. Tiller: Metall. Trans. **3**, 1789 (1972).
17. J.E. Ayers: J. Cryst. Growth **135**, 71 (1994).
18. A. Authier: *Dynamical Theory of X-Ray Diffraction* (Oxford Univ. Press, Oxford 2001).
19. V. Ayvazyan et al.: Eur. Phys. J. **D20**, 149 (2002).
20. L.V. Azároff, R. Kaplow, N. Kato, R.J. Weiss, A.J.C. Wilson, R.A. Young: *X-Ray Diffraction*, (McGraw-Hill, New York 1974).
21. D. Babboneau, I.R. Videnović, M.G. Garnier, P. Oelhafen: Phys. Rev. B **63**, 195401 (2001).
22. D. Bahr, W. Press, R. Jebasinski, S. Mantl: Phys. Rev. B **47**, 4385 (1993).
23. D. Bahr, W. Press, R. Jebasinski, S. Mantl, Phys. Rev. B **51**, 12223 (1995).
24. R.I. Barabash, M.A. Krivoglaz: Sov. Phys. Solid State **29**, 1768 (1987).
25. A.L. Barabási, H.E. Stanley: *Fractal Concepts in Surface Growth* (Cambridge University Press, Cambridge 1995).

26. W.J. Bartels, W. Nijman: J.Cryst. Growth **44**, 518 (1978).
27. W.J. Bartels: J. Vac. Sci. Technol. **B1**, 338 (1984).
28. W.J. Bartels, J. Hornstra, D.J.W. Lobeek: Acta Cryst. A **42**, 539 (1986).
29. G.T. Baumbach, H.-G. Brühl, H. Rhan, U. Pietsch: J. Appl. Cryst. **21**, 386 (1988).
30. G.T. Baumbach, H. Rhan, U. Pietsch: phys. stat. sol.(a) **109**, K7 (1988).
31. G.T. Baumbach, G. Oelgart, H.-G. Brühl, U. Pietsch, E. Tuncel, F. Morier-Genoud, F. Martin, F.K. Reinhart, B. Lengeler: Semicond. Sci. Technol. **7**, 304 (1992).
32. G.T. Baumbach, M. Gailhanou: J. Phys. D: Appl. Phys. **28**, 2321 (1995).
33. G.T. Baumbach, S. Tixier, U. Pietsch, V. Holý: Phys. Rev. B **51**, 16848 (1995).
34. G.T. Baumbach, P. Mikulík: In *Proceedings of the Ecole Française de Reflectivité* (Marseille 1997).
35. G.T. Baumbach, C. Giannini, D. Luebbert, R. Felici, L. Tapfer, T. Marschner, W. Stolz: J. Phys. D **32**, A212 (1999).
36. G.T. Baumbach, D. Lübbert: J. Phys. D: Appl. Phys. **32**, 726 (1999).
37. G.T. Baumbach, D. Lübbert, M. Gailhanou: J. Phys. D: Appl. Phys. **32**, A208 (1999).
38. G.T. Baumbach, D. Lübbert, M. Gailhanou: Jap. Phys. D: Appl. Phys. **32**, A208 (1999).
39. T. Baumbach, P. Mikulík: *X-Ray Reflectivity by Rough Multilayers.* In: *X-Ray and Neutron Reflectivity: Principles and Applications*, ed. J. Daillant, A. Gibaud, Lecture Notes in Physics (Springer, Berlin 1999).
40. G.T. Baumbach, D. Lübbert, M. Gailhanou: J. Appl. Phys. **87**, 3744 (2000),
41. G.T. Baumbach: *X-Ray Scattering by Periodic Nanostructures*, Habilschrift (University of Potsdam 2000).
42. U. Beck, T.H. Metzger, J. Peisl, J.R. Patel: Appl. Phys. Lett. **76**, 2698 (2000).
43. http://www.bede.co.uk/products/ AdvancedXrayTechnologies
44. T. Bedynska: phys. stat. sol. (a) **19**, 365 (1973).
45. C.W.J. Beenakker and H. van Haiten: Solid State Physics **44**, Semiconductor Heterostructures and Nanostructures (Academic, Boston 1991).
46. T. Benabbas, Y. Androussi, A. Lefebvre: J. Appl. Phys. **86**, 1945 (1999).
47. S. Bensoussan, C. Malgrange, M. Sauvage-Simkin: J. Appl. Cryst. **20**, 230 (1987).
48. M. Bhattacharya, M. Mukherjee, M. K. Sanyal, Th. Geue, J. Grenzer, U. Pietsch: J. Appl. Phys. **94**, 2882 (2003).
49. G. Biasol, E. Martinet, F. Reinhardt, A. Gustafsson, E. Kapon: J. Cryst. Growth **170**, 600 (1997).
50. G. Biasol, F. Reinhardt, A. Gustafsson, E. Kapon: Appl. Phys. Lett. **71**, 1831 (1997).
51. G. Biasiol, E. Kapon: Phys. Rev. Lett. **81**, 2962 (1998); J. Cryst. Growth **201/202**, 62 (1999).
52. BICRON web site: http://www.bicrondirect.com/merchant/servlet/merch
53. D. Bimberg, V.A. Shchukin, N.N. Ledentsov, A. Krost, F. Heinrichsdorff: Appl. Surf. Sci.**130-132** 713-718 (1998).
54. D. Bimberg, M. Grundmann, N.N. Ledentsov: MRS Bulletin **23**, 31 (1998);
55. J. Birch, J.-E. Sudgren, P.F. Fewster: J. Appl. Phys. **78**, 6562 (1995).
56. J. Bläsing: private communication.
57. Y. Bodenthin, J. Grenzer, R. Lauter, U. Pietsch, P. Lehmann, D.K. Kurth, H. Möhwald : J. Synchr. Rad. **9**, 206 (2002).

58. U. Bonse, G. Materlik, W. Schr"oder: J. Appl.Cryst. **9**. 232 (1976).
59. J.C. Borkowski, M. Kopp: Rev. Sci. Instr. **39**, 1515 (1968).
60. M. Born, E. Wolf: *Principles of Optics* (Cambridge Univ. Press, Cambridge 1999).
61. D.K. Bowen, B.K. Tanner: *High Resolution X-ray Diffractometry and Topography* (Taylor & Francis, London 1998).
62. Betriebsanleitung für das PSD System PSD-50M (MBRAUN GmbH, Garching 1992).
63. W. Braun, B. Jenichen, V.M. Kaganer, A.G. Shtukenberg, L. Daweritz, K.H. Ploog: Surf. Sci. **525**,136 (2003).
64. W. Braun, B. Jenichen, V.M. Kaganer, A.G. Shtukenberg, L. Daweritz, K.H. Ploog: J. Cryst. Growth **251**, 56 (2003).
65. M. Bruel: Electron. Lett. **31**, 1201 (1995).
66. M. Bruel, World patent No. 99/05711.
67. H.-G. Brühl, G.T. Baumbach, V. Gottschalch, U. Pietsch, B. Lengeler: J. Appl. Cryst. **23**, 228 (1990).
68. H.-G. Brühl, U. Pietsch, B. Lengeler: J. Appl. Cryst. **21**, 240 (1988).
69. O. Brümmer, H.R. Höche, J. Nieber: phys. stat. sol. (a) **37**, 529 (1976).
70. B. Buras, J.S. Olsen, L. Gerward: Nucl. Instr. Methods. **178**, 131 (1980).
71. A. Carlsson, L.R. Wallenberg, C. Persson, W. Seifert: Surf. Sci. **406**, 48 (1998).
72. S.L. Chang: *Multiple Diffraction of X-Rays in Crystals* (Springer, Berlin 1984).
73. E. Chason, T. M. Mayer, Critical Reviews in Solid State and Material Sciences **22**, 1 (1997).
74. J.F. Chen, P.Y. Wang, J.S. Wang, N.C. Chen, X.J. Guo, Y.F. Chen: J. Appl. Phys. **87**, 1251 (2000).
75. R. Cingolani, H. Lage, L. Tapfer, H. Kalt, D. Heitmann, K. Ploog: Phys. Rev. Lett. **67**, 891 (1991).
76. P. Coppens, D. Cox, E. Vlieg, I.K. Robinson: *Synchrotron Radiation Crystallography* (Academic Press, New York 1992).
77. J. M. Cowley: *Diffraction Physics* (North-Holland, Amsterdam 1975).
78. P. Croce, L. Névot: Rev. Phys. Appl. **11**, 113 (1976).
79. J. Daillant, O. Bélorgey: J. Chem. Phys. **97**, 5824 (1992).
80. J. Daillant, A. Gibaud: *X-Ray and Neutron Reflectivity: Principles and Applications.* Lecture Notes in Physics (Springer, Berlin 1999).
81. A.A. Darhuber, E. Koppensteiner, H. Straub, B. Brunthaler, W. Faschinger, G. Bauer: J. Appl Phys. **76**, 7816 (1994).
82. A.A. Darhuber, E. Koppensteiner, G. Bauer, P.D. Wang, Y.P. Song, C.M. Sotomayor Torres, M.C. Holland: J. Phys. D: Appl. Phys. **28**, A195 (1995).
83. A.A. Darhuber, P. Schittenhelm, V. Holý, J. Stangl, G. Bauer, A. Abstreiter: Phys. Rev. B **55**, 15652 (1997).
84. A.A. Darhuber, T. Grill, J. Stangl, G. Bauer, D.J. Lockwood, J.-P. Noel, P. D. Wang, C.M. Sotomayor Torres: Phys. Rev. B **58**, 4825 (1998).
85. A.A. Darhuber, G. Bauer, R.D. Wang, C.M. Sotomayor Torres: J. Appl. Phys. **83**, 126 (1998).
86. N. Darowski, K. Paschke, U. Pietsch, K. Wang, A. Forchel, T. Baumbach, U. Zeimer: J. Phys. D: Appl. Phys. **30**, L55 (1997).
87. N. Darowski, U. Pietsch, Y. Zhuang, S. Zerlauth, G. Bauer, D. Lübbert, T. Baumbach: Appl. Phys. Lett. **73**, 806 (1998).
88. N. Darowski, U. Pietsch, U. Zeimer, V. Smirnitzki, F. Bugge: J. Appl. Phys. **84**, 1366 (1998).

89. N. Darowski: Charakterisierung mesoscopischer Halbleiter-Lateralstrukturen mittels Photoluminescens und Röntgenbeugung unter streifendem Einfall PhD Thesis, (Università Potsdam, Potsdam 1999).
90. I. Daruka, A.L. Barabasi, S.J. Zhou, T.C. German, P.S. Lomdahl, A.R. Bishop: Phys. Rev. B **60**, R1250 (1999).
91. I. Daruka, J. Tersoff, A.L. Barabasi: Phys. Rev. Lett. **82**, 2753 (1999).
92. A.S. Davydov: *Quantum Mechanics* (Oxford, Pergamon Press 1976).
93. D.K.G. deBoer: Phys. Rev. B **44**, 498 (1991).
94. D.K.G. deBoer: Phys. Rev. B **49**, 5817 (1994).
95. D.K.G. deBoer: Phys. Rev. B **53**, 6049 (1996).
96. L. DeCaro, P. Sciacovelli, L. Tapfer: Appl. Phys. Lett. **64**, 34 (1994).
97. L. DeCaro, L. Tapfer, A. Giuffrida: Phys. Rev. B **54**, 10575 (1996).
98. L. DeCaro, C. Giannini, L. Tapfer: Phys. Rev. B **56**, 9744 (1997).
99. P.H. Dederichs, G. Leibfried: Phys. Rev. B **188**, 1175 (1969).
100. P.H. Dederichs: Phys. Rev B. **4**, 1041 (1971).
101. P.H. Dederichs, J. Pollmann: Z. Physik **255**, 315 (1972).
102. P.H. Dederichs: J. Phys. F **3**, 471 (1973).
103. H. deJeu, J.D. Shindler, A.E.L. Mol: J. Appl. Phys. **29**, 511 (1996).
104. S. Di Fonzo, W. Jark, S. Lagomarsino, C. Giannini, L. De Caro, A. Cedola, M. Muller: Nature **403**, 638 (2000).
105. J.W.M. DuMond: Phys. Rev. **52**, 872 (1937).
106. K. Eberl, O.G. Schmidt, S. Schieker, N.Y. Jin-Phillip, F. Phillip: Solid State Electronics **42**, 1593-1597 (1998).
107. H. Ekstein: Phys. Rev **68**, 120 (1945).
108. U. Englisch, T.A. Barberka, U. Pietsch, U.Höhne: Thin Solid Films, **266**, 234 (1995).
109. P. Erhart, H. Trinkaus, B. C. Larson: Phys. Rev. B **25**, 834 (1982).
110. P. Erhart: J. Nucl. Mater. **216**, 170 (1994).
111. A. Erko, D. Roshchupkin, a. Snigerev, A. Smolovich, A. Nikulin, G. Vereshchagin: Nucl. Instr. Meth. A **282**, 634 (1989).
112. A. Erko, F. Schäfers, B. Vidal, A. Yakshin, U. Pietsch, W. Mahler: Rev. Sci. Instrum. **66**, 4845(1995).
113. A.I. Erko, B. Vidal, P. Vincent, Yu.A. Agafonov, V.V. Martynov, D.V. Roschupkin, M. Brunel, Nucl. Instrum. Methods Phys. Res. **A 333**, 599 (1993).
114. J.D. Eshelby: *The Continuum Theory of Lattice Defects*. In: *Solid State Physics*, eds. F. Seitz and D. Turnbull, Vol. 3 (Academic Press, New York 1956).
115. J. Eymery, J.M. Hartmann, G.T. Baumbach: J. Cryst. Growth **184**, 109 (1998).
116. J. Eymery, D. Buttard, F. Fournel, H. Moriceau, T. Baumbach, D. Lübbert: Phys. Rev. B **65**, 165337 (2000).
117. J. Eymery, F. Leroy, F. Fournel, NIM B**200** 73 (2003)
118. A.R. Faruqi, C.C. Bond: Nucl. Instr. Methods **201**, 125 (1982).
119. R. Feidenhansl: Surf. Sci. Rep. **10**, 105 (1989).
120. P.F. Fewster: Semicond. Sci. Technol. **8**, 1915 (1993).
121. P. F. Fewster: *X-Ray Scattering from Semiconductors* (Imperial College Press, London 2001).
122. F. Fournel, H. Moriceau, N. Magnea, J. Eymery, J.L. Rouviere, K. Rousseau, B. Aspar: Mat. Sci. Eng. **B73**, 42 (2000).

123. F. Fournel, H. Moriceau, N. Magnea, J. Eymery, J.L. Rouviere, K. Rousseau, Appl. Phys. Lett. **80** 793 (2002)
124. M. Gailhanou, T. Baumbach, U. Marti, P.C. Silva, F.K. Reinhart, M. Ilegems: Appl. Phys. Lett. **62**, 1623 (1993).
125. H. Gao: J. Mech. Phys. Solids **42**, 741 (1994).
126. P. Gay, P.B. Hirsch, A. Kelly, Acta Metall. **1**, 315 (1953).
127. T.F. Gerhard: *Neue Methoden der Röntgenbeugung mit Hilfe von Synchrotronstrahlung*, Thesis, Julius-Maximilians-Universität (Würzburg 1999).
128. Ch. Gerthsen, H.Vogel: *Physik, ein Lehrbuch zum Gebrauch neben Vorlesungen* (Springer, Heidelberg 1993).
129. Th. Geue, M. Schultz, J. Grenzer, A. Natansohn, P. Rochon: J. Appl. Phys. **87**, 7712 (2000).
130. T. Geue, O. Henneberg, U. Pietsch: Cryst.Res.Technol. **37**, 770 (2002).
131. T. Geue, O. Henneberg, J. Grenzer, U. Pietsch, A. Natansohn, P. Rochon, K. Finkelstein: Colloids and Surfaces A **198-200**, 31 (2002).
132. Th. Geue, M.G. Saphiannikova, O.Henneberg, U.Pietsch: J. Appl. Phys. **93**, 3161 (2003).
133. C. Giannini, L. Tapfer, Y. Zhuang, L. DeCaro: Phys. Rev. B **55**, 5276 (1997).
134. C. Giannini, G.T. Baumbach, D. Luebbert, R. Felici, L. Tapfer, T. Marschner, W. Stolz, N.Y. Jin-Phillipp, F. Phillipp: Phys. Rev. B **61**, 2173 (2000).
135. F. Glas, J. Appl. Phys. **70**, 3556-3571 (1991).
136. M.S. Goorsky and B.K. Tanner: Crystal Res.and Techn. **37**, 647 (2002).
137. J. Grenzer, N. Darowski, U. Pietsch, A. Daniel, S. Rennon, J.P. Reithmaier, A. Forchel: Appl. Phys. Lett. **77**, 4277 (2000).
138. S. Grigorian, J. Grenzer, S. Feranchuk, U. Zeimer, U. Pietsch: J. Phys. D: Appl. Phys. **36**, A222 (2003).
139. S. Grotehans, G. Wallner, E. Burkel, H. Metzger, J. Peisl, H. Wagner: Phys. Rev. B **39**, 8450 (1989).
140. G. Gr"ubel, J. Als-Nielsen, A.K. Freund: J. Phys. (France) IV, Colloque **4**, C9 (1994).
141. A. Guinier, G. Fournet, *Small-Angle Scattering of X-Rays* (Wiley, New York 1955).
142. A. Guinier: *X-Ray Diffraction: In Crystals, Imperfect Crystals, and Amorphous Bodies* (Dover Publ., 1994).
143. J. E. Guyer, P. W. Voorhees, Phys. Rev. B **54**, 11710 (1996).
144. P. Haier, B.A. Herrmann, N. Esser, U. Pietsch, K. Lüders: PROC. EMRS Strasbourg 1997.
145. M. Hart, R.D. Rodriguez: J. Appl. Cryst. **7**, 123 (1978).
146. J. Härtwig: phys. stat. sol. (a) **37**, 417 (1976).
147. J. Härtwig: phys. stat. sol. (a) **42**, 495 (1977).
148. J. Härtwig: Acta Crystallogr. A **37**, 802 (1981).
149. K. Haruta: J. Appl. Phys. **38**, 3312 (1967).
150. H. Hashizume, M. Sauvage, J.F. Petroff, P.Riglet, B. Capelle: *Rapport dactivite* (Lure 1978).
151. H. Heinke, M.O. Möller, D. Hommel, G. Landwehr: J. Cryst. Growth **135**, 41 (1994).
152. O. Henneberg, Th. Geue, M. Saphiannikova, U. Pietsch: J. Phys. D: Appl. Phys. **36**, A241 (2003).
153. A. Hesse, J. Stangl, V. Holý, T. Roch, G. Bauer, O. G. Schmidt, U. Denker, B. Struth: Phys. Rev. B **66**, 085321 (2002).

154. A. Hesse, J. Stangl, V. Holý, G. Bauer, O. Kirfel, E. Müller, D. Grützmacher: Materials Science and Engineering B **101**, 71 (2003).
155. P.B. Hirsch, G.N. Ramachandran: Acta Cryst. **3**, 187 (1950).
156. J.P. Hirth, J. Lothe: *Theory of Dislocations* (Krieger, Malabar 1992).
157. V. Holý, J. Kuběna, E. Abramof, K. Lischka, A. Pesek, E. Koppensteiner: J. Appl. Phys. **74**, 1736 (1993).
158. V. Holý, L. Tapfer, L. Koppensteiner, G. Bauer, H. Laage, O. Brandt, L. Ploog: Appl. Phys. Lett. **63**, 3140 (1993).
159. V. Holý, T. Baumbach: Phys. Rev. B **49**, 10668 (1994).
160. V. Holý, K. Wolf, M. Kastner, H. Stanzl, W. Gebhardt: J. Appl. Cryst. **27**, 551 (1994).
161. V. Holý, T. Baumbach, M. Bessière: J. Phys. D: Appl. Phys. **28**, A220 (1995).
162. V. Holý, A.A. Darhuber, G. Bauer, P.D. Wang, Y.P. Song, C.M. Sotomayor Torres, M.C. Holland: Phys. Rev. B **52**, 8348 (1995).
163. V. Holý: *High-Resolution X-Ray Diffractometry of Thin Layers — Theoretical Aspects*, Habilitation (Masaryk University Brno 1996).
164. V. Holý, C. Giannini, L. Tapfer, T. Marschner, W. Stolz: Phys. Rev. B **55**, 9960 (1997).
165. V. Holý, A.A. Darhuber, J. Stangl, S. Zerlauth, F. Schäffler, G. Bauer, N. Darowski, D. Luebbert, U. Pietsch, I. Vávra: Phys. Rev. B **58**, 7934 (1998).
166. V. Holý, G. Springholz, M. Pinczolits, G. Bauer: Phys.Rev. Lett. 83, **356** (1999).
167. V. Holý, U. Pietsch, T. Baumbach: *High Resolution X-Ray Scattering from Thin Films and Multilayers.* Springer Tracts in Modern Physics 149 (Springer, Berlin 1999).
168. V. Holý: J. Stangl, G. Springholz, M. Pinczolits, G. Bauer: In *Morphological and Compositional Evolution of Heteroepitaxial Semiconductor Thin Films*, eds. J. Mirecki Millunchick, A.L. Barabasi, N.A. Modine, E.D. Jones, Symposium proceedings of Material Res. Society, Vol. 618, 161 (2001).
169. J. Hornstra: J. Phys. Chem. Solids **5**, 129 (1958).
(1978).
170. M.J. Howes, D.V. Morgan (eds): *Charge-coupled devices and systems* (John Wiley, New York 1979).
171. S.M. Hu: J. Appl. Phys. **66**, 2741, (1989).
172. K. Huang: Proc. Roy. Soc. A **190**, 102 (1947).
173. K. Hümmer, E. Weckert: *Determination of Reflection Phases by Three-Beam Diffraction.* In: *X-Ray and Neutron Dynamical Diffraction: Theory and Applications*, eds: A. Authier, S. Lagomarsino and B. K. Tanner, NATO ASI Series B: Physics, Vol. 357 (Plenum Press, New York 1996).
174. V. L. Indenbom, J. Lothe: *Elastic Strain Fields and Dislocation Mobility* (North-Holland Amsterdam, London, New York, Tokyo 1992).
175. R.W. James: *The Optical Principles of X-ray Diffraction* (Bell, London 1950).
176. B. Jenichen, R. Köhler: private communication.
177. B. Jenichen, W. Braun, V.M. Kaganer, A.G. Shtukenberg, L. Daweritz, C.G. Schulz, K.H. Ploog, A. Erko: Rev. Sci. Instr. **74**, 1267 (2003).
178. M. Jergel, P. Mikulík, E. Majková, Š. Luby, R. Senderák, E. Pinčík, M. Brunel, P. Hudek, I. Kostič, A. Konečníková: J. Phys. D: Appl. Phys. **32**, A188 (1999).
179. D.E. Jesson, K.M. Chen, S.J. Pennycook: MRS Bulletin, pp. 31–37 (April 1996).

180. N.Y. Jin-Phillip, F. Phillip: J. Appl. Phys. **88**, 710 (2000).
181. V.M. Kaganer, S.A. Stepanov, and R. Köhler: Phys. Rev. B **52**, 16369 (1995).
182. V.M. Kaganer, R. Koehler, M. Schmidbauer, R. Opitz, B. Jenichen: Phys. Rev. B **55**, 1793 (1997).
183. V.M. Kaganer, K.H. Ploog: Phys. Rev. B **64**, 205301 (2001).
184. V.M. Kaganer, B. Jenichen, G. Paris, K.H. Ploog, O. Konovalov, P. Mikulik, S. Arai: Phys. Rev. B **66**, 035310 (2002).
185. M. Karimi, H. Yates, J.R. Ray, T. Kaplan, M. Mostoller: Phys. Rev. B**58**, 6019 (1998).
186. M. Kaestner, B. Voigtlaender: Phys. Rev. Lett. **82**, 2745 (1999).
187. I. Kegel, T.H. Metzger, J. Peisl, P. Schittenhelm, G. Abstreiter: Appl. Phys. Lett. **74**, 2978 (1999).
188. I. Kegel, T.H. Metzger, J. Peisl, J. Stangl, G. Bauer, D. Smilgies: Phys. Rev. B **60**, 2516 (1999).
189. I. Kegel, T.H. Metzger, P. Fratzl, J. Peisl, A. Lorke, J.M. Garcia, P.M. Petroff: Europhysics Lett. **45**, 222 (1999).
190. I. Kegel, T.H. Metzger, A. Lorke, J. Peisl, J. Stangl, G. Bauer, J.M. Garcia, P.M. Petroff: Phys. Rev. Lett. **85**, 1694 (2000).
191. I. Kegel, T.H. Metzger, A. Lorke, J. Peisl, J. Stangl, G. Bauer, K. Nordlund, W.V. Schoenfeld, P.M. Petroff: Phys. Rev. B **63**, 035318(13) (2001).
192. O. Kienzle, F. Ernst, M. Ruehle, O.G. Schmidt, K. Eberl: Appl. Phys. Lett. **74**, 269 (1999).
193. H. Kiessig: Ann. Phys. **10**, 769 (1931).
194. S. Kishino: J. Phys. Soc. Japan **31**, 1168 (1971); S. Kishino, K. Kohra: Jap. J. Appl. Phys. **10**, 551 (1971); S. Kishino, A. Noda, K. Kohra: J. Phys. Soc. Japan **33**, 158 (1972).
195. Ch. Kittel: *Introduction to Solid State Physics* (J. Wiley, New York 1996).
196. W. Kleber: *Einführung in die Kristallographie* (Verlag der Technik, Berlin 1980).
197. U. Klemradt: *Thesis* (Jülich 1994).
198. E.E. Koch: *Handbook of Synchrotron Radiation* (North-Holland, Amsterdam 1983).
199. F. Kohlrausch: *Praktische Physik: Zum Gebrauch für Unterricht, Forschung und Technik*, (Teubner, Stuttgart 1996).
200. K. Kohra, S. Kikuta: Acta Cryst. **24**, 200 (1968).
201. T. Kojima, M. Tamura, H. Nakaya, S. Tanaka, S. Tamura, S. Arai: Jpn. J. Appl. Phys. **37**, 261 (1998).
202. E.A. Kondrashkina, S.A. Stepanov, R. Opitz, M. Schmidbauer, R. Köhler, R. Hey, M. Wassermeier, D.V. Novikov: Phys. Rev. B **56**, 10469 (1997).
203. E. Koppensteiner, P. Hamberger, G. Bauer, V. Holý, E. Kasper: Appl. Phys. Lett. **64**, 172 (1994).
204. J.B. Kortright: J. Appl. Phys. **70**, 3620 (1991).
205. Z. Kovats, T.H. Metzger, J. Peisl, J. Stangl, M. Muehlberger, Y. Zhuang, F. Schaeffler, G. Bauer: Appl. Phys. Lett. **76**, 3409 (2000).
206. J. Krim and G. Palasantzas: Int. J. Mod. Phys. **9**, 599 (1995).
207. M.A. Krivoglaz: *Diffraction of X-Rays and Neutrons for Non-Ideal Crystals* (Springer, Berlin, Heidelberg 1996).
208. J. Krug, H. Spohn: *Solids far from Equilibrium*, ed. C. Godreche (Cambridge University Press, Cambridge 1992).

209. C. Kunz: *Synchrotron Radiation, Techniques and Application* (Springer, Heidelberg 1979).
210. R.N. Kyutt, P.V. Petrashen, L.M. Sorokin: phys. stat. sol. (a) **60**, 381 (1980).
211. S. Labat, P. Gergaud, O. Thomas, B. Gilles, A. Marty: J. Appl. Phys. **87**, 1172 (2000).
212. B.C. Larson, W. Schmatz: Phys. Rev. B **10**, 2307 (1974).
213. M. v. Laue: *Röntgenstrahlinterferenzen* (Akademische Verlagsgesellschaft Geest & Portig K.-G. Leipzig, 1937).
214. O. Leifeld, D. Gruetzmacher, B. Mueller, K. Kern: Mat. Res. Soc. Symp. Proc. Vol. 533 (1998).
215. W. Leitenberger, S.M. Kuznetsov, A. Snigiriev, Optics Comm. **191**, 91 (2001).
216. C.S. Lent, P.I. Cohen: Surf. Sci. **139**, 121 (1984).
217. L. Leprince, A. Talneau, G.T. Baumbach, M. Gailhanou, J. Schneck: Appl. Phys. Lett. **71**, 3227 (1997).
218. E. Lidorikis, M.E. Bachlechner, R.K. Kalia, A. Nakao, P. Vashishta, G.Z. Voyiadjis: Phys. Rev. Lett. **87**, 086104 (2001).
219. R. Lipperheide, G. Reiss, H. Leeb, H. Fiedeldey, S.A. Sofianos: Phys. Rev. B **51**, 11032 (1995); G. Reiss, R. Lipperheide: Phys. Rev. B **53**, 8157 (1996).
220. F. Liu, J. Tersoff, M. G. Lagally: Phys. Rev. Lett. **80**, 1268 (1998).
221. Ch.-P. Liu, J. Murray Gibson, D.G. Cahill, T.I. Kamins, D.P. Basile, R. Stanley Williams: Phys. Rev. Lett. **84**, 1958 (2000).
222. A.A. Lomov, P. Zaumseil, U. Winter: Acta Cryst. A **41**, 223 (1985).
223. D. Lübbert, *Strain and Lattice Distortion in Semiconductor Structures: A Synchrotron Radiation Study*, Thesis, (Universität Potsdam, Potsdam 1999).
224. D. Lübbert, G.T. Baumbach, S. Ponti, U. Pietsch, L. Leprince, J. Schneck, A. Talneau: Europhys. Lett. **46**, 479 (1999).
225. D. Lübbert, J. Arthur, M. Sztucki, T.H. Metzger, P.B. Griffin, J.R. Patel: Appl. Phys. Lett. **81**, 3167 (2002).
226. A.T. Macrander, S.E.G. Slusky: Appl. Phys. Lett. **56**, 443 (1990).
227. O.Madelung: *Numerical data and functional relationships in science and technology /Landoldt-Börnstein* (Springer, Heidelberg 1993).
228. R. Magalhaes-Paniago, G. Medeiros-Ribeiro, A. Malachias, S. Kycia, T.I. Kamins, R. Stan Williams: Phys. Rev. B **66**, 245312 (2002).
229. W. Mahler, T.A. Barberka, U. Pietsch, U. Höhne, H.-J. Merle: Thin Solid Films **256**, 198 (1995).
230. M.A. Makeev, A. Madhukar: Phys. Rev. Lett. **86**, 5542 (2001).
231. V.I. Marchenko, A. Ya. Parshin: Sov. Phys. JETP **52**, 129 (1980).
232. V.I. Marchenko: JETP Lett. **33**, 381 (1981).
233. G. Margeritondo: *Introduction to Synchrotron Radiation Research* (Oxford University Press, Oxford 1988).
234. W.C. Marra, P. Eisenberger, A.Y. Cho: J. Appl. Phys. **50**, 6927 (1979).
235. R.A. Masumura, G. Sines: J. Appl. Phys. **41**, 3930 (1970).
236. G. Medeiros-Ribeiro, A.M. Bratkowski, T.I. Kamins, D.A.A. Ohlberg, R. Stanley Williams: Science **279**, 353 (1998).
237. M. Meixner, E. Schoell, M. Schmidbauer, H. Raidt, R. Koehler: Phys. Rev. B **64**, 245307 (2001).
238. H. Metzger, C. Luidl, U. Pietsch, U. Vierl: Nucl. Instr. Meth. A**350**, 398 (1994).
239. T.H. Metzger, U. Pietsch, E. Gartstein, J. Peisl: phys. stat. sol. (a) **174**, 395 (1999).

240. P. Mikulík: *Thesis* (University J. Fourier, Grenoble and Masaryk University, Brno 1997).
241. P. Mikulík, T. Baumbach: Physica B **248**, 381 (1998).
242. P. Mikulík, T. Baumbach: Phys. Rev. B **59**, 7632 (1999).
243. P. Mikulík, M. Jergel, T. Baumbach, E. Majková, E. Pinčík, Š. Luby, L. Orega, R. Tucoulu, P. Hudek, I. Kostič: J. Phys. D: Appl. Phys. **34**, A188 (2001).
244. S. Milita, M. Servidori: J. Appl. Cryst. **28**, 666 (1995).
245. P.D. Miller, Chuan-Pu Liu, W.L. Henstrom, J. Murray Gibson, Y. Huang, P. Zhang, T.I. Kamins, D.P. Basile, R. Stanley Williams: Appl. Phys. Lett. **75**, 46 (1999).
246. Z.H. Ming, A. Krol, Y.L. Soo, Y.H. Kao, J.S. Park, K.L. Wang: Phys. Rev. B **47**, 16373 (1993).
247. A. Nefedov: private communication.
248. L. Névot, P. Croce: Rev. Phys. Appl. **15**, 761 (1980).
249. A.Yu. Nikulin, A.W. Stevenson, H. Hashizume: Phys. Rev. B **53**, 8277 (1996).
250. K. Nordlund, U. Beck, T.H. Metzger, J.R. Patel: Appl. Phys. Lett. **76**, 846 (2000).
251. K. Nordlund, P. Partyka, R.S. Averback, I.K. Robinson, P. Erhart: J. Appl. Phys. **88**, 2278 (2000).
252. R. Nötzel, N.N. Ledentsov, L. Däweritz, M. Hohenstein, K. Ploog: Phys. Rev. Lett. **67**, 3812 (1991).
253. R. Noetzel: Semicond. Sci. Technol. **11**, 1365 (1996).
254. http://www-cxro.lbl.gov/optical_constants
255. A. Ougazzaden and F. Devaux: Appl. Phys. Lett. **69**, 4131 (1996).
256. G. Palasantzas, J. Krim: Phys. Rev. B **48**, 2873 (1993).
257. G. Palasantzas: Phys. Rev. B **48**, 14472 (1993).
258. T. Panzner, W. Leitenberger, J. Grenzer, Y. Bodenthin, T. Geue, U. Pietsch, H. Möhwald: J. Phys. D: Appl. Phys. **36**, A93 (2003).
259. L.G. Parrat: Phys. Rev. **95**, 359 (1954).
260. W. Parrish: In *International Tables for Crystallography*, Part C (Kluwer Academic, Bristol 1995), pp. 539–543.
261. A.P. Payne, B.M. Clemens: Phys. Rev. B **47**, 2289 (1993).
262. Y.-H. Phang, R. Kariotis, D.E. Savage, N. Schimke, M.G. Lagally: J. Appl. Phys. **71**, 4627 (1992).
263. Y. H. Phang, C. Teichert, M.G. Lagally, L.J. Peticolas, J.C. Bean, E. Kasper: Phys. Rev. B **50**, 14435 (1994).
264. U. Pietsch, W. Borchard: J. Appl. Cryst. **20**, 8 (1987).
265. U. Pietsch, W. Seifert, J.-O. Fornell, H. Rhan, H. Metzger, S. Rugel, J. Peisl: Appl. Surf. Sci. **54**, 502 (1992).
266. U. Pietsch, H. Metzger, S. Rugel, B. Jenichen, I.K. Robison: J. Appl. Phys. **74**, 2381 (1993).
267. U. Pietsch, U. Zeimer, L. Hofmann, J. Grenzer, S. Gramlich: MRS Proc. **617** 41 (2000).
268. U. Pietsch, J. Grenzer, Th. Geue, F. Neissendorfer, G. Brezesinski, Ch. Symietz, H. Möhwald, W. Gudat: Nucl. Instr. Meth. A**467-468**, 1077 (2001).
269. U. Pietsch: Phys. Rev. B **66** 155430 (2002).
270. Z.G. Pinsker: *Dynamical Scattering of X-Rays in Crystals* (Springer, Heidelberg 1978).
271. P. Poloucek, U. Pietsch, T. Geue, Ch. Symietz, G. Brezesinski: J. Phys. D: Appl. Phys. **34** 450 (2001).

272. A. Ponchet, D. Lacombe, L. Durand, D. Alquier, J.-M. Cardona: Appl. Phys. Lett. **72**, 2984 (1998).
273. S. Ponti, G.T. Baumbach: Phys. Lett. A. **251**, 61 (1999).
274. S. Ponti: private communication.
275. C. Pryor, J. Kim, L.W. Wang, A.J. Williamson, A. Zunger: J. Appl. Phys. **83**, 2548 (1998).
276. P. R. Pukite, C.S. Lent, P.I. Cohen, Surf. Sci. **161**, 39 (1985).
277. M. Rauscher, T. Salditt, H. Spohn: Phys. Rev. B **52**, 16855 (1995).
278. M. Rauscher, R. Paniago, H. Metzger, Z. Kovats, J. Domke, J. Peisl, H.-D. Pfannes, J. Schulze, I. Eisele: J. Appl. Phys. **86**, 6763 (1999).
279. G. Renaud, R. Lazzari, C. Revenant, A. Barbier, M. Noblet, O. Ulrich, F. Leroy, J. Jupille, Y. Borensztein, C.R. Henry, J.-P. Deville, F. Scheurer, J. Mane-Mane, O. Fruchart: Science **300**, 1416 (2003).
280. H.R. Ress: *Neue Messmethoden in der Hochauflösenden Röntgendiffraktometrie*, Julius-Maximilians-Universität, Thesis (Würzburg 1998).
281. H. Rhan, U. Pietsch, H. Metzger, S. Rugel, J. Peisl: J. Appl. Phys. **74**, 146 (1993).
282. J.P. Reithmaier, A. Forchel, IEEE J. Sel. Top. Quantum Electron. **4**. 595 (1998).
283. F. Rieutord: Acta Cryst. A **46**, 526 (1990).
284. I. K. Robinson, D. J. Tweet, Rep. Prog. Phys. **55**, 599 (1992).
285. I.K. Robinson, I.A. Vartanyants, G.J. Williams, M.A. Pfeifer, J.A. Pitney: Phys. Rev. Lett. **87**, 195505 (2001).
286. T. Roch, V. Holý, A. Hesse, J. Stangl, T. Fromherz, G. Bauer, T.H. Metzger, S. Ferrer: Phys. Rev. B **65**, 245324(11) (2002).
287. C. Rocke, S. Zimmermann, A. Wixforth, J. Kotthaus, G. Bhm, A. Weizmann: Phys. Rev. Lett. textbf78 4099 (1997).
288. R.S. Rodberg, R.M. Thaler: *Introduction to the Quantum Theory of Scattering* (Academic Press, New York 1967).
289. Roentec Holding AG homepage : www.roentec.com
290. D. Rose: *Thesis* (University of Potsdam 1996).
291. D. Rose, U. Pietsch, U. Zeimer: J. Appl. Phys. **81**, 2601 (1997).
292. D. Rose, U. Pietsch, A. Förster, H. Metzger: Physica B **198**, 256 (1994).
293. A. Rosenauer, D. Gehrtsen, D. van Dyck, M. Artzberger, G. Boehm, G. Abstreiter: Phys. Rev. B **64**, 245334 (2001).
294. D. Roshchupkin, M. Brunel: Rev. Sci. Instr. **64**, 379 (1993).
295. D. Roshchupkin, M. Brunel, F. de Bergevin, A. Erko, Nucl. Instr. Meth. B **72**, 471 (1992).
296. D. Roshchupkin, R. Tucoulou, A. Masclet, M. Brenel, I. Schelokov, A. Kondakov: Nucl. Instr. Meth. B **142**, 432 (1998).
297. K. Rousseau, J.L. Rouviere, F. Fournel, H. Moriceau, Mater. Sci. Semicond. Process. textbf4, 101 (2001).
298. K. Rousseau, L.J. TRouviere, F. Fournel, H. Moriceau, Appl. Phys. Lett. **80** 4121 (2002)
299. S. Rubini, B. Bonanni, E. Pelucchi, A. Franciosi, A. Garulli, A. Parisini, Y. Zhuang, G. Bauer, V. Holý: J. Vac. Sci. Technol. B **18**, 2263 (2000).
300. T. Salditt, T.H. Metzger, J. Peisl: J. de Physique IV Suppl. **4**, C9-171 (1994).
301. T. Salditt, T.H. Metzger, J. Peisl: Phys. Rev. Lett. **73**, 2228 (1994).

302. T. Salditt, T.H. Metzger, Ch. Brandt, U. Klemradt, J. Peisl: Phys. Rev. B **51**, 5617 (1995).
303. T. Salditt, D. Lott, T.H. Metzger, J. Peisl, G. Vignaud, P. Høghøj, J.O. Schärpf, P. Hinze, R. Lauer: Phys. Rev. B **54**, 5860 (1996).
304. A. Sammar, J.-M. André, B. Pardo, Opt. Commun. **86**, 245 (1991).
305. A. Sammar, J.-M. André, J. Opt. Soc. Am. A **10**, 2324 (1993).
306. W. Sauer, M. Streibl, T. Metzger, A. Haubrich, S. Manus, J. Pesl, J. Mazuelas, J. Hrtwig, J. Baruchel: Appl.Phys.Lett. **75** 1709 (1999).
307. D.E. Savage, J. Kleiner, N. Schimke, Y.-H. Phang, T. Jankowsky, J. Jacobs, R. Kariotis, M.G. Lagally: J. Appl. Phys. **69**, 1411 (1991).
308. D.E. Savage, N. Schimke, Y.-H. Phang, M.G. Lagally: J. Appl. Phys. **71**, 3283 (1992).
309. C. Schelling, G. Springholz, F. Schaeffler: Phys. Rev. Lett. **83**, 995 (1999).
310. J.-P. Schlomka, M. Tolan, L. Schwalowsky, O.H. Seeck, J. Stettner, W. Press: Phys. Rev. B **51**, 2311 (1995).
311. M. Schmidbauer, Th. Wiebach, H. Raidt, M. Hanke, R. Koehler: Phys. Rev. B **58**, 10523 (1998).
312. M. Schmidbauer, M. Hanke, R. Koehler: Cryst. Res. Technol. **36**, 1 (2002).
313. O.G. Schmidt, C. Lange, K. Eberl, O. Kienzle, F. Ernst: Appl. Phys. Lett. **71**, 2340 (1997).
314. O.G. Schmidt, K. Eberl: Appl. Phys. Lett. **73**, 2790 (1998).
315. O.G. Schmidt, S. Schieker, K. Eberl, O. Kienzle, F. Ernst: Appl. Phys. Lett. **73**, 659 (1998).
316. O.G. Schmidt, C. Lange, K. Eberl, O. Kienzle, F. Ernst: Thin Solid Films **321**, 70 (1998).
317. O.G. Schmidt, K. Eberl: Phys. Rev. B **61**, 13721 (2000).
318. T. Schülli: *Anomalous X-ray Diffraction from Semiconductor Heterostructures* PhD Thesis, Johannes-Kepler Universität Linz (Linz 2003).
319. T. U. Schülli, M. Sztucki, V. Chamard, T.H. Metzger, D. Schuh: Appl. Phys. Lett. **81**, 448 (2002).
320. T.U. Schülli, J. Stangl, Z. Zhong, R.T. Lechner, M. Sztucki, T.H. Metzger, G. Bauer: Phys. Rev. Lett. **90**, 066105 (2003).
321. M. Schuster, L. Müller, K.E. Mauser, R. Straub: Thin Solid Films **157**, 325 (1988).
322. M. Schuster, H. Göbel: J. Phys. D: Appl. Phys. **28**, A270 (1995).
323. J. Schwinger: Phys. Rev. **70**, 798 (1946).
324. R. L. Schwoebel: J. Appl. Phys. **40**, 614 (1969).
325. S.J. Shaibani, P.M. Hazzledine: Phil. Mag. A **44**, 657 (1981).
326. V. A. Shchukin, D. Bimberg, V. G. Malyshkin, N. N. Ledentsov: Phys. Rev. B **57**, 12262 (1998).
327. V. A. Shchukin, D. Bimberg: Rev. Mod. Physics **71**, 1125 (1999).
328. Q. Shen, S.W. Kycia, E.S. Tentarelli, W.J. Schaff, L.F. Eastman: Phys. Rev. B **54**,16382 (1996).
329. Q. Shen, S.W. Kycia: Phys. Rev. B **55**, 15791 (1997).
330. Q. Shen, C.C. Umbach, B. Weselak, J.M. Blakely: Phys. Rev. B **48**, 17967 (1993).
331. Q. Shen, C. C. Umbach, B. Weselak, J. M. Blakely, Phys. Rev. B **53**, R4237 (1996).
332. S.K. Sinha, E.B. Sirota, S. Garoff, H.B. Stanley: Phys. Rev. B **38**, 2297 (1988).

333. S.K. Sinha, M. Tolan, A. Gibaud: Phys. Rev. B **57**, 2740 (1998).
334. http://www.slac.stanford.edu
335. M. Sonada, M. Takano, J. Miyahara, H. Kato: Radiology **148**, 833 (1983).
336. B. J. Spencer, J. Tersoff: Phys. Rev. Lett. **79**, 4858 (1997).
337. E. Spiller: Rev. Phys. Appl. **23**, 1687 (1988).
338. E. Spiller, D. Stearns, M. Krumrey: J. Appl. Phys. **74**, 107 (1993).
339. G. Springholz, V. Holý, M. Pinczolits, G. Bauer: Science **282**, 734 (1998).
340. G. Springholz, M. Pinczolits, P. Mayer, V. Holý, G. Bauer, H.H. Kang, L. Salamanca-Riba: Phys. Rev. Lett. **84**, 4469 (2000).
341. D. J. Srolovitz: Acta Metall. **37**, 621 (1989).
342. J. Stangl, V. Holý, P. Mikulík, G. Bauer, I. Kegel, T.H. Metzger, O.G. Schmidt, C. Lange, K. Eberl: Appl. Phys. Lett. **74**, 2785 (1999).
343. J. Stangl, V. Holý, T. Roch, A. Daniel, G. Bauer, J. Zhu, K. Brunner, G. Abstreiter: Phys. Rev. B **62**, 7229 (2000).
344. J. Stangl, T. Roch, G. Bauer, I. Kegel, T.H. Metzger, O.G. Schmidt, K. Eberl, O. Kienzle, F. Ernst: Appl. Phys. Lett. **77**, 3953 (2000).
345. J. Stangl, A. Daniel, V. Holý, T. Roch, G. Bauer, I. Kegel, T.H. Metzger, Th. Wiebach, O.G. Schmidt, K. Eberl: Appl. Phys. Lett. **79**, 1474 (2001).
346. J. Stangl, V. Holý, G. Bauer, Rev. Mod. Phys, in press (2004).
347. D.G. Stearns: J. Appl. Phys. **65**, 491 (1989).
348. A.J. Steinfort, P.M.L.O. Scholte, A. Ettema, F. Tuinstra, M. Nielsen, E. Landemark, D.-M. Smilgies, R. Feidenhansl, G. Falkenberg, L. Seehofer, R.L. Johnson: Phys. Rev. Lett. **77**, 2009 (1996).
349. S.A. Stepanov, R. Köhler: J. Phys. D: Appl. Phys. **27**, 1922 (1994).
350. S.A. Stepanov, U. Pietsch, G.T. Baumbach: Z. Phys. B: Cond. Matter **96**, 341 (1995).
351. S.A. Stepanov, E.A. Kondrashkina, M. Schmidbauer, R. Köhler, J.-U. Pfeiffer: Phys. Rev. B **54**, 8150 (1996).
352. S.A. Stepanov, E.A. Kondrashkina, R. Köhler, D.V. Novikov, G. Materlik, S.M. Durbin: Phys. Rev. B **57**, 4829 (1998).
353. R. Stömmer, J. Grenzer, J. Fischer, U. Pietsch: J. Phys. D: Appl. Phys. **28**, A216 (1995).
354. R. Stömmer, U. Pietsch: J. Phys. D: Appl. Phys. **29**, 3161 (1996).
355. R. Stömmer, T. Metzger, M. Schuster, H. Göbel: Il Nuovo Cimento **19D**, 465 (1997).
356. R. Stömmer, A.R. Martin, T. Geue, H. Göbel, W. Hub, U. Pietsch: Adv. X-Ray Analysis **41**, 223 (1998).
357. S. Takagi: J. Phys. Soc. Japan **26**, 1239 (1969).
358. A. Talneau, N. Bouadma, S. Slempkes, A. Ougazzaden, S. Hansmann: IEEE Photonics Technol. Lett. **9**, 1316 (1997).
359. A. Talneau, S. Slempkes, A. Ougazzaden: to be published in IEEE, Journal of selected Topics in Quantum Electronics.
360. L. Tapfer, K. Ploog: Phys. Rev. B **33**, 5565 (1986).
361. L. Tapfer and P. Grambow: Appl. Phys. A **50**, 3 (1990).
362. L. Tapfer, G. C. La Rocca, H. Lage, R. Cingolani, P. Grambow, A. Fischer, D. Heitmann, K. Ploog, Surface Science **267**, 227 (1992).
363. L. Tapfer, P. Sciacovelli, L. DeCaro: J. Phys. D: Appl. Phys. **28**, A179 (1995).
364. L. Tapfer, L. DeCaro, C. Giannini, H.-P. Schönherr, K.H. Ploog: Solid State Commun. **98**, 599 (1996).

365. M. Taylor, J. Wall, N. Loxley, M. Warmington, T. Laffort: Mat. Sci. and Engn.B **80**, 95 (2001).
366. C. Teichert, Y.H. Phang, L.J. Peticolas, J.C. Bean, M.G. Lagally: *Surface Diffusion: Atomistic and Collective Processes*, ed. M.C. Tringides, NATO-ASI Series (Plenum Press 1997).
367. C. Teichert: Physics Reports **365**, 335 (2002).
368. J. Tersoff, Y.H. Phang, Z. Zhang, M.G. Lagally: Phys. Rev. Lett. **75**, 2730 (1995).
369. J. Tersoff, C. Teichert, M.G. Lagally: Phys. Rev. Lett. **76**, 1675 (1996).
370. http://tesla.desy.de, http://tesla.desy.de/tdr-update.
371. G. Thorkildsen, H.B. Larsen: Acta Cryst. **55**, 1000 (1999).
372. J.G. Timothy, R.P. Madden: In *Handbook of Synchrotron Radiation* (North-Holland, Amsterdam 1983) pp. 323–325.
373. M. Tolan, G. König, L. Brügemann, W. Press, F. Brinkop, J.P. Kotthaus: Europhys. Lett. **20**, 223 (1992).
374. M. Tolan, W. Press, F. Brinkop, J.P. Kotthaus: J. Appl. Phys. **75**, 7761 (1994).
375. M. Tolan, W. Press, F. Brinkop, J.P. Kotthaus: Phys. Rev. B **51**, 2239 (1995).
376. M. Tolan: *X-Ray Scattering from Soft-Matter Thin Films – Materials Science and Basic Research* (Springer, Heidelberg 1999).
377. J. Trenkler: private communication.
378. H. Trinkaus: phys. stat. sol. (b) **51**, 307 (1972).
379. R.D. Twesten, D.M. Follstaedt, S.R. Lee, E.D. Jones, J.L. Reno, J.Mirecki Millunchick, A.G. Norman, S.P. Ahrenkiel, A. Mascarenhas: Phys. Rev. B **60**, 13619 (1999).
380. A.P. Ulyanenkov, S.A. Stepanov, U. Pietsch, R. Köhler: J. Phys. D: Appl. Phys. **28**, 2522 (1995).
381. A. Ulyanenkov, G.T. Baumbach, N. Darowski, U. Pietsch, K.H. Wang, A. Forchel, T. Wiebach: J. Appl. Phys. **85**, 1524 (1999).
382. D. Vaknin, K. Kjaer, H. Ringsdorf, R. Blankenburg, M. Piepenstock, A. Diedrich, M. Lösche: Langmuir **9**, 1171 (1993).
383. I.A. Vartanyants, M.V. Kovalchuk: Rep. Prog. Phys. **64**, 1009 (2001).
384. G. H. Vineyard: Phys. Rev. B **26**, 4146 (1982).
385. R.F. Voss: In *Scaling Phenomena in Disordered Systems*, NATO ASI Series B: Physica, **133**, ed. R. Pynn, A. Skjeltorp (Plenum, New York 1985).
386. B.E. Warren: *X-Ray Diffraction* (Dover Publ., NewYork 1990).
387. E. Weckert, K. Hümmer: Acta Cryst. A **37**, 625 (1990).
388. R. Whatmore, P. Goddard, B. Tanner, G. Clark: Nature **299**, 44 (1982).
389. Th. Wiebach, M. Schmidbauer, M. Hanke, H. Raidt, R. Köhler, H. Wawra: Phys. Rev. B **61**, 5571 (2000).
390. K. Wille: private communication.
391. E.R. Wölfel: J. Appl. Cryst. **16**, 341 (1983).
392. J. Wollschläger, M. Larsson: Phys. Rev. B **57**, 14937 (1998).
393. M. Wormington, I. Pape, T.P.A. Hase, B.K. Tanner, D.K. Bowen: Phil. Mag. Lett. **74**, 211 (1996).
394. Q. Xie, A. Madhukar, P. Chen, N.P. Kobayashi: Phys. Rev. Lett. **75**, 2542 (1995).
395. Y. Yoneda: Phys. Rev. **131**, 2010 (1963).
396. W.H. Zachariasen: *Theory of X-ray Diffraction in Crystal* (J.Wiley and Sons, New York 1945).

397. P. Zaumseil, U. Winter, F. Cembali, M. Servidori, Z. Šourek: phys. stat. sol. (a) **100**, 95 (1987).
398. R. Zaus: J. Appl. Crystallogr. **26**, 801 (1993).
399. U. Zeimer: *Thesis* (University of Potsdam 1998).
400. U. Zeimer, J. Grenzer, S. Grigorian, J. Fricke, S. Gramlich, F. Bugge, U. Pietsch, M. Weyers, G. Tränkle: phys. stat. sol (a) **195**, 178 (2003).
401. Y.W. Zhang, S.J. Xu, C.-H. Chiu: Appl. Phys. Lett. **74**, 1809 (1999).
402. K. Zhang, C. Heyn, W. Hansen, T. Schmidt, J. Falta: Appl. Phys. Lett. **76**, 2229 (2000).
403. Y. Zhuang, V. Holý, J. Stangl, A.A. Darhuber, P. Mikulík, S. Zerlauth, F. Schaeffler, G. Bauer, N. Darowski, D. Lübbert, U. Pietsch: J. Phys. D: Appl. Phys. **32**, A224 (1999).
404. E. Zolotoyabko, D. Shilo, W. Sauer, E. Pernot, J. Baruchel: Appl.Phys.Lett. **73**, 2278 (1998).

Index